Power Electronic Control of AC Motors

J. M. D. MURPHY
Department of Electrical Engineering and Microelectronics
University College, Cork, Ireland

and

F. G. TURNBULL
General Electric Company
Corporate Research and Development
Schenectady, New York, USA

PERGAMON PRESS
OXFORD · NEW YORK · BEIJING · FRANKFURT
SÃO PAULO · SYDNEY · TOKYO · TORONTO

U.K.	Pergamon Press plc, Headington Hill Hall, Oxford OX3 0BW, England
U.S.A.	Pergamon Press, Inc., Maxwell House. Fairview Park, Elmsford. New York 10523, U.S.A.
PEOPLE'S REPUBLIC OF CHINA	Pergamon Press, Room 4037, Qianmen Hotel. Beijing, People's Republic of China
FEDERAL REPUBLIC OF GERMANY	Pergamon Press GmbH, Hammerweg 6. D-6242 Kronberg, Federal Republic of Germany
BRAZIL	Pergamon Editora Ltda. Rua Eça de Queiros, 346, CEP 04011, Paraiso, São Paulo, Brazil
AUSTRALIA	Pergamon Press Australia Pty Ltd., P.O. Box 544, Potts Point, N.S.W. 2011, Australia
JAPAN	Pergamon Press, 5th Floor, Matsuoka Central Building, 1-7-1 Nishishinjuku, Shinjuku-ku, Tokyo 160, Japan
CANADA	Pergamon Press Canada Ltd., Suite No. 271, 253 College Street, Toronto, Ontario, Canada M5T 1R5

Copyright © 1988 J. M. D. Murphy and F. G. Turnbull

First edition 1988

Library of Congress Cataloging in Publication Data
Murphy, J. M. D.
Power electronic control of AC motors.
Includes index.
1. Electric motors, Alternating current—Automatic control. 2. Power electronics. I. Turnbull, F. G.
II. Title.
TK2781.M87   1988   621.46'2   88–6031

British Library Cataloguing in Publication Data
Murphy. J. M. D.
Power electronic control of AC motors.
1. Alternating current machines. Control devices
I. Title. II. Turnbull, F. G.
621.31'33
ISBN 0–08–022683–3

Printed in Great Britain by A. Wheaton & Co. Ltd., Exeter

Power Electronic Control
of AC Motors

Other Pergamon Titles of Interest

CEGB
Advances in Power Station Construction

CORNELLIE
Microprocessors

CRISTOL
Solid State Video Cameras

DA CUNHA
Planning and Operation of Electric Energy Systems 1985

HAMMOND
Electromagnetism for Engineers, 3rd edition

HINDMARSH
Electrical Machines and Drives, 2nd edition
Electrical Machines and Their Applications, 4th edition

HOLLAND
Illustrated Dictionary of Microelectronics and Microcomputers
Integrated Circuits and Microprocessors
Microcomputers and Their Interfacing

YORKE
Electric Circuit Theory, 2nd edition

Pergamon Related Journals (sample copy gladly sent on request)

Computers and Electrical Engineering

Electric Technology USSR

Microelectronics and Reliability

Solid-State Electronics

Preface

It is now some thirty years since the introduction of the thyristor or silicon controlled rectifier heralded the advent of solid-state power electronics and revolutionized the traditional methods of electric power conversion and control. More recently, spectacular advances in power semiconductor devices, integrated electronics, and microprocessors have dramatically reduced the size and cost of power electronic converters. These developments have had a significant impact on the performance and cost-effectiveness of the modern ac drive system. As a result, there is now a rapid expansion in industrial application of the ac drive, and this growth is being accelerated by a greater recognition of the enhanced performance capability. The microcomputer permits a universal hardware design with flexible software control — the resulting ac drive is readily integrated into a computer-based system for overall factory automation. The computational capability of the microcomputer can also be used to provide new levels of sophistication in drive control systems — with traditional induction and synchronous motors, or with nontraditional brushless dc motors and variable-reluctance machines.

This book is intended as an introduction to the basic theory and practice of modern power electronics and, in particular, it deals with the application of power electronic techniques for ac motor control. The text is suitable for degree-level and postgraduate courses in Power Electronics and/or Electrical Machines and Drives. It is also intended for practicing engineers who wish to acquaint themselves with recent developments in these rapidly expanding fields. An extensive set of key references is listed at the end of each chapter, and it is expected that the text will serve as a useful reference work for engineers already active in the field.

Modern ac motor control technology is an interdisciplinary study, and the relevant aspects are treated in this comprehensive text. The authors have attempted to provide an integrated self-contained treatment so that readers, coming from such diverse fields as electrical machines, electronics, or control engineering can become familiar with modern electrical drive systems. Theoretical analysis has been included to the extent necessary for a rational understanding of the principles involved.

Chapter 1 introduces the adjustable-frequency ac drive and reviews the basic principles of ac motor operation. Power electronic devices and circuits are studied in detail in Chapters 2 through 5 and in Chapter 12; these chapters are appropriate for an introductory course in modern power electronics. Chapter 2 discusses the various power semiconductor devices available today; device structure and physical operation are described, and typical terminal characteristics are shown. AC-DC converters are studied in Chapter 3, and the relevant circuit relationships are derived. Chapter 4 is a comprehensive treatment of dc-ac inverters in which the various voltage-fed and current-fed inverter circuits are studied, and some typical thyristor forced-commutating circuits are investigated. Chapter 5 introduces the phase-controlled cycloconverter. Power semiconductor rating, protection, and cooling are discussed in Chapter 12.

Power electronic control of ac motors is treated in Chapters 6 through 11. Chapter 6 examines the influence of nonsinusoidal voltage and current supplies on conventional ac motor operation; harmonic losses are examined and harmonic torque effects are analyzed. Chapter 7 discusses control strategies for adjustable-frequency induction motor drives and includes a detailed treatment of field-oriented (vector) control. Chapter 8 discusses the control of the adjustable-frequency synchronous motor drive. Induction motor speed control by stator voltage adjustment and slip-energy recovery are treated in Chapters 9 and 10, respectively. Brushless dc motor systems, stepping motors, and variable- or switched-reluctance motor drives are treated in Chapter 11. In the final chapter, Chapter 13, induction motor specification and design considerations are discussed, together with ac drive simulation; the text concludes with a review of typical application areas for adjustable-speed ac drive systems.

The authors wish to express gratitude to numerous colleagues at University College, Cork, and at the GE Research and Development Center. Special thanks must go to Dr. James W. A. Wilson of GE who encouraged us to undertake this text. We are also indebted to our GE editors, Dr. Catharine L. Fisher and Russell S. Clark and to Lydia E. Plaskett who typeset the manuscript. We also wish to thank Dr. Michael Egan of University College, Cork, who reviewed the entire text. In addition, we thank our Pergamon editors Annika Corkill and Keith Lambert. Last, but not least, we wish to record our gratitude to our wives for their patience and understanding throughout the preparation of this text.

JOHN M. D. MURPHY
FRED G. TURNBULL

Contents

CHAPTER 1

Principles of Adjustable-Speed AC Drives

1.1. INTRODUCTION

In many modern adjustable-speed drives the demand is for precise and continuous control of speed, torque, or position with long-term stability, good transient performance, and high efficiency. The dc motor has satisfied some of these requirements, but its mechanical commutator is often undesirable because of the regular maintenance required. Maintenance causes difficulty when interruptions cannot be tolerated or when the motor is used at inaccessible locations. AC motors such as the cage-rotor induction motor and the synchronous reluctance and permanent magnet synchronous motors are brushless and have a robust rotor construction which permits reliable maintenance-free operation at high speed. The simple rotor construction also results in a lower cost motor and a higher power/weight ratio. Unfortunately, the induction motor and synchronous motors are inflexible in speed when operated on a standard constant-frequency ac supply. The synchronous motors operate synchronously at a speed which is determined by the supply frequency and the number of poles for which the stator is wound. The induction motor runs slightly below synchronous speed. For intermittent operation at reduced speeds, stator voltage control of the induction motor is satisfactory (Chapter 9). Subsynchronous speed control of a wound-rotor induction motor is obtained by means of a converter cascade for slip-energy recovery (Chapter 10). However, efficient wide-range speed control of the synchronous-type motors or the cage-rotor induction motor is only possible when an adjustable-frequency ac supply is available. Consequently, in this text, attention is largely concentrated on the adjustable-frequency method of obtaining adjustable-speed ac drive systems.

In the past, the various techniques for speed control of ac machines often required the use of auxiliary rotating machines. These auxiliary machines have now been supplanted by static ac drive systems using various types of power semiconductor operating as electrically controlled switches. High efficiency is attained because of the low "on-state" conduction losses when the power semiconductor is conducting the load current and the low "off-state" leakage losses when the power semiconductor is blocking the source, or load, voltage. The transition times between blocking and conduction, and vice versa, depend on the type of power semiconductor used; but times range from $150\,\mu s$ for a large thyristor to 50 ns for a field-effect transistor. The characteristics of various types of power semiconductor used for the control of ac motors are discussed in Chapter 2.

Recently, manufacturers have assembled more than one power semiconductor in a single package, or module. The use of such a module reduces the number of necessary heat sinks and electrical interconnections. Various combinations of devices—for example, thyristors and diodes—are assembled in a common package. For low currents, complete power circuits, made up of, for example, six transistors and six feedback diodes, have been arranged into a three-phase bridge circuit. The selling price of these modules and of the basic power semiconductors themselves has been declining as the market expands.

Another major factor in ac drives technology is the availability of microprocessors for the control of ac drive systems. Microprocessors operate at an adequately high clock frequency (5 to 20 MHz) to complete their calculations in sufficient time to directly control the firing of the power semiconductors in a three-phase bridge circuit operating from the utility supply frequency. In addition to the direct calculation of the power semiconductor firing times, the microprocessor can perform lower priority tasks, such as diagnostics, self-test, start-up and shut-down sequencing, and fault monitoring. Several of the control strategies described in Chapters 7 and 8 are feasible with analog hardware but only practical with a microprocessor and digital controllers.

These rapid technical advancements and declining prices for power semiconductors and microprocessors, coupled with a demand for a high-efficiency, adjustable-speed control for both existing and newly installed equipment, have led to the world-wide application of adjustable-frequency controllers for ac motors.

References 1 and 2 provide reprints of selected technical papers related to power semiconductor applications and ac adjustable-speed drives.

1.2. SELECTING AN ADJUSTABLE-SPEED DRIVE

The problem in selecting an adjustable-speed drive for a particular application is to choose the system that can most economically provide the required range of speed or torque or position control with the desired accuracy and speed of response. The ac commutator motor has been used in the past, because it can be supplied directly from the ac utility supply, but for a reversible drive with continuous speed control over a very wide range, the dc motor has been the most popular solution. The separately excited dc motor can be rapidly and efficiently controlled by variation of the armature voltage and field current. In recent years, the dc supply has been obtained from the ac utility network by means of static converters, which permit the controlled rectification of the alternating voltage, so that an adjustable direct voltage is provided for the armature. Precise speed control is achieved by the adoption of closed-loop feedback methods.

However, the dc motor is not the ideal solution to the problem of adjustable-speed motor operation. The commutator consists of a large number of copper segments separated by thin sheets of mica insulation. This elaborate construction increases the cost of the dc motor and reduces its power/weight ratio. Brush and commutator wear are accentuated by sparking, and the mica insulation limits the voltage between segments. The total armature voltage is therefore limited to a maximum of about 1500 V. The magnitude of the armature current and its rate of change are restricted by commutation difficulties, and the speed of rotation of the dc machine is limited. The cage-rotor, or squirrel-cage, induction motor, on the other hand, has a rotor circuit consisting of a short-circuited winding that can often be made from a single casting. There is no need to insulate the rotor bars from the surrounding laminations, and the cage rotor has a low inertia and can operate at high temperature and high speed for prolonged periods without maintenance. In addition, the cost of the cage-rotor induction motor is about one-tenth

Many adjustable-speed drives require a constant-torque output, which can be achieved if the airgap flux in the motor is maintained constant by operating on a fixed volts/hertz supply. However, the analysis above assumes negligible winding resistance, whereas, in practice, at low frequencies the resistive voltage drop becomes significant compared with the induced emf. This voltage drop causes a reduction in the airgap flux and motor torque. In order to maintain the low-speed torque, the volts/hertz ratio must be increased at frequencies below about 20 Hz for a machine rated at 60 Hz.

1.4. GENERATION OF ADJUSTABLE-FREQUENCY AC POWER

The idea of using an adjustable-frequency supply to control the speed of ac motors is not new; rotating frequency converters were employed for many years. They were used principally in multimotor mill drives and in special applications where a high operating frequency was chosen in order to permit the use of compact ac motors. More recently, the rotating machine methods of adjustable-frequency ac power generation have been supplanted by static conversion methods.

1.4.1. *Static Frequency Converters*

In order to obtain high efficiency in a static frequency converter, it is essential to use solid-state switching devices which are either on or off. In the on-condition, the device approximates to an ideal closed switch having zero voltage drop across it and a current that is determined by the external circuit. In the off-condition, the device approximates to an ideal open switch that has infinite impedance and blocks the flow of current in the circuit. If the solid-state switch can be triggered from the off condition into the on-condition by a low-power control signal, the device can be used in converter circuits for the generation of adjustable-frequency alternating voltages and currents. The grid-controlled mercury arc rectifier and the thyratron—a gas-filled discharge tube with heated cathode—were used in this manner, and many of the circuits used today were originally developed to use these devices. However, these early circuits were not sufficiently attractive technically or economically and were not adopted commercially. The development of the thyristor, which is the semiconductor counterpart of the thyratron, and its availability in high-power ratings have given a new boost to static methods of adjustable-frequency ac generation; many of the earlier circuits have been revived and improved.

The thyristor can be triggered into conduction by an external signal on the gate electrode, but the conducting thyristor cannot be switched off by an external signal. Turn-off, or commutation, as it is called, can only be achieved by interruption of the anode current of the thyristor. If a thyristor circuit operates from an ac supply, the natural reversal of the alternating voltage every half-cycle may be used to reduce the thyristor current to zero and so produce natural commutation. In a dc-operated circuit the current does not go to zero naturally, and forced commutation must be used. This technique usually employs auxiliary charged capacitors to force the current in a conducting thyristor to zero. Chapter 4 describes some of the techniques for providing forced commutation in thyristor circuits supplied from a dc source.

The power transistor, field-effect transistor, and gate turn-off thyristor can be both turned on and turned off by an electrical signal to their control (base or gate) terminals.

This feature allows these three devices to be used readily in circuits supplied from a dc source. They do not require the auxiliary components (inductors, capacitors, and sometimes auxiliary thyristors) that are required for thyristor circuits supplied with dc. The two types of transistor are rated at 1200 V, or less, and therefore find their greatest use on rectified industrial ac voltage systems (120, 240, 380, or 460 V) at power levels of 100 horsepower or less. The gate turn-off thyristor, available with higher voltage and current ratings is used in drive systems up to 1000 horsepower. The thyristor is used exclusively in the very high-power drive systems up to 30 000 horsepower. Of course, increased power can be obtained by the operation of semiconductors in parallel, circuits in parallel, or motor windings in parallel. Harmonic current reduction can be achieved by an appropriate phase shift between inverter circuits.

There are three basic types of static frequency converter, the dc link converter, the cycloconverter, and the high-frequency ac link converter. A block diagram of the dc link converter is shown in Fig. 1.1. The three-phase ac supply is rectified to dc in a controlled

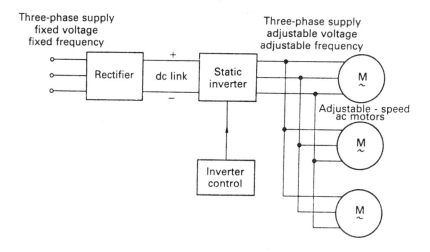

FIG. 1.1. Block diagram of a dc link converter for a multimotor ac drive.

or uncontrolled rectifier, and the resulting dc power is fed to a static inverter. An inverter is a circuit which converts dc power to ac power, and the static inverter uses thyristors or other power devices which are switched sequentially, so that an alternating voltage waveform is delivered to the ac motor. Many suitable inverter circuits have been developed and are described in Chapter 4. The output frequency is determined by the rate at which the inverter devices are triggered into conduction. Triggering can be controlled by the reference oscillator and logic circuits, which generate and distribute firing pulses in the correct sequence to the various devices. An alternate technique is to provide the firing pulses from the rotation of the motor shaft in a synchronous motor drive system. In this way the machine can never be out of synchronism with its inverter and the system behaves like a converter-fed dc drive. In general, at the end of its conduction period, each thyristor must be turned off by an auxiliary commutating circuit. With transistors and other devices having turn-off capability, the auxiliary commutating circuit is not necessary.

The inverter output frequency may be controlled from zero to several hundred hertz by variation of the reference oscillator frequency. The output voltage waveform is usually nonsinusoidal, but external filter circuits are not employed due to the difficulty of obtaining effective operation over a wide range of frequencies. The inverter output is therefore supplied directly to the ac motor, and the harmonic effects must be considered in specifying the motor (Chapter 6). Normally it is found that the distorted waveform does not impose any serious limitations apart from a slight reduction in the rating and efficiency of the motor. Incorrectly chosen pulse-width-modulated voltage waveforms that contain high percentages of harmonic voltage can cause localized heating in the rotor circuit of induction motors operated below their base speed. Some inverter-operated ac motors also exhibit a nonuniform stepping motion at very low speeds due to the nonuniform rotation of the flux wave. The output volts/hertz ratio of the inverter can be controlled by variation of the direct voltage input to the inverter. Alternately, the input voltage may be held constant while the volts/hertz ratio is adjusted within the inverter circuit. Various voltage control techniques are considered in Chapter 4.

In the discussion to date, it has been assumed that the dc link converter of Fig. 1.1 is operating as a voltage-source converter which delivers a load-independent voltage. However, the dc link converter can be modified so that it delivers a load-independent current. It is then classified as a current-source dc link converter, and its operation is studied in Chapter 4.

The dc link converter involves a double power conversion, but efficiencies of 85 to 97 percent are achieved in practice, and the output frequency is chosen to suit the application. The current-source dc link converter is inherently capable of regenerative operation. This means that the direction of energy flow can be reversed, so that energy is returned from the load to the ac utility network. This energy return permits rapid motor deceleration by returning the kinetic energy of the rotating parts to the ac supply. The basic voltage-source dc link converter cannot operate regeneratively unless additional converter circuitry is fitted, and this adds to the cost and complexity of the system. In many applications, regenerative braking is not necessary, and the voltage-source dc link converter is widely used in adjustable-speed ac drives that are highly competitive with more traditional systems.

The second basic form of static frequency converter is the cycloconverter, which has also been used in adjustable-speed ac drives. In a cycloconverter, the supply frequency is converted directly to a lower output frequency without intermediate rectification. Thyristors are used to selectively connect the load to the supply source, and the low-frequency output voltage waveform is fabricated from segments of the supply voltage waveform. With a three-phase input and three-phase output, the output frequency must be less than about one-third of the input frequency; the drive is therefore suitable for low-speed motor operation if the input is at normal utility frequency. However, the output voltage waveform approximates closely to a sine wave, particularly at low output frequencies. As in the static inverter, the output frequency is determined by an independent reference oscillator or shaft position sensor, and the output volts/hertz ratio is also varied by the control circuitry. If a cycloconverter feeds a synchronous machine, the system can be operated so that the motor emf provides the commutation for the cycloconverter thyristors. In this case, the output frequency can exceed the input frequency. Forced commutating circuits can also be provided for thyristor commutation, but they are complex because the direction of current to be commutated is dependent on the load power factor. One of the principal advantages of the cycloconverter is that it is inherently capable of regenerative operation. The cycloconverter is treated more fully in Chapter 5.

The third type of adjustable-frequency converter is the high-frequency ac link converter. This converter also employs a double power conversion process. First, the utility polyphase supply at fixed voltage and fixed frequency is converted to a single-phase, high-frequency ac voltage. Second, the single-phase, high-frequency ac voltage is converted to a polyphase, adjustable voltage and adjustable frequency supply for an adjustable-speed ac motor drive. A high-frequency isolation transformer can be incorporated into the converter as shown in Fig. 1.2.[3] The first converter, which converts

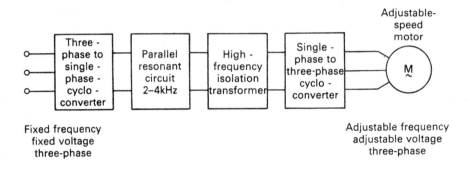

FIG. 1.2. Block diagram of an ac high-frequency link converter for a single ac motor drive.

from utility frequency to a much higher frequency, can use a resonant type of circuit to achieve thyristor commutation. The second conversion, from high frequency to a lower frequency, can use the reversal of the high-frequency voltage to achieve thyristor commutation. An alternate approach is to use power semiconductors with turn-off capability. The high-frequency ac link allows transformer isolation between input and output and provides voltage matching with the selection of the appropriate transformer turns ratio. Several windings can be incorporated into the transformer to provide for multiple isolated sources and loads. An uninterruptible power supply with both an ac and a dc input can be provided to supply a single ac load. Since there are no dc voltages in this type of circuit, the power semiconductor devices must be able to block both polarities of voltage and carry current in both directions. This requirement is generally met with inverse-parallel connected thyristors or power transistors connected to the dc terminals of a diode bridge circuit. With multiple input and output phase numbers, a large number of power devices are required.

All three types of power converter can provide reverse rotation of any motor load by electronically reversing the turn-on sequence of the power devices. No switching of power leads to the motor is necessary.

1.4.2. *Advantages of Static Frequency Conversion*

In a static frequency converter, as already explained, the output frequency is determined solely by the independent reference oscillator. The output frequency is therefore completely independent of fluctuations in the ac supply frequency and voltage, and is also independent of load variations. Consequently, zero frequency regulation is obtained. When the frequency converter supplies an induction motor, the shaft speed departs from synchronous speed by an amount dependent on the load torque, but when a synchronous motor or reluctance motor is used in a simple open-loop system, the motor speed is determined solely by the reference oscillator. Extremely high accuracy and stability are obtained, and the speed regulation on load is zero, although sudden changes in load can produce momentary deviations from synchronous speed.

Other advantages of static frequency conversion are —

1. Static converters have low installation costs because they do not require elaborate foundations or careful machine alignment. Static equipment also has small space requirements and a low noise level.

2. The operating costs of a frequency converter are low due to its high efficiency and the absence of moving parts which deteriorate with time and require periodic replacement.

3. The static converter has great ease of control because the output voltage and frequency can be independently varied over a wide range, and because closed-loop feedback methods are readily applied. The output volts/hertz ratio can be adjusted to suit the adjustable-speed motors and, if necessary, a large starting torque can be developed.

1.5. ADJUSTABLE-FREQUENCY OPERATION OF AC MOTORS

The static frequency converter usually supplies either a cage-rotor induction motor or one of the synchronous types of motor. Conventional ac motor operation at fixed frequency is now briefly reviewed and compared with adjustable-frequency operation. The basic induction motor formulae for both fixed- and adjustable-frequency operation are treated more fully in Chapters 6 and 7. Synchronous motor operation is studied in greater detail in Chapter 8.

1.5.1. The Cage-Rotor Induction Motor[4]

In the induction motor, the rotor winding is short-circuited and receives its supply by induction from the stator. The stator carries a three-phase winding which is directly connected to a three-phase ac supply. This design establishes an airgap magnetic field of constant amplitude which rotates at a synchronous speed, n_1, of $60 f_1/p$ rev/min (p is the number of pole-pairs for which the stator is wound, and f_1 is the stator frequency in hertz). The synchronous speed of the motor can therefore be controlled by variation of the supply frequency, f_1. The amplitude of the airgap flux is determined by the stator volts/hertz ratio, as explained in Section 1.3. When the induction motor is supplied at fixed voltage and frequency, the airgap flux is approximately constant for all normal loading conditions. The phase sequence of the stator supply determines the direction of rotation of the airgap field and hence determines the direction of rotation of the motor. The motor runs at a speed, n, which is usually only a few percentage points less than the synchronous speed, n_1. As a result, the rotor emf is at a frequency, f_2, which equals $s f_1$, where the fractional slip, s, is f_2/f_1, or $(n_1-n)/n_1$. The rotor current, I_2, lags the emf by the rotor phase angle, ϕ_2, and the motor torque is proportional to the in-phase rotor current, $I_2 \cos \phi_2$. The output torque is also proportional to the fundamental airgap flux, Φ_1, and, in general

$$T = K\Phi_1 I_2 \cos \phi_2 \qquad (1.1)$$

where K is a constant of proportionality.

Figure 1.3 shows the torque-speed characteristic of a typical cage-rotor induction

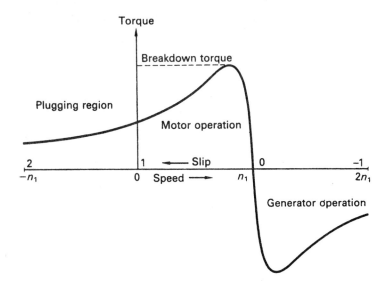

FIG. 1.3. Torque-speed characteristic of the polyphase induction motor at constant voltage and frequency.

motor operating on a standard fixed-frequency supply. At synchronous speed, the motor develops zero torque, and the speed decreases linearly with torque when the slip is small. As the rotor slip frequency increases, the rotor leakage reactance becomes significant and increases the rotor impedance and also the phase angle, ϕ_2. With increasing slip, the motor torque reaches a maximum and then diminishes. The maximum torque is called the breakdown torque, because the motor stalls if the load torque exceeds this peak value. The rotor frequency at the breakdown point is called the rotor breakdown frequency.

If the motor rotates backwards against the forward-rotating field, the slip is greater than unity and a counter-torque is developed which opposes the rotation. The motor operates in this region of the characteristic if two stator leads are suddenly reversed when the machine is operating normally as a motor. The phase sequence of the stator currents is changed, and hence the direction of rotation of the airgap field is reversed. The machine is brought rapidly to rest, and, if the supply is not disconnected, the rotor then speeds up in the opposite direction. This method of braking, or rapidly reversing the induction motor, is known as plugging.

The torque-speed curve may also be extended into the generator region above synchronous speed. In this region, the slip is negative and the machine converts mechanical energy at the shaft into electrical energy which is returned to the ac network. The machine operates under these conditions when the rotor is driven mechanically at supersynchronous speeds. Generator operation also occurs in an adjustable-frequency system when the supply frequency is rapidly reduced. The motor is then overhauled by the load, and the kinetic energy of the rotating masses is converted into electrical energy, giving regenerative braking. If the motor is supplied by a suitable static frequency converter, the regenerated power may be returned to the supply through the converter.

At standstill, or zero speed, the stator current of a typical cage-rotor induction motor supplied from the ac utility network is five or six times rated current, but the starting torque is small because of the low rotor power factor at high rotor frequencies. In an adjustable-frequency system, the supply frequency is reduced for starting. This frequency reduction improves the rotor power factor and so increases the torque per ampere at starting. The airgap flux can also be increased by adjustment of the stator volts/hertz ratio. In this manner, rated torque is available at standstill, and the induction motor is rapidly accelerated to its operating speed by increasing the supply frequency. With this method of starting, there is no danger of the low-frequency crawling which sometimes occurs when induction motors are started on a fixed-frequency supply.

Figure 1.4 shows the torque characteristics of the motor at several different stator fre-

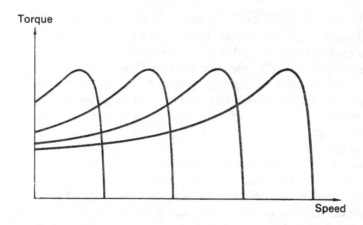

FIG. 1.4. Induction motor torque-speed characteristics at different supply frequencies and constant airgap flux.

quencies. The breakdown torque is maintained constant by adjustment of the stator volts/hertz ratio so that the airgap flux is constant. These motor characteristics are suitable for driving a constant-torque load at an adjustable speed. The difference between the synchronous speed, n_1, and the shaft speed, n, is the slip speed in revolutions per minute, which is practically constant over the speed range for a given load torque. Thus, if the motor has a 40 rev/min slip at rated load and 60 Hz, it will also have approximately a 40 rev/min slip at 30 Hz. The shaft speed is, therefore, not quite proportional to the supply frequency but is near enough for many applications.

If the stator voltage remains constant as the frequency is varied above base or rated motor speed, the airgap flux and breakdown torque decrease with increasing frequency, as shown in Fig. 1.5. These characteristics are suitable for traction applications where a large torque is required below base speed and a reduced torque is sufficient for high-speed running.

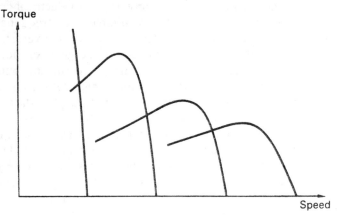

Fɪɢ. 1.5. Induction motor torque-speed characteristics at different supply frequencies and constant terminal voltage.

The shape of the induction motor torque-speed characteristics is largely determined by the rotor resistance, and for normal operation at a fixed frequency, the value of rotor resistance is a compromise. The rotor losses are proportional to slip. A low-resistance rotor gives a small slip and high efficiency at normal loads, but it also permits excessive starting current and yields a low starting torque. A high-resistance rotor, on the other hand, improves the starting conditions, but the full-load slip is large and operating efficiency is reduced. When the induction motor operates on an adjustable-frequency supply, the stator frequency can be reduced for starting, so that the normal starting conditions at fixed utility frequency are not applicable. A low value of rotor resistance is therefore chosen to improve the operating efficiency. A low-resistance rotor is also desirable to reduce the harmonic losses on nonsinusoidal supplies, because the rotor copper loss is one of the main sources of additional harmonic losses (Chapter 6).

In general, induction motor operation takes place at a high power factor with high efficiency, provided the rotor breakdown frequency is not exceeded. Beyond the breakdown point, the motor currents are large, the power factor is low, and the torque per ampere and efficiency are poor. The induction motor can be operated in a closed-loop feedback system in which the rotor slip frequency is directly or indirectly controlled, so that operation always takes place below the breakdown frequency. This type of induction motor drive has speed control and transient performance characteristics which are comparable, if not superior, to those of a dc motor supplied by a thyristor converter. These sophisticated induction motor control techniques are treated in Chapter 7.

1.5.2. *The Synchronous Reluctance Motor*

The elementary synchronous reluctance motor has an unexcited ferromagnetic rotor with polar projections. The reluctance motor, therefore, differs from the conventional synchronous machine, which has either a dc-excited winding or permanent magnets on the rotor. In general, reluctance torque is developed by the tendency of a ferromagnetic material to align itself with a magnetic field. Figure 1.6 shows a two-pole rotor construction that could be used in a reluctance motor.

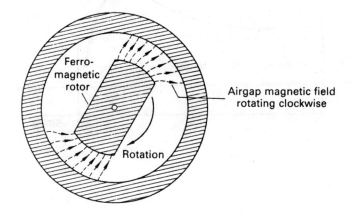

FIG. 1.6. Elementary two-pole synchronous reluctance motor.

If the stator establishes a stationary magnetic field, the salient-pole rotor experiences a torque tending to rotate the polar, or direct rotor, axis into alignment with the stator field axis. This is the position of minimum reluctance for the stator flux. In practice, the stator carries a conventional three-phase winding excited by a three-phase ac supply. A synchronously rotating stator field is established, and the rotor runs in exact synchronism with this field as the salient poles try to maintain the minimum reluctance position relative to the stator flux. The rotor is also fitted with a cage-rotor winding, and the machine accelerates from standstill by induction motor action.

As the rotor speed approaches synchronism, the reluctance torque is superimposed on the induction motor torque, and as a result, the rotor speed oscillates above and below its average value. Provided the load torque and inertia are not excessive, the instantaneous rotor speed can increase to synchronous speed and the rotor locks into synchronism with the stator field. Figure 1.7 shows a typical torque-speed characteristic at a fixed supply frequency and voltage. At subsynchronous speeds the torque is pulsating. Average values are indicated in Fig. 1.7. Once the rotor is synchronized, the cage winding rotates synchronously with the stator field; thus, the rotor winding plays no part in the steady-state synchronous operation of the motor. The machine continues to operate synchronously, provided the pull-out torque of the motor is not exceeded. This is the load torque required to pull the rotor out of synchronism. The *pull-in torque* is defined as the *maximum load torque which the rotor can pull into synchronism with a specified load inertia.* The pull-in torque may be increased at the expense of a larger starting current, but it is always less than the pull-out torque.

When the supply voltage and frequency are varied, a family of torque-speed curves is obtained as in the case of the induction motor. For constant-torque applications, the volts/hertz ratio must again be held approximately constant but a voltage boost is necessary below 20 Hz. Reluctance motors have been widely used in adjustable-speed mul-

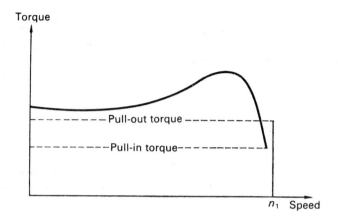

FIG. 1.7. Torque-speed characteristic of the synchronous reluctance motor at constant voltage and frequency.

timotor drives requiring exact speed coordination between individual motors. If all motors are accelerated simultaneously from standstill by increasing the supply frequency, the machines operate synchronously at all times, and they can be designed for optimum synchronous performance without regard to the pull-in torque requirements. Unfortunately, the reluctance motor also exhibits a tendency towards instability at lower supply frequencies (Chapters 6 and 8), but it forms a low-cost, robust, and reliable synchronous machine.

1.5.3. The Permanent Magnet Synchronous Motor

The permanent magnet synchronous motor has the same stator construction as the polyphase induction and reluctance motors. The rotor structure is composed of permanent magnets magnetized in a radial direction. Figure 1.8 shows a cross-section of a

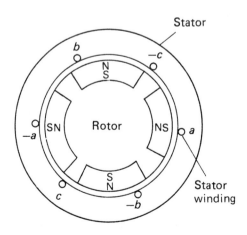

FIG. 1.8. Elementary surface-mounted permanent magnet synchronous machine.

surface-mounted permanent magnet motor in which the distributed three-phase stator winding is represented by three concentrated coils. The magnets are configured in the form of circular arcs and are mounted on the rotor surface. The motor operates at synchronous speed as the rotor flux crosses the airgap and attempts to align itself with the synchronously rotating stator flux. With a fixed stator frequency, the motor speed is independent of torque. As the load torque is increased, the angle between the rotor flux provided by the rotating magnets and the rotating stator flux provided by the stator currents increases. If the load torque is further increased beyond the pull-out torque value, the rotor pulls out of step with the stator flux and the motor stops. Because permanent magnet machines operate at higher efficiencies and power factors than do synchronous reluctance motors, the kVA rating of the inverter is reduced. As the cost of the permanent magnets decreases, these motors are replacing some of the synchronous reluctance motors in industrial applications. They are discussed in greater detail in Chapter 8.

1.5.4. *The Wound-Field Synchronous Motor*

In a wound-field synchronous motor, the rotor flux is produced by electromagnets attached to the rotor structure. The dc current to excite the electromagnets is provided by slip-rings and a remotely located static exciter or by a brushless system with rotating rectifier diodes. In the latter scheme, an ac supply is obtained from a directly coupled ac generator with its armature winding on the rotor, or from the rotor winding of a directly coupled wound-rotor induction motor. The ac power from the rotor winding is rectified to dc by a rectifier bridge mounted on the rotating shaft and is fed directly to the field winding of the synchronous motor without any sliding contacts. The advantage of a wound-field synchronous motor is that the field excitation can be varied to operate the motor at any desired value of lagging, unity, or leading power factor. Leading power factor operation is normal as this helps to overcome the normally lagging power factor associated with induction motors and transformers. Starting torque is provided by adding amortisseur bars to the rotor. These act in a manner similar to the rotor bars of a cage-rotor induction motor, providing starting torque and damping torque for oscillations in load angle caused by sudden changes in load. The torque-speed characteristic of the synchronous motor is similar to that shown in Fig. 1.7 for the reluctance motor. There are two main types of wound-field synchronous machine, the round-rotor machine and the salient-pole machine. They are studied in Chapter 8.

The most important forms of frequency converter and the main types of ac motor have now been introduced. At the present time, the bulk of adjustable-frequency ac drives are a combination of one of the converters and one of the ac motors described in this chapter.

1.6. FUNDAMENTALS OF ELECTRIC DRIVES

An adjustable-speed electric drive system consists of a controllable power converter and an electric motor which drives a mechanical load at an adjustable speed. The main elements of a drive system are shown in the block diagram of Fig. 1.9. The power converter receives ac or dc supply voltages from the main power supply and feeds the motor with appropriately conditioned voltages and currents. In a closed-loop speed control sys-

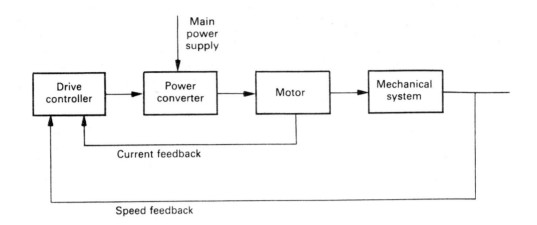

FIG. 1.9. Block diagram of an adjustable-speed drive system.

tem, the drive controller receives actual speed information from the load, which is compared with the reference, or setpoint, value. The difference is used to control the system in such a manner as to reduce the difference. The actual speed should follow the reference value as accurately as possible and without overshoot. Drive control systems usually have a cascaded or hierarchical structure with an inner current control loop and a superimposed outer speed control loop, as indicated in Fig. 1.9. A fast-acting current loop with a preset current limit provides good dynamic performance and effective protection against overcurrents. Position control can be readily added by superimposing a position loop on the existing speed control loop. An inner acceleration control loop may be introduced, if deemed necessary.

The mechanical load has a torque-speed characteristic representing the countertorque which must be overcome by the drive motor. The intersection of the torque-speed curves of drive and load is the point of equilibrium at which the system normally operates in the steady state. The application engineer must know the torque-speed characteristic of the load when selecting and sizing the electric drive. The duty cycle, or variation of load torque with time, is also of great importance in drive selection. If load inertia is high and rapid acceleration is specified, then a significant excess of drive torque over load torque is necessary.

In general, machine torque may act with or against the direction of rotation, and hence there are four operating quadrants in the torque-speed diagram of Fig. 1.10. For consistency, with the earlier ac motor characteristics, speed is plotted horizontally and torque vertically. This form of torque-speed characteristic is preferred by the machine specialist, but for application engineering, the diagram is often drawn as a speed-torque curve with torque plotted horizontally and speed vertically.[5,6] The quadrants of Fig. 1.10 are numbered 1 to 4 in the conventional manner. In the first quadrant, machine torque and speed are both positive, indicating motor operation in the forward direction of rotation. In the third quadrant, torque and speed are both negative, indicating motor operation in the reverse direction. In the second and fourth quadrants, speed and torque

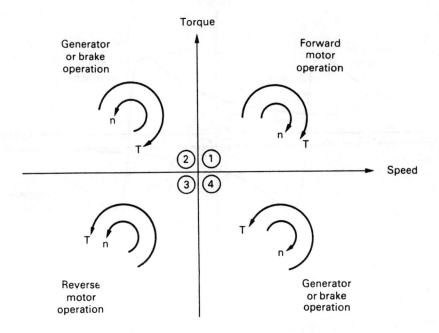

Fig. 1.10. The four-quadrant torque-speed diagram.

have opposite signs and the machine is operating as a generator or brake. A single-quadrant drive is suitable for motor operation only in the first or third quadrants. A two-quadrant drive is capable of driving or braking in a single direction, whereas a four-quadrant drive can operate in all four quadrants, thereby allowing motoring or braking with forward or reverse rotation.

When the machine operates in quadrants 2 and 4, it develops a torque that opposes rotation and retards the load with a braking action. Consequently, the machine absorbs mechanical power from the load and converts it into electrical power. Regenerative braking implies the return of this energy to the electric supply system, with the machine operating as a generator. The electrical power may also be dissipated in an external resistor to give dynamic braking. Alternatively, the electrical power may be dissipated within the machine itself while more electrical power is drawn from the supply and is also dissipated in the machine. This braking action is usually known as plugging.

Four-quadrant operation of the three-phase induction motor is illustrated in Fig. 1.11. The torque-speed characteristic of Fig. 1.3 is repeated, and a second torque-speed curve is added. The latter characteristic is obtained when the stator phase sequence is reversed by interchanging two stator leads. Clearly, motor operation occurs in quadrants 1 and 3, while generator or brake operation occurs in quadrants 2 and 4.

In many drives, operation in the second and fourth quadrants occurs only for brief periods, while the kinetic energy of the rotating elements is dissipated as heat or returned to the supply. However, in the case of a transportation drive or in a crane or hoist application, more prolonged braking operation is common. Thus, when an electric locomotive is running downhill, motor speed tends to increase above its no-load value. This causes the machine to operate as a generator and develop a braking torque which

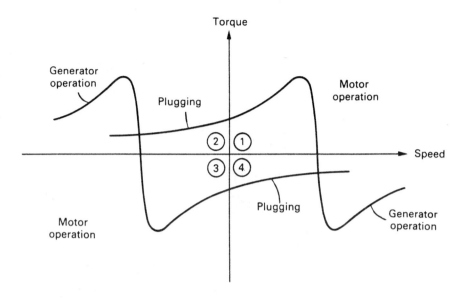

Fig. 1.11. Four-quadrant operation of the three-phase induction motor.

limits locomotive speed. Similarly, when lowering a loaded hoist or elevator, machine speed rises and generator action ensues. These are usually termed overhauling loads because they tend to accelerate, or overhaul, the drive motor, causing it to run faster than the ideal no-load speed and producing a braking action that limits speed.

1.6.1. *Drive Characteristics*

In an adjustable-frequency ac drive, the torque-speed characteristic is altered as shown in Fig. 1.4 by varying stator voltage and frequency. The base speed of the ac motor is its speed at rated frequency and voltage. The drive commonly has a constant-torque capability below base speed, indicating that it is inherently matched to applications requiring a constant drive torque throughout the speed range.

Speed control is possible above base speed by increasing stator frequency, but stator voltage is usually held constant to avoid exceeding the voltage ratings of the motor and converter. This results in a field-weakening or constant-horsepower range of operation in which the available torque decreases with increasing speed. This constant-horsepower characteristic is appropriate for certain industrial loads and for vehicle applications where the available power is limited. In a constant-horsepower drive, the motor operates over a wide range of speed and torque. The physical size of the motor is determined by the peak torque requirement. This occurs at the lowest speed in the constant-horsepower range.

The speed range of a drive system is the ratio of maximum to minimum speeds. The maximum speed may be limited by the power converter or by centrifugal forces in the motor. The restriction on minimum speed may be set by irregular shaft rotation or reduced motor cooling at low speeds. Statement of a speed range figure should clearly indicate whether the motor is developing full-load torque. If the motor is on no-load, the speed range figure has little meaning. Speed regulation, which refers to the ability of the drive to maintain the preset speed for varying loads, is generally defined as the change in speed from no-load to full-load, expressed as a percentage of base speed.

Jogging or threading controls are often fitted in commercial drives to allow small movements or very low speed rotation during setting-up operations. The maximum torque available from the drive is usually set by a torque limit in the drive controller. This limit is essentially a current limit, to avoid overloading of the power converter and motor.

Crane, hoist, and elevator drives operate in all four quadrants of the torque-speed diagram. A wide speed range is required with good speed holding for large variations in load torque from full-load motoring to full-load generating. Process industries frequently require drives in which a number of motors operate at the same speed or at coordinated speeds that are maintained in fixed ratios to each other. For servo drives and position control systems, motor speed must be controllable down to standstill, and a fast reversal is often required. For robotic applications and feed drives in machine tools, a high-performance servo drive is needed with smooth rotation at very low speeds, and torque is required at zero speed to give torsional stiffness to the stationary shaft. These demanding specifications can now be met by ac drive systems. Industrial application of adjustable-speed ac drives is considered further in Chapter 13.

1.6.2. *Load Characteristics*

The load torque may vary with speed in a variety of different ways. A constant-torque load has a torque requirement that is relatively independent of speed, and hence the load power increases linearly with speed. This is characteristic of many industrial loads in which frictional forces predominate—for example, in a belt conveyor with a fixed loading. Other examples of constant-torque loads are extruders, hoisting machines, and mine lifts. Some drives exhibit an initial sticking friction, or stiction, requiring extra breakaway drive torque at standstill, but the running friction is nearly constant for wide variations in speed.

In an eddy-current brake, the torque is viscous and increases linearly with speed so that load power increases as the square of the speed. In fans and centrifugal pumps, the load torque varies almost exactly as the square of the speed; hence the power developed varies as the cube of the speed.

A constant-horsepower load has a torque requirement which decreases with speed so that the power required remains practically constant over a given range of speed. Reel winders, take-up rolls, and spindle drives for machine tools are typical examples of this type of load. Many loads can be classified under one of the headings listed above, but in general, the load torque may be a combination of these characteristics in varying proportions.

1.7. REFERENCES

1. HARNDEN Jr., J.D., and GOLDEN, F.B. (Editors), *Power Semiconductor Applications*, Vols. 1 and 2, IEEE, New York, NY, 1972.

2. BOSE, B.K. (Editor), *Adjustable Speed AC Drive Systems*, IEEE, New York, NY, 1981.

3. ESPELAGE, P.M., and BOSE, B.K., High frequency link power conversion, *IEEE Trans. Ind. Appl.*, **IA-13**, 5, Sept./Oct. 1977, pp. 387-394.

4. ALGER, P.L., *Induction Machines*, Gordon and Breach, New York, NY, 1970.

5. PILLAI, S.K., *A First Course on Electrical Drives*, Wiley Eastern Ltd., New Delhi, 1982.

6. DEWAN, S.B., SLEMON, G.R., and STRAUGHEN, A., *Power Semiconductor Drives*, Wiley-Interscience, New York, NY, 1984.

CHAPTER 2

Power Semiconductors for AC Motor Drives

2.1. INTRODUCTION

A variety of power semiconductor devices is now available for ac motor control, each device with its own particular capabilities and features.[1-6] Advances in manufacturing techniques and process technology have spawned new devices and enhanced the characteristics of existing power semiconductors. In this chapter, the principal devices available to the circuit designer are described and basic circuit considerations relevant to these devices are discussed.

It should be emphasized at the outset that, in power engineering applications, power semiconductors are used as switching devices that are either on or off. This mode of operation is essential because the power engineer is concerned with the conversion and utilization of large amounts of energy, and low-efficiency circuit techniques used in signal electronics are not feasible at the high energy levels which prevail in power electronics.

Switching techniques can be used for the control of dc and ac power, for the rectification of ac to dc, and for the inversion of dc to ac. Thus, dc power control can be achieved by the rudimentary chopper circuit shown in Fig. 2.1. When the mechanical

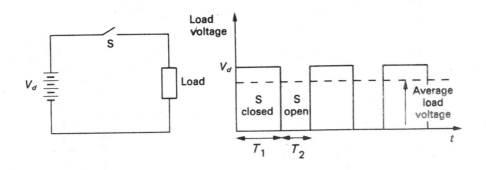

Fig. 2.1. Elementary dc chopper circuit.

switch, S, is closed, the battery voltage, V_d, is applied across the load for the duration of the on-time, T_1. The switch is then opened to remove the load voltage for the off-time, T_2. The average load voltage and load power can be varied by controlling the ratio of on- to off-time. An elementary single-phase inverter circuit is shown in Fig. 2.2. If switches S1 and S2 are alternately opened and closed in antiphase, then a square-wave alternating voltage is developed across the load with a frequency determined by the switching rate.

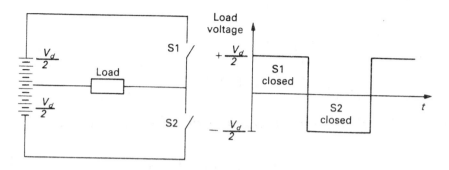

FIG. 2.2. Elementary single-phase inverter circuit.

Obviously, the two switches must not be closed simultaneously; otherwise the battery is short-circuited. It is unrealistic, however, to employ mechanical switches in these circuits, because static switching devices can provide higher speed, better reliability, and longer life. Mercury arc valves and thyratrons were used originally for this purpose, but power semiconductor devices have now supplanted these gas discharge tubes.

An ideal switch does not dissipate energy because it has zero voltage drop in the on-state and zero leakage current in the off-state. It can also switch instantaneously between the zero-resistance on-condition and the infinite-resistance off-condition. The power semiconductor approximates the ideal switch by having a low on-state voltage, or forward voltage drop, resulting in a small conduction loss; and a small leakage current in the off-state, corresponding to a large but finite resistance. It has turn-on and turn-off times of the order of microseconds, or less, permitting sustained high-frequency operation and giving very fast transient response. Fast switching speeds are also important because they imply reduced switching losses and high efficiency.

The basic dc chopper circuit of Fig. 2.1 is redrawn in Fig. 2.3(a) with a transistor switch that is opened and closed by the application of appropriate base currents. When the load is inductive, a free-wheeling diode, D1, is necessary to allow load current to circulate during the off-period of the transistor. The presence of the free-wheeling diode allows the inductive load current to decay with its natural time constant and so avoids destruc-

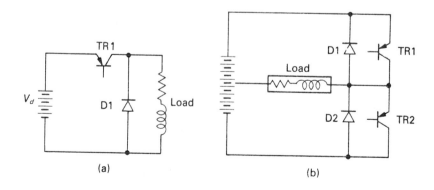

FIG. 2.3. (a) Basic transistor dc chopper; (b) basic single-phase transistor inverter.

tive overvoltages when the transistor is turned off. Similarly, in the single-phase transistor inverter circuit of Fig. 2.3(b), each transistor must have an inverse parallel, or antiparallel, diode when the inverter feeds an inductive load such as an ac motor. During the on-period of TR1, load current flows from right to left in Fig. 2.3(b). When TR1 is turned off, inductive load current continues to flow in the same direction for a time, and must be carried by diode D2 until the load current reverses and TR2 comes into conduction. Diodes D1 and D2 are the so-called "feedback" diodes that are commonly used in voltage-fed inverter circuits.

The transistors in the above circuits may be replaced by other semiconductor switching devices, such as power MOSFETs or thyristors. One of the principal distinctions between the various devices is whether or not a device has a self-turn-off capability. The transistor is turned off by removal of its base drive signal, but the conventional thyristor cannot be turned off at the control electrode when current has been established in the device. In ac-powered circuits such as rectifiers, thyristor turn-off or commutation is accomplished by the natural reversal of the alternating supply voltage, which reduces the thyristor current to zero; but in dc-powered circuits, auxiliary commutating circuitry is required to force the thyristor current to zero.

When the power semiconductor is switching on or off, device voltage and current are both large, resulting in high instantaneous power dissipation and high energy loss. Figure 2.4 shows idealized waveforms of switch voltage, current, and power when current is

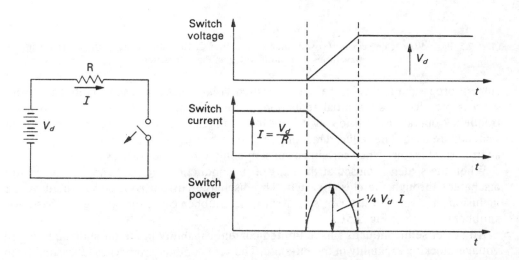

Fig. 2.4. Interrupting current flow in a resistive circuit.

interrupted in a resistive load. A linear variation of switch voltage and current is assumed, and the peak power dissipation is one-quarter of the product of maximum voltage and maximum current. If the circuit is inductive, however, the peak power dissipation is appreciably larger, because inductance tends to prolong current flow.

In general, turn-off losses can be reduced by means of a capacitor connected across the switch, as shown in Fig. 2.5(a). This is the main component of the so-called "snubber" network which snubs, or limits, switch voltage and its rate of rise. As the switch opens, current is diverted into the snubber capacitor, which is initially uncharged. As shown in Fig. 2.5(c), this current diversion slows the build-up of switch voltage, and as a result,

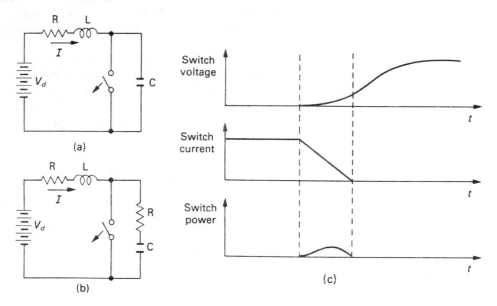

FIG. 2.5. Basic dc switching circuit: (a) with snubber capacitor C; (b) with RC snubber; (c) typical switch waveforms at turn-off with a snubber capacitor.

current drops to a low value before the switch voltage has increased significantly. This process produces a substantial reduction in the switching energy loss, so that high-frequency operation with low switching losses is feasible when power semiconductor switches are used. The reduced rate of rise of switch voltage, or dv/dt, at turn-off, is also an important feature when certain power semiconductor devices are used.

When the switch is closed at the start of its conduction cycle, the snubber capacitor discharges through the closed switch. The discharge current is usually limited by the inclusion of a small resistor in series with the snubber capacitor, giving the basic RC snubber network of Fig. 2.5(b).

All power semiconductors have limited current capability in the on-state and limited voltage-blocking capability in the off-state. The semiconductor device is fabricated from a small silicon chip with a low thermal capacity, which, when conducting, heats up rapidly due to on-state conduction loss. The maximum allowable junction temperature is typically of the order of 125 °C to 200 °C. The maximum temperature must not be exceeded, or device damage will result. The maximum allowable current is determined by the internal power dissipation in the device and the efficiency with which heat is removed from the temperature-sensitive junction. Normally, the power semiconductor is mounted on a metal heat sink with a large surface area which conducts heat away from the device and dissipates it to the surrounding air. Thermal design considerations are discussed in more detail in Chapter 12.

2.2. BASIC SEMICONDUCTOR THEORY

It is well known that semiconductors are substances whose electrical properties lie between those of good conductors and good insulators. Power semiconductor devices use silicon or, less commonly, germanium as the basic material. Both elements have atoms with four orbital, or valence, electrons, and each crystallizes into a lattice with each atom forming strong links, or covalent bonds, with four adjacent atoms, and sharing a valence electron with each of them. The semiconductor is electrically neutral, but mobile charges, or carriers, are generated by overcoming the bonding forces between the valence electrons and the atoms. This separation produces two kinds of carriers, electrons and holes. When the covalent bond is broken, the valence electron is free to move. This free electron creates a vacancy in the lattice which may be filled by an adjacent electron, thereby causing a movement of the vacant electron position. This electron vacancy is called a hole, and it behaves as if it were an independent mobile positive charge. If an electric field is applied to the semiconductor, holes and electrons move in opposite directions, but produce currents in the same direction because of their opposite charges. If a hole and electron collide, then recombination occurs, with the electron completing the covalent bond and the hole disappearing.

The intrinsic, or pure, semiconductor has relatively few carriers at normal temperatures, and consequently, its properties resemble those of an insulator. The conductivity is increased by doping the pure silicon with a small, carefully controlled impurity content. The impurity atoms replace the silicon atoms at various points in the crystal lattice. Doping elements having five valence electrons, such as phosphorus, arsenic, or antimony, are called "donors" because they donate their fifth valence electron as a free electron. The resulting semiconductor material conducts predominately by electron flow, and because the electron is negatively charged, the material is known as an n-type semiconductor. Doping elements with three valence electrons, such as boron, aluminum, gallium, or indium, create holes in the lattice and are called "acceptors" because they accept an electron to complete the covalent bond. Since the semiconductor material conducts mainly by the movement of positively charged holes, it is known as a p-type semiconductor. If the p- or n-type material is heavily doped, it is known as a p^+ or n^+ semiconductor. Lightly doped silicon is designated as p^- or n^- material.

At normal temperatures, the silicon atoms are in a state of vibration. The vibration may result in the breaking of some covalent bonds linking the silicon atoms, thereby generating a limited number of hole-electron pairs. This thermal ionization phenomenon occurs in both the undoped, or intrinsic, semiconductor and in the doped, or extrinsic, semiconductor. In p-type silicon, holes due to impurity atoms constitute the dominant, or majority, carrier; free electrons due to thermal ionization constitute a minority carrier. Conversely, in n-type silicon, electrons constitute the majority carrier and thermally generated holes constitute the minority carrier.

2.3. THE RECTIFIER DIODE

A junction between p- and n-type silicon in a single crystal forms the basis of the p-n junction rectifier, or diode. It may be regarded as a two-terminal electronic switch which is automatically on, or closed, when the anode (p-side) is positive with respect to the cathode (n-side), and is off, or open, when the anode is negative with respect to the cathode. Because the p-n junction is also a basic building block in many other semiconductor devices, an understanding of its behavior is vital.

In the simplified representation of the p-n junction diode in Fig. 2.6, the n-type semi-conductor on the right is depicted solely in terms of donor impurity atoms. These are indicated in the form of an encircled, positively charged atom and an associated free electron. The atom is part of the lattice and is immobile, whereas the electron is a free carrier. Similarly, the p-type semiconductor on the left is depicted in terms of the accep-

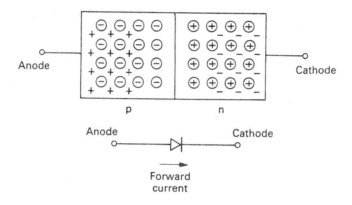

FIG. 2.6. Physical structure and electrical symbol of a p-n rectifier diode.

tor impurity atoms, with encircled immobile negative ions and mobile positive carriers or holes. Consequently, there is a high concentration of electrons on the right of the junction and a high concentration of holes on the left. Even in the absence of an applied electric field, there is a tendency for carriers to move by diffusion from a region of high concentration to a region of lower concentration because of the mutual repulsion between like charges. Thus, free electrons and holes diffuse across the junction and recombine, thereby depleting the region on either side of the junction of mobile carriers and exposing the immobile charged atoms. This depletion, or space charge, region produces an electric field at the junction, which establishes a potential barrier opposing the diffusion of additional majority carriers.

The diode is forward biased by the application of an external voltage which makes the p-side of the junction positive and the n-side negative. The electric field resulting from this voltage narrows the depletion region and reduces the potential barrier. Majority carriers with sufficient energy diffuse across the reduced potential barrier, resulting in a forward current flow across the junction. This forward current increases as the external voltage is increased, until the depletion layer vanishes completely and a very large forward current is possible.

When the diode is reverse biased by an external voltage, which makes the anode negative with respect to the cathode, mobile electrons in the n-region and holes in the p-region are drawn away from the junction. Thus, the depletion layer is widened, increasing the potential barrier and effectively suppressing the diffusion of majority carriers across the junction. However, some minority carriers, due to thermal ionization of the silicon atoms, are also present on either side of the junction (holes in n-region, electrons in p-region). The electric field at the junction assists the flow of these minority carriers across the junction, and this drift current constitutes a small leakage current in the reverse-biased p-n junction.

The static characteristic of a diode is shown in Fig. 2.7. In the forward conduction mode, the diode has a forward voltage drop that is practically independent of current and is typically about 1 V for a silicon diode. The power dissipation is determined by the product of the forward voltage drop and the current. The maximum allowable current is governed by the ability of the device to dissipate heat without exceeding the maximum allowable junction temperature.

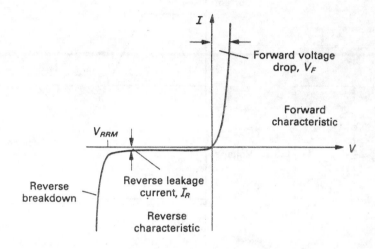

FIG. 2.7. Steady-state characteristic of a p-n rectifier diode.

The reverse characteristic of Fig. 2.7 shows the small leakage current in the reverse blocking mode due to minority carrier flow. If excessive reverse voltage is applied, these minority carriers acquire sufficient velocity and kinetic energy to dislodge additional carriers upon collision with fixed atoms of the crystal in the depletion layer. These carriers, in turn, accelerate, and the process is repeated. A rapid increase in current and an avalanche breakdown of the reverse-biased junction results. The reverse avalanche condition is avoided by operating the diode below the specified repetitive peak reverse voltage, V_{RRM}. Commercial rectifier diodes are generally manufactured by the diffusion process and are now available with forward current ratings from less than 1 A to several thousand amperes, and with reverse voltage ratings from 50 V to 3000 V, or more.

Zener and avalanche diodes are p-n junction diodes in which the reverse breakdown voltage is well defined and which operate in the reverse breakdown region without damage, provided the power rating of the device is not exceeded. These properties make Zener and avalanche diodes ideal for supplying reference voltages in electronic circuit applications and for protecting other semiconductor devices from overvoltage conditions.

2.3.1. *Fast-Recovery Diodes*

In a forward-biased diode, majority carriers from the p- and n-regions cross the junction to become minority carriers on the other side. The diode cannot block reverse voltage until these carriers have been removed or have recombined. Consequently, when reverse voltage is suddenly applied to a diode which has been conducting, the minority carriers in the p- and n-regions are swept back across the junction, and there is a brief interval of high reverse current before the diode reverts to the blocking mode.[7]

Typical current and voltage waveforms are shown in Fig. 2.8. Initially, the diode car-

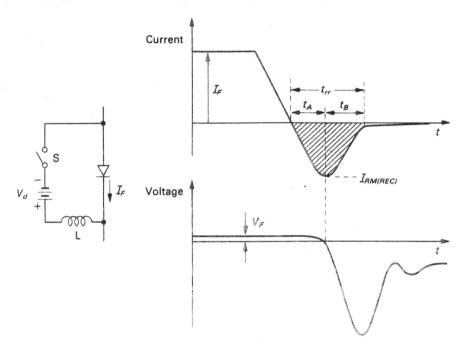

FIG. 2.8. Reverse recovery of a p-n rectifier diode.

ries a forward current, I_F, and has a forward voltage drop, V_F, of about 1 V. When switch S is closed to reverse-bias the diode, the applied voltage, $-V_d$, causes the diode current to decrease at a rate $di/dt = -V_d/L$, where L is the loop inductance. The current reverses direction and the reverse recovery current rises to a peak value, $I_{RM(REC)}$, before decreasing to zero. The ampere-seconds area associated with this recovery current is shaded in Fig. 2.8. This shaded area represents the reverse recovered charge, or "stored charge," the magnitude of which is dependent on the forward current, I_F, prior to turn-off. The total reverse recovery time, t_{rr}, is composed of segments t_A and t_B. If t_B is small, the reverse recovery current falls very rapidly and may generate large transient overvoltages due to high values of $L\,di/dt$. The ratio t_B/t_A, or S-factor, is an indicator of the likelihood of excessive voltage transients when the diode recovers. A "soft" recovery diode has an S value of about unity, signifying low oscillatory overvoltages, whereas a "snappy" recovery diode has a smaller S-factor and larger voltage overshoot.

Reverse recovery performance is not critical in rectifier circuits operating on 50 or 60 Hz ac supplies. In chopper and inverter circuits, however, fast-recovery diodes must be used in conjunction with semiconductor switching devices. In these applications, peak reverse recovery current is an important diode parameter because this current, which must be carried by a semiconductor device elsewhere in the circuit, increases the switching burden on that device. A soft recovery characteristic is desirable to avoid large transient overvoltages that could cause device damage.

For voltage ratings below about 400 V, the sophisticated epitaxial process for the growth of silicon crystals may be used to fabricate diode structures with doping profiles that give fast recovery and low forward voltage drop. At higher voltages, diffusion techniques are used, with gold or platinum doping, to shorten the recovery phase by assisting carrier recombination; but the forward voltage drop increases. Alternatively, electron irradiation of the silicon crystal can be employed to create dislocations in the lattice that speed recombination; but this method tends to give a snappy recovery characteristic. Reverse recovery times for low-voltage epitaxial diodes are measured in tens of nanoseconds. Recovery time rises to several microseconds for high-voltage diffusion diodes rated at 2000 to 3000 V.

The advantages of very fast recovery and low forward voltage drop are combined in the Schottky diode, which employs a rectifying metal-to-semiconductor contact instead of a p-n junction. Current flow is by majority carriers only, giving low conduction voltage and eliminating the turn-off delay associated with minority carrier recombination. However, reverse voltage ratings are limited to about 100 V.

2.4. THE POWER TRANSISTOR AND POWER DARLINGTON[8-12]

The power transistor is a larger version of the normal bipolar junction transistor (BJT) and is used in a switching mode for power conversion and control. The transistor must be held in the on-state by a continuous current signal at the control, or base, terminal. When this signal is removed, the transistor automatically switches off.

As is well known, the bipolar junction transistor is a three-layer device with emitter (E), base (B), and collector (C) regions. The term "bipolar" indicates that current flow consists of a movement of both positive and negative charges — that is, holes and electrons. As indicated in Fig. 2.9(a), the outer elements may be of n-type material, sepa-

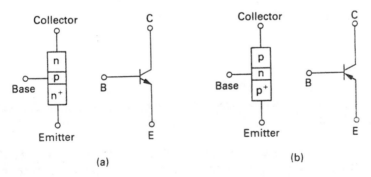

FIG. 2.9. Simplified structure and circuit symbol for (a) an n-p-n transistor and (b) a p-n-p transistor.

rated by a p-type base region. This is the n-p-n transistor, which has the circuit symbol shown. Alternatively, an n-type base region separates p-type elements, giving the p-n-p transistor, with the circuit symbol of Fig. 2.9(b). In each case, the base and emitter form a p-n junction that conducts like a diode in the direction indicated by the arrowhead. For high-voltage and high-current applications, n-p-n transistors are more widely used because they are easier to manufacture and cheaper to buy.

For normal transistor operation, the base-emitter junction is forward biased and the emitter acts as a source of mobile carriers which enter the base region. These injected carriers are electrons in the n-p-n transistor and holes in the p-n-p transistor. In general, the emitter region is made of heavily doped material to increase the number of injected carriers, which become minority carriers when they enter the base. Most of these minority carriers diffuse through the base region, which is very narrow, and arrive at the collector-base junction. This junction is reverse biased by an external voltage, and hence the minority base carriers injected by the emitter are swept into the collector region by the electric field at the collector-base junction. Figure 2.10 is a diagrammatic representa-

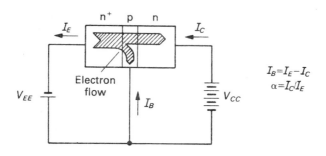

FIG. 2.10. Electron flow in an n-p-n transistor.

tion of the electron current flow in an n-p-n transistor. Some electrons recombine in the base region and do not reach the collector. Hence, collector current, I_C, is slightly less than emitter current, I_E. The difference between these currents is responsible for the small base current, I_B. Thus,

$$I_B = I_E - I_C. \tag{2.1}$$

The ratio of collector to emitter current is defined as the forward current gain, α. Thus,

$$\alpha = I_C / I_E \tag{2.2}$$

and is typically in the range 0.95 to 0.99.

If the current I_E in Fig. 2.10 is reduced to zero by open-circuiting the emitter terminal, there is a small collector leakage current, I_{CBO}, in the reverse-biased collector-base junction.

In practice, the transistor is usually connected in the common-emitter configuration, in which the emitter serves as a common terminal for the input at the base and the output at the collector. This configuration has a high current and power gain. Figure 2.11 shows the common-emitter connections of the n-p-n and p-n-p transistors. As explained above, operation in the active, or conducting, region requires that the base-emitter junction be forward biased. This implies a base supply voltage, V_{BB}, with the polarity indicated in Fig. 2.11. Simultaneously, the collector-base junction is reverse biased by a collector supply voltage, V_{CC}, of appropriate polarity. Normally, V_{CC} is much greater than V_{BB}. As shown in Fig. 2.11 (a), both base-emitter and collector-emitter voltages are positive for n-p-n transistors. Conversely, the base-emitter and collector-emitter voltages are both negative for p-n-p transistors. The transistor has a very low voltage capability for reverse collector-to-emitter voltages and should not be operated in the reverse mode.

In the common-emitter connection, base current, I_B, is the input current which controls collector current, I_C, at the output. The ratio of dc collector current to dc base current is the forward current transfer ratio, or current gain, β (or h_{FE}). Thus,

$$\beta = \frac{I_C}{I_B}. \tag{2.3}$$

On combining Equations 2.1, 2.2, and 2.3, it is readily shown that

$$\beta = \frac{\alpha}{1-\alpha}. \tag{2.4}$$

Consequently, for values of α between 0.95 and 0.99, β falls approximately in the range 20 to 100.

If the convention is adopted that currents flowing into the transistor are regarded as positive, as shown in Fig. 2.11, then I_E is negative for the n-p-n transistor, while I_B and I_C are negative for the p-n-p transistor.

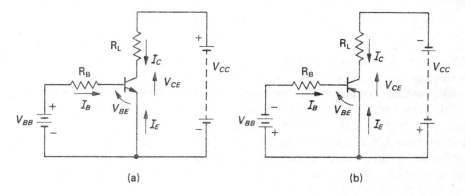

(a)   (b)

FIG. 2.11. Common-emitter connection of (a) an n-p-n transistor and (b) a p-n-p transistor.

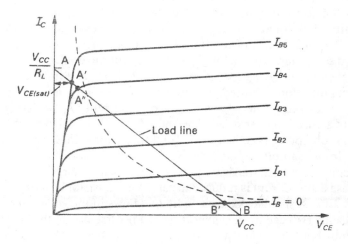

FIG. 2.12. Collector characteristics for a transistor in the common-emitter connection.

2.4.1. *Common-Emitter Characteristics*

The output, or collector, characteristics of a transistor in the common-emitter connection plot collector current, I_C, against collector-emitter voltage, V_{CE}, for fixed values of base current, I_B. Typical characteristics for an n-p-n transistor are shown in Fig. 2.12, where it is evident that collector-emitter voltage exerts little control over collector current. For zero base current, the collector current is very small and the transistor is said "to operate in the cut-off region," corresponding to the off-state. At point A' on the rising portion of the collector characteristics, the collector-emitter voltage is small, and the transistor is said "to operate in the saturation region," corresponding to the on-state. Between these two extremes, there is the active, or transition, region where the transistor can operate as a linear amplifier. The product $I_C V_{CE}$ gives the power dissipation in the transistor, assuming the base power is negligible, and the maximum allowable power dissipation is indicated by the dotted curve in Fig. 2.12. If junction overheating is to be avoided, the transistor must normally operate in the region below this curve.

For a particular load resistor, R_L, as in Fig. 2.11, the collector current, I_C, is given by

$$I_C = \frac{V_{CC} - V_{CE}}{R_L}. \tag{2.5}$$

This is the equation of a load line which may be superimposed on the collector characteristics and which represents the locus of all possible operating points. Ideally, when the transistor is fully conducting, there is negligible collector-to-emitter voltage drop, and the full supply voltage, V_{CC}, appears across R_L, producing a current of V_{CC}/R_L. This collector current value locates point A on the vertical axis of Fig. 2.12. Conversely, when the ideal transistor is in the nonconducting, or cut-off, region, the supply voltage, V_{CC}, appears across the transistor and there is zero collector current. This value locates point B on the horizontal axis of Fig. 2.12. For a purely resistive load, the straight line connecting points A and B is the load line.

For linear operation in an amplifier circuit, the transistor is biased to an appropriate operating point on the load line. In a switching circuit, however, the transistor is often required to operate as close as possible to the idealized operating points, A and B, corresponding to the saturated and cut-off conditions, respectively. In practice, a large base current will cause the transistor to operate in the saturated condition at point A' with a small saturation voltage, $V_{CE(sat)}$, between the collector and emitter. This value represents the on-state voltage of the transistor and is typically less than 1 V. Full conduction, or hard saturation, gives a low $V_{CE(sat)}$ and minimizes on-state losses. Hard saturation is achieved by supplying excess base current. Thus, as explained above, the collector current at saturation is limited to a value $I_{C(sat)} = V_{CC}/R_L$, and excess base current has no influence on this value. By forcing a base current, I_B, which is greater than $I_{C(sat)}/h_{FE}$, hard saturation is ensured. The ratio $I_{C(sat)}/I_B$ is called the "forced current gain," β_F, and is obviously less than the natural transistor gain β, or h_{FE}. It should also be noted that, in the saturated state, the base-emitter voltage, $V_{BE(sat)}$, is greater than $V_{CE(sat)}$, and, as a result, both transistor junctions are forward biased.

When the base drive current is reduced to zero, the transistor is in the cut-off region at point B' in Fig. 2.12. A small leakage current, I_{CO}, flows in the collector circuit, but this may be significantly reduced by the application of a reverse bias to the base-emitter junction. As a result, both transistor junctions are reverse biased and the operating point moves closer to the ideal operating point, B.

In switching applications, the load line may pass through the region of excessive power dissipation, as indicated in Fig. 2.12. When switching between the saturated and cut-off states occurs, the operating point must traverse the active region, and instantaneous power dissipation is high. However, the average power dissipation is not excessive if the operating points A' and B' lie outside the region of excessive power dissipation and if the transistor switches rapidly from one state to the other at moderate repetition frequencies.

2.4.2. Switching Performance [8, 9, 11]

The switching speed of the transistor is limited by certain inherent delays in its response. Figure 2.13 shows idealized switching waveforms for an n-p-n transistor feed-

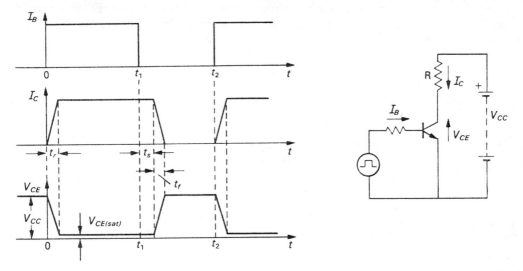

FIG. 2.13. Transistor switching waveforms for a resistive load.

ing a purely resistive collector load. These waveforms indicate that collector current does not respond immediately to changes in input base current. Initially, the base current is zero and the transistor is cut off. Consequently, the collector current is very small and the collector supply voltage, V_{CC}, appears across the transistor. At time zero, the base-emitter junction is forward biased and the base current rises instantaneously to a positive value. However, collector current, I_C, does not increase instantaneously, but rises to its maximum value in a time t_r, which is defined as the rise time. Simultaneously, the collector-emitter voltage falls to $V_{CE(sat)}$. The transistor remains in this saturated state while sufficient forward base current is maintained.

Subsequently, at time t_1, the base drive is removed instantaneously, but the collector current does not respond until the transistor comes out of saturation after a time t_s, known as the storage time. The collector current then starts to fall. Simultaneously, voltage builds up across the transistor and reaches the supply voltage, V_{CC}, when I_C has decreased to the cut-off value, I_{CO}, which is assumed negligible in Fig. 2.13. This interval is defined as the fall time, t_f. The sum $(t_s + t_f)$ is the transistor turn-off time, t_{off}. A typical power transistor rated at 300 to 400 V and having a continuous collector current rating of 15 A might have a rise time of 1 μs, a storage time of 4 μs, and a fall time of 1 μs.

Storage time, usually the major factor in determining the turn-off time of the transistor, is related to a carrier recombination process in the device. When the transistor is in the saturated state, excess minority carriers are present in the base and collector regions; the storage time represents the interval required for recombination to the level corresponding to the boundary between the active and saturation regions. However, the excess, or stored charge, is a function of the excess base current in the on-state. Consequently, storage time can be substantially reduced if the excess charge is minimized by holding the forward base current at the minimum value required to hold the transistor in saturation. The transistor is then said to operate in "quasi-saturation," or "soft saturation," at an operating point such as A″ in Fig. 2.12. Quasi-saturation can be achieved by using an antisaturation, or clamping, circuit such as the well-known Baker clamp of Fig. 2.14, in which the transistor is held in soft saturation by a diode network. If V_{D1},

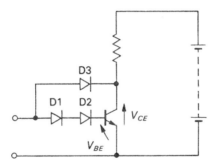

FIG. 2.14. Transistor antisaturation circuit using diodes (Baker clamp).

V_{D2}, and V_{D3} are the diode forward voltage drops, then in the on-state

$$V_{CE} + V_{D3} = V_{BE} + V_{D1} + V_{D2}. \tag{2.6}$$

Assuming similar diodes are used, so that $V_{D1} = V_{D2} = V_{D3}$, then V_{CE} is clamped at a potential of

$$V_{CE} = V_{BE} + V_{D1} \tag{2.7}$$

and storage time is significantly reduced. However, the reduced storage time is obtained at the expense of a larger $V_{CE(sat)}$ and higher on-state losses, as compared with hard-saturation designs.

The rise time of the transistor can be reduced by overdriving the base with forward-bias current at turn-on. For optimum switching performance, the transistor base current should peak rapidly at turn-on; when the collector current reaches its maximum value, the base current is reduced by the antisaturation circuitry. The application of a reverse base bias at turn-off reduces both storage and fall times but also increases the possibility of transistor breakdown, particularly in the case of inductive loads. Chapter 12 examines transistor switching performance on inductive loads and discusses snubber circuits for transistor protection.

2.4.3. *The Power Darlington*[8,9,12]

One of the disadvantages of the power transistor is its relatively low current gain in the saturated state, necessitating substantial base drive current throughout the on-period. Thus, a power transistor with a collector current of 100 A may require a continuous base current of 10 A in the saturated mode. However, two discrete transistors can be interconnected so that the input transistor drives the base of the other, and the resulting Darlington configuration has an enhanced dc current gain. Figure 2.15(a) shows the clas-

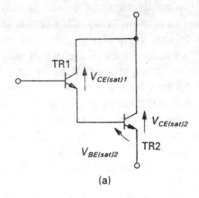

(a)

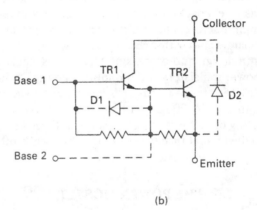

(b)

Fig. 2.15. (a) Basic Darlington connection of two transistors; (b) typical configuration of monolithic Darlington transistor.

sical Darlington connection in which the input base current requirement of the driver transistor is typically 300 to 500 mA for a collector current of 100 A in the output transistor. In the modern monolithic power Darlington, the two transistors are fabricated on a single silicon chip. Leakage current stabilization resistors may be added, as shown in Fig. 2.15(b), and the resulting three-terminal device has the characteristics of a high-current, high-gain transistor.

A disadvantage of the Darlington connection is that the collector-to-emitter saturation voltage is somewhat higher than that of a single transistor of the same rating, resulting in a higher on-state power dissipation. This fact can be explained with reference to

Fig. 2.15(a). For a single transistor in the fully saturated condition, the collector-emitter voltage, $V_{CE(sat)}$, is less than the base-emitter voltage, $V_{BE(sat)}$. However, it is evident from Fig. 2.15(a) that the Darlington connection imposes the constraint $V_{CE(sat)2} = V_{CE(sat)1} + V_{BE(sat)2}$. Thus, the $V_{CE(sat)}$ of the output transistor is equal to the $V_{CE(sat)}$ of the input transistor, plus the $V_{BE(sat)}$ of the output transistor. For this reason, the output transistor cannot be driven into hard saturation, and because it operates in quasi-saturation, it has an increased collector-emitter saturation voltage. However, this on-state voltage is not unduly high compared with other power semiconductor devices.

In the basic Darlington connection, the driver transistor must turn on or off before the output transistor begins to turn on or off. This serial, or sequential, operation inevitably means slow device switching, particularly at turn-off, when the storage time of the Darlington is the sum of the storage times of the two transistors. If the switching process is to be accelerated, the output transistor must be switched simultaneously with the input transistor. Faster turn-off may be achieved by the introduction of a speed-up diode, such as D1 in Fig. 2.15(b). When the base terminal of TR1 is reverse biased at turn-off, diode D1 provides a path for the reverse base current for TR2, and so speeds up the turn-off process. In the modern monolithic Darlington, the base of the output transistor is often made accessible at an external terminal, as indicated in Fig. 2.15(b). The output transistor can then be turned on or off simultaneously with the input transistor to optimize the switching performance.[8]

The monolithic Darlington transistor also has a parasitic diode, D2, in inverse parallel with the device, as shown in Fig. 2.15(b). However, this diode has a long recovery time and cannot be used as a feedback diode in high-frequency inverter circuits. For such applications, several manufacturers have introduced a power Darlington with a very fast recovery diode on a separate chip within the device package.

Aided by advances in design and processing techniques, the voltage and current ratings of large bipolar power transistors and monolithic power Darlingtons have increased steadily. Individual devices are now available having a voltage rating of 1200 V and a continuous-current rating of 300 A. With a reduced voltage rating of 100 V, a continuous-current rating of 750 A is possible. Some of these semiconductors use multiple parallel-connected devices fabricated on a single chip, while others have a single large junction.

2.5. THE POWER MOSFET [13-16]

The basic principle of the field-effect transistor (FET) has been known for many years, but the metal-oxide-semiconductor field-effect transistor (MOSFET) is a more recent development of the field-effect concept, based upon MOS technology, which was originally developed for integrated circuits.

MOSFET devices, as shown in the circuit symbol of Fig. 2.16(a), have three external terminals designated the drain, source, and gate. These correspond to the collector, emitter, and base, respectively, of a conventional bipolar junction transistor. Low-power MOSFETs usually have a planar structure, as shown in Fig. 2.16(b). Fabrication begins with a substrate of p-type silicon into which two n^+ regions are diffused. An insulating silicon dioxide layer, grown on the surface, is etched to allow the metallic source and drain connections to the n^+ regions. Metal is also deposited on the silicon dioxide layer to form the gate contact. In the absence of gate bias, current cannot flow between the drain and source because of the two back-to-back p-n junctions. However, when the insulated

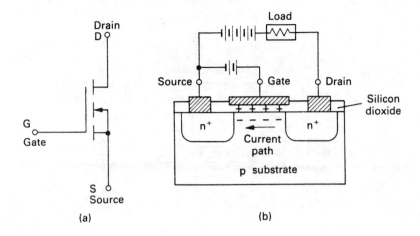

FIG. 2.16. (a) MOSFET circuit symbol; (b) simplified structure of an n-channel planar MOSFET.

gate is made positive with respect to the source, the electric field draws free electrons to the surface of the p substrate. This process forms an n-type channel, which allows electrons to flow from source to drain and causes a lateral current flow, as indicated by the arrow in Fig. 2.16(b). The current flow is enhanced by increasing the magnitude of the gate voltage to form a deeper conducting channel. Consequently, the n-channel enhancement-mode planar MOSFET of Fig. 2.16(b) is a voltage-controlled current device in which conduction is entirely due to the movement of electrons. Conversely, it is possible to construct a p-channel MOSFET in which conduction is entirely due to the movement of holes.

In general, MOSFET conduction is by majority carriers only, as distinct from the conventional bipolar junction transistor, in which hole and electron movement contribute to current flow. Consequently, time delays for removal or recombination of minority carriers are eliminated, and the MOSFET is capable of switching at frequencies in the megahertz range. It should be noted that the gate contact is electrically insulated and draws a negligible dc leakage current, but the input capacitance of the device must be charged and discharged during switching. The main disadvantage of the planar construction shown in Fig. 2.16(b) is that it entails a long conducting channel between the drain and source, giving a large value of on-resistance. The high power dissipation associated with this high resistance limits the use of the planar MOSFET to power levels of about 1 W.

The modern power MOSFET retains the high input impedance and high switching speed of the planar MOSFET, but overcomes its power-handling limitations by allowing current to flow vertically rather than laterally through the device. This vertical DMOS (double-diffused MOS) technology results in the structure shown in Fig. 2.17 for an n-channel device. This has an n^+ substrate on which a high resistivity n^- layer is epitaxially grown. The thickness and resistivity of this epitaxial layer are determined by the required blocking voltage capability of the device. As shown in Fig. 2.17, p^- regions are then diffused into the epitaxial layer, and within these regions n^+ layers are diffused; and an insulated gate contact is added. Source metallization is also deposited, and drain current is collected from the device substrate. A practical power MOSFET has a cellular construction in which this basic cell is repeated thousands of times on a single chip of sili-

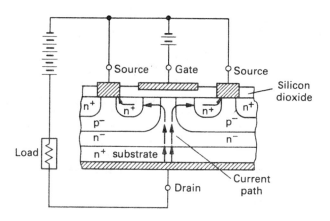

FIG. 2.17. Simplified structure of a vertical channel DMOS power MOSFET.

con. As in the planar MOSFET, a positive gate voltage causes a conducting channel of electrons to be induced in the p^- material under the gate oxide layer. Current flows vertically from the drain electrode and then laterally through the channel region, as indicated by the arrows in Fig. 2.17. Channel lengths are determined by the diffusion process and can be accurately controlled to give short length and low on-resistance.

For normal operation, the n-channel MOSFET drain-to-source voltage is positive, but if the applied voltage is reversed, the device appears as a forward-biased p-n junction. However, this internal reverse diode does not have the fast recovery characteristics needed to function as an antiparallel feedback diode in high-frequency inverter circuits.

The basic switching circuit for an n-channel power MOSFET is shown in Fig. 2.18 with appropriate supply voltages, and an n-p-n bipolar transistor circuit is included for comparison. In a MOSFET, drain current is controlled by gate voltage. The MOSFET output characteristics of Fig. 2.19 show drain current (I_D) as a function of drain-to-source voltage (V_{DS}) with gate-to-source voltage (V_{GS}) as a parameter. At low values of V_{DS}, the curves are approximately linear, indicating a constant value of on-resistance, $R_{DS(on)} = V_{DS}/I_D$. As V_{DS} is increased, the operating characteristics move into a constant-current region, where the characteristics are relatively flat. As in the case of the bipolar transistor, a load line may be superimposed on these output characteristics, and for power switching applications, operation is in the fully on- or off-condition, with a rapid transition from one state to the other.

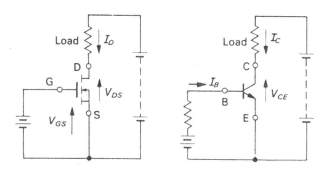

FIG. 2.18. Comparison of a voltage-driven MOSFET switch and a current-driven bipolar transistor switch.

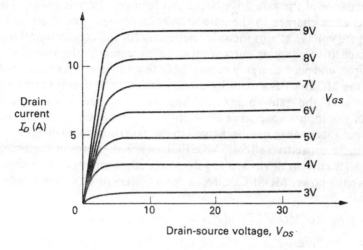

FIG. 2.19. Typical output characteristics of a power MOSFET.

2.5.1. *MOSFET Characteristics*

The transfer characteristic of Fig. 2.20 shows the relationship between drain current (I_D) and gate-to-source voltage (V_{GS}). It is evident that drain current is negligible unless the device is turned on by application of a gate voltage that exceeds a certain threshold value, $V_{GS(th)}$. Above the threshold voltage, the characteristic is nearly linear, indicating that the change in drain current, due to a 1 V change of gate voltage, is constant. This ratio is termed the transconductance (g_m) and is typically between 1 and 10 siemens, or amperes per volt, while the threshold gate voltage is in the region of 2 to 3 V. The gate voltage required to hold the MOSFET in the fully on-condition is of the order of 10 V.

The switching behavior of the power MOSFET is largely determined by the time constant of the series RC circuit, consisting of the input capacitance of the device and the

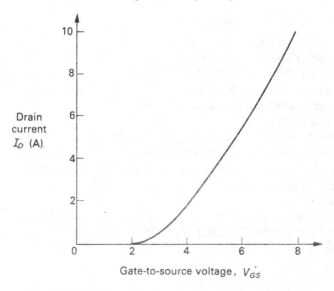

FIG. 2.20. Typical transfer characteristic of a power MOSFET.

source impedance of the gate drive circuit. At turn-on, there is an initial delay while the input capacitance charges to the gate threshold voltage, and there is a further delay before the gate voltage attains the value necessary for full conduction. To reduce turn-on time, the capacitive charging current must be increased by the use of a low-impedance driving source, and peak charging currents of a few amperes may be required.

Because the MOSFET is a majority-carrier transistor, it begins to turn off immediately upon removal of the gate voltage, but the switching speed is limited by the rate of discharge of the input capacitance through the drive circuit. Turn-off is not completed until the input capacitance has discharged to the gate threshold voltage.

Despite these capacitive effects, MOSFET switching times of the order of 100 ns are feasible in high-current devices, and gate power requirements are so small that it is possible to drive the power MOSFET directly from CMOS or TTL integrated circuit logic.

2.5.2. Comparison of MOSFET and Bipolar Transistors

The power MOSFET has several features which make it an attractive candidate for power-switching applications. Its very fast switching times vary little with temperature, as opposed to those of the bipolar transistor. The higher switching speed of the MOSFET also gives significantly lower switching losses. Unlike the bipolar transistor, which is current controlled, the MOSFET is voltage controlled, and its high input impedance results in low drive requirements and compact control circuitry. Paralleling of MOSFETs is facilitated by the positive temperature coefficient of resistance due to majority-carrier operation. Consequently, there is a natural tendency to equalize the current flow among parallel devices. The bipolar transistor, on the other hand, has a negative temperature coefficient, so that current-sharing resistors are necessary for parallel operation. The bipolar transistor is also vulnerable to the second breakdown phenomenon, in which current crowding and localized hot spots are liable to occur. Because of the negative temperature coefficient, extra current is concentrated into a small area, resulting in destructive breakdown of the die. This particular second breakdown failure mechanism is absent in a MOSFET, and consequently, it has a wider safe-operating area than the bipolar transistor.

If the total power loss of a MOSFET and a bipolar transistor of the same die area are compared, it is found that the switching losses of the MOSFET are considerably less than those of the bipolar, but the on-resistance and conduction loss of the MOSFET are greater. At high switching frequencies, the bipolar switching losses are excessive and the MOSFET is more efficient. At low frequencies, conduction losses are dominant and the bipolar is superior. Although the crossover frequency is a matter of some debate, it lies in the region of 10 to 30 kHz.[17]

The major limitation of the power MOSFET, particularly at high blocking voltages, is the large on-resistance, $R_{DS(on)}$. This parameter is obviously crucial for determining the on-state power dissipation of the device. Manufacturers' data sheets quote the value corresponding to a junction temperature of 25 °C. In low-voltage devices rated at less than 100 V and with current ratings of the order of 30 A, $R_{DS(on)}$ values of about 30 or 40 mΩ have become practical. However, at the maximum operating junction temperature of 150 °C, the on-state resistance may be twice the listed value. A typical device with a voltage rating of 100 V and a current rating of 30 A has an on-state voltage of 2 to 3 V at rated current and normal operating temperature. In the case of a 500 V, 10 A MOSFET, however, the voltage drop at rated current and maximum junction temperature is about

8 V. At higher voltage ratings, the on-state voltage is unacceptably large (particularly in high-current devices); thus, the power MOSFET is best suited to high-frequency switching applications at voltages below about 500 V.

2.6. THE THYRISTOR[7,18-24]

The thyristor, or silicon controlled rectifier (SCR), was introduced in 1957 by the General Electric Company (U.S.A.). It quickly became the workhorse of the expanding power electronics industry. The basic properties of the thyristor are now well known. It is essentially a solid-state switch that can be triggered from an open, or off-condition, into a closed, or on-condition. Like the diode, the thyristor is a unidirectional conducting device that blocks current flow in the reverse, or cathode-to-anode, direction. Unlike the diode, however, the thyristor can also block current flow in the forward, or anode-to-cathode, direction until it is triggered into conduction by a low-power pulse applied between the gate electrode and the cathode.

As shown in Fig. 2.21, the thyristor is a four-layer arrangement of p-n-p-n silicon. The

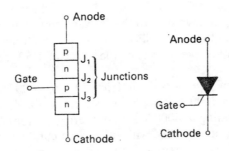

FIG. 2.21. Thyristor structure and circuit symbol.

main terminal electrodes are attached to the two end layers, known as the emitter layers. The p-type emitter is the anode and the n-type emitter is the cathode. The two inner layers are known as the base layers; the control electrode, or gate, is attached to the p-type base. When negative anode-to-cathode voltage is applied, the center junction is forward biased, but the two outer junctions are reverse biased. Consequently, the reverse characteristic is similar to that of a silicon diode, where a small leakage current flows until the critical reverse breakdown voltage is exceeded, when a destructive breakdown occurs.

Unlike the diode, the thyristor draws a small leakage current when positive anode-to-cathode voltage is applied, because the center junction, J_2, is now reverse biased. If the applied voltage is increased to the critical forward breakover voltage, $V_{(BO)}$, the thyristor switches rapidly into the fully conducting condition. But this triggering mechanism is not normally used. Gate triggering, the usual method, is achieved by injecting a small external gate current which forward-biases the gate-cathode junction, J_3. This current reduces the forward breakover voltage, as shown by the steady-state characteristics of Fig. 2.22. Consequently, the thyristor may be triggered by the application of an external forward voltage that is less than the breakover value and then, at the desired instant, the injection of sufficient gate current to reduce the breakover voltage below the applied voltage.

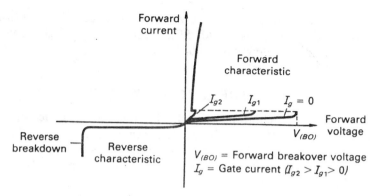

Fɪɢ. 2.22. Anode-to-cathode characteristics of the thyristor.

Once the thyristor has been gated, it latches in the on-state and the gate electrode loses control. The forward-blocking condition cannot be regained unless the gate signal is removed and the forward current is reduced below a certain minimum value known as the holding, or sustaining, current. The peak repetitive forward and reverse voltage ratings are denoted by V_{DRM} and V_{RRM}, respectively, and are approximately equal for the common reverse-blocking thyristor. In the forward-conducting region, the thyristor has characteristics similar to those of a silicon diode of approximately the same current rating. Thus, the forward voltage drop is of the order of 1.5 V and increases slightly with current.

The voltage ratings of thyristors range from 50 V to 4000 V, and current ratings vary from a few amperes to 3000 A. Single devices are available for operation at line frequency with a voltage rating of 4000 V and an average current rating of 1000 A, or more. Since these ratings are not possible with other semiconductor devices, the thyristor is paramount in high-power applications up to tens of megawatts.

2.6.1. *The Two-Transistor Analogy*

Since the thyristor is a p-n-p-n device, it can be visualized as consisting of two interconnected complementary transistors, a p-n-p and an n-p-n, having a common collector-base junction, as shown in Fig. 2.23. The various mechanisms of thyristor triggering can be explained with the use of this two-transistor analogy.

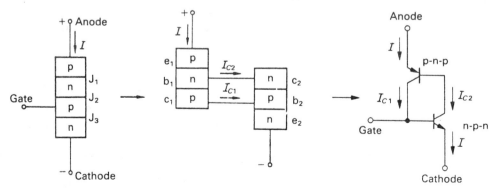

Fɪɢ. 2.23. The two-transistor analog of the thyristor.

Figure 2.23 shows that the collector of each transistor is cross-coupled to the base of the other; consequently, there is a positive feedback action that allows each transistor to drive the other into saturation. Thus, a positive gate current causes the n-p-n transistor to conduct, and as a result, the collector current, I_{C2}, flows. This current becomes a base current for the p-n-p transistor, causing the collector current, I_{C1}, to flow. Because I_{C1} is also a base current for the n-p-n transistor, the thyristor is latched into the on-state, even if the external gate drive is removed. In general, when the thyristor has a positive anode-to-cathode voltage, each transistor is biased in the conventional manner and has a common-base current gain, α, which expresses the ratio of collector to emitter current. Thus, for the p-n-p transistor, α_1 is the fraction of the hole current injected at emitter e_1 that reaches collector c_1. If I is the external current entering emitter e_1, the collector current is given by $I_{c1} = \alpha_1 I + I_{1co}$, where I_{1co} is the collector leakage current of the transistor.

Similarly, the n-p-n transistor has a current gain, α_2, which is the fraction of electron current injected at emitter e_2 that reaches collector c_2. Thus, $I_{c2} = \alpha_2 I + I_{2co}$, where I_{2co} is the leakage current of the n-p-n transistor.

The sum of the two collector currents is equal to the external circuit current, I. Therefore,

$$I = I_{c1} + I_{c2}$$

$$= (\alpha_1 + \alpha_2) I + I_{co} \qquad (2.8)$$

where I_{co} is the total leakage current of the thyristor.
Solving for I gives

$$I = \frac{I_{co}}{1 - (\alpha_1 + \alpha_2)}. \qquad (2.9)$$

If the current gains α_1 and α_2 are small, the thyristor current is only slightly greater than the leakage current, I_{co}, and hence the device is in its forward-blocking state. The thyristor remains in this off-condition unless the sum of the current gains, $\alpha_1 + \alpha_2$, is increased by one of the methods about to be described. When $\alpha_1 + \alpha_2$ approaches unity, the denominator of Equation 2.9 approaches zero, indicating that regenerative action takes place and the thyristor current becomes infinitely large. In practice, the resistance of the external circuit must limit the flow of current, because both transistors are saturated and all junctions are forward biased. Under these conditions, the p-n-p-n device has a very low impedance and is in the forward-conducting state, or on-state.

Gate triggering is explained by the fact that current gain, α, of a transistor is a function of emitter current (Fig. 2.24); consequently, if the base-emitter junction of either

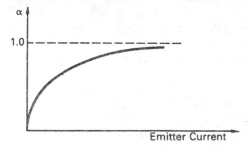

FIG. 2.24. Variation of transistor current gain with emitter current.

transistor is forward biased, the sum of the current gains, $\alpha_1 + \alpha_2$, will increase, causing the thyristor to turn on. Usually the base of the n-p-n transistor is used as the gate terminal into which current is injected in order to switch the thyristor into the conducting state. As explained above, the thyristor continues to conduct when the gate current is removed, provided the load current exceeds the holding, or sustaining, current. When load current is less than the holding current, $(\alpha_1 + \alpha_2) < 1$, and the device cannot sustain regeneration. Consequently, the thyristor reverts to the forward-blocking mode.

An avalanche breakdown occurs in a transistor when the collector-to-emitter voltage is increased excessively. This is due to the leakage current carriers which arrive at the collector junction with sufficient energy to dislodge additional carriers. These carriers, in turn, dislodge more carriers, and the avalanche multiplication causes a sharp increase in transistor current; hence the current gain, α, also increases. The forward-blocking voltage of a thyristor can be gradually increased until avalanche breakdown occurs at the common collector junction, J_2. The sum of the current gains then reaches unity and the thyristor switches on. The forward-blocking voltage just prior to switching is the forward breakover voltage, $V_{(BO)}$, as already defined. In practice, this method of firing a thyristor will destroy some devices. Gate triggering is the mechanism normally employed.

Static dv/dt capability. A high rate of increase of forward anode-to-cathode voltage (dv/dt) can turn on a thyristor, even though there is no gate signal and the forward breakover voltage is not exceeded. When the thyristor is in the forward-blocking state, or off-state, the applied voltage appears as a reverse bias on junction J_2 of Fig. 2.21, because junctions J_1 and J_3 are forward biased. The reverse-biased p-n junction is equivalent to a capacitor, because the mobile current carriers are removed from the vicinity of J_2, leaving exposed charges on either side of the junction, as explained earlier. When the applied voltage is increased, there is a further movement of mobile charges which is equivalent to the charging current of the capacitor. If the forward-blocking voltage has a high rate of increase, there is a significant charge movement away from the central junction, J_2. Additional carriers are injected from the end regions to neutralize the excess charge. This represents a current flow across the gate-cathode junction, J_1, which is equivalent to an externally supplied gate current. Thus, if the applied dv/dt exceeds a critical value, the thyristor turns on. Spurious dv/dt triggering will not normally damage the thyristor, but it will cause circuit maloperation which may result in a destructive overcurrent condition.

Thyristors have been developed with a "shorted emitter" structure which has a superior dv/dt withstand capability.[7] These devices have a partial gate-to-cathode short circuit that diverts some of the dv/dt-induced capacitive charging current around the gate-cathode junction. An enhanced dv/dt capability results, but some gate triggering sensitivity is sacrificed.

Commercial thyristors have off-state dv/dt ratings that are typically between 200 and 1000 V/μs. It is normal practice to limit the maximum dv/dt applied to the thyristor by placing an RC snubber network across the terminals of the device, as discussed in Chapter 12.

2.6.2. *Gate Triggering*

Gate triggering is the normal mechanism for thyristor firing, as explained above. Figure 2.25 shows the internal structure of a conventional center-gate thyristor in which the

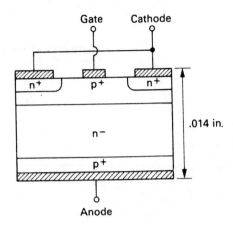

FIG. 2.25. Structure of a conventional center-gate thyristor.

gate electrode lies within the circular cathode electrode. Steep-fronted gate pulses are desirable for proper device operation and to ensure that thyristors with different gate sensitivities turn on at precisely the same time. The gate pulse must not be removed until the anode current has increased to a certain minimum value known as the latching current; otherwise the thyristor will turn off when the gating signal is removed. Once the latching current has been exceeded, the anode current can be reduced to a somewhat lower value before the thyristor reverts to the forward-blocking state, or off-state. This is the holding current, which has already been mentioned. With resistive loads, anode current builds up rapidly, and short gate pulses are satisfactory. Inductive loads delay the build-up of anode current, and longer gate pulses are necessary to ensure that the thyristor will not unlatch when the gating pulse is terminated. If the load is highly inductive or if the current flow can become intermittent, then a low-level continuous gating signal or a train of gating pulses should be applied.

Because the gate-cathode circuit of the thyristor is a p-n junction, it has a characteristic similar to that of a diode. The voltage-current characteristic may vary considerably between nominally identical devices, and the gate firing circuitry, therefore, must be designed to accommodate this variation. Figure 2.26 shows typical gate characteristics as included in the thyristor data sheet. The spread in the V/I characteristic is indicated by the limit lines A and B, and the superimposed load line shows that an appropriate gate firing circuit for the device could have an open-circuit voltage of 6 V and a source impedance of 6 Ω, corresponding to a short-circuit current of 1 A. The indicated load line avoids the cross-hatched area, where firing cannot be guaranteed for all units, and it also lies within the thyristor gate ratings for maximum instantaneous voltage, current, and power. Reference 19 contains more detailed information.

Rate of rise of anode current (di/dt). When a thyristor is triggered into conduction by a gate signal, there is a finite turn-on time during which the anode-to-cathode voltage decreases from the off-state value to the on-state value. Simultaneously, the anode current increases. The product of voltage and current represents the instantaneous power loss in the thyristor during turn-on. If the anode current increases rapidly, the power loss at turn-on becomes significant, and over-heating may occur at high switching frequencies. The problem is aggravated by the fact that the anode current is initially confined to a

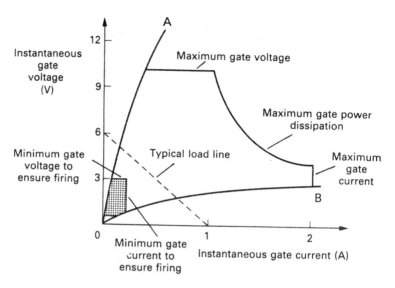

FIG. 2.26. Gate characteristics of a thyristor.

small area near the gate electrode. The conducting area then expands over the rest of the junction. If the anode current increases rapidly, the power loss is initially dissipated in a localized hot spot near the gate electrode. Thus, an excessive rate of rise of forward current (di/dt) will degrade or destroy the thyristor; a maximum di/dt rating is specified in the data sheet.

Excessive di/dt may occur when a thyristor which blocks a high forward voltage is suddenly triggered into a resistive or capacitive load, but the slow build-up of current in an inductive load prevents excessive in-rush current. When thyristors are fitted with shunt capacitors for transient voltage suppression, a small series resistor is included to limit the in-rush current when the thyristor is triggered. Small series inductors are sometimes included in high-power inverter circuits to limit the turn-on di/dt. Saturable reactors may also be used to delay the build-up of current for a few microseconds. When the reactor saturates, it no longer affects the circuit operation.

Turn-on time can be reduced and di/dt capability enhanced by overdriving, or "hard-firing," the thyristor with a gate current that is significantly in excess of the minimum value required. A gate firing waveform that is widely used has an initial high value of current with a very fast rise time; the initial current is then quickly reduced to a lower value and remains at that level for several microseconds. Most high-power thyristors now use the amplifying gate structure, which gives improved di/dt capability without excessive gate current. This capability is achieved by the fabrication of a pilot thyristor on the same silicon wafer as the main thyristor. The gate pulse from the firing circuit initiates conduction in the pilot thyristor, which then drives a large-area gate for the main thyristor and rapidly turns it on. New device structures have also been developed with interdigitated gate electrodes that initiate conduction over a large area and so enhance the di/dt capability.

2.7. PRINCIPLES OF THYRISTOR COMMUTATION

As already mentioned, the thyristor gate electrode loses control when the device conducts. The process of interrupting current flow and restoring the thyristor to its nonconducting, or blocking, state is known as commutation. The technique used to achieve commutation is one of the main distinctions between thyristor circuits. [25-30]

When the thyristor is in the conducting state, as shown in Fig. 2.27, the base regions

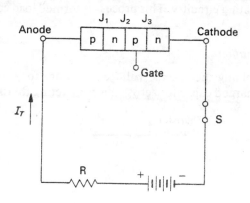

FIG. 2.27. Thyristor in conducting state with gate open-circuited.

on either side of the center junction, J_2, contain a high concentration of holes and electrons, and each of the three junctions is forward biased. The simplest way to commutate the thyristor is to interrupt the current by a mechanical switch. The charges within the thyristor then recombine and the thyristor regains its forward-blocking ability. Mechanical switching is impractical for high-frequency operation, and more effective static commutation techniques have been developed. In such circuits, the thyristor current is interrupted by the application of a reverse anode-to-cathode voltage across the thyristor. This procedure reduces the forward current to zero and then produces a brief pulse of reverse recovery current as the holes and electrons in the vicinity of the two end junctions, J_1 and J_3, diffuse to these junctions. While the reverse current flows, the anode-to-cathode voltage of the thyristor remains at about $+1.5$ V. When the current carriers at the end junctions have been removed, the reverse current falls to zero and the reverse-biased end junctions are able to block the applied inverse voltage. However, the thyristor is not able to block forward voltage because a high concentration of current carriers is still present in the vicinity of the center junction, J_2. A further time interval must be allowed for these carriers to recombine naturally. If forward voltage is reapplied before recombination is complete, the thyristor will immediately revert to the conducting state.

The turn-off time, t_q, is defined as *the time interval between the reduction of anode current to zero and the regaining of the forward-blocking capability.* The amount of forward current before commutation influences the turn-off time, because the trapped charge at junction J_2 is directly proportional to this current. For present-day thyristors, the turn-off time is typically between 3 and 100 μs, depending upon the design and construction of the particular thyristor.

Thyristor circuits may be classified in terms of the method used to achieve commutation. In circuits which operate from a dc supply, the thyristor current must be forced to zero by the application of a reverse anode-to-cathode voltage produced by auxiliary components in the circuit. This process is called forced commutation, and the auxiliary circuit is called the forced commutating circuit. In ac-powered circuits, commutation is achieved by the natural reversal of the alternating supply voltage, which reduces the thyristor current to zero and then applies the reverse bias. This method is termed natural, or phase, commutation or ac line commutation. In some dc-powered thyristor circuits, the nature of the load is such that thyristor commutation occurs naturally, without any auxiliary commutating circuitry. This process is termed load commutation.

2.7.1. *Forced Commutation*

One method of applying the reverse voltage necessary for forced commutation is to switch a previously charged capacitor across the conducting thyristor (Fig. 2.28). Switch

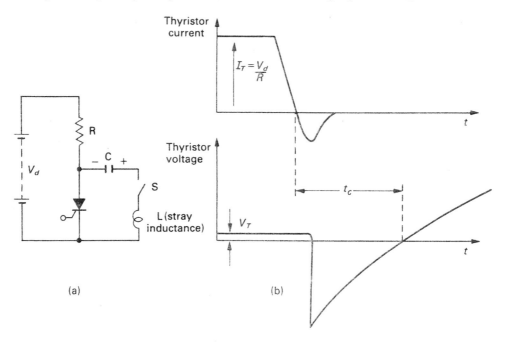

FIG. 2.28. Forced commutation by means of a charged capacitor: (a) basic circuit; (b) current and voltage waveforms.

S represents a transistor or auxiliary thyristor which is triggered at the desired commutation instant. The current and voltage waveforms for the main thyristor are shown in Fig. 2.28(b). Before the commutation is initiated, the anode-to-cathode voltage has a value of approximately 1.5 V, corresponding to the on-state voltage drop, V_T. The turn-off behavior of the thyristor exhibits a reverse recovery phenomenon which is similar to that of a rectifier diode: following the switching of the charged capacitor across the thyristor there is a period during which the forward current drops to zero and a brief pulse of reverse current flows. The reverse recovery time is typically only a few microseconds, but the sudden cessation of reverse current may produce damaging voltage transients, as in the case of the rectifier diode. The reverse recovery current, which is limited by the stray inductance of the circuit, removes the free charge from the end junctions, resulting in a corresponding reduction in the charge on the capacitor, C. The drop in capacitor voltage is small, however, and the remaining capacitor voltage is blocked by the two end junctions and appears as a reverse bias on the thyristor.

The dc supply now charges up the capacitor with the opposite polarity. When its voltage reverses, the thyristor is again subjected to a forward voltage, but it will not conduct, provided recombination is complete at junction J_2 and the thyristor has regained its forward-blocking capability. The time between the reduction of the anode current to zero and the reapplication of forward voltage is called the circuit turn-off time, or commutating time, t_c. This interval must be equal to or greater than the turn-off, or recovery, time, t_q, required by the thyristor; otherwise a commutation failure will result. In practice, the thyristor turn-off time is measured by reducing the capacitance, C, thereby reducing the commutating time provided by the circuit, until a commutation failure occurs.

In general, the thyristor current before commutation determines the amount of free charge at junction J_2 and, hence, determines the recovery time required by the thyristor. The commutating circuit must provide sufficient turn-off time for the largest current to be commutated. The provision of adequate turn-off time determines the capacitance required.

Practical forced commutating circuits usually contain a capacitor and an inductor. Turn-off of the conducting thyristor may be initiated by the triggering of another main thyristor in the power converter circuit or by the triggering of an auxiliary thyristor in the forced commutating circuit. Figure 2.29 shows a single-phase inverter circuit which

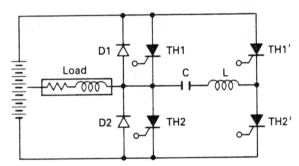

FIG. 2.29. Single-phase thyristor inverter employing the McMurray commutation method.

employs the McMurray method of forced commutation by an auxiliary resonant circuit. [31] If this circuit is compared with the corresponding transistor circuit of Fig. 2.3(b), it is seen that transistors TR1 and TR2 are now replaced by main thyristors, TH1 and TH2, with their associated forced commutating circuitry consisting of auxiliary thyristors TH1' and TH2' and an LC circuit. Devices with a controlled turn-off capability, such as transistors, do not require forced commutating circuitry and obviously give more compact power converter equipment. In a thyristor circuit, the size, weight, and cost of the auxiliary components required for forced commutation are typically greater than the size, weight, and cost of the main thyristor itself, but in medium- and high-power inverter applications, the thyristor may be the only available device with the necessary power-handling capability. The McMurray forced commutating circuit is studied in more detail in Chapter 4.

Reapplied dv/dt. In most forced-commutated circuits, the thyristor is reverse biased for a specified time (of the order of 10 to 30 μs) and is then required to block forward voltage. The rate at which this forward voltage is reapplied must not be excessive; otherwise the thyristor will fail to turn off. The reapplied dv/dt capability should not be confused with the static, or off-state, dv/dt capability, which is applicable when the thyristor is already in the forward-blocking state, or off-state. Because the reapplied dv/dt capability is a turn-off time condition, the maximum allowable value is generally included in the turn-off time specification.

2.7.2. *Natural Commutation*

Thyristors are widely used in such ac-powered circuits as phase-controlled converters, where the natural reversal of the ac supply voltage commutates the conducting thyristor, so that forced commutating circuitry is not required. The simplest thyristor circuit employing natural, or phase, commutation is the single-phase half-wave rectifier. Figure 2.30 shows the basic circuit with a resistive load. During the negative half-cycle of

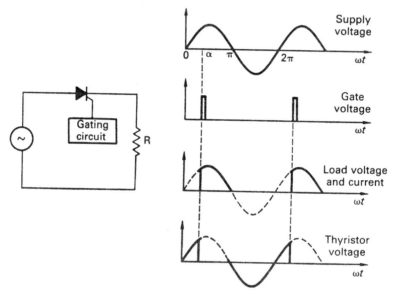

FIG. 2.30. Single-phase half-wave thyristor rectifier circuit with a resistive load.

the ac supply voltage, the thyristor blocks the flow of load current. During the positive half-cycle, the thyristor is forward biased but will not conduct until a gating pulse is applied. The thyristor then quickly turns on and the supply voltage appears across the load for the remainder of the positive half-cycle (assuming that the on-state voltage of the thyristor is small compared with the supply voltage, as is usually the case). The load voltage and current waveforms are identical in the case of a purely resistive load. At the end of the positive half-cycle, the thyristor current falls to zero and the supply voltage reverses. The resulting reverse bias voltage applied across the thyristor rapidly turns it off. (The brief pulse of reverse recovery current is ignored in Fig. 2.30.) The thyristor is now reverse biased for the complete negative half-cycle of the supply voltage, a length of time that provides more than adequate thyristor recovery time unless the supply frequency is increased to tens of kilohertz.

When a rectifier diode is used in this circuit, the full positive half-cycle of supply voltage appears across the load. The angle by which the thyristor retards the start of conduction is called the firing angle, or delay angle, α. Variation of α from zero to 180 degrees reduces the average dc output voltage from a maximum value to zero. This process is known as phase control.

When the load is inductive, the current builds up gradually at the start of conduction, and a prolonged gating pulse is necessary to allow the inductive load current to increase above the latching value of the thyristor. Subsequently, the decaying inductive current holds the thyristor in conduction after ωt equals 180 degrees. The negative supply voltage then opposes the inductive current, reducing it to zero, and the thyristor turns off and is reverse biased for the remainder of the negative half-cycle. Additional single-phase and polyphase rectifier circuits are described in greater detail in Chapter 3.

AC power control can be obtained if a second thyristor is introduced into the half-wave rectifier circuit as shown in Fig. 2.31. The thyristor is a unidirectional device, and

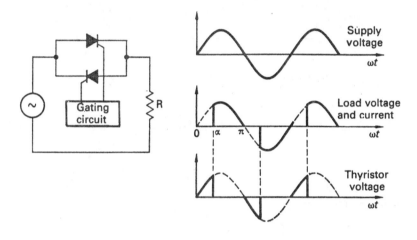

FIG. 2.31. Phase control of alternating voltage in a single-phase back-to-back thyristor circuit with a resistive load.

hence alternating current flow requires a pair of devices connected in inverse parallel as shown. The thyristors are triggered alternately at identical points in their anode-to-cathode voltage cycles, and a symmetrical alternating voltage is applied across the load. This voltage waveform has no dc component, but odd harmonics are present, principally the third. For a purely resistive load, the voltage and current waveforms are as shown in Fig. 2.31.

Variation of the thyristor delay angle, α, controls the average power delivered to the load by the ac supply. Thus, this circuit can be used as an ac power regulator for heating or lighting loads. The circuit is also widely used for controlling the speed of single-phase ac series or universal motors.[32] In these schemes, motor speed is controlled by variation of the applied voltage, but applied frequency is not varied. A typical motor control scheme is shown in Fig. 2.32. Three-phase versions of this system are studied in Chapter 9. The bidirectional thyristor, or triac, which is described in Section 2.8.1 of this chapter, may be used in place of the two antiparallel thyristors.

2.7.3. *Load Commutation*

This form of commutation is achieved in a dc-powered circuit when the nature of the load is such that there is a natural tendency for current to fall to zero sometime after the thyristor has been gated. If the load is a series RLC circuit, as in Fig. 2.33, and if R is

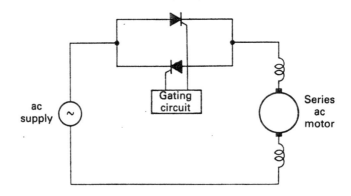

FIG. 2.32. Basic circuit for speed control of a single-phase ac series motor.

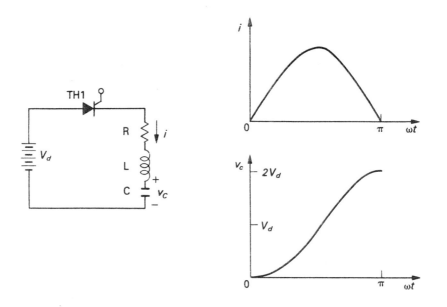

FIG. 2.33. Thyristor load commutation by means of a series LCR circuit.

small, the load current is a near-sinusoidal half-cycle pulse. When the load current falls to zero at the end of the half-cycle, thyristor TH1 turns off, leaving capacitor C charged to a voltage approaching $2V_d$, which applies a reverse bias to the thyristor. This load-commutated circuit can be used for the resonant overcharging of a capacitor. However, the commutation technique is classified in some textbooks as a particular form of forced commutation.

Load commutation of the inverter is also achieved when the inverter feeds a synchro-nous motor. Above a certain speed, the generated emfs in the windings of the synchro-nous machine are able to effect the commutation of thyristor current without any auxil-iary forced commutating circuitry. The commutation process is essentially the same as that in a naturally commutated thyristor converter fed from an ac network, and the volt-age and current waveforms are identical, as shown in Chapter 4. The absence of a forced commutating circuit improves inverter efficiency and reliability; consequently, the syn-chronous motor powered by a load-commutated inverter (LCI) is widely used in high-power pump and compressor drives. Below about 10 percent of full speed, however, the generated emf in the synchronous machine is insufficient for load commutation; alterna-tive forced commutation techniques must be introduced for starting and low-speed run-ning.

2.8. OTHER MEMBERS OF THE THYRISTOR FAMILY

The term thyristor is used to describe all four-layer p-n-p-n devices with regenerative latching action. The reverse-blocking triode thyristor, or silicon controlled rectifier (SCR), described above, is the most important member of the thyristor family, but there are several other devices that are of interest in ac motor control applications. The triac is a bidirectional thyristor widely used for ac power control of heating and lighting loads and for the fixed-frequency control of ac motors. The inverter-grade thyristor is a con-ventional reverse-blocking thyristor with enhanced dynamic characteristics to give improved performance in forced-commutated inverter circuits.

Several new thyristor devices developed in recent years are also described in this sec-tion. These are the asymmetric thyristor (ASCR), the reverse conducting thyristor (RCT), the gate-assisted-turn-off thyristor, and the field-controlled thyristor (FCT). The most important new thyristor device to gain acceptance recently is the gate turn-off thyristor (GTO), which is discussed in detail in Section 2.9.

2.8.1. *The Triac*

The conventional thyristor, or SCR, has a reverse-blocking characteristic that pre-vents current flow in the cathode-to-anode direction. However, there are many applica-tions, particularly in ac circuits, where bidirectional conduction is required. Two thyris-tors may be connected in inverse parallel, as shown in Fig. 2.31, but at moderate power levels the two antiparallel thyristors can be integrated into a single device structure, as shown in Fig. 2.34(a). This device, commonly known as a triac (*triode ac* switch), is represented by the circuit symbol shown in Fig. 2.34(b).

The vertical region between the main terminals, MT1 and MT2, can be visualized as a p-n-p-n switch in parallel with an n-p-n-p switch or, in other words, as an inverse-parallel pair of reverse-blocking thyristors with a single external gate connection. For MT2 posi-tive or negative with respect to MT1, the triac can be triggered at its gate terminal by pos-itive or negative gate current. This capability provides a cheap, compact ac power switch requiring only one firing circuit.

Figure 2.35 shows the volt-ampere characteristic of the triac with terminal MT1 as reference point. If the triac is triggered by a gate signal or by an applied voltage in excess of $V_{(BO)}$, it will remain in the conducting state until the current drops below the holding current. In an ac circuit, the triac must turn off in the brief interval during which the load

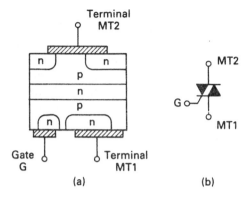

FIG. 2.34. Triac structure and circuit symbol.

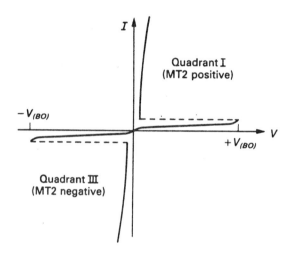

FIG. 2.35. Volt-ampere characteristic of the triac.

current is passing through zero. This is readily accomplished in a resistive circuit, but with an inductive load there is a phase shift between the supply voltage and current. As a result, when the triac turns off, the voltage across the device suddenly rises to the instantaneous ac supply voltage, causing it to resume conduction immediately unless the rate of rise of voltage is limited by an RC snubber circuit in parallel with the triac.

Because of the interaction between the two halves of the device, triacs are limited in voltage, current, and frequency ratings as compared with conventional thyristors, but devices are available with rms current ratings up to 300 A and voltage ratings of 1200 V. The triac finds widespread use in consumer and light industrial appliances operating from 50 or 60 Hz ac supplies at moderate power levels. The plastic-encapsulated triac is a particularly cheap and compact device and is widely used for controlling the speed of single-phase ac series or universal motors, in such consumer appliances as food mixers and portable drills.

2.8.2. *The Inverter-Grade Thyristor*

Rectifier circuits use conventional thyristors that are naturally commutated by the normal reversal in polarity of the ac supply voltage; consequently, turn-off time is not critical. For such applications, a standard, or "converter-grade," thyristor is satisfactory. In a chopper or inverter circuit operating from a dc supply, the thyristor is turned off by a forced commutating circuit, using the energy stored in a charged capacitor. Each forced commutation involves the discharging and recharging of the capacitor with a consequent energy loss. At high switching frequencies, the resulting power loss may be unacceptably high. In these applications, thyristor turn-off time is a critical parameter that must be minimized to limit the size and weight of the commutating capacitor and to reduce commutation losses. To meet this requirement, semiconductor manufacturers have developed a range of fast, or "inverter-grade," thyristors in which the silicon crystal is doped with gold atoms to reduce the lifetime of charge carriers in the n-base region. By accelerating the recombination of electron-hole pairs during turn-off, the turn-off time, t_q, is reduced to a value of 5 to 50 μs, depending on the voltage rating. Electron irradiation of the crystal can also be used to achieve this accelerated recombination and reduced turn-off time.

A good di/dt capability is often important in high-frequency inverter applications because it permits fast turn-on with low power dissipation and may also eliminate the need for a di/dt limiting inductor. A high di/dt capability can be achieved by adoption of the amplifying gate structure described in Section 2.6.2. However, some design trade-offs are involved in optimizing thyristor performance to give fast turn-off. In particular, the voltage-blocking capability and dv/dt rating of the device are reduced, and the on-state voltage is increased, resulting in higher conduction losses. Nevertheless, large inverter-grade thyristors are now available having a voltage-blocking capability of 1400 V, an rms current rating of 900 A, a turn-off time of 25 μs at a reapplied dv/dt of 400 V/μs, and a di/dt rating of 800 A/μs.

2.8.3. *The Asymmetric Thyristor (ASCR)*

As indicated above, thyristor design involves a trade-off among various device parameters, such as forward and reverse voltage-blocking capability, turn-on and turn-off times, and on-state voltage drop. The conventional thyristor may have a reverse-blocking capability of thousands of volts, but this capability is not required for every application. In particular, the voltage-fed inverter circuit, which converts dc power to ac, usually has a rectifier diode connected in antiparallel across each thyristor to conduct reactive load currents and excess commutating current. In such circuits, the anti-parallel diode clamps the thyristor reverse voltage to 1 or 2 V under steady circuit conditions.

If a high reverse voltage rating is unnecessary, the remaining thyristor characteristics can be optimized. The asymmetric thyristor, or ASCR, is specifically designed for applications where reverse-blocking capability is unimportant. Typically, the reverse voltage rating is about 20 or 30 V and the forward voltage rating is in the range of 400 V to 2000 V. The switching times and on-state voltage drop of the ASCR are smaller than those of a conventional thyristor of the same rating. As already indicated, fast turn-off is important because it minimizes the size, weight, and cost of commutating circuit components, and permits operation at switching frequencies of 20 kHz, or more, with high efficiency.

The conventional center-gate thyristor shown in Fig. 2.25, and repeated in Fig. 2.36(a), has a thick, lightly doped n-base region. Base thickness must be large

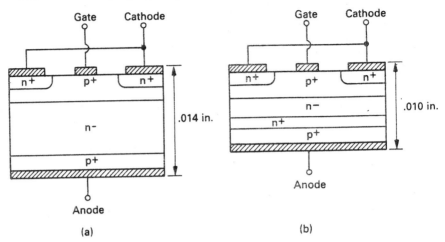

FIG. 2.36. Comparison of (a) conventional center-gate thyristor and (b) asymmetric thyristor (ASCR).

enough to prevent the spreading of the depletion region, during forward blocking, into the anode p^+ region. When this spreading occurs due to the application of excessive forward voltage, a punch-through condition occurs and the thyristor immediately turns on. Unfortunately, because the on-state voltage drop is also proportional to n-base thickness, a thyristor with a high peak repetitive forward voltage rating, V_{DRM}, will also have high on-state losses.

A narrower n-base is possible if a highly doped n^+ layer is introduced adjacent to the p^+ emitter, as shown in Fig. 2.36(b). This new n^+ layer acts as a buffer, preventing the depletion region from extending into the p^+ layer and allowing a higher average electric field in the lightly doped n-region. Consequently, a thinner n-base is obtained for the same forward-blocking voltage capability.

In the reverse-blocking mode, the n^+p^+ junction quickly avalanches at less than 50 V, so that the peak repetitive reverse voltage rating, V_{RRM}, is somewhat less than this value. However, the reduction in crystal thickness can be used to give a reduction in on-state voltage and/or a faster turn-off time. If the amount of gold doping is the same as that in a conventional fast thyristor, turn-off time is unaltered, but the on-state voltage drop is considerably reduced. If the gold doping level is increased to a value which gives the same on-state voltage as that in a conventional thyristor, the turn-off time is substantially reduced. Doping levels may also be adjusted to give lesser gains in both on-state voltage and turn-off time. Thus, the turn-off time can be halved and the on-stage voltage reduced to allow more efficient operation at double the switching frequency of a conventional thyristor.

2.8.4. *The Reverse-Conducting Thyristor (RCT)*

The reverse-conducting thyristor is simply an asymmetric thyristor with a monolithically integrated, antiparallel diode in a single silicon chip.[2.33] The thyristor is turned off by passing a current pulse through the diode part of the chip. By combining the ASCR and the diode in one device, a more compact circuit layout is obtained and heat sinking is simplified. Stray loop inductance between the ASCR and the diode is also minimized, avoiding the generation of reverse voltage transients across the ASCR and making its turn-off behavior more predictable. Isolation of thyristor and diode functions is important to ensure that charge carriers present in the diode during commutation do not diffuse into the thyristor part of the chip to cause retriggering when forward voltage is reapplied.

A disadvantage of the RCT is that it is inflexible compared with two discrete devices, because the current ratio between thyristor and diode parts of the chip is fixed for a given design. If the diode carries a commutation current pulse only, the greater part of the chip can be devoted to the thyristor to maximize its current capability. In the voltage-source inverter, the load current is controlled by the thyristor and flows freely in the other direction through the diode. For such circuits, the RCT must have equal current ratings for thyristor and diode sections. Purpose-designed RCT devices are now being manufactured for high-performance inverter and chopper circuits. Device ratings presently available are similar to those of ASCRs.

2.8.5. *Other Thyristor Devices*

The field-controlled thyristor (FCT), or static induction thyristor (SIT), a new, more complex power device that is normally in the on-state, is still under development. It is a $p-n^- -n^+$ rectifier with a fine-grid pattern of p^+ silicon overlaid on the n^+ surface. The application of a negative drive to the grid will turn off the normally conducting device, but it is necessary to hold this negative grid bias to maintain the off-state. Prototype devices have been reported indicating that the FCT has a high voltage-blocking capability, a low on-state voltage, and excellent dv/dt and di/dt capabilities. [34,35]

The gate-assisted turn-off thyristor is a conventional p-n-p-n structure having a negative-bias voltage applied between the gate and the cathode during the turn-off interval. This removes current carriers from the p-base region and accelerates the turn-off process. The technique can be applied to conventional thyristors or to asymmetric thyristors to increase their upper frequency capability. In order to have an appreciable effect on turn-off time, the gate-cathode junction must be highly interdigitated, so that stored charge is effectively removed over the whole cross-section of the base region.

The logical extension of gate-assisted turn-off is actual gate turn-off without any forced commutating circuitry. The gate turn-off thyristor was originally developed in the late 1960s, but this early device had performance problems and limitations that prevented its industrial acceptance. With modern technology, the GTO has acquired improved characteristics and is used in many commercial inverter drives. The GTO is now discussed in detail.

2.9. THE GATE TURN-OFF THYRISTOR (GTO)

The gate turn-off thyristor (GTO) incorporates many of the advantages of the conventional thyristor and the high-voltage switching transistor. It is a p-n-p-n device which, like the regular thyristor, or SCR, can be triggered into a conducting state by a pulse of positive gate current. However, the GTO is more versatile because it can be turned off by a brief pulse of negative current at the gate terminal. This facility allows the construction of inverter circuits without the bulky and expensive forced commutating components associated with conventional thyristor circuitry.[36–39] The GTO also has a faster switching speed than the regular thyristor, and it can withstand higher voltage and current than the power transistor or MOSFET.

The GTO is a three-terminal device with anode, cathode, and gate terminals. The circuit symbol is shown in Fig. 2.37(a), indicating the bidirectional gate current capability.

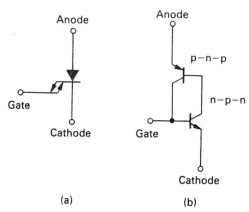

FIG. 2.37. (a) GTO circuit symbol and (b) two-transistor analog of the GTO.

Since the GTO is a four-layer p-n-p-n device, it can be modeled by the two-transistor circuit of Fig. 2.23, which is repeated in Fig. 2.37(b). Like the conventional thyristor, the GTO switches regeneratively into the on-state when a positive gating signal is applied to the base of the n-p-n transistor. In a regular thyristor, the current gains of the p-n-p and n-p-n transistors are large in order to maximize gate sensitivity at turn-on and to minimize on-state voltage drop. But this pronounced regenerative latching effect means that the thyristor cannot be turned off at the gate. Internal regeneration is reduced in the GTO by a reduction in the current gain of the p-n-p transistor, and turn-off is achieved by drawing sufficient current from the gate. The turn-off action may be explained as follows. When a negative bias is applied at the gate, excess carriers are drawn from the base region of the n-p-n transistor, and the collector current of the p-n-p transistor is diverted into the external gate circuit. Thus, the base drive of the n-p-n transistor is removed and this, in turn, removes the base drive of the p-n-p transistor, and stops conduction.

The reduction in gain of the p-n-p transistor can be achieved by the diffusion of gold or other heavy metal to reduce carrier lifetime, or by the introduction of anode to n-base short-circuiting spots, as in Fig. 2.38, or by a combination of these two techniques. Device characteristics are influenced by the particular technique used. Thus, the gold-doped GTO retains its reverse-blocking capability but has a high on-state voltage drop. The shorted anode emitter construction has a lower on-state voltage, but the ability to block reverse voltage is sacrificed. Large GTOs also have an interdigitated gate-cathode structure in which the cathode emitter consists of many parallel-connected n-type fingers diffused into the p-type gate region, as in Fig. 2.38. This configuration ensures a simultaneous turn-on or turn-off of the whole active area of the chip.

2.9.1. Switching Performance

The simplified gate drive circuit of Fig. 2.39 shows separate dc supplies for turn-on and turn-off. The GTO is gated into conduction by means of transistor TR1 in the turn-on

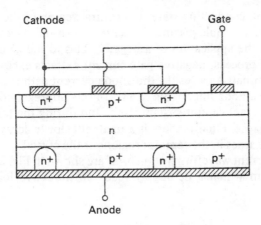

Fig. 2.38. Basic GTO structure showing anode to n-base short-circuiting spots.

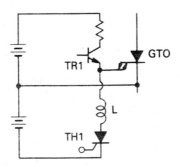

Fig. 2.39. Basic gate-drive circuit for a GTO.

circuit. The switching device in the turn-off circuit should have a high peak current capability. An auxiliary thyristor or MOSFET is appropriate for this duty. Figure 2.39 shows an auxiliary thyristor, TH1, which is gated to initiate the turn-off process. Turn-off performance may be enhanced by the presence of some series inductance, L, as indicated.[40] The voltage supply for the turn-off circuit is in the region of 10 to 20 V and the gate current at turn-off, applied for some microseconds, is typically about one-fifth of the anode current prior to turn-off. Consequently, the energy required to turn off the GTO is much less than that needed to turn off a conventional thyristor.

Gate turn-on. The gate turn-on mechanism of the GTO is similar to that of a conventional thyristor. A steep-fronted pulse of gate current turns on the device, and gate drive can be removed without the loss of conduction when the anode current exceeds the latching current level. Turn-on time is reduced by increasing forward gate current, as in a conventional thyristor, but because the regenerative effect is reduced in the GTO, the gate drive current required for turn-on is larger. To ensure conduction of all cathode fingers and a reduction in on-state voltage, some manufacturers recommend a continuous gating current during the entire conduction period.

Gate turn-off. In the conducting state, the central region of the GTO crystal is filled with a conducting electron-hole plasma. To achieve turn-off, excess holes in the p-base must be removed by the application of a negative bias to the gate. During the storage phase of the turn-off process, negative gate current extracts excess holes in the p-base through the gate terminal. As a result, the anode current path is pinched into a narrow filament under each cathode finger (Fig. 2.38). In this nonregenerative three-layer section of the crystal, current cannot sustain itself, and during the fall period, the current filaments quickly collapse. Finally, there is a small but slowly decaying tail of anode current due to residual charges in the remoter regions of the crystal.

The voltage and current waveforms at turn-off are shown in Fig. 2.41 for the simplified circuit of Fig. 2.40, in which a dc supply feeds an inductive load through a series-

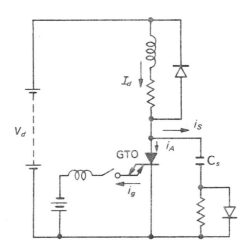

FIG. 2.40. Basic GTO switching circuit with a clamped inductive load.

connected GTO. A free-wheeling diode is connected across the load to allow circulation of load current during the off-period of the GTO. The snubber capacitor reduces the rate of rise of forward voltage at turn-off, thereby improving the current interrupt capability of the GTO and also limiting the turn-off losses in the device.

Assume the GTO is conducting a steady load current, I_d, when gate turn-off is initiated at time zero by the application of a reverse bias between gate and cathode. A reverse gate current, i_g, builds up at a rate determined by the inductance of the gate circuit, but the anode current remains constant throughout the storage time, t_s. Anode current, i_A, then decreases rapidly to the residual, or tail, current during the fall time, t_f, and the load current is diverted into the snubber capacitance, C_s. As a result, forward voltage v_A, builds up across the device at a rate dv_A/dt equal to I_d/C_s, and the anode tail current decays to zero to complete the turn-off process. Similar waveforms are obtained in inverter applications, but in practical circuits, stray inductance will cause some departure from the idealized waveforms described. In particular, supply inductance will cause a transient overshoot in GTO voltage, as indicated in Fig. 2.41. Excessive overvoltage can be avoided with an appropriate snubber circuit design.[41,42]

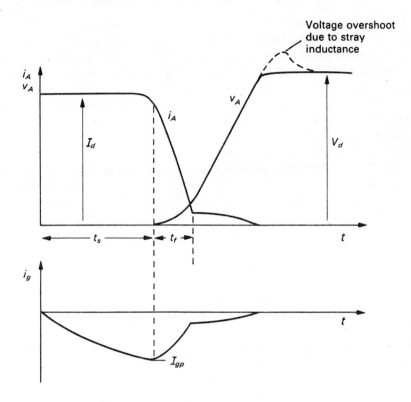

FIG. 2.41. Voltage and current waveforms during turn-off of a GTO.

2.9.2. *GTO Characteristics*

The peak value of off-gate current, I_{gp}, is a function of the anode current, I_d, prior to turn-off. The ratio I_d/I_{gp}, which is termed the turn-off gain, is typically between 3 and 5. The controllable current is the maximum anode current that can be interrupted by gate turn-off. It is highly dependent on device structure and gate drive conditions. It is also seriously reduced by an excessive rate of rise of anode voltage at turn-off. [43,44] The controllable current is typically about three times the rms on-state current. Device damage will occur if an attempt is made to turn off an anode current that is greater than the maximum controllable current. Consequently, auxiliary snubber circuits are imperative to slow the build-up of reapplied anode voltage, but snubberless GTOs are also being developed. [45]

In order to realize the gate turn-off capability of the GTO, basic design trade-offs are necessary, and therefore, some device characteristics are inferior to those of a conventional thyristor of comparable rating. Because of the low internal regeneration, there is an increase in latching and holding current levels, and there is also an increase in the on-state voltage drop, and the associated power loss. In a GTO with a shorted anode emitter, the reverse voltage rating is appreciably less than the forward-blocking voltage, but many inverter circuits do not require a reverse voltage withstand capability. However, the GTO retains many of the advantages of the thyristor and has a faster switching speed. Its surge current capability is comparable to that of a conventional thyristor, so that device protection is possible with a fast semiconductor fuse. Because of the interdigitated gate-cathode structure of the GTO, the di/dt limitation at turn-on is less stringent than that in a conventional thyristor. In general, the GTO has the high blocking voltage and large current capability that are characteristic of thyristor devices. Consequently, the GTO can be used in equipment operating directly from three-phase ac supplies at 460 V and above. This voltage capability is difficult to achieve with bipolar power transistors. A wide range of GTO devices is now available, and a single device can be obtained with a peak voltage rating of 2500 V and an average on-state current of 800 A.

The GTO inverter has a number of advantages over the conventional thyristor inverter. In particular, the GTO circuit has about 60 percent of the size and weight of the thyristor unit and has a higher efficiency because the increase in gate drive power and on-state power loss is more than compensated by the elimination of forced commutation losses. Several manufacturers have adopted GTO devices as the switching elements in a range of packaged adjustable-frequency inverter drives,[46] and the use of GTOs in inverters is growing rapidly.

2.10. COMPOSITE POWER SEMICONDUCTOR DEVICES

Power semiconductor devices can be combined to give a hybrid device with enhanced characteristics. Thus, as explained earlier, the MOSFET has a high input impedance and can be driven directly from integrated circuit logic, but it is best suited to voltage ratings below about 500 V. Many applications require a greater device voltage capability, which can be provided by the bipolar junction transistor (BJT) and the reverse-blocking thyristor, or silicon controlled rectifier (SCR). If the MOSFET is used as a driver for the bipolar transistor or thyristor, rather than as an output device, the resulting combination has the high input impedance of the MOSFET and the voltage, current, and power capabilities of a bipolar transistor or thyristor. Both hybrid and monolithic combinations of these bipolar and MOS devices are now being marketed.[1,47] The MOS-BJT offers the low drive requirements of the MOSFET with the low saturation losses of the high-voltage bipolar transistor. The MOS-SCR combines the advantages of a MOSFET input with the regenerative latching action and low on-state voltage of a thyristor. The insulated gate transistor is an example of the monolithic combination of MOS and bipolar technologies in a composite device.

2.10.1. *The Insulated Gate Transistor (IGT)*

A recent development in power MOS technology is the insulated gate transistor (IGT) or conductivity-modulated field-effect transistor (COMFET).[48–50] These devices retain the high input impedance of the MOSFET and, in addition, have a low on-state voltage drop, which is comparable to that of a bipolar transistor and which varies only moderately with junction temperature. However, the turn-off time of an IGT can be significantly greater than that of a power MOSFET. Figure 2.42(a) shows the IGT circuit symbol with its collector, emitter, and gate terminals. The device is turned on by applying a positive gate-emitter voltage and is turned off by reducing this voltage to zero.

The IGT structure, as shown in Fig. 2.42(b) is similar to that of a power MOSFET. The key difference is that the n^+ substrate of the MOSFET is replaced by a p^+ collector region which floods the epitaxial n^- layer with injected minority carriers during forward

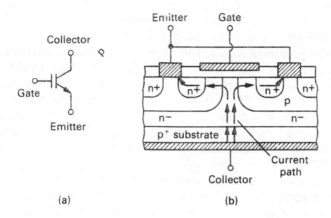

FIG. 2.42. Circuit symbol and physical structure of the insulated gate transistor (IGT).

conduction. This condition drastically reduces the resistance of the epitaxial layer as compared with that of a power MOSFET and allows the IGT to operate at much higher current densities. But this use of bipolar conduction imposes a penalty in turn-off speed. When the on-state voltage of the IGT exceeds about 1 V, its current capability is superior to that of a MOSFET or bipolar transistor. At a forward voltage drop of 2 V, the IGT will operate with a current density which is twenty times that of a MOSFET, provided one accepts an IGT turn-off time of 15 to 20 μs. Operation with a reduced current density gives a higher on-state voltage but a shorter turn-off time. Thus, IGT characteristics can be tailored to give minimum conduction losses and slow turn-off times (tens of microseconds) or fast turn-off times (a microsecond or less) with a resultant increase in conduction losses. An IGT with a voltage-blocking capability of 500 V can have a turn-off time of 0.25 μs and an on-state voltage drop still substantially better than that of a 500 V power MOSFET. For blocking voltages below 100 V, however, the power MOS-FET has lower conduction losses and is superior to the IGT. At higher blocking voltages, the on-state resistance of the MOSFET increases sharply. Detailed calculations have shown that the IGT is superior for motor drive applications at operating voltages above 200 V and switching frequencies below 50 kHz.[51]

The IGT has the high input impedance and voltage-controlled characteristics of the power MOSFET, but the gate turn-off capability is lost at high collector current levels.[52] At its present state of development, the IGT cannot achieve the voltage-blocking capability of the GTO.

2.10.2. Integrated Power Modules

Much development work has been concerned with the packaging of power semiconductor devices for increased economy, reliability, and user convenience. At low power levels, cheap, plastic-encapsulated devices are appropriate for light industrial and consumer markets. At high current levels, above several hundred amperes, heat removal considerations are predominant and double-sided cooling of disk-type thyristors is usual. At current levels up to 100 or 150 A, discrete semiconductors with electrically live metal housings have traditionally been employed. However, because the mounting and interconnection of these devices is labor intensive and costly, semiconductor manufacturers moved toward modular assemblies in the late 1970s.[53,54]

In these integrated power modules, two or more semiconductor chips are mounted on a common metal base plate. Electrical isolation between the chips and the base plate is achieved by means of a material such as beryllium oxide, which is electrically insulating but thermally conductive, so that heat is effectively removed from the chip to the metal base. Hermetically sealed or glass-passivated silicon chips are used for improved long-term voltage-blocking stability. The insulation voltage between the device and the metal base is 2500 V rms; hence, a number of power modules can be mounted directly on a common heat sink or on a convenient cooling surface, such as an equipment frame or an enclosure wall. Connection is made to the power module by means of convenient, top-mounted connectors.

Compact integrated diode and thyristor modules for rectifier circuit applications have been available for several years, and more sophisticated modules are now available. These may incorporate one or two Darlington transistors with discrete-chip, fast-recovery, antiparallel diodes for inverter applications. Six-Darlington modules are also available, which incorporate the six Darlington transistors required for a three-phase bridge inverter and the six associated feedback diodes. Such integrated power modules offer enhanced reliability and significant reductions in inverter size, weight, and assembly costs.

2.11. REFERENCES

1. PELLY, B.R., Power semiconductor devices — a status review, *IEEE Int. Semicond. Power Converter Conf., 1982*, pp. 1-19.

2. SITTIG, R., and ROGGWILLER, P. (Editors), *Semiconductor Devices for Power Conditioning*, Plenum, New York, NY, 1982.

3. GHANDHI, S.K., *Semiconductor Power Devices: Physics of Operation and Fabrication Technology*, Wiley-Interscience, New York, NY, 1977.

4. BALIGA, B.J., Switching lots of watts at high speeds, *IEEE Spectrum*, 18, 12, Dec. 1981, pp. 42-48.

5. ADLER, M.S., WESTBROOK, S.R., and YERMAN, A.J., Power semiconductor switching devices — an assessment, *Conf. Rec. IEEE Ind. Appl. Soc. Annual Meeting, 1980*, pp. 723-728.

6. FOSTER, A., Trends in power semiconductors, IEE Conf. Publ.No. 234, *Power Electronics and Variable-Speed Drives*, 1984, pp. 1-6.

7. LOCHER, R.E. and SMITH, M.W. (Editors), *Electronic Data Library, Thyristors-Rectifiers*, General Electric Co., 1982.

8. SMITH, M.W. (Editor), *Electronic Data Library, Transistors-Diodes*, General Electric Co., 1982.

9. PETER, J.M. (Editor), *The Power Transistor in Its Environment*, Thomson-CSF Sescosem Semiconductor Division, 1978.

10. CLEARY, J.F. (Editor), *Transistor Manual*, General Electric Co., 1964.

11. PELLY, B.R., and SHEN, P., Power transistors for choppers and inverters — an application review, *Conf. Rec. IEEE Ind. Appl. Soc. Annual Meeting, 1978*, pp. 1097-1106.

12. SMITH, M.W., Application of high power Darlington transistors, *Power Conversion International*, Sept./Oct. 1979, pp. 33-40.

13. *Power MOSFET HEXFET Databook*, Third Edition, International Rectifier Corp., 1985.

14. *MOSPOWER Applications Handbook*, Siliconix Ltd., 1985.

15. TIEFERT, K.H., TSANG, D.W., MYERS, R.L., and LI, V., The vertical power MOSFET for high-speed power control, *Hewlett-Packard J.* 32, 8, Aug. 1981, pp. 18-23.

16. SEVERNS, R., Using the power MOSFET as a switch, *Intersil Application Bulletin A036*, 1981.

17. HOWER, P.L., A comparison of bipolar and field-effect transistors as power switches, *Conf. Rec. IEEE Ind. Appl. Soc. Annual Meeting, 1980*, pp. 682-688.

18. GENTRY, F.E., GUTZWILLER, F.W., HOLONYAK, Jr., N., and VON ZASTROW, E.E., *Semiconductor Controlled Rectifiers: Principles and Applications of P-N-P-N Devices*, Prentice-Hall, Englewood Cliffs, NJ, 1964.

19. GRAFHAM, D.R., and GOLDEN, F.B. (Editors), *SCR Manual*, Sixth Edition, General Electric Co., 1979.

20. RICE, L.R. (Editor), *Silicon Controlled Rectifier Designers Handbook*, Second Edition, Westinghouse Electric Corp., 1970.

21. HOFT, R.G. (Editor), *SCR Applications Handbook*, International Rectifier Corp., 1974.

22. FINNEY, D., *The Power Thyristor and Its Applications*, McGraw Hill, U.K., 1980.

23. HEUMANN, K., and STUMPE, A.C., *Thyristoren*, Teubner, Stuttgart, 1969.

24. ZACH, F., *Leistungselektronik*, Springer Verlag, Wien, 1979.

25. BEDFORD, B.D., and HOFT, R.G., *Principles of Inverter Circuits*, J. Wiley, New York, NY, 1964.

26. DAVIS, R.M., *Power Diode and Thyristor Circuits*, Cambridge University Press, London, 1971.

27. MAZDA, F.F., *Thyristor Control*, J. Wiley, New York, NY, 1973.

28. DEWAN, S.B., and STRAUGHEN, A., *Power Semiconductor Circuits*, Wiley-Interscience, New York, NY, 1975.

29. MOTTO, J.W. (Editor), *Introduction to Solid State Power Electronics*, Westinghouse Electric Corp., 1977.

30. BIRD, B.M., and KING, K.G., *Power Electronics*, J. Wiley, U.K., 1983.

31. McMURRAY, W., SCR inverter commutated by an auxiliary impulse, *IEEE Trans. Commun. and Electron.*, **83**, 75, Nov. 1964, pp. 824-829.

32. ADEM, A.A., *Speed Controls for Universal Motors*, Application Note 200.47, General Electric Co., 1966.

33. VITINS, J., and DeBRUYNE, P., Modern power semiconductors for traction applications, *Brown Boveri Rev.*, **69**, 7/8, July/Aug. 1982, pp. 279-283.

34. WESSELS, B.W., and BALIGA, B.J., Vertical channel field-controlled thyristors with high gain and fast switching speeds, *IEEE Trans. Electron. Devices*, **ED-25**, 10, Oct. 1978, pp. 1261-1265.

35. TERASAWA, Y., MIYATA, K., MURAKAMI, S., NAGANO, T., and OKAMURA, M., High power static induction thyristor, *IEEE IEDM Tech. Dig.*, 1979, pp. 250-253.

36. KISHI, K., KURATA, M., IMAI, K., and SEKI, N., High power gate turn-off thyristors (GTOs) and GTO-VVVF inverter, *IEEE Power Electron. Spec. Conf.*, 1977, pp. 268-274.

37. MATSUDA, Y., FUKUI, H., AMANO, H., OKUDA, H., WATANABE, S., and ISHIBASHI, A., Development of PWM inverter employing GTO, *Conf. Rec. IEEE Ind. Appl. Soc. Annual Meeting, 1981*, pp. 1208-1216.

38. BURGUM, F.J., EBBING, W., and HOULDSWORTH, J.A., A new gate turn-off switch (GTO) used in a PWM ac motor drive, *Proc. Motorcon 1981*, pp. 3B.6-1 to 3B.6-9.

39. ISHIBASHI, A., FUJII, H., and MATSUDA, Y., GTO inverter for general purpose motor drive, *Hitachi Rev.* **31**, 4, Aug. 1982, pp. 185-188.

40. FUKUI, H., MATSUDA, Y., NAGANO, T., and SAKURADA, S., *Switching Characteristics of a Gate Turn-off Thyristor*, Application Note, Hitachi Ltd.

41. STEIGERWALD, R.L., Application techniques for high power gate turn-off thyristors, *Conf. Rec. IEEE Ind. Appl. Soc. Annual Meeting, 1975*, pp. 165-174.

42. OHASHI, H., Snubber circuit for high-power gate turn-off thyristors, *IEEE Trans. Ind. Appl.*, **IA-19**, 4, July/Aug. 1983, pp. 655-664.

43. MATSUDA, Y., and FUKUI, H., Application engineering of gate turn-off thyristors, *Hitachi Rev.* **31**, 4, Aug. 1982, pp. 173-178.

44. WOODWORTH, A., Understanding GTO data as an aid to circuit design, *Electron. Compon. and Appl.* **3**, 3, May 1981, pp. 159-166.

45. NAGANO, T., FUKUI, H., YATSUO, T., and OKAMURA, M., A snubber-less GTO, *Conf. Rec. IEEE Power Electron. Spec. Conf., 1982*, pp. 383-387.

46. PAICE, D.A., and MATTERN, K.E., Application of gate-turn-off thyristors in 460 V, 7.5-250 hp ac motor drives, *Conf. Rec IEEE Ind. Appl. Soc. Annual Meeting, 1982*, pp. 663-669.

47. RUGGLES, T.E., Power semiconductors extend control horizons, *Electron. Power*, **29**, 3, Mar. 1983, pp. 233-236.

48. BALIGA, B.J., ADLER, M.S., GRAY, P.V., LOVE, R.P., and ZOMMER, N., The insulated gate rectifier: A new power switching device, *IEEE IEDM Tech. Dig.*, Abst. 10.6, 1982, pp. 264-267.

49. BALIGA, B.J., Fast-switching insulated gate transistors, *IEEE Electron. Device Lett.*, **EDL-4**, 12, Dec. 1983, pp. 452-454.

50. RUSSELL, J.P., GOODMAN, A.M., GOODMAN, L.A., and NEILSON, J.M., The COMFET — A new high conductance MOS-gated device, *IEEE Electron. Device Lett.*, **EDL-4**, 3, March 1983, pp. 63-65.

51. BALIGA, B.J., The new generation of MOS power devices, *Proc. Conf. on Drives/Motors/Controls, Harrogate* (U.K.), *Oct. 1983*, pp. 139-141.

52. SMITH, M., SAHM, W., and BABU, S., Insulated-gate transistors simplify ac-motor speed control, *EDN*, Feb. 9, 1984, pp. 181-201.

53. BALTHASAR, P.P., and REIMERS, E., The integrated power switch, *IEEE Trans. Ind. Appl.* **IA-12**, 2, Mar./Apr. 1976, pp. 179-191.

54. WETZEL, P., Integration of power semiconductor devices, *Electron. Power*, **26**, 9, Sept. 1980, pp. 712-714.

CHAPTER 3

AC-DC Converter Circuits

3.1. INTRODUCTION

A rectifier converts ac power to dc, whereas an inverter converts dc power to ac. The general term "converter" is used to denote a rectifier or inverter but is often used to signify a circuit capable of both rectifier and inverter operation. This chapter describes the circuit operation and provides design equations for the application of semiconductor rectifier diodes and thyristors in both single-phase and polyphase converter circuits. These circuits are used to convert single-phase or polyphase ac voltages to a constant or adjustable dc voltage as the first step in an ac-dc-ac adjustable-speed ac motor drive system, as the power converter for an ac-dc adjustable-speed dc motor drive system, and as the power converter in a wound-rotor motor slip-power recovery system. Such circuits are also used as field current controllers for wound-field synchronous motors. For the most part, these converter circuits use the reversal of the incoming utility supply voltage to achieve natural thyristor commutation. Improved forced-commutated circuits that overcome some of the disadvantages of line voltage commutation are also presented. The characteristics of each of the various circuits when used as the input converter in a dc link adjustable-speed drive system (ac-dc-ac) are emphasized.

3.2. SINGLE-PHASE CONVERTERS

The most common single-phase rectifier circuit is the full-wave bridge shown in Fig. 3.1.[1-5] A supply transformer may be required to provide electrical isolation or to alter the ac supply voltage to the rectifier. The rectifier voltage and current waveforms are shown in Fig. 3.2. Circuit operation is conveniently demonstrated by drawing the ac

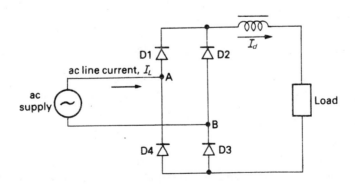

FIG. 3.1. The single-phase bridge rectifier circuit.

69

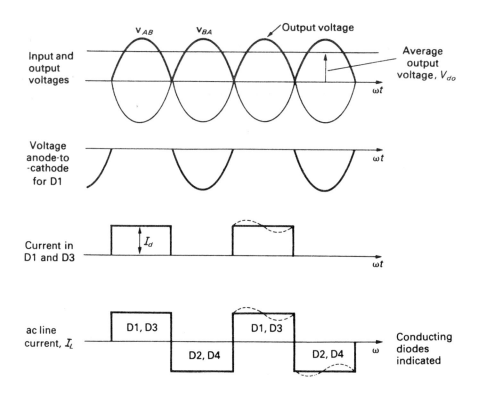

FIG. 3.2. Voltage and current waveforms for a single-phase bridge rectifier with a highly inductive load.

supply voltage waveforms, v_{AB} and v_{BA}, where obviously, v_{BA} is $-v_{AB}$. Since the diodes cannot block forward voltage, each diode starts conduction when the anode-to-cathode voltage on the device changes from negative to positive. Consequently, the diodes conduct in diagonal pairs, D1, D3 and D2, D4, and each pair of devices conducts for 180 degrees, or one half-cycle, of the ac supply. When the source voltage, v_{AB}, is positive, diodes D1 and D3 conduct, and v_{AB} appears at the output terminals, assuming the diode forward voltage drop is negligible. Similarly, when v_{BA} is positive, it appears at the output terminals via diodes D2 and D4. Thus, a full-wave rectified voltage is developed across the load circuit. The load current, I_d, has negligible ripple if the load circuit is highly inductive. However, load inductance cannot affect the flow of dc current; hence the magnitude of the load current is equal to the average output voltage divided by the resistance of the load circuit.

Examination of the waveforms in Fig. 3.2 indicates that the peak reverse voltage applied to the rectifier devices is equal to 1.414 times the rms supply voltage. The peak, average, and rms diode currents are, respectively, 1.0, 0.5, and 0.707 times the dc output current. The ac line current has no dc component but does contain all the odd harmonics of the input frequency. The amplitude of the harmonics decreases as $1/k$, where k is the order of the harmonic: that is, there is a third harmonic of one-third of the fundamental, a fifth harmonic of one-fifth of the fundamental, etc. It is readily shown that the dc voltage has an average value, V_{do}, equal to $2\sqrt{2}/\pi = 0.900$ times the rms source voltage, together with all even harmonics of the input frequency. Therefore, the lowest ripple frequency in the dc output voltage is the second harmonic of the supply frequency. For industrial 50 or 60 Hz sources, the ripple frequency is either 100 or 120 Hz, and a large reactor is required in the dc circuit to keep the current waveforms as shown in Fig. 3.2. A finite reactor (less than infinite inductance) will allow some ac ripple current to exist in the diode and ac line currents. Dotted waveforms on the second cycle of the two current waveforms in Fig. 3.2 show the effect of a finite inductance. If the load inductance is small enough, the current will drop to zero (discontinuous conduction). For discontinuous current flow, the load voltage is determined by the nature of the load (dc motor, dc filter capacitor, etc.) rather than the supply voltage.

If the dc load is composed of an L-C filter as shown in Fig. 3.3, then care must be exercised in the selection of the L-C filter components. Resonance must be avoided at harmonics of the source frequency, in order to prevent ripple voltage amplification rather

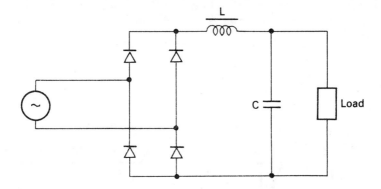

Fig. 3.3. Single-phase bridge rectifier circuit with L-C filter.

than reduction. Various handbooks contain filter design nomographs to aid in this procedure.[6,7] Low-power converters sometimes dispense with the reactor and provide only a capacitor for the dc filter. In such a case, the diode and ac line current waveforms are composed of a pulse of current occurring before each peak of the supply voltage waveform. This shortened conduction interval requires that the diodes have a higher peak-to-average current ratio than the value of 2 to 1 given for the circuit, shown in Fig. 3.1, with an infinite inductance.

The full-wave bridge circuit is used as the input converter for low-power (less than 3 to 5 hp) ac-dc-ac adjustable-speed ac drives. As will be explained in subsequent chapters, the inverter (dc to ac) in this case must provide both voltage and frequency control for the ac motor load because the uncontrolled ac to dc rectifier provides a fixed dc link voltage. Single-phase rectifiers at high power are used in electric locomotives supplied from an overhead ac catenary. These systems are more complex and usually are provided with a voltage stepdown transformer having separate low-voltage secondary windings that allow the use of multiple series-connected bridge circuits. Some systems provide forced commutation of the thyristors in order to improve the line power factor at reduced output voltages.

If the diodes (D1-D4) in Fig. 3.1 are replaced by thyristors (TH1-TH4), then the dc output voltage can be controlled by adjusting the instant at which gate signals are supplied to the four thyristors. If the gating signals are delayed until the voltage has become positive, the thyristors block the forward voltage applied to them and the load current is maintained in the previously conducting thyristor. Figure 3.4 shows the power circuit, and Fig. 3.5 indicates the circuit voltage and current waveforms. The waveforms are similar to those shown in Fig. 3.2 except that the 180-degree current conduction interval

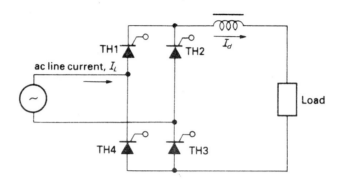

FIG. 3.4. The single-phase bridge phase-controlled converter circuit.

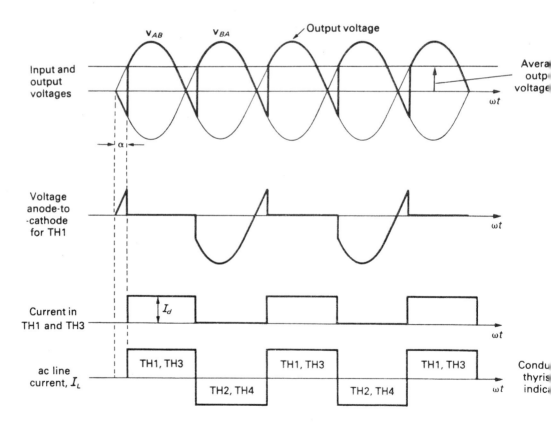

FIG. 3.5. Voltage and current waveforms for a single-phase bridge phase-controlled converter with a highly inductive load.

is phase shifted in time with respect to the voltage zero crossing. The average, rms,and peak thyristor currents have the values given for the diodes of the single-phase uncontrolled rectifier circuit. The peak reverse thyristor voltage is the same as before, but the maximum forward thyristor voltage requirement is also equal to 1.414 times the rms source voltage, occurring when the delay angle, α, is greater than 90 degrees. The output dc voltage varies as a function of delay angle and is equal to $V_{do} \cos \alpha$. The value of V_{do} is, as before, 0.900 times the rms source voltage. Discussion of the derivation of this voltage transfer function between output dc voltage and delay angle is deferred until Section 3.3.1. A discussion of overlap conduction during thyristor commutation is deferred until Section 3.3.2. Free-wheeling operation can be provided by the addition of a diode in parallel with the two series-connected thyristors, as shown in Fig. 3.6. The free-wheeling diode conducts whenever the instantaneous output voltage attempts to become negative. The thyristor current is reduced to zero and all four thyristors are nonconducting. The free-wheeling diode provides the necessary path for the current to flow in the inductor-load loop. This current decays at the circuit's L/R time constant.

The single-phase controlled rectifier converter can be used as the input converter in an ac-dc-ac drive system with an adjustable-voltage or -current dc link. Because this type of drive generally is of higher horsepower, polyphase controlled rectifier converters are normally used. Low-horsepower (less than 3 to 5 hp) adjustable-speed dc motors are supplied from single-phase controlled converters.

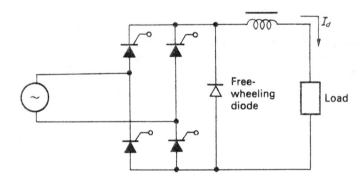

FIG. 3.6. Single-phase bridge phase-controlled converter with free-wheeling diode.

3.3. POLYPHASE CONVERTERS

The single-phase rectifier has both a pulsating output voltage and a pulsating instantaneous power demand. Polyphase ac to dc converter circuits are therefore used in high-power industrial applications.[1-5] For simplicity, the operation of a three-phase half-wave circuit is analyzed. The three thyristors have a common cathode connection, as shown in Fig. 3.7, which forms the positive terminal of the rectified dc output. The transformer neutral constitutes the negative output terminal.

In these polyphase ac circuits, the natural commutation process consists of a cyclic transfer of current from one thyristor to the next. Assume, for the moment, that the thyristors are replaced by diodes and hence no firing delay is possible. Each diode then conducts the full output current for one-third of a cycle while its anode voltage is the

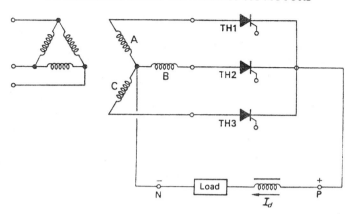

Fig. 3.7. Three-phase half-wave converter circuit.

most positive of the three. Figure 3.8 shows the transformer (or anode) voltages relative to neutral and the rectified output voltage, assuming zero forward voltage drop in the rectifier device. In effect, the positive load terminal, P, is automatically connected by a conducting diode to whichever phase of the secondary is most positive, and the nonconducting diodes are reverse biased relative to their common cathode terminal. These waveforms can also be obtained with thyristors, provided each one is triggered as soon as its anode-to-cathode voltage becomes positive.

When supplying a resistive load, the output current waveform is a replica of the output voltage. If a large inductor is connected in the dc circuit, it will suppress harmonic currents and maintain a steady output current, I_d. Each thyristor carries this current for one-third of the cycle, as shown in Fig. 3.8.

3.3.1. Delayed Commutation

The average output voltage is controlled by the use of a thyristor circuit with delayed commutation. The firing angle, or delay angle, α, is measured from the point where commutation occurs when diodes are used. Figure 3.9 shows the voltage and current waveforms for an inductive load which maintains constant current. The thyristor gating pulses are now delayed and each thyristor blocks the positive anode voltage for an angle α. This forces the preceding phase to continue supplying the load current although it is not the most positive of the three. When the incoming thyristor is triggered, the load current immediately transfers to it and a reverse bias voltage is applied across the outgoing thyristor, causing it to turn off. The output voltage now has the waveform shown in Fig. 3.9 and the average value has been reduced.

A general expression for the average output voltage is obtained as follows: assume an m-phase half-wave circuit in which each phase conducts for $2\pi/m$ electrical radians in one cycle of the supply voltage. The instantaneous phase voltage is given by $v = \sqrt{2}\, V_{ph} \cos \omega t$, where V_{ph} is the rms value of the transformer secondary phase voltage and the time origin is taken at the instant of peak voltage (Fig. 3.10). If the firing angle is zero, conduction takes place from $(-\pi/m)$ to $(+\pi/m)$. If the firing angle is α,

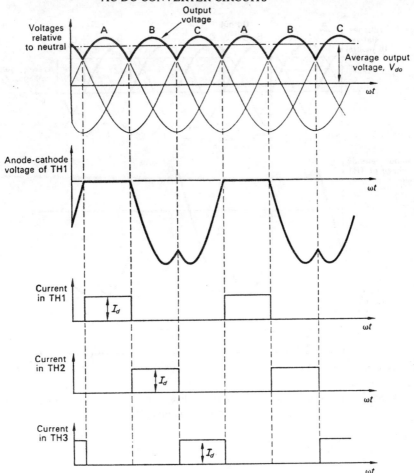

FIG. 3.8. Voltage and current waveforms for a three-phase half-wave converter with a highly inductive load and zero firing angle.

the phase conducts from $[(-\pi/m)+\alpha]$ to $[(+\pi/m)+\alpha]$, and the average output voltage, V_d, equals the average height of the shaded area in Fig. 3.10:

$$V_d = \frac{m}{2\pi} \int_{[-(\pi/m)+\alpha]}^{[(\pi/m)+\alpha]} \sqrt{2}\, V_{ph} \cos \omega t \, d(\omega t) = \sqrt{2}\, V_{ph} \left(\frac{m}{\pi}\right) \sin \left(\frac{\pi}{m}\right) \cos \alpha . \quad (3.1)$$

When the delay angle is zero, V_d has a maximum value:

$$V_{do} = \sqrt{2}\, V_{ph} \left(\frac{m}{\pi}\right) \sin \left(\frac{\pi}{m}\right) . \quad (3.2)$$

For the three-phase, half-wave wye circuit (Fig. 3.7) $m = 3$, and $V_{do} = \left(3\sqrt{6}/2\pi\right) V_{ph}$ $= 1.17\, V_{ph}$. The average output voltage, with phase control, is therefore given by

$$V_d = V_{do} \cos \alpha . \quad (3.3)$$

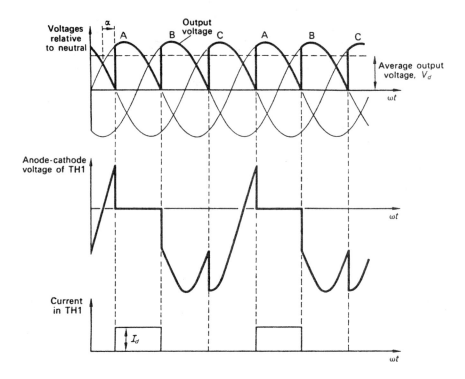

FIG. 3.9. Voltage and current waveforms for a three-phase half-wave converter with a highly inductive load and delayed commutation.

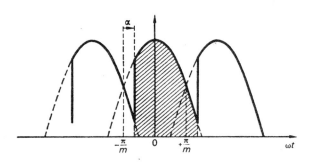

FIG. 3.10. Output voltage waveform for an m-phase half-wave converter with a firing angle, α.

This formula assumes continuous current conduction and is not valid if current flow becomes discontinuous as a result of delayed commutation. The output voltage and current waveforms for a three-phase half-wave controlled rectifier with a firing delay of 60 degrees are shown in Fig. 3.11(a) for a highly inductive load and in Fig. 3.11(b) for a resistive load. When the dc load is highly inductive, it tends to maintain constant output current, even with large delay angles, and the thyristor current retains its rectangular waveform. Consequently, the induced emf in the load inductor maintains current flow, even when the anode polarity has reversed, indicating that energy is being returned from

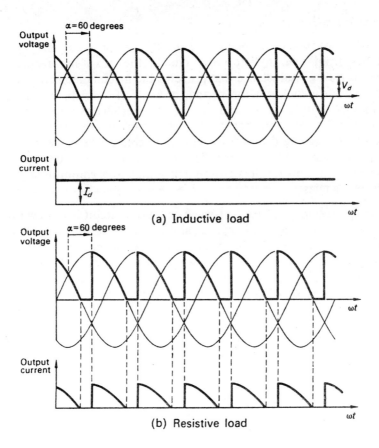

FIG. 3.11. Output voltage and current waveforms when commutation is delayed by 60 degrees: (a) with a highly inductive load; (b) with a resistive load.

the magnetic field of the inductor through the transformer to the supply, and the circuit is temporarily acting as an inverter. However, with the firing delay of 60 degrees, the average output voltage, V_d, is positive and there is a net power flow from the ac supply to the load. With a delay of 90 degrees and a highly inductive load, V_d has zero average value and energy oscillates between the load inductor and the supply, with no net power flow (assuming zero circuit losses). If firing is delayed beyond 90 degrees, V_d is negative and inverter operation may be obtained by introducing a dc source in series with the inductor.

In the case of a resistive load, each thyristor ceases to conduct as soon as its anode voltage becomes negative. In general, the result is a discontinuous current flow in half-wave circuits when the delay angle exceeds $(\pi/2 - \pi/m)$ radians, i.e., 30 degrees in the case of a three-phase circuit. For smaller delay angles, the formula $V_d = V_{do} \cos \alpha$ is valid, but for greater delays the formula is

$$V_d = V_{do} \left\{ \frac{1 - \sin(\alpha - \pi/m)}{2 \sin(\pi/m)} \right\} . \tag{3.4}$$

3.3.2. *Commutation Overlap*

In the analysis to date, it has been assumed that the load current, I_d, commutates instantaneously from one thyristor or diode to the next. In practice, the supply transformer has leakage inductance that prevents the instantaneous transfer of current. As a result, there is a commutation period, or angle of overlap, u, during which two thyristors conduct simultaneously and two transformer phases are short-circuited. The voltage difference between the phases circulates a commutating current which reduces the current in the outgoing thyristor to zero and increases the current in the incoming thyristor to I_d. In a phase-controlled rectifier, the overlap angle, u, decreases as the delay angle, α, is increased. This result is caused by the increased commutating voltage available at large delay angles.

In Fig. 3.12(a), the three-phase half-wave converter circuit is redrawn to show each transformer secondary-phase emf, e_A, e_B, and e_C in series with the effective leakage inductance per phase, L. This inductance is also known as the commutating inductance. A highly inductive load is assumed so that the output current, I_d, is constant. When commutation takes place from thyristor TH1 to TH2, the phase current, i_A, decreases from I_d to zero, and phase current, i_B, increases from zero to I_d, as shown in Fig. 3.12(c). Hence, during this overlap period

$$i_A + i_B = I_d . \tag{3.5}$$

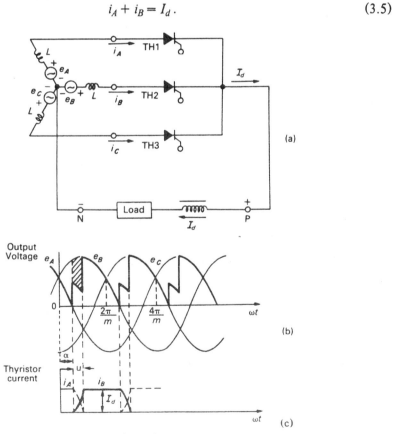

Fig. 3.12. Basic circuit diagram and waveforms for a phase-controlled converter with delayed commutation and finite commutating reactance.

Neglecting thyristor on-state voltage drop, the instantaneous output voltage is given by

$$e_d = e_A - L\, di_A/dt = e_B - L\, di_B/dt. \tag{3.6}$$

Thus,

$$2e_d = e_A + e_B - L\left(di_A/dt + di_B/dt\right). \tag{3.7}$$

But, from Equation 3.5

$$di_A/dt + di_B/dt = 0. \tag{3.8}$$

Combining Equations 3.7 and 3.8 gives

$$e_d = \tfrac{1}{2}\left[e_A + e_B\right]. \tag{3.9}$$

Thus, the output voltage during the overlap period is the average emf of the two conducting phases, as shown in Fig. 3.12(b). When the incoming thyristor is triggered, the output voltage rapidly rises to this average value and remains there until commutation is completed, when it rapidly rises to the incoming phase voltage. Consequently, the average output voltage, V_d, is less than the theoretical value obtained with instantaneous commutation because the volt-seconds, represented by the shaded area in Fig. 3.12(b), subtract from the theoretical output voltage. This diagram shows that the instantaneous reduction in output voltage during the overlap period is $\tfrac{1}{2}\left[e_B - e_A\right]$, and the reduction in average output voltage due to overlap is

$$\Delta V_d = \frac{m}{2\pi} \int_{\alpha}^{\alpha+u} \tfrac{1}{2}\left[e_B - e_A\right] d(\omega t). \tag{3.10}$$

With a time origin at the intersection of e_A and e_B, as shown in Fig. 3.12(b), e_A equals $\sqrt{2}\, V_{ph} \cos{(\omega t + \pi/m)}$ and e_B is $\sqrt{2}\, V_{ph} \cos{(\omega t - \pi/m)}$. On substituting these expressions in Equation 3.10 and performing the integration, it is found that

$$\Delta V_d = V_{do}\left[\frac{\cos\alpha - \cos(\alpha + u)}{2}\right] \tag{3.11}$$

where V_{do} is the output voltage with zero delay and no overlap, as given by Equation 3.2. The mean output voltage, V_d, is therefore

$$V_d = V_{do} \cos\alpha - \Delta V_d$$

$$= V_{do}\left[\frac{\cos\alpha + \cos(\alpha + u)}{2}\right]. \tag{3.12}$$

In Equations 3.11 and 3.12, the voltages, ΔV_d and V_d, are expressed in terms of delay angle, α, and overlap angle, u. It is also possible to express these voltages in terms of the load current, I_d. The required expressions are readily derived by means of the following volt-second method.

The well-known relationship between voltage and current for an inductor, L, is

$$v_L = L \, di/dt .$$
(3.13)

This expression may be rewritten as

$$di/dt = v_L/L$$
(3.14)

and integrating over the time interval from zero to t gives

$$\Delta i = \frac{1}{L} \int_0^t v_L \, dt .$$
(3.15)

This result indicates that the current change Δi is directly proportional to the voltage-time area supported by the inductor during the time t. Thus, the current change Δi requires the application of $L\Delta i$ volt-seconds.

In the ideal phase-controlled converter, there is no overlap and the dc current, I_d, transfers instantaneously from phase to phase at each commutation. If the transformer has a leakage inductance per phase, L, as in Fig. 3.12(a), the transfer of current is delayed and the growth of current from zero to I_d in the incoming phase requires the application of LI_d volt-seconds across the leakage inductance. This voltage-time area, which is shown shaded in Fig. 3.12(b), is absorbed by the leakage inductance of the transformer and does not appear across the load. Consequently, the average load voltage is reduced.

In an m-phase half-wave converter, there are m commutations per cycle, and the total voltage-time area absorbed per cycle is mLI_d. The reduction in average output voltage due to overlap is therefore mLI_d/T, where T is the periodic time of the ac supply waveform. Thus

$$\Delta V_d = \frac{mLI_d}{T} = mfLI_d = \frac{m I_d X}{2\pi}$$
(3.16)

where X equals $2\pi fL$, the commutating reactance per phase at supply frequency, f.

Equation 3.16 shows that the average voltage drop due to overlap is proportional to the load current, I_d. In general, therefore, the average output voltage of a half-wave rectifier with firing delay and overlap is

$$V_d = V_{do} \cos \alpha - \frac{mI_d X}{2\pi} .$$
(3.17)

This equation indicates that the output voltage decreases linearly with load current from the no-load value of $V_{do} \cos \alpha$.

3.3.3. The Phase-Controlled Inverter

In a dc circuit, a reversal in the direction of power flow is normally associated with a reversal of the current direction, but the same effect is obtained if the voltage polarity is reversed and the current direction is maintained. The latter technique must be used in the phase-controlled inverter, which is a unidirectional current device due to the thyristors.

In a normal ac to dc converter circuit, each phase of the ac supply delivers current through a thyristor while the phase or anode voltage is positive. Thyristor current must flow from anode to cathode, and the direction of energy flow can only be reversed if the ac supply delivers current during periods of negative anode voltage. Reverse energy flow is achieved by delaying commutation until the anode voltage becomes negative and then forcing current flow in opposition to the negative phase voltage. The dc source which produces this current delivers energy to the ac supply, and inverter operation is achieved.

For inversion, the commutation instant must be delayed beyond 90 degrees when the average output voltage, $V_d = V_{do} \cos \alpha$, becomes negative. Figure 3.13 shows the voltage and current waveforms, assuming a constant current I_d is maintained by means of the dc smoothing inductor. The transformer neutral is now the positive dc terminal and the dc supply voltage, V, exceeds the average counter emf V_d by the amount necessary to circulate the current I_d through the circuit resistance. The dc smoothing inductor is required to support the instantaneous voltage differences between the dc source and the inverter. Natural commutation by phase voltage reversal is obtained because the incoming thyristor is triggered while its anode is more positive than the anode of the conduct-

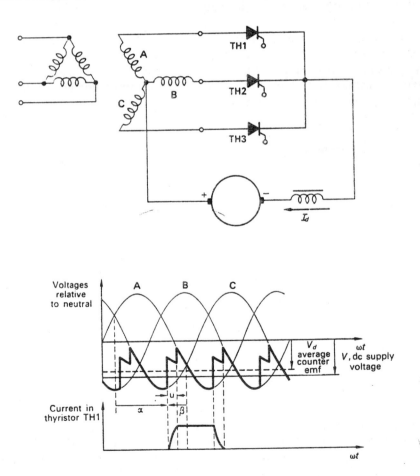

Fig. 3.13. Basic circuit diagram and waveforms for the phase-controlled inverter.

ing thyristor. The voltage difference between phases circulates a commutating current in the usual manner. If the delay angle, α, is greater than 180 degrees, this commutating voltage is no longer available and conduction continues into the positive half-cycle of voltage. The rectified voltage then aids the dc supply voltage and a short-circuit condition is produced. Commutation must, therefore, be completed before the instant at which the phase voltages become equal. The angle by which firing is advanced ahead of this point is denoted by β, where $\beta = \pi - \alpha$. The angle β must always be large enough to provide the volt-second area required to commutate the load current to the incoming thyristor and to allow the outgoing thyristor to recover. For inverter operation, the voltage drop due to overlap has the effect of increasing the magnitude of the negative counter emf of the inverter, which therefore increases with load current. Because the inverter is simply a phase-controlled converter with α greater than 90 degrees, all the equations derived in Section 3.3.2 are valid for inverter operation.

Figure 3.14 shows the anode-cathode voltage of a thyristor in an inverter circuit, assuming negligible overlap. This waveform is readily derived from the transformer (or anode) voltages relative to neutral, as in Figs. 3.8 and 3.9. After the transfer of current, a reverse-bias voltage is applied to the outgoing thyristor for a short interval. This interval, called the extinction angle or margin angle, must be greater than the thyristor recovery, or turn-off, time; otherwise a commutation failure leading to a short-circuit will occur. The phase-controlled inverter, unlike the forced-commutated inverter, needs no auxiliary apparatus for commutation because the commutating voltages are provided by the ac system which the inverter is supplying. This principle is used in ac-dc-ac high-voltage dc power transmission systems and is also the basic commutation process in the naturally commutated cycloconverter (Chapter 5).

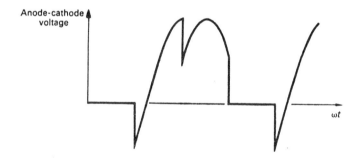

FIG. 3.14. Thyristor anode-cathode voltage in a phase-controlled inverter with zero commutating reactance.

3.3.4. *Reactive Power and Power Factor*

In a phase-controlled rectifier, the introduction of a firing delay causes the thyristor current to lag the anode, or phase, voltage (Fig. 3.9). If the load is highly inductive and overlap is negligible, the thyristor current retains its rectangular waveform but is delayed by an angle α. The displacement angle, ϕ, between the phase voltage and the fundamental component of the phase current is equal to the delay angle, α, and hence the fundamental power factor, cos ϕ, decreases in proportion to the per-unit output voltage

V_d/V_{do}. This phase lag in transformer secondary current is reflected to the primary side, resulting in a reduced input power factor. The presence of overlap tends to increase the phase lag of the fundamental current and, even with zero firing delay, the input fundamental power factor is somewhat less than unity.

In the case of an inverter, the fundamental component of output current would have unity power factor if commutation were instantaneous and could be delayed until α equaled 180 degrees. In practice, firing must be advanced in order to allow sufficient time for overlap and turn-off, and hence the output current always leads the phase voltage. A phase-controlled inverter, therefore, delivers current to the supply at a leading power factor, and the inverter may be regarded either as a generator of leading reactive power or a consumer of lagging reactive power. The inverter delivers real power to the ac supply, but its lagging reactive power requirements must be supplied by the synchronous alternators of the ac system which the inverter is supplying. Alternatively, compensating capacitors may be connected on the ac side of the inverter in order to reduce its reactive power consumption.

In general, when a thyristor converter is connected to the ac supply, a discontinuous nonsinusoidal current flows in the ac system, so that one must distinguish between the real power, the reactive power, and the harmonic power.

The real power per phase, P, is equal to the product of the rms phase voltage V_{ph} and the in-phase component of the fundamental current. If I_1 is the fundamental rms current and $\cos \phi$ is the fundamental power factor, then

$$P = V_{ph} I_1 \cos \phi . \tag{3.18}$$

The reactive power, Q, as usual, is the product of the rms phase voltage and the quadrature component of the fundamental current:

$$Q = V_{ph} I_1 \sin \phi . \tag{3.19}$$

The harmonic power, Q_{har}, is given by the product of the phase voltage, V_{ph}, and the total rms harmonic current, I_{har}:

$$Q_{har} = V_{ph} I_{har} . \tag{3.20}$$

The harmonic power, like the reactive power, oscillates between the load and the supply. If I_{ph} is the total rms current, including harmonics, then

$$I_{ph} = \sqrt{(I_1^2 + I_{har}^2)} \tag{3.21}$$

and the total apparent power, S, is given by

$$S = V_{ph} I_{ph} = \sqrt{(P^2 + Q^2 + Q_{har}^2)} . \tag{3.22}$$

Strictly speaking, when harmonics are present, the fundamental power factor, $\cos \phi$, is called the displacement factor, and the ratio of real to apparent power is called the total power factor, λ. Thus

$$\lambda = \frac{P}{S} = \frac{V_{ph} I_1 \cos \phi}{V_{ph} I_{ph}} = \frac{I_1}{I_{ph}} \cos \phi \tag{3.23}$$

or

$$\lambda = \mu \cos \phi \qquad (3.24)$$

where μ is the ratio of the fundamental rms current to the total rms current and is called the distortion factor.

3.3.5. Free-Wheeling Operation

The operation of a polyphase ac phase-controlled rectifier with an inductive load is modified when a free-wheeling diode is connected across the load (Fig. 3.15). As in the single-phase circuit of Fig. 3.6, this diode diverts the flow of thyristor current by allowing the load current to circulate through it during parts of the cycle when the output voltage tends to become negative. A continuous dc load current is therefore obtained at large delay angles, even though the thyristor current is discontinuous. When the free-wheeling diode is removed from the polyphase circuit, the inductive load maintains a continuous current and forces thyristor current flow during periods of negative output voltage. In the modified circuit, negative output voltage excursions are prevented because the free-wheeling diode immediately conducts when the load voltage tends to reverse polarity. The output current is then maintained by the load emf, and the thyristor current is zero. The inductive load is effectively short-circuited by the conducting diode until the rectified load voltage again becomes positive. The load voltage therefore has the same discontinuous waveform as in Fig. 3.11(b) for a normal three-phase rectifier supplying a resistive load. The presence of the free-wheeling diode prevents a reversal in the polarity of the output voltage and so reduces the ripple content of the output voltage and current waveforms. It also modifies the alternating current waveform at the input to the phase-controlled rectifier so that the reactive power consumption is reduced and the fundamental power factor is improved.[2] The free-wheeling diode must be removed or replaced with a thyristor if inverter operation is required.

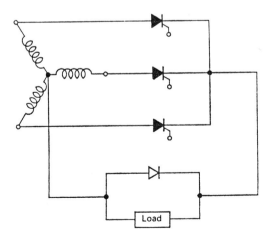

Fig. 3.15. The three-phase half-wave converter with a free-wheeling diode.

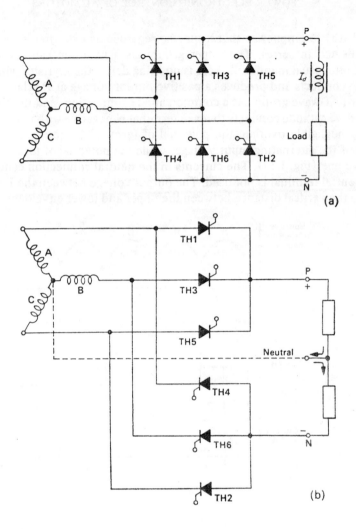

FIG. 3.16. The basic three-phase bridge converter: (a) the usual circuit diagram; (b) the bridge circuit redrawn as two half-wave circuits.

3.3.6. *The Three-Phase Full-Wave Bridge Converter*

The basic principles of phase-controlled rectification and inversion have now been introduced. In practice, a three-phase half-wave circuit would use a rectifier transformer with a zig-zag secondary connection.[1,2] This configuration prevents magnetic saturation of the transformer core by the dc component of the secondary current. However, the intermittent flow of current in each secondary phase winding results in a poor transformer utilization as compared with normal sinusoidal current flow. Other circuits, using interphase transformers have been developed in order to improve transformer utilization. The basic theory presented above is readily extended to these more complex configurations.

A circuit arrangement that is commonly used with thyristors is the three-phase bridge circuit of Fig. 3.16(a). This is a relatively simple circuit using six thyristors. It is redrawn

in Fig. 3.16(b) to emphasize that it may be regarded as two three-phase half-wave circuits connected in series. The upper group has a common cathode connection and
operates in the usual manner. If there is no firing delay, the thyristor with the most positive anode conducts and produces a positive output voltage at P relative to the neutral.
The lower half-wave group has a common anode connection, and the thyristor with the
most negative cathode conducts to make terminal N negative with respect to the neutral.
Assuming negligible overlap, the potential of terminal P therefore follows the upper
envelope of the alternating input voltage, while the potential of terminal N follows the
lower envelope (Fig. 3.17). The currents in the neutral connection cancel one another,
and the neutral terminal is not used. The output voltage between the P and N terminals
is equal to the vertical distance between the upper and lower envelopes in Fig. 3.17, and

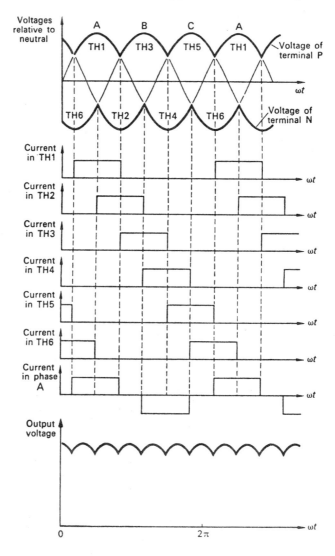

FIG. 3.17. Voltage and current waveforms for a three-phase bridge converter with zero firing angle and
negligible commutating reactance.

therefore it is formed by segments of the three-phase line-to-line voltages and has a six-pulse ripple (six pulses per cycle) as shown. This six-pulse ripple occurs because of the six uniformly spaced commutations in one cycle of the ac supply frequency, and the three-phase bridge converter is therefore classified as a six-pulse converter. The average output voltage is twice that of a three-phase half-wave circuit. Thus, the average output voltage with negligible overlap and zero firing delay is

$$V_{do} = 2 \left(\frac{3\sqrt{6}}{2\pi} \right) V_{ph} = \left(\frac{3\sqrt{6}}{\pi} \right) V_{ph} = 2.34 \ V_{ph} = 1.35 \ V_L \qquad (3.25)$$

where V_{ph} is the rms phase voltage of the wye-connected ac supply, and V_L is the rms line voltage. The actual load voltage is less than the theoretical output voltage of Equation 3.25 by the on-state voltage of two conducting thyristors in series — that is, approximately 3 V.

Each thyristor carries the full direct current, I_d, for 120 degrees in each cycle. Two thyristors always conduct simultaneously in order to complete the circuit, but the conduction periods of devices in the upper and lower groups do not coincide in time. For a phase sequence A, B, C, the thyristors conduct in pairs: TH1 and TH2; TH2 and TH3; TH3 and TH4; TH4 and TH5; TH5 and TH6; TH6 and TH1, and they are consequently numbered in their correct firing sequence in Fig. 3.16. A commutation occurs every 60 degrees and alternate commutations take place in the upper and lower groups. If the dc load has sufficient inductance to maintain a constant dc output current, I_d, the thyristor and ac line currents are as shown in Fig. 3.17 and the line current has no dc component.

When a firing delay is introduced, the voltage waveform is modified and the average output voltage is given by $V_d = V_{do} \cos \alpha$, as before. The presence of overlap causes further modification of the voltage waveform and introduces a voltage drop of $3 I_d X / \pi$, which is twice that for the three-phase half-wave circuit because there are twice as many commutations per cycle. Inverter operation is again obtained when the firing delay angle is greater than 90 degrees, and all previous inverter theory is directly applicable.

In order to facilitate the analysis above, it has been assumed that the three-phase bridge circuit has a wye-connected ac supply. Because the neutral connection is not used, the bridge converter can be supplied from a delta-connected machine or transformer secondary, but circuit operation can be analyzed in terms of an equivalent wye-connected ac supply, as in Fig. 3.17. Converter operation may also be studied by using the ac line-to-line voltage waveforms.[8,9] This method involves working with a set of six sine waves, as shown in Fig. 3.18. If A, B, C are the three ac supply terminals, then the six line-to-line voltages, AB, AC, BC, BA, CA, and CB, are drawn as shown. Clearly, AC = −CA, BA = −AB, and CB = −BC. Consider that all the thyristors in Fig. 3.18(a) are turned on throughout the entire cycle so that they behave like diodes, giving an uncontrolled rectifier. The two devices that are conducting at any instant are those connected to the two ac lines with the highest voltage between them. Thus, for one-sixth of a cycle, the line-to-line voltage, AB, is the most positive, and thyristor TH1 in the upper group is conducting with thyristor TH6 in the lower group, and the output voltage is on the line-to-line wave AB. At each commutation, the output voltage wave switches to a different line-to-line wave, as shown in Fig. 3.18(b). When a firing delay is introduced, the commutation instants are delayed, but the output voltage waveform is again composed of segments of the line-to-line voltages.

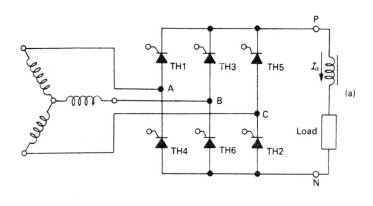

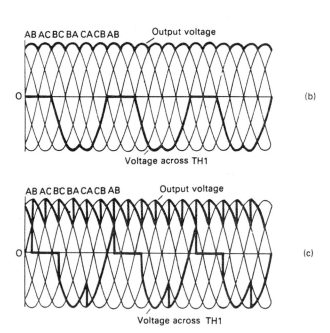

Fig. 3.18. Analysis of a three-phase bridge converter using ac line-to-line voltage waveforms: (a) basic circuit: (b) voltage waveforms with zero delay; (c) voltage waveforms with $\alpha = 30$ degrees.

The anode-to-cathode voltage waveform for a thyristor may be readily derived from the ac line-to-line voltage waveforms, because one terminal of the semiconductor device is always connected to an ac line, while the other terminal is connected to a dc line. Fig. 3.18(b) shows the derivation of the anode-to-cathode voltage waveform for TH1 with zero delay angle. Fig. 3.18(c) shows the output voltage and anode-to-cathode voltage waveforms with a delay angle of 30 degrees. Commutation overlap effects are readily incorporated into these waveforms.

The three-phase bridge circuit is popular for thyristor applications because the thyristors are subjected to only half the peak reverse voltage ($1.045\ V_{do}$) as compared with a six-phase half-wave circuit or a double three-phase circuit with interphase transformer ($2.09\ V_{do}$). For a given direct current, I_d, the thyristors in the bridge circuit must carry twice as much current, but thyristors are more sensitive to voltage requirements and their cost increases sharply with voltage rating. In a bridge rectifier, the output voltage is reduced by the voltage drop of two conducting thyristors in series, but this reduction is not normally excessive because of the small on-state voltage of a thyristor. The bridge circuit has the advantage that there is no dc component in the ac current, making a transformer supply unnecessary. The fact that a neutral connection is not required is often an advantage in ac motor applications because the wye point of the machine winding or the ac source may not be available.

3.3.7. *The Half-Controlled Bridge Converter*

The fully controlled three-phase bridge phase-controlled converter has six thyristors connected as shown in Fig. 3.16(a). As explained above, this circuit operates as a rectifier when each thyristor has a firing angle, α, which is less than 90 degrees and functions as an inverter for α greater than 90 degrees. If inverter operation is not required, the circuit may be simplified by replacing three controlled rectifiers with power diodes, as in Fig. 3.19. This simplification is economically attractive because diodes are considerably less expensive than thyristors, and they do not require firing angle control electronics.

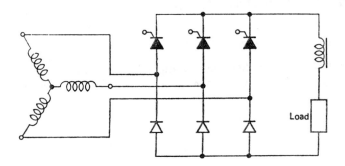

FIG. 3.19. The half-controlled three-phase bridge converter circuit.

Adopting the technique of Section 3.3.6, the half-controlled bridge, or "semiconverter," is analyzed by considering it as a phase-controlled half-wave circuit in series with an uncontrolled half-wave rectifier. The output voltage of the uncontrolled rectifier is constant, but the controlled rectifier circuit delivers an adjustable output voltage which is added to that of the uncontrolled section. By delaying the thyristor firing angle beyond 90 degrees, the controlled section operates as an inverter, and power delivered by the diodes is returned to the ac supply by the thyristors. The inverter counter emf opposes the rectifier voltage and reduces the resultant dc output voltage of the bridge. Theoretically, zero output voltage is obtained when the firing angle is 180 degrees, but in practice, the circuit may malfunction at large delay angles due to the failure of a thyristor to commutate in the limited time available. This malfunction is known as the "half-waving" effect.

The uncontrolled half-wave rectifier develops a constant average output voltage relative to neutral with a value of $(3\sqrt{6}/2\pi)\ V_{ph}$. The output voltage of the controlled half-wave rectifier depends on the delay angle, α, and its average value relative to neutral is $(3\sqrt{6}/2\pi)\ V_{ph}\cos\alpha$. The total average output voltage is therefore

$$V_d = \frac{3\sqrt{6}}{2\pi}\ V_{ph}\ (1+\cos\alpha)\,. \tag{3.26}$$

When the delay angle is zero, V_d has a maximum value, V_{do}, of $(3\sqrt{6}/\pi)\ V_{ph} = 2.34\ V_{ph}$. Thus,

$$V_d = V_{do}\left[\frac{1+\cos\alpha}{2}\right]. \tag{3.27}$$

Clearly, V_d cannot be negative and is zero when α equals 180 degrees.

The output voltage waveforms of half-controlled bridge are similar to those of a fully controlled bridge with a free-wheeling diode. The output voltage ripple gradually changes its character from a six-pulse ripple when α is zero, to a three-pulse ripple when α is large. Consequently, the harmonic content of the load voltage waveform is higher than that of a fully controlled bridge rectifier without a free-wheeling diode.

Where the supply neutral terminal is available, two free-wheeling diodes can be added to the full-wave bridge circuit, as shown in Fig. 3.20. This circuit provides an improved power factor over the conventional three-phase full-wave bridge rectifier. The circuit has been called a "four-legged bridge" in the literature.[10]

3.3.8. Regenerative Converters

The circuits described in Sections 3.2 and 3.3 have achieved regeneration, or return of energy, from the dc side of the converter to the ac supply by reversing the polarity of the dc voltage while maintaining the same direction of dc current. In cases where the load on

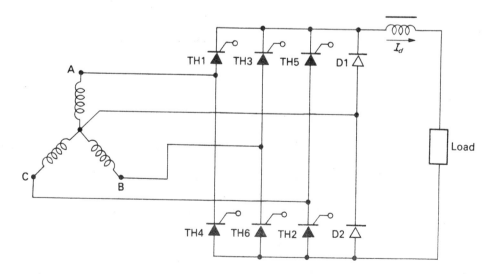

FIG. 3.20. Three-phase full-wave bridge converter with two free-wheeling diode circuits.

the dc side is such that the direction of the power flow can reverse (for example, an overhauling load on a dc motor or a sudden reduction in speed of an ac motor connected to a dc to ac inverter), then provision must be made in the rectifier circuit to prevent overvoltage on the dc bus. Where multiple motors are connected to the dc bus, the regenerated power from one group of motors can supply power to the remaining motors of the drive system that still require power. An alternative is to provide a resistor and dc-dc chopper to control and dissipate the power accepted by the dc bus. In higher power applications and high-performance servo drive systems, it is desirable to return the regenerated power to the ac supply.

As shown in the three-phase half-wave line-commutated inverter of Fig. 3.13, the thyristors have a common cathode connection, and there is no path for dc current flow in the opposite direction to I_d. This circuit and its full-wave bridge counterpart [Fig. 3.16(a)] are used to provide a regenerative dc supply for current-source inverters (Section 4.8) and load-commutated inverters (Section 4.9). These two types of converter system achieve regeneration by reversing the polarity of the dc link voltage.

Voltage-source inverters, however, must achieve reverse power flow by reversing the direction of the dc link current and maintaining the same dc link voltage polarity. This capability can be provided by a double-converter connection in which two thyristor converters are connected in inverse parallel, as shown in Fig. 3.21. For normal motoring operation, one thyristor bridge operates as a rectifier and the second thyristor bridge is blocked. When regeneration occurs, the second bridge accepts the reverse current from the dc link circuit, and energy is returned to the ac supply network if the firing angle is suitably adjusted for inverter operation. The inverse-parallel, or double-converter, connection thus permits direct current and power flow in either direction, but it requires

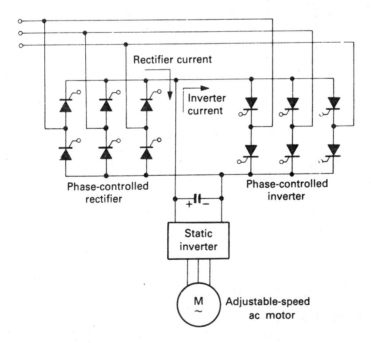

FIG. 3.21. Double-converter connection for regenerative braking of a voltage-source inverter drive.

twice as many thyristors as the single-converter circuit, and the control circuitry is also more complex. The double-converter circuit is often used for speed control and braking of reversible dc motor drives, and it is also a basic element in phase-controlled cycloconverter circuits (Chapter 5).

Many voltage-source inverter drives have a constant dc link voltage obtained by uncontrolled rectification, as described in Sections 3.2 and 3.3.6. The fixed dc link voltage is supplied to a voltage-source inverter that provides both voltage and frequency control for an adjustable-speed ac motor (Chapter 4). For regenerative operation of the diode bridge converter, additional power circuit components can be placed in parallel with the six diodes to provide a path and to control the current returned to the ac supply. For this mode of operation, the dc link voltage is greater than the normal dc voltage when the diodes are rectifying the ac voltage. Consequently, forced commutation must be provided for thyristors,[11] or self-commutated devices (transistors or gate turn-off thyristors) must be used. Figure 3.22 shows a three-phase full-wave bridge circuit with a regenerative rectifier. Transistors TR1 to TR6 provide a current path for $-I_d$ during regeneration, while diodes D1 to D6 allow a current $+I_d$ during motoring operation. Each transistor is supplied with a base-current signal for the 120-degree interval during which conduction occurs in the antiparallel-connected diode during normal rectifier operation. Consequently, the converter is capable of delivering or accepting dc current and can change instantaneously from one mode of operation to the other.

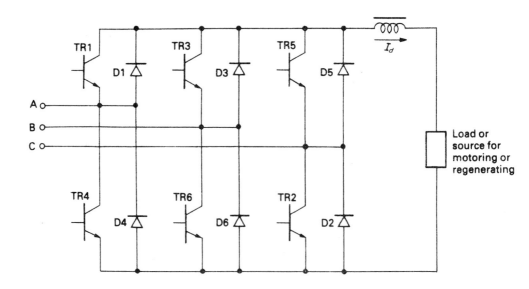

FIG. 3.22. Three-phase bridge converter with transistors for regeneration.

3.4. FORCED-COMMUTATED AC-DC CONVERTERS

The circuits described in Sections 3.2 and 3.3 have generally used the reversal of the source voltage on alternate half-cycles for commutation of the thyristors. Another class of thyristor converter circuit employs forced commutation by using capacitors, reactors, and auxiliary thyristors to provide commutating voltages for the main power thyristors.

These circuits overcome some of the performance disadvantages associated with conventional phase control. A simplified circuit of a single-phase converter with forced commutation is shown in Fig. 3.23.[12] Two main thyristors, TH1 and TH2, together with diodes D1 and D2, comprise the main power converter circuit components. Two auxiliary thyristors, TH3 and TH4, together with commutating capacitors C1 and C2, form the forced commutating circuit for both main thyristors, TH1 and TH2. Resistor R1 and diodes D5 and D8 provide a path for charging capacitors C1 and C2 to the peak of the ac source voltage. Capacitor C3 provides a voltage clamp for the energy stored in the ac supply reactance.

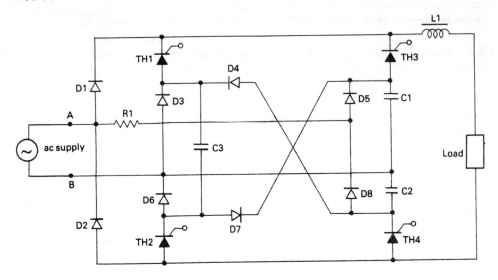

Fig. 3.23. Single-phase ac-dc converter with forced commutation.

To understand the forced-commutated mode, assume that source terminal A is positive with respect to B, that TH2 has been gated on, that C2 was charged positive on its upper plate through R1 and D8 during the preceding half-cycle, and that C3 is discharged. The conduction path for the load current is from A through D1, L1, the load, TH2, and D6 back to B. At some later time in the half-cycle, thyristor TH4 is gated on, which essentially connects C2 in parallel with the series combination of diode D6 and thyristor TH2. Because C2 has been charged to a voltage equal to the peak value of the ac supply during the previous half-cycle, TH2 and D6 become reverse-biased by this voltage. When current flow through the TH2 path is interrupted, the ac supply current remains relatively constant due to the inductance of the ac supply and diverts through the TH4 and C2 path because of the initial polarity of the charge on C2. This current diversion causes the voltage across C2 to decay to zero. In addition, during this period, diodes D4 and D6 are reverse biased because C3 has no charge and does not acquire any charge during this time.

As soon as a slight voltage of negative polarity appears across C2, diodes D4 and D6 become forward biased, causing C3 to be connected in parallel with C2. The ac supply current still remains relatively constant but now divides between the C2 and C3 paths according to the ratio of the capacitance of each capacitor to the total capacitance. The current flow causes C2 and C3 to charge so that the upper plate of C3 is positive and the upper plate of C2 is negative until the capacitor voltage becomes equal to that of the ac supply.

At the time the voltage across C2 and C3 becomes equal to the ac supply voltage, the load inductance starts to take over as the source of current in the dc loop. However, the energy stored in the source inductance has to be dissipated or transferred to another circuit element. The transfer of this energy into C1, C2, and C3 results in a voltage overcharge of C2, of C3, and also of C1 through the R1, D5 path, which results in a voltage transient on the ac supply voltage. During this overcharge period, the ac supply current goes to zero, as, of course, does the current through C1, C2, C3, and TH4. TH4 then recovers and once again blocks forward voltage.

Late in the half-cycle, TH1 and TH2 are both gated on in order to discharge C3 and prepare it for the next half-cycle. If this were not done, the charge would be trapped on C3 by diodes D3 and D6, and the circuit would not operate as described. The circuit operates in a similar manner for the opposite polarity of the ac supply voltage.

In each half-cycle of the ac supply voltage, the flow of ac line current commences after a controllable delay angle, α. With the correct gating sequence of the main and auxiliary thyristors, the ac line current is extinguished at an angle α before the end of the voltage half-cycle. This ensures that the current conduction interval is symmetrical around the maximum voltage point in the ac source voltage waveform, resulting in a unity displacement factor for all values of output voltage. The harmonic currents contribute to a distortion factor; therefore, the circuit power factor is less than unity. The advantages of these forced-commutated circuits over conventional phase-controlled ac-dc converters are as follows:

1.   Higher source current displacement factor at reduced levels of output voltage.

2.   Reduced levels of telephone interference current.

The principal disadvantages are —

1.   Auxiliary components or devices with turn-off capability are required — for example, gate turn-off thyristors.

2.   The main thyristors and dc load are subjected to higher peak voltages.

These systems are used for high-power electric locomotives supplied from a high-voltage, single-phase, low-frequency ac voltage source with considerable series reactance.

The circuit just described operates once per half-cycle of the fundamental frequency. If additional commutations could be provided during each half-cycle of the fundamental, then the line current could approach a sinusoid with minimum harmonic currents. The ac source current would then be sinusoidal and at unity power factor for all values of output voltage. Circuits using pulse-width modulation have been constructed with devices having turn-off capability such as power transistors, field-effect transistors,[13] and gate turn-off thyristors.[14]

3.5. RECTIFIERS AND DC-DC CHOPPERS

The circuits described in Sections 3.2 through 3.4 have achieved the ac to adjustable dc voltage conversion in a single converter, either a line-commutated phase-controlled rectifier or a forced-commutated phase-controlled rectifier. An adjustable direct voltage

input for the inverter may also be obtained by rectifying the three-phase ac supply voltage with an uncontrolled diode rectifier and then adjusting the rectified dc output voltage
by means of a dc chopper. In a series chopper circuit, a thyristor is placed in series with
the dc supply and the load. It is rapidly switched on and off so that the output voltage
consists of a series of rectangular pulses.[15] Figure 3.24(a) shows an ac motor speed control scheme in which the chopper thyristor, TH1, is alternately switched on by a gate
pulse and then turned off by a commutating circuit of the type used in forced-
commutated inverters. The idealized output voltage waveform of the chopper circuit is
shown in Fig. 3.24(b). The average output voltage is usually varied: (1) by keeping the
total cycle time, T, constant and varying the ratio of the on-time of the thyristor, T_1, to
the off-time, T_2, or (2) by holding the on-time, T_1, constant and varying the off-time,
T_2. If the pulsating output voltage is filtered, an adjustable direct voltage is obtained,
which can be used for speed control of a dc machine or as the input voltage to an inverter
for ac motor control.

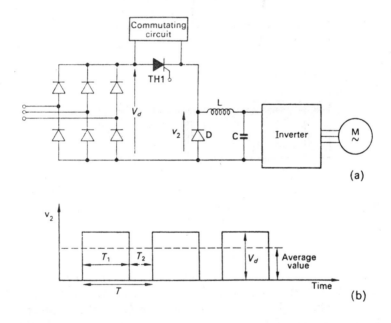

FIG. 3.24. AC motor drive with voltage control by means of a chopper in the dc link: (a) basic circuit;
(b) idealized output voltage waveform for the chopper circuit.

Figure 3.25(a) shows a typical dc chopper circuit with a constant input voltage, V_d.
Thyristor TH1, the main series thyristor, is alternately switched on and off in order to
adjust the output voltage, v_2. Diode D is the free-wheeling diode that allows continuous
output current, i_2, when TH1 is turned off. The series inductor, L, and shunt capacitor,
C, filter the output voltage. An auxiliary thyristor, TH2, is triggered to initiate turn-off of
TH1, and initially, when TH2 is gated, capacitor C1 charges up to the input voltage, V_d,
through the load circuit. Thyristor TH1 is then gated to deliver current to the load, and
C1 discharges around the oscillatory circuit, consisting of L1, C1, and diode D1. After
the first half-cycle of this oscillation, the capacitor is charged to a voltage, V_c, with the

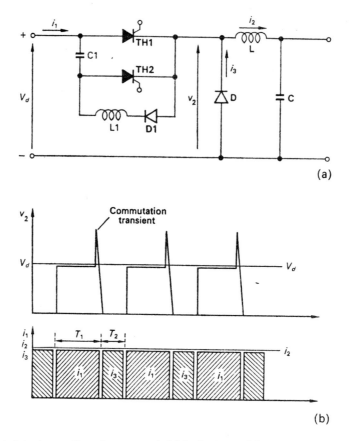

FIG. 3.25. Typical dc chopper for voltage control: (a) basic circuit; (b) voltage and current waveforms with a large filter inductance.

negative plate at the anode of TH1. If the oscillatory circuit is lossless, V_c is equal to the input voltage, V_d. The reversal of current flow for the next half-cycle of the oscillation is prevented by diode D1, and consequently, capacitor C1 retains the voltage V_c. At some subsequent time, when TH2 is gated, the capacitor C1 applies a reverse-bias voltage across TH1 and turns it off. The output current, i_2, is maintained by the filter inductance and flows through C1 and TH2. This current gradually reduces the voltage on C1 to zero and recharges it with the opposite polarity. If C1 is large enough, positive anode-cathode voltage is not reapplied across TH1 until it has regained its forward-blocking capability. When C1 has been fully charged by the dc supply, TH2 turns off. The filter inductance maintains the output current, i_2, by drawing the current through the free-wheeling diode, D, and the output voltage, v_2, is zero until TH1 is again gated on and the cycle of operation is repeated. Figure 3.25(b) shows the voltage and current waveforms for a large filter inductance that maintains a constant output current.

A transistor-based circuit is shown in Fig. 3.26. It utilitizes the same principles as the thyristor circuit of Fig. 3.25 without the need for commutating circuit elements. Also, the load voltage does not have the commutation transient shown in Fig. 3.25(b). The switching frequency is usually higher (up to 50 kHz) to reduce the size of the L-C filter.

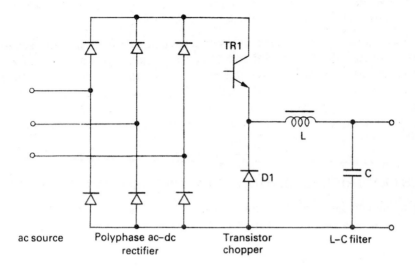

ac source   Polyphase ac–dc   Transistor   L–C filter
                    rectifier   chopper

FIG. 3.26. Diode rectifier bridge and transistor-based chopper.

The advantages of the rectifier-chopper approach over the phase-controlled rectifier approach can be summarized as follows:

1. Higher input displacement factor and power factor are possible at reduced dc output voltages.

2. Only one controlled device is required, rather than two, three, or six controlled devices with their associated control electronics.

3. The dc chopper can operate at a faster repetition rate, which increases the speed of response and reduces the time lag associated with phase-controlled rectifier operation.

4. Because of the higher operating frequency of the chopper, the L-C filter can be smaller for equivalent filtering.

5. With a current sensor, an instantaneous load current limit can be provided.

6. A similar chopper circuit can be used for both ac and dc input conversion systems.

The disadvantages of the rectifier-chopper circuit compared with the phase-controlled rectifier are as follows:

1. Efficiency is reduced due to series connection of power converters.

2. An energy storage filter is required on the fixed dc voltage bus when it is supplied from a single-phase input, to prevent the voltage from falling to zero every half-cycle.

3. The circuit shown in Fig. 3.26 is nonregenerative and only provides a single polarity of voltage and unidirectional load current; additional power circuit elements must be added to provide both the chopper and rectifier with regenerative capabilities.

4.   If two energy storage filters are present, beat frequency effects between the two resonant filter frequencies must be prevented.

These rectifier-chopper conversion systems have been used in multiple ac motor drive systems supplied with utility power and having a storage battery across the fixed-voltage dc link for back-up power. The inverter provides ac motor frequency control, while ac motor voltage control is effected by the chopper. A second application is in low-horsepower, single-quadrant dc motor drives that can utilize a single controllable power transistor operating at a very high frequency instead of a four-semiconductor single-phase ac-dc converter operating at 50 or 60 Hz.

3.6. HARMONIC DISTORTION DUE TO POWER CONVERTER CIRCUITS

The ac line current of a phase-controlled converter is nonsinusoidal, and therefore, the circuit acts as a generator of harmonic currents. These harmonic currents develop harmonic voltage drops across the network source impedance, causing distortion of the supply voltage delivered to the converter itself and to other consumers in the area.[16–19]

In general, the harmonic currents on the ac side of the converter are of order $k = np \pm 1$, where p is the pulse number of the converter and n is an integer; the pulse number is the number of diodes or thyristors which successively begin to conduct during one cycle of the supply frequency. The six-phase half-wave circuit is therefore classified as a six-pulse system, as also is the three-phase bridge circuit. The lowest order harmonics generated by these circuits are the fifth and the seventh. A 12-pulse converter is obtained if two six-pulse groups, with a 30-degree phase displacement, feed a common dc load. Similarly, two 12-pulse circuits with a 15-degree displacement form a 24-pulse system. Increasing the pulse number reduces the total harmonic content by eliminating low-order harmonics, but correct balance between converter groups is essential. Theoretically, the amplitude of the kth harmonic current is given by I_1/k, where I_1 is the fundamental amplitude. However, this relationship assumes infinite dc circuit inductance and instantaneous commutation. In practice, commutation overlap reduces the amplitude of the harmonic current below this theoretical value. Delayed firing reduces the overlap angle and so tends to increase the harmonic amplitude.

Distortion of the ac supply voltage waveform is objectionable, as it may cause overheating of power-factor-correcting capacitors due to large harmonic currents. At the harmonic frequencies, undesirable resonances may also occur between the power-factor-correcting capacitance and the system inductance. A certain amount of telephone interference is also possible. In order to avoid these difficulties, supply authorities specify that the total converter rating must not form an excessive part of the load on the power system. Engineering Recommendation G.5/3 of the Electricity Council of Great Britain is a typical utility specification on the connection of converter equipment to the power system.[20] This document states that smaller items of converter equipment which are in general use are acceptable for connection to the ac supply without any detailed consideration of the system itself. Thus, three-pulse and six-pulse converters can be connected to the three-phase, 415 V, ac supply provided their ratings are not greater than 8 kVA and 12 kVA, respectively. Larger kVA ratings are permissible at higher system voltages. If the equipment size exceeds these limits, then it must be judged against a table of maximum permitted values of harmonic current that a consumer may feed into the system, assuming that the existing harmonic levels on the system are within specified limits. If the converter is intended for connection to the 415 V distribution system, the permissible harmonic current levels are 56 A for the fifth harmonic, 40 A for the seventh harmonic, 19 A for the eleventh harmonic, and 16 A for the thirteenth harmonic. In order to satisfy these limits, the rating of a diode converter should be less than 150 kVA for a six-pulse circuit and less than 300 kVA for a 12-pulse system. Again, larger ratings are permitted at higher system voltages.

At higher power levels, the pulse number of the converter installation may have to be increased in order to reduce the harmonic content to an acceptable value. Transformers are required to produce the correct phase displacement between converter groups, which increases the cost of the installation. Harmonic filters are also used to suppress specific harmonics. These are series resonant circuits connected between the ac lines to provide low impedance paths for particular harmonic currents.

Power converter circuits also cause system voltage distortion known as line notching.[17] During the brief commutation overlap period, two phases of the ac supply are short-circuited, as explained earlier. This produces a notch in the ac line voltage waveform for the duration of the overlap angle, u. Thus, a three-phase bridge circuit produces six notches per cycle of the line voltage, and these notches are rich in high frequencies, which are readily propagated through the power system.

Power converters may also generate radio-frequency interference (RFI) or electromagnetic interference (EMI) due to the sudden changes of current which occur in diode and thyristor circuits. This interference is partly radiated directly but is mainly propagated in the supply lines and may cause interference to broadcast reception or malfunction of thyristor equipment. In general, distortion of the ac system voltage due to a converter load may cause interaction between converter installations by coupling transient emfs into sensitive trigger circuits or by producing excessive dv/dt transients in thyristor anode circuits, or by the introduction of spurious zero crossings in the line voltage waveform. The danger of interaction may usually be eliminated by good circuit design and layout, by shielding sensitive control circuits and by fitting dv/dt suppression circuits.

3.7. REFERENCES

1. RISSIK, H., *The Fundamental Theory of Arc Converters*, Chapman and Hall, London, 1939.

2. SCHAEFER, J., *Rectifier Circuits: Theory and Design*, J. Wiley, New York, NY, 1965.

3. MÖLTGEN, G., *Line Commutated Thyristor Converters*, Pitman, London, 1972.

4. PELLY, B.R., *Thyristor Phase-Controlled Converters and Cycloconverters*, Wiley-Interscience, New York, NY, 1971.

5. DEWAN, S.B., and STRAUGHEN, A., *Power Semiconductor Circuits*, Wiley-Interscience, New York, NY, 1975.

6. LOCHER, R.E., and SMITH, M.W. (Editors), *Electronic Data Library, Thyristors-Rectifiers*, General Electric Co., 1982.

7. JORDAN, E.C. (Editor), *Reference Data for Engineers: Radio, Electronic, Computer, and Communications*, Seventh Edition, H.W. Sams, Indianapolis, IN, 1985.

8. LUDBROOK, A., and MURRAY, R.M., A simplified technique for analyzing the three-phase bridge rectifier circuit, *IEEE Trans. Ind. Gen. Appl.*, **IGA-1**, 3, May/June 1965, pp. 182-187.

9. LYE, R.W. (Editor), *Power Converter Handbook*, Canadian General Electric Co. Ltd., 1979.

10. STEFANOVIC, V.R., Power factor improvement with a modified phase-controlled converter, *IEEE Trans. Ind. Appl.*, **IA-15**, 2, Mar./Apr. 1979, pp. 193-201.

11. WILSON, J.W.A., The forced commutated rectifier as a regenerative rectifier, *IEEE Trans. Ind. Appl.*, **IA-14**, 4, July/Aug. 1978, pp. 335-340.

12. BEZOLD, K.H., FORSTER, J., and ZANDER, H., Thyristor converters for traction dc motor drives, *Conf. Rec. IEEE Int. Semicond. Power Converter Conf., Baltimore, MD, May 8-10, 1972*, pp. 3-8-1 to 3-8-7.

13. KOCHER, M.J., and STEIGERWALD, R.L., An ac to dc converter with high quality input waveforms, *Conf. Rec. IEEE Power Electron. Spec. Conf., Cambridge, MA, June 14-17, 1982*, pp. 63-75.

14. YOSHIOKA, T., MIYAZAKI, K., and OKUYAMA, T., Quick response ac motor drive system, *Hitachi Rev.*, **31**, 4, Aug. 1982, pp. 179-184.

15. HEUMANN, K., Pulse control of dc and ac motors by silicon controlled rectifiers, *IEEE Trans. Commun. and Electron.*, **83**, 1964, pp. 390-399.

16. STACEY, E.M., and SELCHAU-HANSEN, P.V., SCR drives — ac line disturbance, isolation and short-circuit protection, *IEEE Trans. Ind. Appl.*, **IA-10**, 1, Jan./Feb. 1974, pp. 88-105.

17. SCHIEMAN, R.G., and SCHMIDT, W.C., Power line pollution by 3-phase thyristor motor drives, *Conf. Rec. IEEE Ind. Appl. Soc. Annual Meeting, 1976*, pp. 680-690.

18. STRATFORD, R.P., Rectifier harmonics in power systems, *IEEE Trans. Ind. Appl.*, **IA-16**, 2, Mar./Apr. 1980, pp. 271-276.

19. SMITH, R.L., and STRATFORD, R.P., Power system harmonics effects from adjustable-speed drives, *IEEE Trans. Ind. Appl.*, **IA-20**, 4, July/Aug. 1984, pp. 973-977.

20. *Limits for Harmonics in the United Kingdom Electricity Supply System*, Engineering Recommendation G.5/3, The Electricity Council, London, 1976.

CHAPTER 4

DC-AC Inverter Circuits

4.1. INTRODUCTION

In an inverter circuit, dc power is converted to ac power. The phase-controlled converter discussed in Chapter 3 can feed power from a dc source into an existing ac utility network. The circuit uses conventional thyristors, as shown in Fig. 3.13, but natural commutation is achieved because of the ac network voltages.

This chapter is concerned with a type of inverter circuit that can function as an isolated source, delivering ac power to a passive network or a motor. Semiconductor devices with controlled turn-off capability, such as transistors, MOSFETs, and GTOs, can be used when these devices are available with the required voltage and current ratings. If conventional thyristors are employed, forced commutating circuits are essential because the ac system voltages needed for natural commutation are not present.

The output frequency of the static inverter is determined by the rate at which the semiconductor devices are switched on and off by the inverter control circuitry; consequently, an adjustable-frequency ac output is readily provided. However, the basic switching action of the inverter normally results in nonsinusoidal output voltage and current waveforms that may adversely affect motor performance. The filtering of harmonics is not feasible when the output frequency varies over a wide range, and the generation of ac waveforms with low harmonic content is important. When the inverter feeds a transformer or ac motor, the output voltage must be varied in conjunction with frequency to maintain the proper magnetic conditions. Output voltage control is therefore an essential feature of an adjustable-frequency system, and various techniques for achieving voltage control are discussed in this chapter.

The inverter may receive its dc power from a battery, but in most industrial applications it is fed by a rectifier, as already described. This configuration is classified as a dc link converter because it is a two-stage static frequency converter in which ac power at network frequency is rectified and then filtered in the dc link before being inverted to ac at an adjustable frequency. Rectification is performed by standard diode or thyristor converter circuits, as described in Chapter 3, and inversion is achieved by the circuit techniques described in this chapter.

Inverters can be broadly classified either as voltage-source or current-source inverters. The voltage-fed, or voltage-source, inverter (VSI) is powered from a stiff, or low-impedance, dc voltage source such as a battery or a rectifier, the output voltage of which is smoothed by an LC filter. The large filter capacitor across the inverter input terminals maintains a constant dc link voltage. The inverter is therefore an adjustable-frequency voltage source, the output voltage of which is essentially independent of load current.

On the other hand, the current-fed, or current-source, inverter (CSI) is supplied with a controlled current from a dc source of high impedance. Typically, a phase-controlled thyristor rectifier feeds the inverter with a regulated current through a large series inductor. Thus, load current rather than load voltage is controlled, and the inverter output voltage is dependent upon the load impedance.

4.2. THE SINGLE-PHASE BRIDGE VOLTAGE-SOURCE INVERTER

The single-phase inverter introduced in Chapter 2 is shown in two basic circuit configurations in Fig. 4.1. These circuits are suitable for adjustable-frequency operation because an output transformer is not required. (A transformer cannot deliver a dc out-

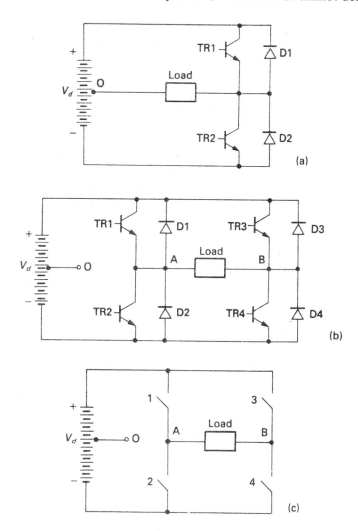

FIG. 4.1. Single-phase inverters: (a) half-bridge circuit; (b) full-bridge circuit; (c) equivalent full-bridge circuit using ideal mechanical switches.

put voltage, and transformer size and cost are prohibitive at low output frequencies.) Transistors are shown in the diagram, but the basic inverter operation is independent of the particular semiconductor device used. Each n-p-n transistor shown in Fig. 4.1 permits current flow from collector to emitter, and an inverse-parallel (or antiparallel) rectifier diode allows current flow in the reverse direction. Such a reactive feedback diode is characteristic of a voltage-source inverter, providing a reverse current path for load current that permits a return of energy from a reactive load, through the inverter, to the dc supply.

The configuration in Fig. 4.1(a), commonly known as a half-bridge inverter, requires a center-tapped dc supply. This supply can be provided by a battery, as shown, or by placing a pair of large series-connected capacitors across a rectified ac supply. Alternate switching of transistors TR1 and TR2 then produces a square-wave alternating voltage across the load.

The half-bridge circuit is the basic building block for single-phase and three-phase bridge inverters. In Fig. 4.1(b), two half-bridges are combined to give a single-phase full-bridge inverter which avoids the center-tapped dc supply at the expense of additional semiconductor devices. Attention will be concentrated on the operation of this circuit.

4.2.1. *Voltage Waveforms*

Normally, transistors TR1 and TR2 in Fig. 4.1(b) are switched on and off for alternate 180-degree intervals by supplying and removing forward base current. Neglecting on-state voltage drop, load terminal A is therefore connected alternately to the positive and negative sides of the dc supply. Likewise, transistors TR3 and TR4 are switched alternately to connect load terminal B to the positive or negative dc rail. When TR1 and TR4 conduct simultaneously, the dc supply voltage, V_d, appears across the load. Similarly, when TR2 and TR3 conduct, the supply voltage is applied across the load with reverse polarity.

If the transistor switching action occupies a negligible fraction of the half-cycle, the voltages at terminals A and B have square waveforms. It is convenient to take the dc center-tap, O, as a reference point, despite the fact that it is not used, and Fig. 4.2 shows the resultant square voltage waveforms of amplitude $V_d/2$ for v_{AO} and v_{BO}. These voltages relative to the dc mid-point will be termed the pole voltages. In Fig. 4.2, the transistors are switched in diagonal pairs, with TR1 and TR4 turned on and off simultaneously, and similarly for TR2 and TR3. The resultant load voltage is the difference of the pole voltages. Thus

$$v_{AB} = v_{AO} - v_{BO} \tag{4.1}$$

giving a square-waveform voltage of amplitude V_d, as shown in Fig. 4.2(c). For a purely resistive load, the current has an identical square waveform, and the feedback diodes are unnecessary because load current reverses instantaneously.

4.2.2. *Current Waveforms*

If the inverter supplies an inductive load, the current lags the applied voltage. Figure 4.2(d) shows the steady-state load current waveform when the square-wave

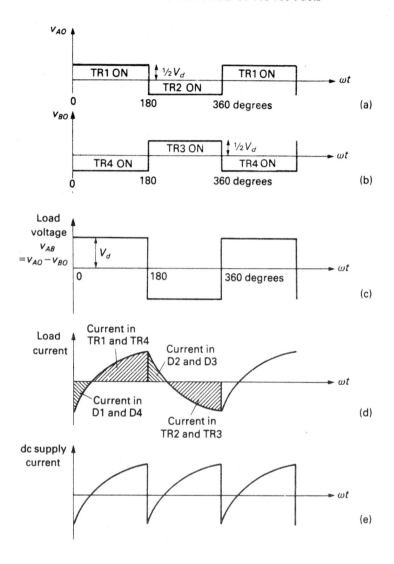

Fɪɢ. 4.2. Voltage and current waveforms for a single-phase bridge inverter with a series RL load: (a), (b) pole voltages; (c) load voltage; (d) load current; (e) dc supply current.

inverter feeds a passive RL circuit. The current is a series of exponentials, and for an interval after the load voltage reverses polarity, the instantaneous power consumption of the load is negative because voltage and current have opposite signs. This signifies that stored energy in the inductive load is being returned to the dc supply through the inverter feedback diodes. The behavior is analogous to that of the simple series RL circuit of classical single-phase ac circuit theory, where voltage and current waveforms are sinusoidal but current lags voltage. In the interval between the zero-crossings of voltage and current, instantaneous power consumption is negative, indicating the return of energy to the ac supply network by the reactive load.

The mechanism of energy feedback through the inverter can be seen by reference to the load current waveform of Fig. 4.2(d), where the conducting devices are indicated for each portion of the ac cycle. At time zero, TR2 and TR3 are turned off and TR1 and TR4 are turned on, but the load current already established in the inductive load flows for a time in the negative direction, from B to A. This negative load current flows through feedback diodes D1 and D4 and the dc supply, thereby returning inductive load energy to the dc source.

Later in the half-cycle, when the load current falls to zero and reverses, the increasing positive current flows through transistors TR1 and TR4. This instant of current reversal is load dependent and can occur at any time in the half-period. Consequently, the transistors must have forward base drive throughout the half-cycle so that they can take up the load current when required. However, conduction of D1 and D4 at time zero results in an immediate reversal of load voltage.

After a half-period, TR1 and TR4 are turned off and TR2 and TR3 are turned on. Positive load current continues to flow through feedback diodes D2 and D3 and the dc power supply, and load energy is again returned to the dc source during the initial portion of this half-cycle.

There is also a return of load energy to the dc supply when the inverter feeds a leading power factor load such as an underdamped series RLC circuit with a resonant frequency which is greater than the inverter output frequency. In this case, the load current is approximately sinusoidal and reverses before the reversal of load voltage. As shown in Fig. 4.3, there is now an interval in the latter portion of each voltage half-cycle when load voltage and current have opposite signs and load energy is returned through the feedback diodes. The leading power factor load causes the transistor current to fall naturally to zero before the end of the voltage half-cycle, and the reverse load current then increases in the associated feedback diode. The forward voltage drop of the diode applies

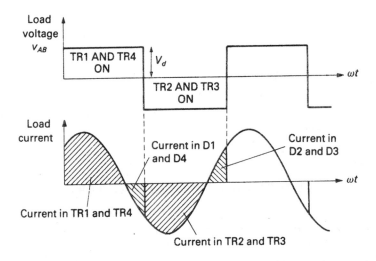

FIG. 4.3. Load voltage and current waveforms for a single-phase bridge inverter with a leading power factor load.

a small reverse bias to the transistor for the remainder of the half-period. For this load condition, the self-turn-off capability of the transistor is not required and conventional thyristors can be used without forced commutating circuits. This is another example of thyristor load commutation, as discussed in Chapter 2.

In general, it should be noted that, during the half-period for which TR1 has forward base drive, load terminal A is effectively connected to the positive dc rail, although feedback diode D1 may conduct for a portion of this half-cycle. Similarly, for the next half-period, TR2 is turned on, and terminal A is connected to the negative dc rail by conduction of TR2 or D2. Thus, the square-wave firing signals for transistors TR1 and TR2 uniquely determine the potential of terminal A. The transistor with its antiparallel diode functions as a bidirectional switch that is closed while a firing signal is applied to the transistor base. This switching action is emphasized by the inverter representation of Fig. 4.1(c), where ideal mechanical switches are shown in place of semiconductors. This diagram also emphasizes that the two devices in an inverter leg or half-bridge must always be switched alternately, with a brief interval between the turning off of one device and the turning on of the other. Spurious firing signals which result in simultaneous conduction of both devices will cause a "shoot-through" fault that short circuits the dc supply.

DC Supply Current. The dc supply current waveform is a replica of the load current waveform except for the polarity reversal in alternate half-cycles, as shown in Fig. 4.2(e). Conduction of the feedback diodes clearly results in intervals of negative dc supply current, confirming that there is a return of load energy to the dc source during a portion of each half-cycle. The average dc supply current is determined by the average power input to the inverter. Because the inverter is nearly lossless, the average dc current is therefore proportional to the mean load power in watts.

The inverter may feed an ac motor which operates transiently as a generator when the motor brakes or suddenly decelerates. Under these conditions, the square-wave firing signals for the inverter still determine the half-cycle closure interval of each inverter switch, but the conduction interval of the feedback diode now exceeds that of the transistor, and the dc supply current has an average value which is negative. This negative current indicates that there is a net return of energy from the ac generator to the dc source. A dc battery can absorb this regenerated energy, but a rectifier can not. Consequently, in a dc link converter, the dc link voltage will rise due to overcharging of the filter capacitor, and the regenerated energy is normally dissipated in a dynamic braking resistor, as discussed in Section 4.6.

4.3. THE THREE-PHASE BRIDGE SIX-STEP INVERTER

The three-phase six-step, or quasi-square-wave, inverter is a voltage-source inverter that has been widely used in commercial adjustable-speed ac motor drives. As shown in Fig. 4.4(a), a third leg, or half-bridge, is added to the single-phase bridge circuit of Fig. 4.1(b), and the output terminals A, B, and C are connected to a three-phase ac motor. A reversal of the direction of motor rotation is readily accomplished by changing the inverter output phase sequence by means of the firing signals.

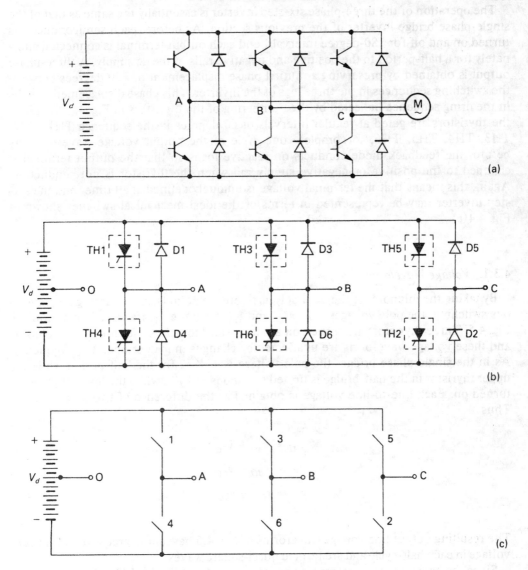

FIG. 4.4. The three-phase bridge inverter: (a) transistor bridge circuit feeding an ac motor; (b) general three-phase bridge circuit; (c) equivalent circuit using ideal mechanical switches.

Conventional thyristors have been used in many three-phase six-step inverters, and the general circuit diagram of Fig. 4.4(b) shows each thyristor device within a dashed rectangle. This circuit symbol is intended to represent the conventional thyristor with its associated forced commutating circuit. The symbol may also be regarded as representing any semiconductor device with a self-turn-off capability such as a GTO, power MOS-FET, or transistor. In general, the semiconductor must have a sustained gating, or firing, signal throughout the half-cycle during which conduction may occur, because the exact instant at which conduction commences is load dependent.

The operation of the three-phase six-step inverter is essentially the same as that of the single-phase bridge inverter of the previous section. As before, each semiconductor is turned on and off for 180-degree intervals, and each output terminal is connected alternately for a half-period to the positive and negative rails of the dc supply. A three-phase output is obtained by preserving a mutual phase displacement of 120 degrees between the switching sequences in the three legs of the inverter. This phase displacement results in the firing sequence indicated by the numbering of the thyristors in Fig. 4.4(b). Thus, the thyristors are gated at regular intervals of 60 degrees in the sequence TH1, TH2, TH3, TH4, TH5, TH6, to complete one cycle of the output voltage waveform. As before, the feedback diode conducts on reactive loads so that the output terminal is clamped to the positive or negative supply rail when the thyristor is not conducting. Again, this means that the terminal voltage is uniquely defined at all times and the six-step inverter may be represented in terms of the ideal mechanical switches shown in Fig. 4.4(c).

4.3.1. *Voltage Waveforms*

By taking the midpoint of the dc supply as reference point O and assuming instantaneous switching, the pole voltages v_{AO}, v_{BO}, and v_{CO} have the square waveforms shown in Fig. 4.5. Three thyristors are turned on at any instant to define the three pole voltages, and these voltage waveforms are unaffected by changes in load or operating frequency. As in the single-phase bridge, the pole voltage is $+ V_d/2$ for the half-period while the upper thyristor in the half-bridge is turned on, and is $- V_d/2$, while the lower thyristor is turned on. Each line-to-line voltage is obtained as the difference of two pole voltages. Thus

$$v_{AB} = v_{AO} - v_{BO}$$

$$v_{BC} = v_{BO} - v_{CO}$$

$$v_{CA} = v_{CO} - v_{AO}. \qquad (4.2)$$

The resulting output line voltage waveforms of Fig. 4.5 have 60-degree intervals of zero voltage in each half-cycle and are termed quasi-square waves.

Since the pole voltage waveform is a square wave of amplitude $V_d/2$, it has the familiar Fourier series expression involving all odd harmonics. Thus

$$v_{AO} = \frac{4}{\pi} \cdot \frac{V_d}{2} \left[\sin \omega t + \frac{1}{3} \sin 3 \omega t + \frac{1}{5} \sin 5 \omega t \right.$$

$$\left. + \frac{1}{7} \sin 7\omega t + \frac{1}{9} \sin 9 \omega t + ... \right]. \qquad (4.3)$$

Similarly, v_{BO} is a square wave displaced by 120 degrees. Its Fourier series expression is

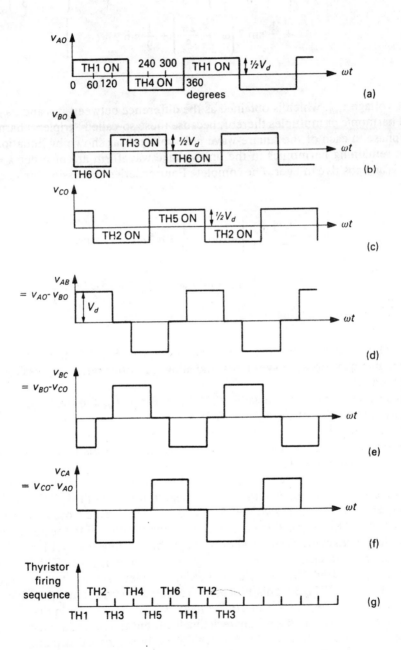

FIG. 4.5. Voltage waveforms for six-step operation of the three-phase bridge inverter: (a), (b), (c) pole voltages; (d), (e), (f) line-to-line voltages; (g) thyristor gating sequence.

$$v_{BO} = \frac{4}{\pi} \cdot \frac{V_d}{2} \left[\sin\left(\omega t - \frac{2\pi}{3}\right) + \frac{1}{3}\sin 3\,\omega t + \frac{1}{5}\sin 5\left(\omega t - \frac{2\pi}{3}\right) \right.$$

$$\left. + \frac{1}{7}\sin 7\left(\omega t - \frac{2\pi}{3}\right) + \frac{1}{9}\sin 9\,\omega t + ... \right]. \qquad (4.4)$$

The line voltage v_{AB}, which is obtained as the difference between v_{AO} and v_{BO}, contains no third harmonic or multiples thereof, because these so-called "triplen" harmonics are in time phase in each of the square-wave pole voltages, as shown by Equations 4.3 and 4.4. The remaining harmonics in the line voltage waveform are of order $k = 6\,n \pm 1$, where n is any positive integer. The complete Fourier series expression for v_{AB} is

$$v_{AB} = \frac{2\sqrt{3}}{\pi} V_d \left[\sin\omega t - \frac{1}{5}\sin 5\,\omega t - \frac{1}{7}\sin 7\,\omega t \right.$$

$$\left. + \frac{1}{11}\sin 11\,\omega t + \frac{1}{13}\sin 13\,\omega t - ... \right]. \qquad (4.5)$$

(This equation assumes that the waveform for v_{AB} in Fig. 4.5(d) is shifted by $+30$ degrees to give quarter-wave symmetry and allow a Fourier series expression involving sine terms only.)

The rms value of the line voltage is $\sqrt{2/3}\ V_d$, or 0.816 V_d, and the fundamental component has an rms value of $\sqrt{6}\ V_d/\pi$, or 0.78 V_d.

If the quasi-square-wave inverter feeds a balanced wye-connected (star-connected) load, the line-to-neutral, or phase voltage, waveform has six distinct steps per cycle and is termed a six-step wave. This is readily demonstrated in the case of a purely resistive load, by drawing the equivalent circuit for each of the six conduction sequences in one cycle of inverter operation. For the first 60-degree interval, TH5, TH6, and TH1 are turned on, or the mechanical switches 5, 6, and 1 of Fig. 4.6 are closed, connecting load terminals A and C to the positive rail and connecting terminal B to the negative rail. The corresponding equivalent circuit is shown under the heading Interval 1, in Fig. 4.6, and the line voltages are clearly $v_{AB} = + V_d$, $v_{BC} = - V_d$, and $v_{CA} = 0$. The line-to-neutral voltages are deduced by simple potential divider action to be $v_{AN} = v_{CN} = + V_d/3$, and $v_{BN} = - 2\ V_d/3$. This process is repeated for the remaining five intervals, and the resulting six-step line-to-neutral voltage, v_{AN}, is also shown in Fig. 4.6.

This analysis has assumed a purely resistive load, but the same six-step phase voltage waveform is obtained with any balanced three-phase linear load. Calculation of the phase voltage waveforms with the use of the general voltage equations derived in Appendix 4.1, at the end of Chapter 4, demonstrates this fact. These general equations, which relate the instantaneous line-to-neutral voltages to the instantaneous pole voltages, are not load dependent. Thus, in general,

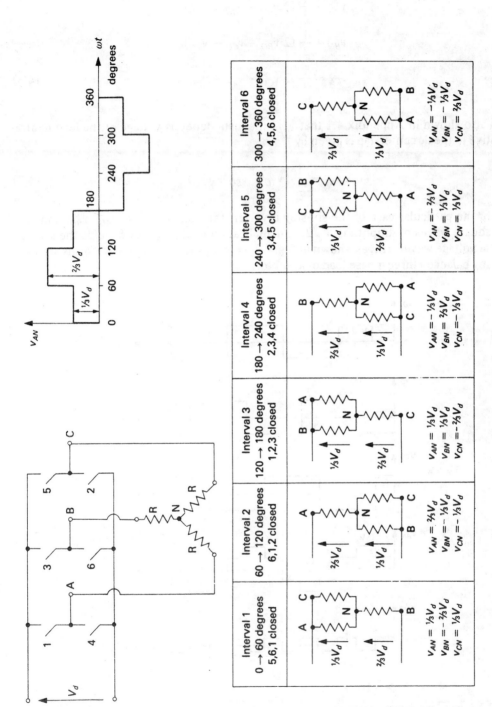

FIG. 4.6. Equivalent circuits for a three-phase six-step inverter with a balanced wye-connected resistive load.

$$v_{AN} = \frac{1}{3} (2 v_{AO} - v_{BO} - v_{CO})$$

$$v_{BN} = \frac{1}{3} (2 v_{BO} - v_{CO} - v_{AO})$$

$$v_{CN} = \frac{1}{3} (2 v_{CO} - v_{AO} - v_{BO}) . \tag{4.6}$$

It is also shown in Appendix 4.1 that v_{NO}, the instantaneous voltage of the load neutral relative to the dc center-tap is given by

$$v_{NO} = \frac{1}{3} (v_{AO} + v_{BO} + v_{CO}) . \tag{4.7}$$

For the particular case of the six-step inverter, the pole voltages, v_{AO}, v_{BO}, and v_{CO}, are square waves of amplitude $V_d/2$, and application of Equation 4.6 yields the six-step phase voltage waveform, as shown in Fig. 4.7. Hence, these voltage waveforms are valid for any balanced three-phase load or ac motor.

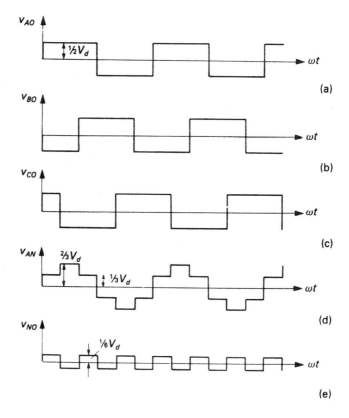

FIG. 4.7. General voltage waveforms for a three-phase six-step inverter with a balanced wye-connected load: (a), (b), (c) pole voltages; (d) line-to-neutral voltage; (e) voltage between load neutral and dc midpoint.

The harmonic content of the six-step line-to-neutral voltage is the same as that of the quasi-square-wave line-to-line voltage, and the difference in waveshape between the two voltages is due solely to a different phase relationship between the fundamental and the harmonics. This difference can be seen by comparing Equation 4.5 with the complete Fourier series expression for v_{AN}, which is

$$v_{AN} = \frac{2}{\pi} V_d \left(\sin \omega t + \frac{1}{5} \sin 5 \omega t + \frac{1}{7} \sin 7 \omega t \right.$$

$$\left. + \frac{1}{11} \sin 11 \omega t + \frac{1}{13} \sin 13 \omega t + ... \right).$$ (4.8)

The square-wave pole voltage waveforms of Fig. 4.7 have a high triplen harmonic content, but the isolation of the load neutral suppresses all triplen harmonics in the line-to-neutral voltages, as discussed in Appendix 4.1. The waveform for v_{NO} in Fig. 4.7(e) shows that all the suppressed triplen harmonic voltages now appear between load neutral and the dc mid-point as a square-wave voltage at three times the inverter output frequency.

The actual voltage waveforms may differ slightly from the ideal waveshapes due to commutation effects and internal voltage drops in the inverter circuit.

4.3.2. Current Waveforms

Commercial six-step inverters are available using transistors, thyristors, and GTOs, but the inverter current waveforms are determined by the load characteristics and are independent of the power semiconductor device used, except during the brief switching or commutation interval. The various inverter current waveforms will now be investigated, assuming a balanced delta-connected RL load and instantaneous semiconductor switching. The positive directions of the line currents i_A, i_B, i_C and the phase currents i_R, i_S, i_T are indicated in Fig. 4.8.

The line voltage waveforms of Fig. 4.5 show that each phase of the delta-connected load is subjected to a quasi-square-wave voltage with levels of $+ V_d$, zero, and $- V_d$. The phase current, therefore, consists of a series of exponential changes in current produced by the step changes in applied voltage. Assume a time origin is taken at the instant when TH4 is turned off and the complementary thyristor, TH1, is turned on, leaving TH5, TH6, and TH1 as the three gated thyristors. The voltage applied across phase R of the delta-connected load now has the line voltage waveform v_{AB} of Fig. 4.9(a) with $+ V_d$ applied for the first 120 degrees of the cycle. This produces the exponentially rising current denoted by i_1 in Fig. 4.9(b). After one-third of a cycle, TH3 is turned on and TH6 is turned off, thereby connecting both terminals of phase R to the positive supply rail and reducing the phase voltage to zero. However, the inductive current can circulate around the circuit composed of TH1 and D3. Thus, the phase current decays exponentially and is denoted by i_2 in Fig. 4.9(b). One half-cycle is completed after 180 degrees, when TH4 is turned on and the voltage V_d is applied across phase R with reverse polarity. The other two phases have similar current waveforms displaced by 120 degrees and 240 degrees, respectively. The line current is obtained in the usual manner as the difference of two

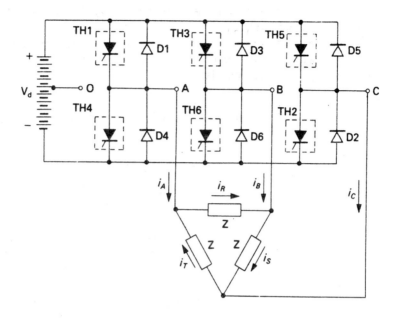

FIG. 4.8. Three-phase bridge inverter with a balanced delta-connected load.

phase currents. Thus, $i_A = (i_R - i_T)$ and is shown in Fig. 4.9(e). Mathematical expressions for the various currents are readily determined by writing the circuit equations for each conducting interval and matching the boundary conditions.[1]

The line current of a wye-connected load is similar to that shown in Fig. 4.9(e), because every wye-connected load has a delta equivalent. If the six-step line-to-neutral voltage, v_{AN} of Fig. 4.7(d), is superimposed on the phase or line current i_A of Fig. 4.9(e), it is seen that the exponential changes in phase current correspond to step changes in the applied six-step phase voltage.

Thyristor and Diode Currents. In Fig. 4.9, a time origin has been chosen at the instant when TH4 is turned off and TH1 is turned on. However, the line current i_A is negative, and must be carried by feedback diode D1, thereby reducing the net current drawn from the dc supply. Subsequently, i_A reverses and TH1 starts to conduct. As stated earlier, the instant of current reversal is load dependent and the thyristor must be able to take up load current when required. Consequently, a continuous gating signal is required for the full 180 degrees during which conduction may occur.

At the end of the half-period, TH1 is turned off, but the line current remains positive for a time and must be carried by diode D4. The line current waveform is repeated in Fig. 4.10(a) with the conducting device indicated for each portion of the ac cycle. Thus, the thyristor and diode current waveforms are as shown in Fig. 4.10(b) and (c). The feedback diodes also conduct for a portion of each half-cycle on leading power factor loads and load-commutated operation of a thyristor inverter is feasible, as discussed in Section 4.2 for the single-phase bridge inverter.

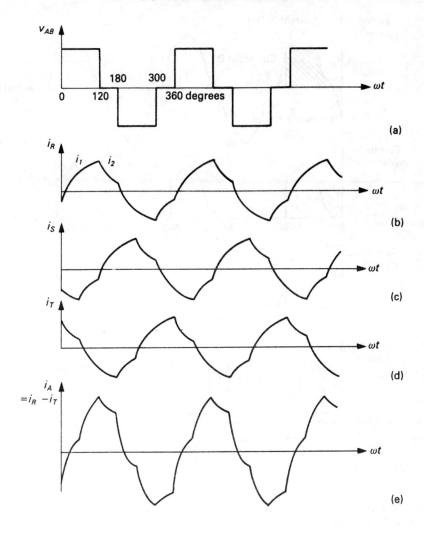

FIG. 4.9. Voltage and current waveforms for a three-phase six-step inverter with a balanced delta-connected RL load: (a) line-to-line voltage; (b), (c), (d) phase currents; (e) ac line current.

DC Supply Current. The waveforms of the currents in phases R, S, and T of the delta-connected load are shown in Fig. 4.9. In each phase, current component i_1 flows through the dc supply but component i_2 does not, as it is simply the natural decay of the phase current circulating through a conducting thyristor and diode. The current delivered by the dc supply is, therefore, the sum of the i_1 components of the phase currents, as in Fig. 4.10(d).

As explained above, the line current i_A may be negative for part of the interval from zero to 60 degrees, thereby indicating that current is flowing through feedback diode D1 and reducing the current drawn from the dc supply. In the interval from 60 degrees to 120 degrees, TH1 is the only gated thyristor which is connected to the positive dc rail, and i_A may still be negative during this interval if the load is very inductive. Under these

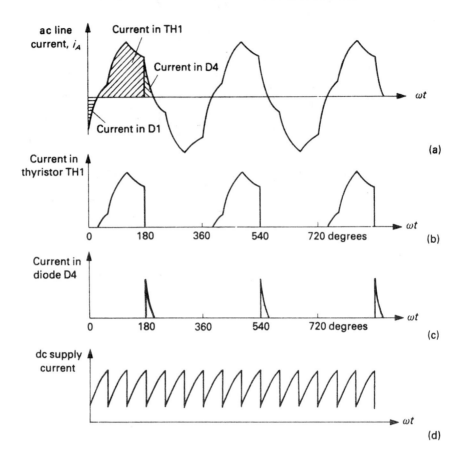

FIG. 4.10. Current waveforms in a three-phase six-step inverter with a balanced delta-connected RL load: (a) ac line current; (b) thyristor current; (c) diode current; (d) dc supply current.

conditions, the dc supply current reverses, indicating the substantial return of inductive load energy through the feedback diodes. In practice, this occurs when the fundamental power factor of the inductive load is less than 0.55. In the dc link converter, the filter capacitor must be large enough to limit the rise in dc link voltage as a result of this intermittent current reversal.

As before, regenerative braking of an ac motor load expands the conduction intervals of the feedback diodes so that the mean dc supply current becomes negative, and there is a net return of energy to the dc link. This energy must be dissipated or returned to the ac supply network, as discussed in Section 4.6.

AC Motor Current. When an induction motor is powered by a six-step VSI, the motor current waveforms are very similar to those shown above for a passive RL load. These currents have a significant harmonic content whose influence on motor behavior is studied in detail in Chapter 6. For the present, it can be noted that the fifth and seventh harmonic currents both react with the fundamental airgap flux to produce a sixth harmonic pulsating torque which causes an irregular stepping, or cogging rotation, of the motor shaft at low speeds. This sets a lower limit to the useful speed range of the six-step inverter drive, and if smooth low-speed rotation is required at frequencies less than 5 Hz, then an alternative form of converter is required that can deliver a near-sinusoidal current waveform at low frequencies.

4.3.3. *Output Voltage Control*

As already explained in Chapter 1, many adjustable-speed drives are required to have a constant-torque capability throughout the speed range, and this is achieved if the airgap flux in the motor is maintained constant at all speeds. To a good approximation, constant flux is obtained by operating the motor with a constant voltage/frequency ratio (constant volts/hertz), but a voltage boost is necessary at low frequencies to overcome the effects of winding resistance. To satisfy these requirements, the static inverter must deliver an adjustable-frequency alternating voltage whose magnitude is suitably related to the output frequency.

In a six-step inverter, the output voltage waveform does not vary with frequency, and there is a fixed ratio between the magnitudes of the dc link voltage and the ac output voltage. In some early drives, voltage control was achieved by means of a variable-ratio output transformer, but modern six-step inverter drives vary the dc link voltage to control the ac output voltage. The adjustable dc link voltage is commonly provided by a phase-controlled thyristor converter with an LC filter on the output, as shown in Fig. 4.11 (a). The filter smooths the rectified voltage but introduces a time lag which slows the transient response of the system.

The adjustable dc link voltage can also be obtained by using an uncontrolled diode rectifier bridge to deliver a constant dc voltage that is then regulated by means of a dc chopper, as explained in Chapter 3. Figure 4.11 (b) shows a six-step inverter drive in which the chopper thyristor TH1 is alternately gated on and then turned off by the forced commutating circuit. The pulsating output voltage from the chopper is filtered and fed to the six-step inverter, and the average dc voltage is adjusted by controlling the ratio of on-to off-time for the chopper thyristor. At high power levels, a large forced-commutated thyristor or GTO is required to regulate the full dc link current, but a single power transistor is adequate at reduced power levels. The relative merits of the two circuits shown in Fig. 4.11 have already been discussed in Section 3.5 of Chapter 3.

4.3.4. *Basic Drive Control Strategies*

Control of the output voltage/frequency relationship of the six-step inverter drive can be achieved (a) by direct control of output frequency with voltage adjusted accordingly, or (b) by direct control of output voltage with frequency following voltage.

Direct control of frequency is appropriate when the six-step inverter feeds one or more permanent magnet or reluctance-type synchronous motors in applications requiring precise speed holding but with unexacting transient performance requirements. A precise and stable output frequency is obtained by controlling the inverter switching rate with a stable reference oscillator, as in Fig. 4.12. The precise output frequency is con-

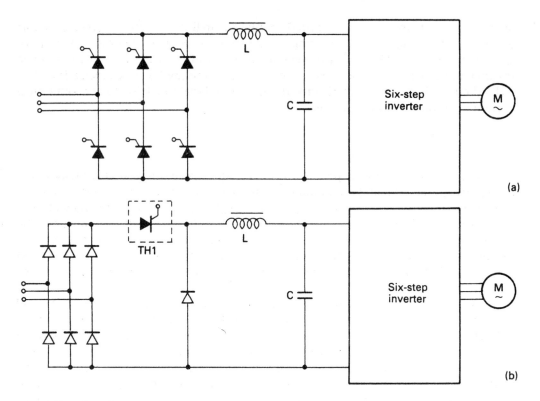

FIG. 4.11. Adjustable-voltage dc supply for a six-step inverter using (a) a phase-controlled thyristor rectifier and (b) a diode bridge rectifier and dc chopper.

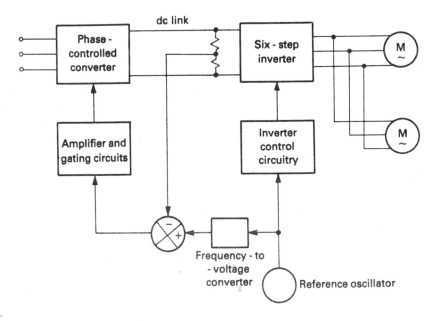

FIG. 4.12. Operation of a six-step inverter with direct control of inverter frequency.

verted to a precise shaft speed in the synchronously rotating machines, giving excellent speed holding and accurate speed matching without tachometer feedback. Direct control of frequency ensures that frequency and speed will not deviate when the load is altered or individual motors are started. The dc link voltage is varied in sympathy with the demanded frequency to give the desired motor voltage/frequency relationship. A closed-loop system of voltage control may be employed, as in Fig. 4.12. The reference voltage from the frequency-to-voltage converter is compared with a certain fraction of the dc link voltage, and the difference is amplified and used to control the rectifier delay angle so that the required volts/hertz output is obtained. A similar control scheme is used with the dc chopper. The control system must also ensure that the reference frequency is not increased at a rate which the motors are unable to follow or reduced at a rate which causes excessive regeneration of energy to the dc link.

Direct control of voltage, with the inverter frequency dependent on voltage, is a suitable control strategy for an induction motor drive, where very precise speed-holding accuracies are unnecessary. Many commercial six-step inverter drives use this control technique for industrial applications where a fast dynamic response is not required. The phase-controlled converter or dc chopper regulates the dc link voltage in response to the speed reference signal in a closed-loop voltage regulator. The dc link voltage in turn determines the output frequency and hence the shaft speed.

The control loop for the thyristor rectifier may incorporate an inner current loop, as used in dc motor drives. Figure 4.13 shows how the voltage error defines a set current and the current error controls the rectifier firing angle. This technique gives good control over dc link current and permits clamping of the maximum current at some predetermined limit, typically the full-load value. Similarly, in a dc chopper system, the dc link current is monitored and, if it exceeds the preset limit, then current control takes over and reduces the output voltage from the chopper to prevent excessive current.

The output power from the induction motor is proportional to the product of shaft torque and speed. The input power to the inverter is given by the product of dc link voltage and current. Shaft speed is roughly proportional to frequency and hence to inverter dc volts, with this control strategy. Consequently, the dc link current is approximately proportional to shaft torque, and the current limit functions as a torque control.

When starting from standstill, the current control acts to limit current to the full-load value by reducing the inverter input voltage. This, in turn, reduces the stator excitation frequency so that the induction motor has a low rotor frequency and a high input power factor, and develops a large starting torque at rated current. As the motor speeds up, the current control allows voltage and frequency to rise gradually, and the motor continues to operate with a low rotor frequency as it accelerates to the set speed. By restricting motor operation to low slip frequencies, in this manner, the output torque per ampere is maximized and the inverter rating is closely matched to that of the motor.

Under normal load conditions, current is less than the limiting value, and the system is under voltage control with the speed reference determining the terminal voltage, and thereby the output frequency and shaft speed. If the load torque ever exceeds the torque limit, the system reverts to current control. The inverter voltage and frequency are then reduced to limit motor current and torque to the full-load values.

The speed regulation of the induction motor from no load to full load can be improved by incorporating a slip compensation technique. This consists of adding a small current-dependent term to the speed reference signal so that the link voltage and output fre-

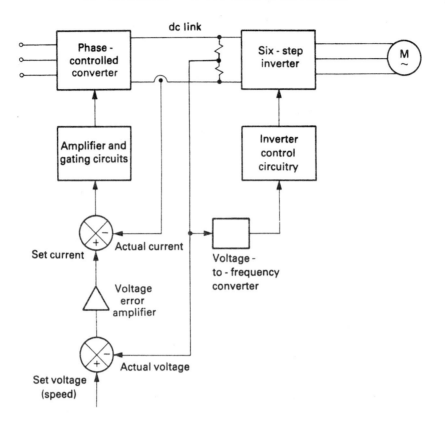

FIG. 4.13. Operation of a six-step inverter with direct control of dc link voltage.

quency increase with load, and compensate for the increasing slip of the motor. Speed holding to within 1 percent of top speed is readily obtained in this manner.

Using the inverter voltage to determine frequency, as described, has the advantage that magnetic saturation cannot occur, even under transient operating conditions, because frequency rapidly follows voltage. A sudden reduction in the speed reference signal will not produce a sudden reduction in frequency, because energy is regenerated to the dc link, causing a rise in link voltage and a consequent increase in inverter frequency. Rapid deceleration requires the removal or dissipation of this regenerated energy. Closed-loop control of the induction motor drive is discussed in more detail in Chapter 7.

4.4. THE SINGLE-PHASE BRIDGE PULSE-WIDTH-MODULATED INVERTER

As shown in Section 4.2 of this chapter, the single-phase bridge inverter delivers a square-wave output voltage of amplitude V_d. However, this presumes that the power semiconductors are turned on and off in diagonal pairs to give the voltage waveforms of Fig. 4.2. The bridge circuit is repeated in Fig. 4.14, and the pole voltages, v_{AO} and v_{BO}, are also shown. As before, these are square-wave alternating voltages of amplitude $V_d/2$,

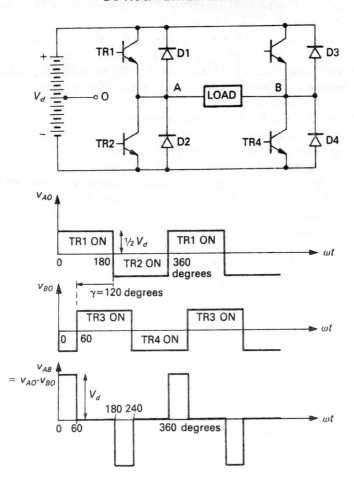

FIG. 4.14. Voltage control in a single-phase bridge inverter by simple pulse-width modulation.

but a phase displacement is now introduced between the pole voltages by phase-shifting the firing signals of one half-bridge relative to the other. In Fig. 4.14, the firing of TR3 and TR4 is advanced by an angle, γ, equal to 120 degrees. The resulting load voltage, v_{AB}, equals $(v_{AO} - v_{BO})$ and is a quasi-square wave having zero-voltage intervals of 120 degrees duration in each half-cycle. These intervals obviously correspond to the times when terminals A and B are connected simultaneously to the same dc supply rail, and load current circulates through a transistor and a feedback diode.

   The importance of this technique is that the fundamental output voltage can be varied from a maximum value to zero by advancing the firing of TR3 and TR4 from zero to 180 degrees. Thus, voltage control is achieved by operating with a constant inverter supply voltage and modifying the output voltage waveform. This general method of voltage control is termed pulse-width modulation (PWM), and the technique described above is an elementary form of PWM known as single-pulse modulation, or pulse-width control.

   In general, the PWM process modifies the harmonic content of the output voltage waveform, and it may be used to minimize objectionable harmonic effects in the load. In the single-pulse modulation technique above, certain values of γ eliminate troublesome low-order harmonics. Thus, when γ equals 60 degrees, there is no third harmonic or multiples thereof, and when γ is 72 degrees, the fifth harmonic is eliminated. For large values of γ, however, the output waveform is a series of very narrow pulses and the low-order harmonic content is excessive.

An alternative form of PWM, known as multiple-pulse modulation or square-wave PWM, involves the removal of a series of notches from the square-wave output voltage, leaving a series of equal-width pulses in each half-cycle, as shown in Fig. 4.15(a). This waveform can be generated by switching one half-bridge at the required fundamental frequency, as before, and switching the other half-bridge at a multiple of this frequency. The ratio, $T_1/(T_1 + T_2)$, is termed the duty cycle of the PWM waveform, and the magnitude of the fundamental output voltage is controlled by variation of the duty cycle. At reduced output voltages, the low-order harmonic content of this waveform is much less than that of the earlier single-pulse modulation technique of Fig. 4.14.

In more refined PWM waveforms, the pulse width is varied throughout the half-cycle in a sinusoidal manner and low-order harmonics are eliminated. Strictly speaking, the pulses should be positioned at regular intervals throughout the cycle, as shown in

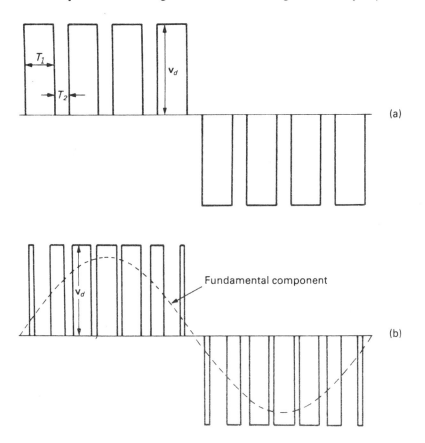

FIG. 4.15. Output voltage waveform for a single-phase bridge PWM inverter: (a) multiple-pulse PWM; (b) sinusoidal PWM.

Fig. 4.15(b), and the pulse width at a particular angle should be proportional to the sine of that angle. Voltage control is obtained by varying the widths of all pulses without upsetting the sinusoidal relationship.

Sinusoidal PWM is approximated by use of an inverter control circuit in which a high-frequency triangular carrier wave, v_c, is compared with a sinusoidal reference wave, v_r, of the desired frequency, and the crossover points are used to determine the inverter switching instants. In Fig. 4.16, an asymmetrical triangular wave is mixed with a sine wave reference to produce a PWM waveform in which the width of each pulse is defined by the interval between successive intersections. Consequently, the pulse width is approximately proportional to the mean sine wave ordinate in a pulse interval. When the frequency of the triangular wave is much greater than that of the sine wave, the variation in sine wave magnitude is insignificant between adjacent intersections, and the resulting PWM waveform approaches closely the ideal waveform in which pulse width is a sinusoidal function of angular position. This PWM technique will be discussed in more detail in the next section, in connection with the three-phase bridge inverter, because the three-phase circuit is more relevant to ac motor control applications.

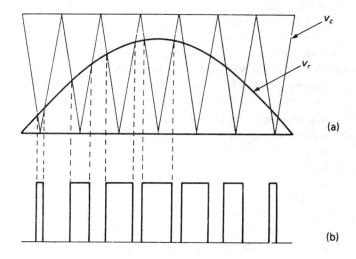

FIG. 4.16. Control circuit waveforms for a sinusoidal PWM inverter: (a) comparator input voltages; (b) comparator output voltage.

4.5. THE THREE-PHASE BRIDGE PWM INVERTER

The three-phase bridge inverter circuit for PWM waveform generation is identical to the six-step inverter circuit of Fig. 4.4, but the switching sequence is more complex. Voltage control is achieved by modulating the output voltage waveform within the inverter, and an adjustable-voltage dc link is not required. Consequently, an uncontrolled diode bridge rectifier is used, as shown in Fig. 4.17. The fast switching speed of the power transistor is advantageous in PWM inverters, and Fig. 4.17 shows a transistor inverter circuit, but thyristors or GTOs are necessary at high power levels.

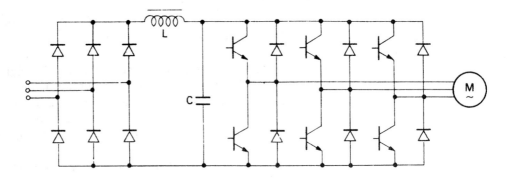

FIG. 4.17. Diode rectifier bridge feeding a three-phase transistor PWM inverter.

A wide variety of three-phase PWM techniques has been devised, but only some of the more important strategies will be discussed here. As in the single-phase circuit, the idea is to introduce notches in the basic square-wave pole voltage, thereby reducing the half-cycle average value and the fundamental component.[2] It is desirable to have several notches in each half-cycle so that the dominant switching harmonics are at high frequency, and motor inductance can effectively limit the harmonic current flow.

Output frequency control is accomplished, as usual, by varying the switching rate of the inverter devices. Thus, output voltage and frequency can be rapidly altered within the inverter circuit, and the PWM inverter drive has a transient response which is much superior to that of the six-step inverter drive. When a sophisticated PWM strategy is adopted, there are no low-order harmonics in the motor current, and low-speed torque pulsations and cogging effects are eliminated. However, the switching frequency of the PWM inverter is substantially higher than that of the six-step inverter and, as a result, inverter switching losses may be significant.

The three-phase inverter must develop a balanced supply with identical phase voltage waveforms having a mutual phase displacement of 120 degrees. It is essential, therefore, that the firing signal patterns for each inverter leg, or half-bridge, are identical and have a constant phase displacement of 120 degrees between them. This precludes the use of the single-pulse modulation technique, as used in the single-phase bridge inverter, because this approach involves a variable phase displacement between the half-bridges.

In general, the half-bridge pole voltage is a replica of the firing signal waveform, being positive while the upper device is turned on and negative while the lower device is turned on. Consequently, the operation of the PWM inverter is conveniently studied in terms of the techniques used to generate the firing signal waveforms. Traditionally, these waveforms have been produced by comparing a triangular carrier wave of high frequency with a low-frequency modulating reference wave. The latter may be square, trapezoidal, or sinusoidal.

4.5.1. Square-Wave PWM[3-6]

The square-wave PWM inverter has a control circuit in which a high-frequency symmetrical triangular carrier wave is compared with a synchronized square-wave modulating, or reference, wave with the required output frequency. The control circuit waveforms for one of the inverter half-bridges are shown in Fig. 4.18, and the switching instants of the semiconductor devices are determined by the intersections of the two waveforms. When the instantaneous square-wave reference voltage, v_r, exceeds the carrier voltage, v_c, the comparator output is "high," and the upper device in the half-bridge is switched on. When v_r is less than v_c, the comparator output is "low," and the lower device is turned on. The comparator output voltage is a pulse-width-modulated waveform, as shown in Fig. 4.18(b), and if the power semiconductor switching times are negligible, this is also the inverter pole voltage waveform. Clearly, this pole voltage has the required notched waveform, and its fundamental frequency equals that of the square-wave reference. The number of notches in a half-cycle is determined by the ratio of the carrier to reference frequency, or carrier ratio, p. In Fig. 4.18, p has a value of nine.

In the three-phase inverter, each half-bridge has a separate comparator which is fed with the same triangular carrier wave. However, the three square-wave reference voltages are displaced by 120 degrees, forming a balanced three-phase system. If the carrier ratio is a multiple of three, the triangular wave has an identical phase relationship with each of the three modulating square waves, and the resultant switching signals and pole voltage waveforms are also identical but have a mutual phase displacement of 120 degrees. Thus, for balanced three-phase operation, it is essential to operate with a carrier ratio that is a multiple of three.

Figure 4.19 shows the reference square waves for phases A, B, and C, and the common triangular carrier wave, for a carrier ratio of six. The resulting pole voltages v_{AO}, v_{BO}, and v_{CO} are also shown. As usual, the line voltage $v_{AB} = v_{AO} - v_{BO}$, giving a series of uniformly spaced, equal-width pulses of amplitude V_d in each half-cycle, with a half-width pulse at either end, as shown in Fig. 4.19(g). The corresponding line-to-neutral voltages for a wye-connected load are obtained by application of the general voltage equations (4.6), and the waveform for v_{AN} is shown in Fig. 4.19(h).

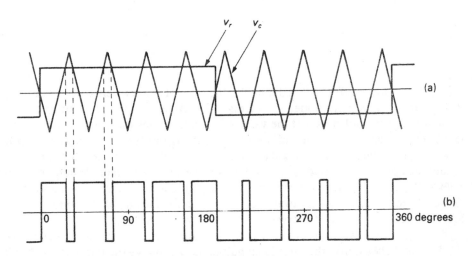

FIG. 4.18. Control circuit waveforms for a square-wave PWM inverter: (a) comparator input voltages; (b) comparator output voltage and pole voltage.

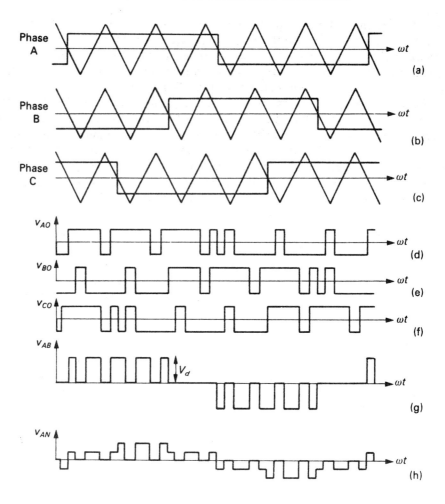

FIG. 4.19. Voltage waveforms for a three-phase square-wave PWM inverter: (a), (b), (c) comparator input voltages; (d), (e), (f) pole voltages; (g) ac line voltage; (h) line-to-neutral voltage.

The modulation index, M, is defined as the ratio of the reference wave amplitude, V_r, to the carrier wave amplitude, V_c. In Fig. 4.19, M equals 0.6, and an examination of these waveforms indicates that the value of M determines the notch width in the modulated pole voltage waveform and therefore controls the fundamental output voltage of the inverter. Usually, the triangular carrier amplitude is fixed, and the reference wave amplitude controls the modulation index and the output voltage. When M is zero, the pole voltages of Fig. 4.19 are symmetrical unmodulated square waves and the instantaneous line voltage, v_{AB}, is always zero. For small values of M, the line voltage pulses are very narrow but, as M increases, the pulse width increases proportionally, thereby increasing the volt-seconds area per half-cycle and the fundamental voltage amplitude. As M approaches unity, the individual pulses merge to give the familiar six-step line-to-line and line-to-neutral voltage waveforms. Thus, square-wave PWM produces voltage waveforms that are pulse-width modulated within a six-step envelope, and the harmonic content of such waveforms cannot decrease below the level associated with a six-step wave.

The line voltage waveform of Fig. 4.19(g) has a series of equal-width pulses in each half-cycle, flanked by two half-width pulses. A somewhat different line voltage waveform is obtained when the comparator voltages are as shown in Fig. 4.20. The square-wave reference is now shifted by one quarter-cycle of the carrier, as compared with Fig. 4.19, and the resulting line voltage waveform contains equal-width pulses only, within a six-step envelope. In the corresponding line-to-neutral voltage waveform of Fig. 4.20(h), the number of pulses per cycle is twice the carrier ratio, p. This PWM waveform has been used in many commercial inverter drives, but the modulation strategy has often been implemented by a different technique, in which the square-wave pole voltage is only modulated during the middle 60-degree segment of each half-cycle, as shown in Fig. 4.21. This strategy limits the number of switchings per cycle and so reduces the inverter switching losses. In Fig. 4.21, the pole voltages are shown with two

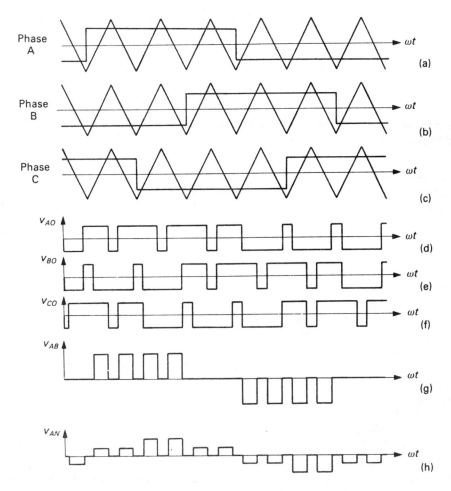

FIG. 4.20. Voltage waveforms for a three-phase square-wave PWM inverter when the carrier wave is phase shifted by one quarter-cycle, as compared with Fig. 4.19.

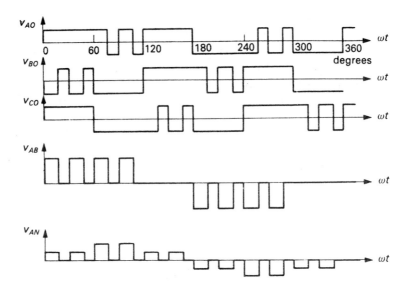

FIG. 4.21. Alternative modulation technique for the generation of the PWM voltage waveforms of Fig. 4.20.

notches removed from the mid-60-degree segment of each half-cycle, giving line-to-line and line-to-neutral voltages that are the same as those of Fig. 4.20.

Harmonic Analysis. As shown above, square-wave PWM produces an output voltage waveform which is a pulse-width-modulated six-step wave. This waveform contains all the harmonics of the six-step wave with additional high-frequency switching harmonics. In a three-phase system, one naturally selects the carrier ratio, p, to be a multiple of three to give a balanced three-phase output. The pole voltage waveform has a particularly large harmonic at the carrier frequency, but because p is a multiple of three, this carrier harmonic is a triplen harmonic and does not appear across the three-phase, three-wire load.

The dominant switching harmonics in the output voltage are sidebands of twice the carrier frequency, and multiples thereof. Thus, there are major odd harmonic voltage components of order $k = 2p \pm 1$, with somewhat smaller components of order $2p \pm 5$. (Components of order $2p \pm 3$ are present in the pole voltage waveform but, being triplen harmonics, are suppressed in the load.) If $p = 12$, for example, the major switching harmonics are of order 23 and 25, with lesser harmonics of order 19 and 29. Similar odd harmonics occur as sidebands of $4p, 6p, 8p \ldots$.

At high carrier ratios, the switching harmonics and the low-order six-step harmonics can be clearly distinguished, and the harmonic amplitudes are practically independent of p. In Fig. 4.22, the dominant harmonic voltages are plotted as a function of the modulation index, M. The fundamental voltage component is a maximum in the unmodulated six-step wave, and in Fig. 4.22(a), the harmonic voltage magnitude, V_k, is expressed as a fraction of this maximum fundamental voltage, $V_{1\,max}$. It is clear that there is a linear relationship between the magnitude of the fundamental voltage and the modulation

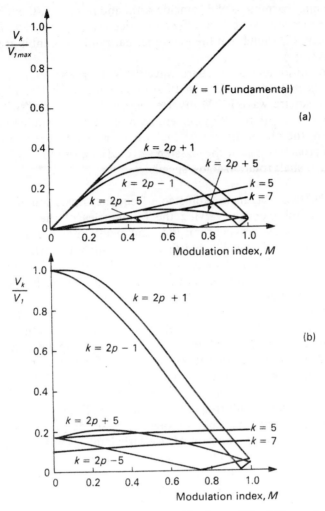

FIG. 4.22. Harmonic content of the square-wave PWM voltage of Fig. 4.19, as a function of the modulation index: (a) harmonic amplitude relative to maximum fundamental amplitude; (b) harmonic amplitude relative to actual fundamental amplitude.

index. At low values of M, the dominant switching harmonics of order $(2p \pm 1)$ are comparable in size to the fundamental, indicating large relative harmonic distortion. This fact is shown more clearly in the normalized curves of Fig. 4.22(b), where the harmonic voltage, V_k, in a particular waveform is expressed as a fraction of the corresponding fundamental magnitude, V_1. As M approaches unity, the harmonic content is naturally that of a six-step wave with harmonics of order $k = 5, 7, 11, 13...$ $(6n \pm 1)$, and each harmonic has an amplitude of $1/k$ of the fundamental.

The results in Fig. 4.22 were computed for the output voltage waveforms of Fig. 4.19, assuming a carrier ratio of 12, but somewhat similar results are obtained for the output waveforms of Fig. 4.20, and the curves do not change significantly at higher carrier ratios. However, the curves of Fig. 4.22 are *not* valid for lower values of p, when some of the major switching harmonics will coincide with, and reinforce, the inherent low-order harmonics of the six-step wave. This effect is particularly pronounced for carrier ratios of three or six, when it is found that the low-order harmonic content is greater than that of the six-step waveform.

In general, motor losses with a square-wave PWM supply are greater than those with a six-step supply because of the increased harmonic content.[7] The advantage of a high carrier ratio in a square-wave PWM inverter is that the dominant switching harmonics are at high frequency, with resulting current harmonics more readily filtered by the leakage inductance of the motor. Even with high carrier ratios, however, the harmonic content, at best, approaches that of the six-step wave, giving rise to the same torque pulsations and irregular shaft rotation at low speeds.

Constant Volts/Hertz Operation. Adjustable-frequency operation of the PWM inverter is achieved by simultaneously varying the frequencies of the carrier and reference waveforms while preserving synchronization and maintaining the correct phase relationship. These conditions are readily satisfied when both waveforms are generated by a common reference oscillator. As already explained, a set of three-phase square-wave reference voltages of adjustable amplitude is required, and each square wave is compared with a common fixed-amplitude triangular wave.

An examination of Fig. 4.19 or Fig. 4.20 shows that the duration of an individual voltage pulse in the modulated line-to-line voltage is proportional to the modulation index, M, and the periodic time of the reference wave, T. Thus, the pulse duration, T_p, is proportional to MT, or M/f, where f is the frequency of the reference wave. If the reference wave amplitude is varied linearly with frequency, then M/f is constant and the pulse duration, T_p, is independent of frequency. Consequently, the volt-seconds area per half-cycle is the same at all frequencies, which implies constant volts/hertz operation, as required. From another point of view, because the fundamental voltage amplitude is a linear function of M, constant M/f clearly implies constant V_1/f.

At higher frequencies in the constant volts/hertz range of operation, low carrier ratios are required to limit the switching frequency and associated inverter losses. The low-order harmonic content may be large, but this should not cause problems at these high speeds.

Figure 4.23 shows the manner in which the line voltage waveform changes with frequency. At a certain base frequency, the modulation index is nearly unity, and the output voltage approaches the six-step waveform, corresponding to the maximum fundamental voltage available. Below base frequency, the voltage pulses separate, as shown, but the pulse duration remains constant, as explained above, giving constant volts/hertz. At low frequencies, the modulation index is small and the pulses are widely separated, resulting in high harmonic distortion. The machine inductance is also less effective in smoothing the current waveform at these low frequencies, and it is usual to increase the order of the dominant carrier-related harmonics by incrementing the carrier ratio in a discrete number of steps as the fundamental frequency is reduced. This technique is termed "gear-changing." If required, a volts/hertz boost is implemented at low frequencies by increasing the modulation index. Above base frequency, constant-voltage operation is possible with a six-step output voltage waveform. These aspects of square-wave PWM inverter operation, common to the sinusoidal PWM inverter, will be discussed in the next section.

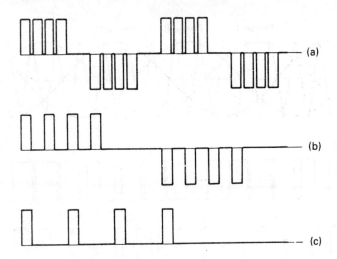

FIG. 4.23. The production of a constant volts/hertz output with a three-phase square-wave PWM inverter. AC line voltage waveform: (a) near base frequency; (b) at half-frequency; (c) at quarter-frequency.

4.5.2. Sinusoidal PWM

In the square-wave PWM strategy just described, the modulating reference waveform is a three-phase square wave. The frequency, amplitude, and harmonic content of this reference signal are reproduced in the inverter pole voltages; consequently, the low-order harmonics of the reference wave appear in the output waveform. However, most ac motors are designed to operate on a sine wave supply, and the inverter output voltage should be as nearly sinusoidal as possible. It seems obvious, therefore, that the three-phase square-wave reference should be replaced by a three-phase sine wave reference, to give a PWM waveform in which the pulse width is sinusoidally modulated throughout the half-cycle. This is termed sinusoidal PWM, or sine wave PWM, or subharmonic PWM.[8]

As before, each inverter phase or half-bridge has a comparator which is fed with the reference voltage for that phase and a symmetrical triangular carrier wave which is common to all three phases, as shown in Fig. 4.24(a). Again, the ratio of carrier to reference frequency or carrier ratio, p, must be a multiple of three to ensure identical phase voltage waveforms in a three-phase system. The triangular carrier has a fixed amplitude, and the ratio of sine wave amplitude to carrier amplitude is termed the modulation index, M.

Output voltage control is achieved by variation of the sine wave amplitude. This variation alters the pulse widths in the output voltage waveform but preserves the sinusoidal modulation pattern. In Fig. 4.24, the carrier ratio is nine and the modulation index is almost unity. The corresponding pole voltages v_{AO}, v_{BO}, v_{CO}, and the resultant line-to-line voltage, v_{AB}, are shown in Fig. 4.24(b), (c), (d), and (e), respectively.

Adjustable-frequency operation of a sine wave PWM inverter for ac motor control requires the generation of a set of three-phase sine wave reference voltages of adjustable amplitude and frequency. If the motor is to operate at very low speeds down to standstill, the reference oscillator must have a corresponding low-frequency capability, down to

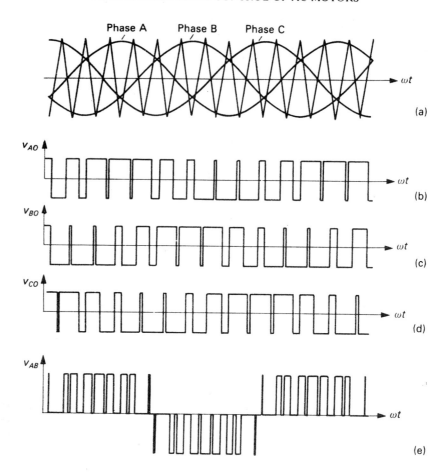

FIG. 4.24. Voltage waveforms for a three-phase sinusoidal PWM inverter: (a) comparator voltages; (b), (c), (d) pole voltages; (e) ac line voltage.

zero frequency. With traditional analog circuit techniques, it is difficult to generate such a sine wave reference without encountering problems of dc offset and parameter drift. Consequently, many of the early PWM inverter drives adopted the square-wave PWM strategy because the electronic circuit design of an adjustable-frequency square-wave oscillator is relatively straightforward. However, the implementation of sinusoidal PWM has been facilitated by modern digital circuit techniques utilizing programmed memory or custom-designed large-scale integrated (LSI) circuits.

For large carrier ratios, the sinusoidal PWM inverter delivers a high-quality output voltage waveform in which the dominant harmonics are of a high order, being clustered around the carrier frequency and its harmonics. Smooth motor rotation is obtained, even at very low speeds, because the undesirable low-order harmonics and troublesome torque pulsations, which are characteristics of the six-step and square-wave PWM waveforms, are eliminated with a sinusoidal PWM supply.

Harmonic Analysis. The carrier ratio, p, determines the order of the predominant harmonics in the sinusoidally modulated pole voltage waveform. Analytical techniques based on Fourier series methods can be used to derive a mathematical expression for the harmonic spectrum of the PWM waveform, but the resulting expression is complex and involves Bessel functions.[9-11] The analysis shows that the harmonics occur as sidebands of the carrier frequency and its multiples and, in general, the harmonic order is given by $k = np \pm m$, the mth sidebands of the nth carrier harmonic.

For even values of n, there is an odd sideband spectrum because the harmonics are nonexistent when n and m are both even. Thus, for $n = 2$, there are harmonics of order $2p \pm 1, 2p \pm 3, 2p \pm 5...$ in the pole voltage waveform, but harmonic amplitude diminishes rapidly with increasing values of m. It is also found that odd values of n possess an even sideband spectrum, because the harmonics are nonexistent when n and m are both odd. Thus, for $n = 1$, there are harmonics of order $p \pm 2, p \pm 4...$ and other smaller harmonics. Large harmonics are also present in the pole voltage waveform at odd multiples of p, but since p is a multiple of three, these carrier harmonics, and certain other harmonics such as those of order $2p \pm 3$, are triplen harmonics, and are not applied to the three-phase load.

A detailed harmonic analysis shows that the harmonic magnitudes are independent of the carrier ratio, p (within 1 part in 10^4), provided p is greater than nine. The magnitudes of the major harmonic components are plotted as a function of the modulation index, M, in Fig. 4.25. As before, there are two sets of curves, Fig. 4.25(a) with harmonic magnitude V_k expressed as a fraction of the maximum fundamental voltage V_{1max}, and Fig. 4.25(b) showing the normalized harmonic voltage V_k / V_1. The curves are plotted for values of M between zero and unity.

It is evident from Fig. 4.25(a) that there is a linear relationship between fundamental voltage magnitude and modulation index. Harmonic analysis shows that this is true provided p is greater than three. The fundamental line-to-neutral voltage amplitude for a wye-connected load is given by

$$V_1 = M \, V_d/2 \text{ for } 0 \leqslant M \leqslant 1 \qquad (4.9)$$

where V_d is the dc link voltage.[10]

Figure 4.25 shows that the harmonics of order $(2p \pm 1)$ are dominant over most of the range of M, but harmonics of order $(p \pm 2)$ are also significant. Thus, for $p = 15$, say, the principal harmonics are of order 29 and 31, with lesser harmonics of order 13 and 17. In general, the lowest harmonic of appreciable magnitude is of order $p - 2$, and for large values of p, it is of a much higher order than the lower order harmonics in the square-wave PWM waveform. Consequently, most commercial PWM inverters now implement sinusoidal PWM.

Voltage Utilization. The fundamental output voltage of the sinusoidal PWM inverter is increased to its maximum value by raising the modulation index to a value of unity. For this condition, it is evident that the modulated pole voltage waveform has a reduced volt-seconds area per half-cycle, as compared with a six-step inverter, where the pole voltage is an unmodulated square wave. This fact implies that the maximum fundamental output voltage of a sine wave PWM inverter is less than that of a six-step inverter.

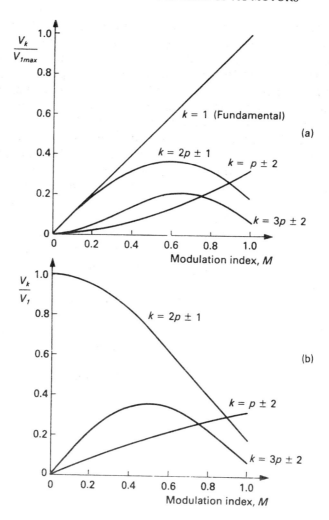

FIG. 4.25. Harmonic content of the sinusoidal PWM voltage as a function of modulation index:
(a) harmonic amplitude relative to maximum fundamental amplitude; (b) harmonic amplitude relative
to actual fundamental amplitude.

For a dc link voltage, V_d, and unity modulation index, the amplitude of the fundamental line-to-neutral voltage for a wye-connected load is given by Equation 4.9 as $V_d/2$. For a six-step line-to-neutral voltage waveform, Equation 4.8 gives the fundamental amplitude as $2V_d/\pi$ or $0.636\ V_d$. The corresponding fundamental rms line-to-line voltages are $0.61\ V_d$ for sinusoidal PWM and $0.78\ V_d$ for a six-step waveform. Thus, for a given dc link voltage, the sine wave PWM inverter has a fundamental voltage capability which is only 78 percent of that for a six-step inverter.

Improved utilization of the available dc link voltage is possible in a sine wave PWM inverter when M is increased above unity, as shown in Fig. 4.26. This increase in M overrides the normal sinusoidal modulation process and is termed over-modulation. Some of the intersections between reference and carrier waves are lost, and as a result,

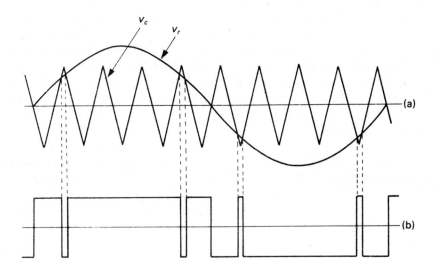

v_c

v_r

(a)

(b)

FIG. 4.26. Over-modulation in a sinusoidal PWM inverter: (a) comparator input voltages; (b) comparator output voltage and pole voltage.

pulses are dropped in the output voltage waveform. For large values of M, the only intersections are those at the zero crossings of the sine wave. This means that the pole voltage waveform is now an unmodulated square wave, and the inverter has transitioned from sine wave PWM to the six-step mode of operation, which develops the maximum possible value of fundamental voltage.

The penalty for adopting over-modulation is that the true sinusoidal modulation strategy is abandoned, and low-order harmonics reappear in the output voltage waveform. The earlier Fig. 4.25 shows the harmonic content in the normal region of operation, where the modulation index is less than unity and pulse dropping does not occur.

The output voltage capability of the inverter can also be extended by employing a non-sinusoidal reference. A trapezoidal reference wave, for example, gives good utilization of the available dc link voltage and introduces smaller low-order harmonics than a square-wave reference. Alternatively, a third harmonic component can be added to a sine wave reference to flatten the peaks of the sine wave.[12,13] This modified reference wave gives some increase in fundamental output voltage as compared with a pure sine wave reference, and there is no increase in harmonic content, because the third harmonic component introduces triplen harmonic pole voltage components that are in time phase and consequently do not appear across the three-phase load.

Minimum Dwell Time. In the discussion of over-modulation in the preceding paragraphs, it was tacitly assumed that the sine wave PWM inverter can operate with a modulation index that equals or exceeds unity, thereby facilitating a transition to the unmodulated six-step output voltage waveform. Figure 4.24 shows that, as the modulation index approaches unity, the interval between consecutive switchings in an inverter half-bridge becomes very short near the positive and negative peaks of the sine wave reference, and the pole voltage waveform has some notches of very brief duration. This fact implies that a power semiconductor which has just been turned off must be quickly switched on again, and vice versa. In practice, the duration of the interval between switchings, or dwell time, cannot be allowed to fall below a certain minimum value, t_{min}, which is related to the turn-off time of the device. In the case of a forced-commutated thyristor inverter, the commutating circuit may require significant time intervals for redistribution of charge, or over-charge dissipation, and t_{min} values of several hundreds of microseconds may be required to safeguard the commutating ability. Any violation of this minimum dwell time constraint may result in an inverter shoot-through fault that short circuits the dc link.

Practical inverter control circuits must incorporate a minimum dwell time clamp or pulse "lockout" feature, which prevents the notch width from decreasing below a certain preset minimum. As the sine wave reference is increased in amplitude and the modulation index approaches unity, the sine-triangle intersections call for a notch width that is less than t_{min}. However, the minimum dwell time clamp overrides this request, and the relevant notch is held at t_{min} until a further increase in the reference wave amplitude eliminates the intersection altogether and the notch is abruptly dropped.[14] An alternative technique can be used in which the notch is removed as soon as the sine-triangle intersections call for a notch width that is less than t_{min}.[15] In either case, the sudden removal of a notch of width t_{min}, results in a step change in the fundamental output voltage that may produce objectionable transient currents in the motor and inverter, and may cause stability problems in closed-loop systems. A number of these discontinuities will occur as the modulation index is increased until a value is reached at which all the intersections disappear (except those at the sine wave zero-crossings) and the inverter delivers a six-step wave.

Over-modulation may result in unacceptably large jumps in fundamental voltage, particularly in the case of thyristor inverters with long dwell times. Various intermediate PWM strategies have been suggested that are designed to achieve a smooth transition from sine wave PWM to six-step operation without unduly large steps in voltage.[14–20] The choice of transition PWM strategy is also important with regard to harmonic losses in the motor, because the harmonic content may be high.[14,18] It is also desirable to select PWM waveforms that minimize current ripple in the motor and thereby utilize fully the limited instantaneous current capability of the inverter.[20]

Modern power semiconductor devices such as the transistor and MOSFET have fast switching times which offer significant advantages in PWM inverter applications. The minimum dwell time, t_{min}, is appreciably smaller than that in a forced-commutated thyristor inverter; hence, fundamental voltage discontinuities are usually negligible. The power MOSFET permits ac motor control at switching frequencies of 20 kHz.[21] As a result, harmonic acoustic noise in the motor and inverter is outside the audible range and noiseless operation is achieved, but the power handling capability of the MOSFET inverter is rather limited.

Constant Volts/Hertz Operation. The sinusoidal PWM inverter can readily provide the constant volts/hertz supply required for constant-torque operation of an ac motor. As in the square-wave PWM inverter, the modulation index is varied linearly with the frequency of the reference wave to give a fundamental output voltage that is proportional to output frequency. As before, the carrier ratio is reduced at the higher output frequencies to limit the switching frequency of the inverter and the associated inverter switching losses.

A high carrier ratio improves waveform quality by raising the order of the principal harmonics. At low fundamental frequencies, very large carrier ratios are feasible, and the resulting near-sinusoidal output current waveforms account for one of the main attributes of the sine wave PWM inverter drive — the extremely smooth rotation at low speeds. As the fundamental output frequency is reduced, the carrier ratio is increased by discrete integer values at certain preset output frequencies. Normally, these changes in carrier ratio, or gear-changes, are designed so that the inverter switching frequency lies between predetermined upper and lower limits: The upper limit avoids excessive inverter switching losses, and the lower limit maintains good waveform quality without excessive current ripple. A large number of gear-changes are necessary when operation is required down to very low output frequencies, but waveform quality is excellent. Careful matching of the fundamental voltage components of the PWM waveforms used on either side of a gear-change is necessary so as to minimize the current transients produced by the sudden change in carrier ratio.

The above discussion assumes that the carrier and reference waves have an integer triplen carrier ratio and are always synchronized in phase, as shown in Fig. 4.24. For high carrier ratios, however, waveform quality is unaffected by the adoption of a free-running strategy in which carrier and reference waveforms are unsynchronized. Many commercial inverters have adopted this approach. In a forced-commutated thyristor inverter, the choice of carrier frequency is restricted by the need to limit commutation losses. If it is decided to limit the inverter switching frequency to 500 Hz, for example, then a triangular carrier wave with a fixed frequency of 500 Hz is compared with an adjustable-frequency sine wave reference having the required output frequency. This unsynchronized, or asynchronous, approach gives very high carrier ratios at very low speeds without gear-changing, and maintains a constant inverter switching frequency of 500 Hz.

As the reference frequency increases, the carrier ratio falls, and the unsynchronized waveforms produce low-frequency subharmonic beat components in the inverter output voltage.[22] These components arise because the frequency ratio between the carrier and reference waves has a noninteger value and, as a result, the PWM waveform may vary slightly from cycle to cycle. For large carrier ratios, these beat-frequency subharmonics are negligible, but when the carrier ratio is less than nine, the beating effect can cause intolerable low-frequency pulsations in motor current, torque, and speed. These objectionable beat-frequency effects are eliminated when the carrier and reference waves are synchronized, as in Fig. 4.24. Consequently, as the output frequency is increased and the carrier ratio falls below nine, the PWM inverter must transition to a synchronized mode of operation (with p as a multiple of three). In a thyristor inverter, this transition typically occurs at an output frequency in the range from 30 to 50 Hz. Careful circuit design is necessary to ensure a smooth transition between the asynchronous and synchronous modes of operation with minimum switching transients.

As the operating frequency is increased further, the carrier ratio is reduced to six, and possibly three, to limit the switching frequency and ease the transition to the six-step mode of operation. Over-modulation is then implemented, or special transition PWM strategies are adopted that allow a gradual transition to six-step operation at base frequency without excessively large voltage jumps. The merits of PWM and six-step operation are combined in this fashion.[14–18,22] Below base frequency, PWM techniques are used to give the linear voltage/frequency relationship required for constant-torque operation. Above base frequency, the motor has a high-speed range of operation with a fixed-amplitude six-step voltage. This mode of operation gives the motor a so-called constant-horsepower characteristic, and a torque capability which decreases with speed. The overall drive characteristic is similar to that obtained in a standard dc motor drive with armature voltage control below base speed and field weakening above base speed.

Advantages of Sinusoidal PWM. The sinusoidal PWM inverter drive offers an enhanced low-speed performance as compared with the square-wave PWM and six-step inverter drives. As described in Section 4.3.3, the six-step inverter drive operates in the constant volts/hertz range below base speed when the dc link voltage is varied proportionally with output frequency, but there is a substantial time lag in the voltage response due to the dc link filter, and this prejudices the low-speed stability of the drive. The two types of PWM inverter have a better transient response, as voltage and frequency control are both accomplished within the inverter. The dc link filter components are also smaller and less costly in a PWM inverter drive, but drive losses may be high, particularly in the case of square-wave PWM.

The sinusoidal (or square-wave) PWM inverter operates with a fixed dc link voltage, which offers a number of advantages. A constant link voltage can be provided by an uncontrolled diode rectifier bridge, and the absence of delayed firing means that the fundamental power factor presented to the ac utility supply is always high (of the order of 0.96) and is independent of motor power factor. In forced-commutated thyristor inverters in which the commutating capacitors are charged by the inverter supply voltage, the fixed dc link voltage maintains a constant current-commutating capability. A constant-torque drive requires approximately constant motor current at all speeds. The PWM inverter drive with its fixed link voltage can commutate rated current, even at very low fundamental output voltages (and frequencies), whereas the six-step inverter drive requires an auxiliary fixed-voltage dc supply for the commutating circuit if it is to develop rated torque at low speeds. The presence of a constant-voltage dc link also permits the parallel operation of several independent PWM inverters on the same rectified supply, with a significant saving in rectifier costs. In addition, a standby battery or motor-generator set can be switched in for emergency operation during ac supply failures.

Recent Developments. Recent developments in PWM inverter technology have primarily been in the area of digital control circuitry and real-time microprocessor-based waveform generation. The classical sine wave modulation strategy of Fig. 4.24 is known as natural sampled PWM. For digital and microprocessor-based systems, a modified strategy known as "regular" sampling has certain advantages, as discussed later in Section 4.5.5.

In a digital PWM modulator, the waveform appropriate to a particular fundamental voltage can be stored in a programmable read-only memory (PROM) which is accessed by a microprocessor or hardwired digital logic. This approach eliminates many of the inherent difficulties in the analog implementation of a synchronized sinusoidal PWM strategy, and in particular, it facilitates a smooth transition from PWM to six-step operation. A custom-designed LSI chip has been marketed which implements regular sampled PWM and a transition to six-step operation.[23] This integrated circuit has been adopted by a number of commercial PWM inverter manufacturers.

4.5.3. Current-Controlled PWM

The current-controlled PWM inverter consists of a conventional PWM voltage-source inverter fitted with current-regulating loops to provide a controlled current output. If the inverter has a high switching frequency, the stator currents of the synchronous or induction motor can be rapidly adjusted in magnitude and phase. As in a dc drive, high-quality dynamic control of motor current is particularly important for the implementation of a high-performance, single-motor ac drive. Servo drives, in particular, must satisfy exacting performance specifications with respect to dynamic response and smoothness of rotation down to zero speed. These characteristics are highly dependent on the quality of the current control, and the techniques described in this section are widely used in such drives.

The current controller can take a number of different forms.[24–29] Usually, a sinusoidal reference current waveform is generated and fed to a comparator, together with the actual measured current of the motor. The simplest approach uses the comparator error to switch the devices in the inverter half-bridge so as to limit the instantaneous current error. Figure 4.27(a) shows the control for one inverter leg. If the motor phase current is more positive than the reference current value, the upper device is turned off and the lower device is turned on, causing the motor current to decrease, and vice versa. The comparator has a deadband, or hysteresis, that determines the permitted deviation of the actual phase current from the reference value before an inverter switching is initiated. Thus, the actual current tracks the reference current without significant amplitude error or phase delay. In a three-phase system, there is usually an independent current controller for each inverter phase. Standard lockout circuitry is required to ensure sufficient dwell time between successive switchings in an inverter leg.

Figure 4.27(b) illustrates the type of output current waveform obtained with the simple hysteresis or on/off current controller. A small deadband gives a near-sinusoidal motor current with a small current ripple, but requires a high switching frequency in the inverter. However, the switching frequency is not constant for a given deadband but is modulated by the variations in motor inductance and back emf. When the back emf of the motor is low, the switching frequency may rise excessively. It has been shown that in a three-phase system without a neutral connection, the instantaneous current error can reach double the hysteresis band.[25,28] Also, low current levels cannot be achieved because the modulation vanishes when the reference current lies within the hysteresis band. In addition, the variable switching frequency produces objectionable acoustic noise, and despite its simplicity, this on/off or "bang-bang" technique is seldom used in practice.

In commercial applications, a fixed switching frequency is preferred because acoustic noise is less annoying and inverter switching losses are more predictable. Figure 4.27(c) shows the most common current control technique in which the current error is compared with a fixed-frequency triangular carrier wave. Thus, the current error is essentially the reference or modulating signal in a conventional, asynchronous, sine-triangle, pulse-width modulator. The resulting PWM signal, whose duty cycle is proportional to the current error, controls the inverter switching as before. If the reference current is more positive than the actual current, the resulting error is positive, and the on-period of the upper device exceeds that of the lower device. Consequently, the inverter leg is switched predominantly in the positive direction to increase the ac line current. Con-

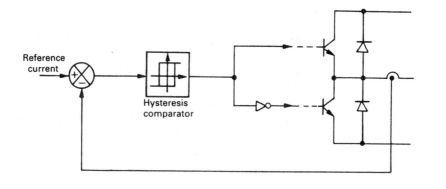

(a)

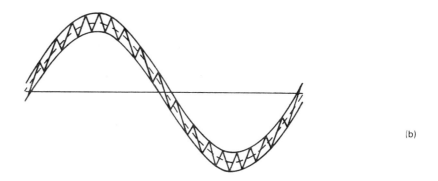

(b)

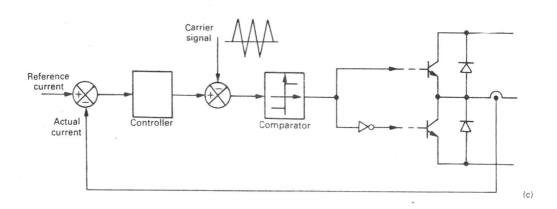

(c)

FIG. 4.27. The current-controlled PWM inverter: (a) hysteresis control for one inverter leg; (b) sinusoidal current waveform generated by hysteresis control; (c) fixed-frequency PWM control for one inverter leg.

versely, if the current error is negative, the inverter leg is switched predominantly in the negative direction. Again, a three-phase system has three current controllers, but the high-frequency triangular carrier signal is common to all three phases, and each inverter leg switches at the carrier frequency. With a sine-wave reference and a high carrier ratio, a near-sinusoidal motor current waveform is produced, containing high-order harmonics only. The current comparison in Fig. 4.27(c) is performed in a linear operational amplifier current controller with proportional or proportional-integral compensation. The compensation is adjusted to minimize the magnitude and phase errors in the ac line currents. This PWM technique is also widely used in dc servo motor drives where a high-frequency dc chopper, or "PWM servo amplifier," operates under closed-loop current control to give rapid adjustment of dc motor current and torque.

Fixed-frequency or on/off PWM current-control techniques, as described above, can provide a high-quality, controlled-current ac source. A high-frequency inverter with fast current-control loops allows rapid adjustment of motor current in magnitude and phase, despite the speed-dependent back emf in the stator windings. By feeding the ac motor with impressed stator currents, the effects of stator resistance and leakage reactance are eliminated, and the control of the ac motor is greatly simplified. In steady-state operation, the sinusoidal reference currents are accurately tracked to give smooth rotation at very low speeds. Servo drives for machine tools and robotics are usually rated at less than 10 kW, and consequently, a transistor inverter can be used with a switching frequency of 10 kHz or more. Operation above 20 kHz will place the switching frequency outside the audible range. Servo control of the ac motor is discussed in detail in later chapters with the help of block diagrams. In these diagrams, the current-controlled PWM inverter is represented as in Fig. 4.28, showing the three current-regulating loops which cause the actual line currents, i_A, i_B, and i_C, to track the reference currents, i_A^*, i_B^*, and i_C^*. The block labeled "PWM GEN" represents the hysteresis controller or fixed-frequency modulator that generates the PWM firing signals for the inverter.

In a wye-connected stator winding with isolated neutral, the three ac line currents must obey the relationship $i_A + i_B + i_C = 0$, and consequently, it is sufficient to sense and control only two of the line currents. In practice, however, each of the three currents is usually controlled for reasons of symmetry. For proper operation as a controlled-current source, the dc link voltage must be sufficiently large to force the ac line current to vary as required. When the motor speed rises, the back emf increases and approaches the dc link voltage. Because the net voltage available is insufficient, the output current is unable to track the reference current. Thus, at high motor speeds, the PWM operation will transition to the six-step, voltage-source mode of operation. However, rapid variation in the magnitude of the current error may cause random pulse dropping or multiple switchings in the transition region.

4.5.4. *Optimum PWM*

Voltage-source PWM inverters for ac motor control have conventionally employed the square-wave and sinusoidal PWM strategies already described. However, in controlling the fundamental voltage output, the PWM strategy introduces additional harmonic components whose presence is detrimental to motor performance and efficiency. In recent years, more sophisticated switching patterns have been suggested in which specific low-order harmonics are suppressed or total harmonic content is mini-

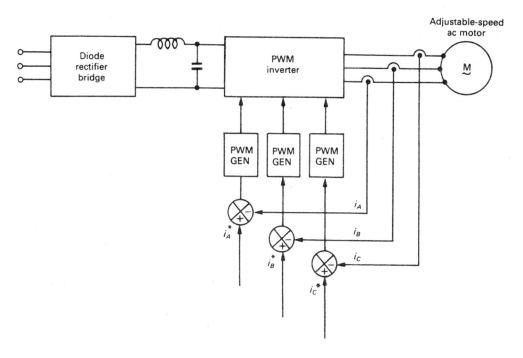

FIG. 4.28. Block diagram of a three-phase current-controlled PWM inverter feeding an ac motor.

mized.[30–33] These optimized modulation strategies are extremely difficult to realize with conventional analog circuitry, but they can be effectively implemented with modern microprocessor-based control techniques.

In general, the basic three-phase bridge inverter configuration develops a pole voltage waveform as in Fig. 4.29. Fourier analysis shows that the amplitudes of the harmonic voltages may be expressed in terms of the switching or commutation angles $\alpha_1, \alpha_2, \alpha_3....$ If the waveform has quarter-wave and half-wave symmetry, as shown, the amplitude of the kth harmonic voltage is

$$V_k = \frac{2 V_d}{k \pi} \left[1 + 2 \sum_{i=1}^{m} (-1)^i \cos k \, \alpha_i \right] \tag{4.10}$$

where m is the number of switchings per quarter-cycle.

Equation 4.10 can be used to determine the switching angles necessary to set the fundamental voltage at some specific magnitude and simultaneously suppress certain selected harmonics, but numerical techniques are necessary to solve the nonlinear equations of the problem. If there are m switchings in a quarter-cycle, the fundamental can be controlled and $(m-1)$ harmonics suppressed. This is the harmonic elimination PWM technique.

In a symmetrical three-phase, three-wire system, the triplen harmonics are automatically suppressed in the load voltage waveform, and there are no even harmonics when the phase voltage symmetry is as in Fig. 4.29. Consequently, the four lowest order harmonics of significance are the fifth, seventh, eleventh, and thirteenth, and these can

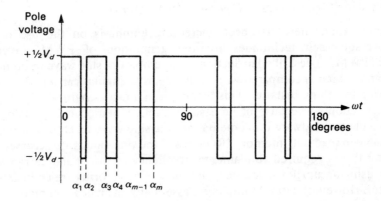

FIG. 4.29. General pole voltage waveform in a PWM inverter.

cause torque pulsations and speed fluctuations in ac motor drives at low rotational speeds. Elimination of these four harmonics produces a waveform in which there is no harmonic of order lower than the seventeenth. Control of the fundamental voltage and the suppression of four harmonics requires five switchings per phase per quarter cycle, or 22 switchings per cycle, including the switchings at zero and 180 degrees.

An alternative approach is to define a performance index related to the undesirable effects of the harmonics and to select the switching angles so that the fundamental voltage is controlled and the performance index is minimized.[33–40] This is classified as a programmed waveform, or distortion minimization, PWM technique. It has been suggested[33] that an appropriate performance index for ac motor control applications is the total harmonic voltage distortion as defined by

$$\sigma = \left[\sum_{k \neq 1}^{\infty} \left(V_k / k \right)^2 \right]^{1/2}. \tag{4.11}$$

This quantity is proportional to the total rms harmonic current, if skin effect is negligible, and hence σ^2 is proportional to the harmonic copper loss in the machine. Minimization of this quantity and control of the fundamental voltage are achieved by appropriate positioning of the switching angles, but again an off-line, mainframe computer solution is necessary.

Practical implementation of these complex PWM strategies cannot be achieved by the simple waveform-mixing techniques used for square-wave or sinusoidal PWM, but economic realization is possible with modern microprocessors, read-only memory, and digital hardware. These modern circuit techniques also facilitate the selection of different modulation strategies for different portions of the speed range to give optimum drive performance. It has been shown that the optimum PWM strategies are particularly attractive when there are only a few switchings per cycle. Thus, in the transition PWM region immediately prior to the change to six-step operation, the use of optimum PWM strategies can result in a significant reduction in harmonic content. A number of authors have compared the optimum PWM techniques with the more conventional PWM strategies.[39–42]

4.5.5. *Digital and Microprocessor Control of PWM Inverters*

In recent years, there has been increasing emphasis on the use of digital and microprocessor-based techniques for the generation of PWM waveforms.[43-64] Sinusoidal PWM, as described in Section 4.5.2, employs a sine wave reference, or modulating, signal which is compared with an isosceles triangular carrier wave to determine the inverter switching instants. This technique, known as natural sampled PWM, has been widely adopted because of its ease of implementation using analog control circuitry. In a digital hardware implementation, the sine wave reference may be stored as a look-up table in read only memory (ROM), and the sine values are accessed at a rate corresponding to the required fundamental frequency. A triangular carrier wave is generated by using an up/down counter, and the two waveforms are compared in a digital comparator. However, natural sampling is essentially an analog technique, and this form of digital implementation is not very effective.

In a microprocessor-controlled PWM inverter, it is difficult to calculate the pulse widths of the naturally sampled waveform because they cannot be defined by an analytic expression. It is possible to simulate the process of natural sampling in software but, again, this is not an effective technique.

Regular Sampled PWM. An alternative approach which is essentially digital in nature and which is more appropriate for digital hardware or microprocessor implementation is illustrated in Fig. 4.30. The sinusoidal modulating wave is now sampled at regular intervals corresponding to the positive peaks (or both positive and negative peaks) of the synchronized carrier wave. The sample-and-hold circuit maintains a constant level until the next sample is taken. This process results in a stepped, or amplitude-modulated, version of the reference waveform. This stepped waveform is compared with the triangular carrier, and the points of intersection determine the inverter switching instants. The sample-hold version of the modulating wave has a constant step magnitude while a pulse width is being defined. Hence, the width of a pulse is proportional to the step height, and the centers of the pulses occur at uniformly spaced sampling times. This technique is known as uniform sampled, or regular sampled, PWM.[9]

In a ROM-based implementation of regular sampled PWM, there is a substantial reduction in the amount of memory required, as compared with natural sampling. Figure 4.30 shows that the number of sine values needed to fully define one cycle of the sample-hold version of the modulating wave is equal to the carrier ratio used, whereas in the natural sampled scheme, a normal sine wave is required and the definition of a complete cycle at intervals of 0.5 degrees, for example, requires 720 values.

Recent advances in digital large-scale integrated (LSI) circuitry have resulted in a commercial digital PWM modulator based on regular sampling.[23] This modulator is available in a single integrated circuit package covering the full frequency range required in an industrial induction motor drive. The integrated circuit has four clock inputs which define motor frequency, motor volts/hertz, inverter switching frequency, and minimum pulse width. At low output frequencies, the carrier ratio has a value of 168, giving torque smoothness at low speeds. As the fundamental frequency is increased, the carrier ratio is reduced in integer steps at discrete points in the frequency range. Approaching 50 Hz, the carrier ratio is reduced to 15, and thereafter pulses are dropped in a manner that maintains quarter-wave symmetry until six-step operation is achieved.

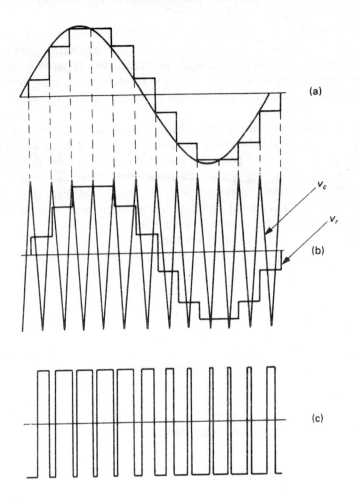

FIG. 4.30. Regular sampled PWM: (a) modulating sine wave and sample-hold version of modulating signal; (b) comparator input voltages; (c) comparator output voltage and pole voltage.

A microprocessor-based technique is also more effective for implementing regular sampling. The microprocessor can now be used to calculate the PWM waveforms directly, because the pulse width duration is defined by the analytic expression[9]

$$t_{pw} = \frac{T}{2}\left[1 + M \sin (\omega_m T_1)\right]$$ (4.12)

where T is the sampling interval, T_1 is the sample instant, and $M \sin (\omega_m t)$ is the original sine wave modulating signal of Fig. 4.30(a). The various options available for microprocessor-based implementation of regular sampled PWM are discussed in Reference 58, and the feasibility of using a software-based calculation over a frequency range of 0 to 100 Hz has been demonstrated.

Optimum PWM Implementation. The optimum PWM waveforms of Section 4.5.4 cannot be generated with analog control circuitry, and the switching angles cannot be computed in real time. The usual technique adopted involves storing the precomputed switching information for a number of different fundamental voltage levels as a look-up table in ROM and accessing this information by hardwired digital logic or by a microprocessor.

The memory requirement is related to the number of discrete levels of inverter output voltage and the look-up table format. One approach uses an 8-bit or 16-bit word in which three bits are used to store the polarity status (high or low) of the three inverter output terminals. The remaining bits specify the number of interrupts to the next change in inverter conduction pattern. Interrupts are generated at uniform intervals throughout the ac cycle, and the binary word which indicates the number of interrupts to the next switching is reduced by one at each interrupt until a reading of zero indicates that a change in the inverter switching pattern is due.

An alternative technique uses a look-up table giving the polarity status of the three output terminals at each interrupt. Information regarding one quarter-cycle is sufficient to allow the switching pattern for the entire cycle to be developed. Typically, the first quarter-cycle of phase A is chosen, and information on the status of phases A, B, and C is stored for this 90-degree interval. The second quarter-cycle is generated by reading the look-up table in the reverse order, interchanging the status of phases B and C. The remainder of the cycle is generated by reading the look-up table in the forward and reverse directions, as before, and complementing the output. This format lends itself to reversing the direction of rotation of the ac motor by altering the phase sequence.

Improved microcomputer-based controllers for optimum PWM have also been described in which real-time, continuous-amplitude control of the fundamental voltage is achieved by the use of an interpolation algorithm. [65,66]

4.5.6. *Drive Control*

The PWM inverter drive is commonly used in industrial drive applications. Constant volts/hertz control gives a constant-torque region of operation below base frequency, as already explained. A constant-horsepower range of operation is usually implemented above base frequency. For routine applications, where good dynamic performance is unimportant, open-loop control is adopted without tachometer feedback. Precise control of shaft speed is obtained when the PWM inverter is used to power a synchronous motor or synchronous reluctance motor, as discussed in Section 4.3.4 on the six-step inverter drive.

The PWM inverter has a very fast transient response, as already emphasized. In fact, the rate of reduction of PWM inverter output voltage must not be excessive in an induction motor drive; otherwise the motor may develop large fault-type currents. Hence, the control system must limit the rate of voltage reduction for stable current regulation. [67]

The fast response of the PWM inverter drive and its capability for smooth low-speed rotation down to standstill make it attractive for a high-performance industrial drive. This application requires the adoption of the more sophisticated closed-loop control strategies discussed in Chapters 7 and 8; further consideration of this topic will be deferred until then.

4.6. BRAKING OF VOLTAGE-SOURCE INVERTER DRIVES

DC power for the six-step and PWM voltage-source inverters is obtained by rectification of the alternating supply voltage with a thyristor or diode rectifier circuit. The dc link supply is filtered as shown in Fig. 4.11 and Fig. 4.17. The large capacitor across the inverter input terminals smooths the direct voltage supply for the inverter, and reduces the source impedance.

If the adjustable-speed ac motor is overhauled by the load, regeneration occurs, and power is returned to the dc link through the feedback diodes of the voltage-source inverter. This occurs when the inverter output frequency is suddenly reduced and in crane and hoist applications when a load is lowered. In a multiple-motor drive in which a single rectifier feeds several inverters simultaneously, regenerated power that is returned to the dc link by one inverter will reduce the rectifier load. In a single-motor drive, regenerated power charges up the smoothing capacitor across the inverter input terminals and raises the dc link voltage. When the dc link converter has a diode rectifier input, this regenerated energy cannot be returned to the ac supply network because the diode rectifier circuit is incapable of inverter operation. If a fully controlled thyristor bridge converter is employed, then phase-controlled inverter operation is feasible, but the polarity of the applied dc voltage must be reversed. Hence, the connections between capacitor and rectifier must be reversed and the firing angle increased beyond 90 degrees to allow regeneration to the ac supply, as discussed in Chapter 3.

Switching of power leads is undesirable and may be avoided by connecting another thyristor bridge in inverse parallel, giving the regenerative double-converter connection described in Chapter 3 and illustrated in Fig. 3.21. This converter circuit allows direct current and power flow in either direction, permitting full four-quadrant operation of the induction motor drive. However, the double-converter connection requires additional thyristors and more sophisticated control circuitry. A transistor bridge may also be fitted to a diode bridge rectifier circuit to provide a path for regenerated current, as shown in Fig. 3.22, but again there is an increase in circuit complexity.

The provision of regenerative braking obviously adds significantly to the cost and complexity of the voltage-source inverter drive. In many applications, the rapid deceleration offered by regenerative braking is not essential, and dynamic braking gives a satisfactory transient performance for less capital expenditure. In the dynamic method, the regenerated power is dissipated in a resistor which is switched across the dc link when the link voltage reaches a certain level due to regenerative charging of the filter capacitor. The switching may be achieved automatically by means of an auxiliary thyristor or transistor which is triggered from a voltage-sensing circuit.

There are many undemanding practical drive applications which do not require the fast deceleration provided by regenerative or dynamic braking. In such cases, it is sufficient to ensure that the inverter control circuitry limits the demanded rate of deceleration and the resulting rise in dc link voltage so that destructive overvoltages do not occur.

4.7. THE SINGLE-PHASE BRIDGE CURRENT-SOURCE INVERTER

All the single-phase and three-phase inverters discussed in the preceding sections of this chapter have a dc voltage-source supply, and the amplitude of the inverter output voltage is essentially independent of load. Consequently, each of these circuits is classified as a voltage-source inverter (VSI). If the inverter is fed from a controlled current source, the output current amplitude is essentially independent of load, giving a current-source inverter (CSI).[68,69] The output voltage of the CSI is determined by the response of the load to the applied current. The controlled current supply for the inverter can be obtained from a thyristor phase-controlled rectifier which is given the characteristics of a current source by operating it in a current regulation loop and placing a large series inductor in the dc output. This aspect is discussed further in the next section.

The current-source inverter is mainly used at medium to high power levels in a circuit configuration that employs conventional thyristor devices. The single-phase version of the common autosequentially commutated inverter (ASCI) circuit is shown in Fig. 4.31. It should be noted that the forced commutating circuit components are included. The feedback diodes which are characteristically present in a voltage-source inverter are now removed, and two capacitors store the energy necessary for commutation. The four series diodes effectively prevent the capacitors from discharging through the load in the interval between inverter commutations.

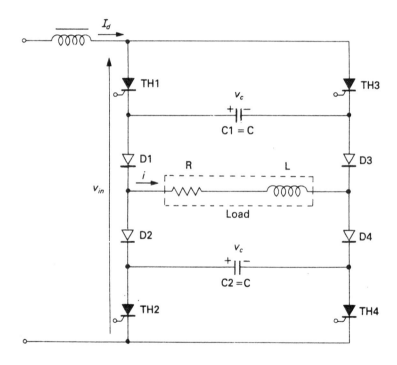

FIG. 4.31. The single-phase bridge autosequentially commutated current-source inverter with a series RL load.

The dc current source delivers a steady current of amplitude I_d to the single-phase bridge CSI, and the thyristors are switched in diagonal pairs so that the load current is a square wave of amplitude I_d. As usual, the triggering frequency of the thyristors determines the output frequency. The circuit behavior during the commutation interval is highly load dependent. If the inverter feeds a passive RL load, the voltage and current waveforms are as shown in Fig. 4.32. Circuit operation proceeds as follows.

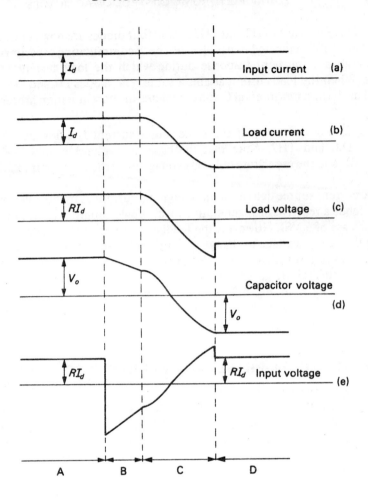

FIG. 4.32. Voltage and current waveforms for the single-phase bridge current-source inverter.

Interval A. In this interval between commutations, the constant source current I_d is flowing in the load, and hence the load voltage is RI_d. Thyristors TH1 and TH4 are conducting, and the current I_d is established in the circuit consisting of TH1, D1, the load, D4, and TH4. The capacitors C1 and C2 are charged to a voltage $v_c = + V_o$ from the previous half-cycle. Note that V_o is greater than the load voltage RI_d, as shown in Fig. 4.32.

Interval B. When TH2 and TH3 are gated on, the capacitors C1 and C2 apply a reverse bias to TH1 and TH4, respectively; causing them to turn off. However, the load current I_d continues to flow in the same direction as before, through TH3, C1, D1, the load, D4, C2, and TH2. The capacitors are in series with the load and are discharged by the constant current I_d. When the capacitor voltages have fallen from V_o to RI_d, diodes D2 and D3 conduct and interval B is terminated.

Interval C. Thyristors TH2 and TH3 and all four diodes are now conducting so that the load is effectively connected in parallel with both commutating capacitors. This RLC circuit undergoes a transient response during which the load current falls to zero and reverses. When the load current attains a value $-I_d$, diodes D1 and D4 become reverse biased. This terminates interval C and completes the commutation process.

Interval D. For the next half-cycle, the source current I_d is flowing through TH3, D3, the load, D2, and TH2. Note that the capacitor voltage is now $-V_o$, as shown in Fig. 4.32(d), and the capacitors hold this voltage until the next commutation.

The dc current source delivers a unidirectional current I_d to the inverter, and when the inverter feeds a reactive or regenerative load, the dc supply current cannot reverse as it did in the case of a VSI. However, the feedback diodes of the VSI have been removed, allowing the dc link voltage to change polarity and permitting a return of energy to the dc link by virtue of a reversal of dc link voltage rather than a reversal of dc link current. When TH2 and TH3 are gated to initiate Interval B, the inverter input voltage, v_{in}, goes negative, as shown in Fig. 4.32(e), and remains negative throughout this interval. Hence, the input power $v_{in}I_d$ is also negative, indicating that power is being returned to the dc current source.

4.8. THE THREE-PHASE BRIDGE CURRENT-SOURCE INVERTER

The three-phase bridge current-source inverter is formed by adding a third leg, or half-bridge, to the single-phase circuit of Fig. 4.31. The resulting self-commutated or autosequentially commutated inverter (ASCI) is used to deliver adjustable-frequency ac power to a cage-rotor induction motor or synchronous motor, giving a rugged ac motor drive.[70-78] The ASCI circuit is the most common inverter configuration, but other commutating circuits are available.[79-83] In a synchronous motor drive, the auxiliary commutating circuits may be employed for starting the motor, but when the motor runs at more than 10 percent of rated synchronous speed, the generated emfs of the synchronous machine are adequate for load commutation of the inverter thyristors, and the commutating circuit is rendered inoperative.

The dc link converter is given the characteristics of a controlled current source by removing the shunt capacitor in the dc link and employing a current regulation loop to control the output dc current from the phase-controlled rectifier, as shown in Fig. 4.33. A signal representing the desired current is compared with the actual current, as measured on the dc or ac side of the rectifier. The difference is amplified and used to control the rectifier delay angle so that the required current value is obtained. A large series inductor in the dc link circuit filters the output current from the controlled rectifier. If the polarity of the dc link voltage reverses, the phase-controlled converter can function as an inverter, returning power to the ac supply network. Consequently, this converter configuration, in which a thyristor rectifier feeds a current-source inverter, has an inherent regenerative capability that is advantageous in many drive applications. A diode rectifier bridge, followed by a current-regulated dc chopper, is sometimes used as a nonregenerative current source for the CSI.

Figure 4.34 shows the three-phase autosequentially commutated inverter feeding a balanced wye-connected RL load. The path of the regulated input current, I_d, through

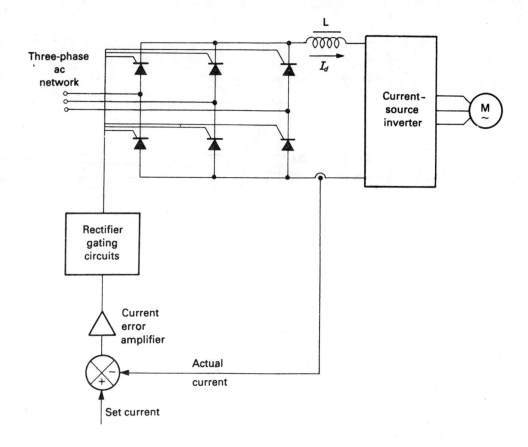

FIG. 4.33. Phase-controlled thyristor rectifier operating as a controlled current source for a CSI.

the inverter and load phases is governed by the particular inverter thyristors that are gated into conduction. In Fig. 4.34, the six thyristors, TH1 to TH6, are numbered in the sequence in which they are gated, and each thyristor conducts for 120 degrees of the output period. The gating of one thyristor causes an adjacent conducting thyristor of another phase to turn off, and so establishes the 120-degree conduction pattern. The two banks of delta-connected capacitors store the energy necessary for commutation, and the six blocking diodes, D1 to D6, isolate the capacitors from the load.

The inverter conduction sequence is such that the regulated source current, I_d, is directed through a pair of conducting thyristors, one connected to the positive dc rail and the other connected to the negative rail. There is a 60-degree interval in each half-cycle during which both thyristors in the same half-bridge are turned off, and the corresponding ac line current is zero. Assuming instantaneous commutation, the ac line current has the quasi-square waveform shown in Fig. 4.34, which is identical to the line-to-line output voltage waveform of a six-step VSI. Fourier analysis of the current waveform gives the expression

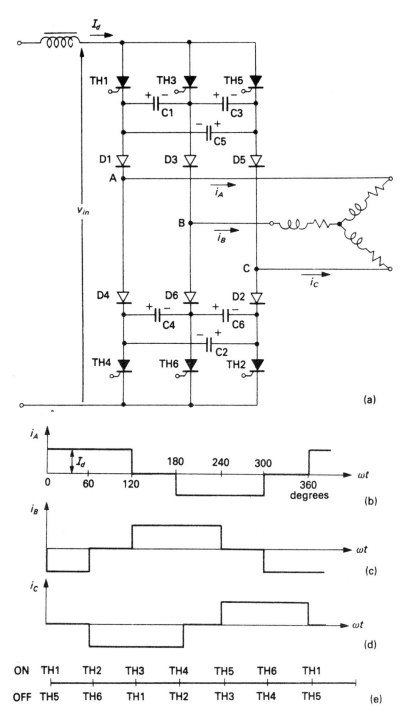

FIG. 4.34. The three-phase bridge autosequentially commutated current-source inverter with a wye-connected RL load: (a) circuit diagram; (b), (c), (d) idealized ac line current waveforms; (e) thyristor gating sequence.

$$i_A = \frac{2\sqrt{3}}{\pi} I_d \left(\sin \omega t - \frac{1}{5} \sin 5\,\omega t - \frac{1}{7} \sin 7\,\omega t \right.$$

$$\left. + \frac{1}{11} \sin 11\,\omega t + \frac{1}{13} \sin 13\,\omega t - \ldots \right). \qquad (4.13)$$

Thus, the fundamental component of ac line current has an amplitude of $2\sqrt{3}\,I_d/\pi$, and an rms value of $\sqrt{6}\,I_d/\pi$, or $0.78\,I_d$.

The unique relationship between input and output voltage magnitudes for the six-step VSI has its counterpart in the unique relationship between input and output current magnitudes for the current-fed inverter. If the CSI feeds the balanced delta-connected load of Fig. 4.35(a), the phase current has the six-step waveform of Fig. 4.35(b), which

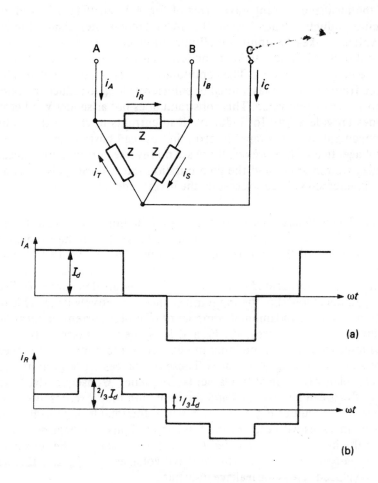

FIG. 4.35. Idealized current waveforms in a balanced delta-connected load fed by a six-step current-source inverter: (a) line current; (b) phase current.

is characteristic of the line-to-neutral voltage for a balanced wye-connected load fed by a six-step VSI. By Fourier analysis:

$$i_R = \frac{2}{\pi} I_d \left(\sin \omega t + \frac{1}{5} \sin 5\,\omega t + \frac{1}{7} \sin 7\,\omega t \right.$$

$$\left. + \frac{1}{11} \sin 11\,\omega t + \frac{1}{13} \sin 13\,\omega t + ... \right). \qquad (4.14)$$

Clearly, the current waveforms delivered by the six-step CSI have pronounced low-order harmonics. In an ac motor drive, these harmonics give rise to pulsating torques and irregular shaft rotation at low speeds, as in the case of a six-step VSI drive.

When an electrical load is fed from an ac current source, the terminal voltage waveform is determined by the response of the load to the applied current. The voltage across an inductor L is $L\,di/dt$, where di/dt is the rate of change of current. Consequently, the idealized current waveforms of Fig. 4.34 and Fig. 4.35 cannot be realized with practical inductive loads because the instantaneous step changes in current would develop voltage spikes of infinite amplitude. In practical circuits, the rate of change of load current must be limited to keep the terminal voltage within the peak voltage capability of the inverter thyristors. The commutation interval, during which load current is transferred from phase to phase, must be sufficiently long to reduce the rate of change of current to an acceptable value. This constraint does not arise in a VSI because the feedback diodes provide a path for inductive load currents to charge the dc link capacitor. This arrangement prevents rapid interruption of load current and clamps the inverter output voltage. In a CSI, however, there are no reverse current paths, because the feedback diodes are removed, and the commutation interval can only be shortened at the expense of increased voltage stresses on the inverter devices.

Induction Motor Voltage Waveforms. For an induction motor load, the terminal voltage waveform is determined by the impedance presented to the fundamental and harmonic components of the inverter output current. The relevant fundamental and harmonic equivalent circuits are derived in Chapter 6. At fundamental frequency, the induction motor is represented by the conventional equivalent circuit of Fig. 6.1, and for normal operation at low slip, the input impedance is relatively large. This impedance, in conjunction with the fundamental component of motor current, determines the fundamental line-to-neutral voltages, V_a, V_b, and V_c, at the motor terminals.

At harmonic frequencies, the input impedance of the motor is effectively the sum of the stator and rotor leakage reactances. These reactances are large at harmonic frequencies but are relatively small at fundamental frequency. Consequently, fundamental and harmonic effects can be separated and the induction motor represented by the approximate equivalent circuit of Fig. 4.36(a), in which the total leakage inductance per phase, L, is placed in series with each of the fundamental phase voltage sources, V_a, V_b, and V_c. When the induction motor is fed from a current source, the terminal voltages are obtained by superposition of the fundamental voltages, V_a, V_b, and V_c, and the $L\,di/dt$ voltages developed across the leakage inductances.

In the case of a six-step current supply, $L\,di/dt$ is zero except when a step change in current occurs. Consequently, the terminal voltage is a fundamental sine wave with

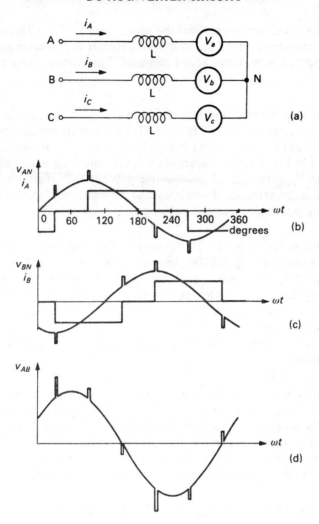

FIG. 4.36. Idealized voltage and current waveforms for an induction motor fed by a six-step current-source inverter: (a) approximate induction motor equivalent circuit; (b), (c), (d) voltage and current waveforms.

superimposed voltage spikes, as shown in the phase and line voltage waveforms of Fig. 4.36.[72,84] The position of the voltage spikes on the motor voltage waveform clearly depends on the fundamental power factor. These commutation voltage spikes must not cause the blocking voltage capability of the inverter thyristors to be exceeded. This voltage restraint is ensured by employing an induction motor with a small leakage reactance and extending the commutation interval so that the rate of change of current is not excessive.

4.8.1. *Commutating Circuit Analysis*

The autosequentially commutated CSI has a relatively simple circuit, but an analysis of the commutating action is complicated by the fact that the commutating phase interacts with the other two phases and the motor load.[72,85–91] Figure 4.37 shows the CSI feeding a wye-connected induction motor. Inverter operation is assumed to be in a steady-state condition with motor speed constant. The commutating cycle proceeds as follows.

Interval A. In this interval, the inverter is in the normal operating mode between commutations. Assume thyristors TH1 and TH2 have been conducting for some time, so that phases A and C carry current but phase B does not. The resulting current flow path is indicated in Fig. 4.37(a). Capacitors C1, C3, and C5 are assumed to be charged with the voltages V_o, zero, and $-V_o$, respectively. When the inverter is first switched on, the capacitors must be precharged with a voltage distribution of this nature, but the auxiliary precharging circuit is not needed subsequently.

Interval B. When TH3 is fired, TH1 is reverse biased by the voltage on capacitor C1, and turns off. As shown in Fig. 4.37(b), the current I_d, which was flowing in TH1, is now flowing through TH3, the capacitor bank formed by C1 in parallel with C3 and C5, and diode D1. During this charging interval, the constant source current, I_d, linearly charges the capacitor bank. The outgoing thyristor, TH1, is reverse biased until the voltage on capacitor C1 changes polarity. Diode D3 is reverse biased and the motor phase currents have the same values as existed during interval A. This charging mode ends when diode D3 starts to conduct.

Interval C. This is the current transfer interval during which motor current transfers from phase A to phase B, as shown in Fig. 4.37(c). When diode D3 conducts, the upper capacitor bank is connected in parallel with the motor through diodes D1 and D3. The resulting LC circuit resonates, and in one-quarter of the resonant period the oscillatory current reduces the phase A current from I_d to zero, and increases the phase B current from zero to I_d. Diode D1 then blocks, and the commutation cycle is complete.

Interval D. The source current I_d is now feeding phases B and C through thyristors TH3 and TH2, as shown in Fig. 4.37(d). This condition lasts until TH4 is gated to initiate the next commutation. Because D3 is the only conducting diode in the upper group, the upper capacitor group retains its charge until the next upper group commutation.

The duration of transfer interval C determines the rate of change of motor phase current and the magnitudes of the resulting $L \, di/dt$ voltage transients developed across the motor leakage inductances. These voltage spikes appear in the output voltage waveform, as discussed previously and as shown in Fig. 4.36. A large value of commutating capacitance prolongs transfer interval C and thereby limits the voltage stresses on the inverter devices. The increased commutating time is therefore of adequate duration to permit the use of low-cost, converter-grade thyristors rather than expensive, inverter-grade devices with fast turn-off characteristics.

The commutation cycle described above is applicable at low output frequencies, when commutations in the upper and lower groups do not overlap. As the output frequency of the inverter is increased, partial overlap occurs when the charging interval B commences in the upper (lower) group before the current transfer interval C is completed in the

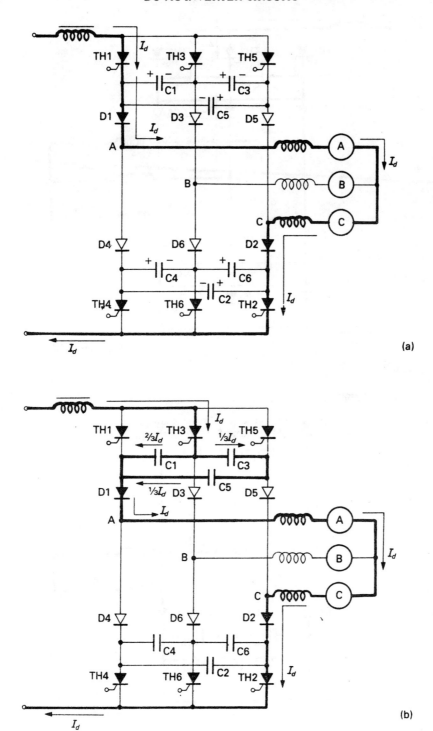

FIG. 4.37. The commutation process in the three-phase autosequentially commutated current-source inverter: (a) Interval A; (b) Interval B.

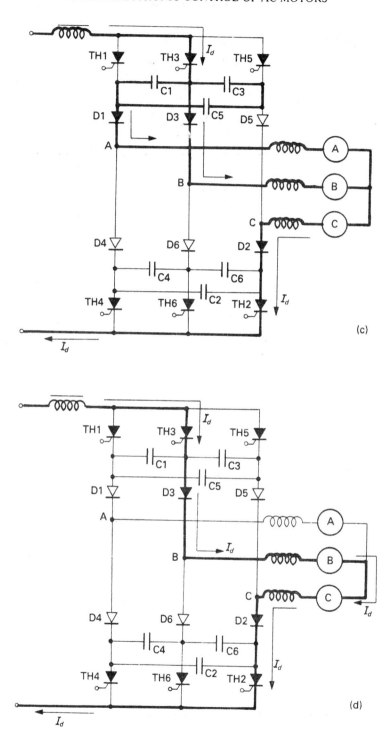

Fig. 4.37. The commutation process in the three-phase autosequentially commutated current-source inverter: (c) Interval C; (d) Interval D.

lower (upper) group. As the inverter frequency is increased further, the current transfer mode may occur simultaneously in both upper and lower groups. This so-called multiple commutation effect, or full commutation overlap, results in a current bypass effect in which a portion of the source current, I_d, is diverted from the motor by two conducting diodes in the same inverter leg. [86,87] This diversion of current causes a reduction in the power input to the motor for a given source current and adversely affects the output torque capability, efficiency, and stability of the drive.

Large commutating capacitors have a beneficial effect in reducing inverter voltage stresses, but they also lower the operating frequency at which the current bypass effect occurs, and so limit the useful frequency range of the inverter. Consequently, it may be desirable to reduce the commutating capacitance and limit voltage stresses by means of a voltage-clamping circuit. This may consist of a three-phase diode bridge rectifier connected in parallel with the CSI across the motor terminals and feeding a series string of Zener diodes that determine the clamping voltage level.

4.8.2. Regenerative Braking

In a CSI drive, the phase-controlled converter and filter inductor deliver a regulated dc current to the inverter. The average dc voltage at the inverter terminals varies with inverter loading so that, neglecting losses, the inverter power input balances the load power. With an unloaded motor, the inverter dc voltage is near zero, while at full load, the dc link voltage has a maximum value. This behavior contrasts with that of the six-step VSI drive, where the dc voltage is the regulated quantity and dc current is a function of the power demand of the load.

When regeneration occurs in a CSI drive, energy is returned to the dc link. The dc link current is unidirectional, and as explained earlier, a power reversal is achieved by a change in polarity of the mean dc link voltage rather than a reversal of dc link current. The removal of the feedback diodes, which are characteristically present in a VSI, ensures that the dc link current is unidirectional whether the machine is motoring or regenerating, and also permits a polarity reversal of the mean link voltage. The phase-controlled converter and current control loop of Fig. 4.33 seek to maintain the set current, and consequently, the change in voltage polarity at the input terminals of the CSI automatically results in a corresponding change in polarity at the output terminals of the phase-controlled converter, as shown in Fig. 4.38(a) and (b) for motor and generator operation, respectively. When the drive regenerates, the delay angle of the phase-controlled converter is increased beyond 90 degrees by the control loop to give a dc link voltage reversal, allowing the circuit to function as a phase-controlled inverter that returns energy from the dc link to the ac utility network. Thus, the phase-controlled converter and filter inductor, together with the current feedback loop, act as a regenerative current source, and a second phase-controlled converter is unnecessary for regeneration of energy to the ac supply. Hence, the CSI drive, which has a relatively simple power circuit, employing only 12 converter-grade thyristors, has the inherent capability of regenerating to the input ac power lines.

A reversal in the direction of motor rotation is readily obtained by electronically reversing the output phase sequence of the current-source inverter. Consequently, the CSI drive is capable of full four-quadrant operation and rapid speed reversal without any additional components. It is obviously cost competitive with the standard double-

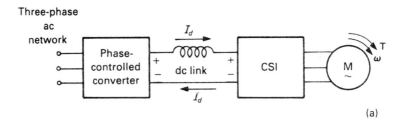

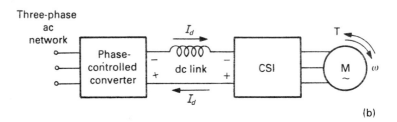

FIG. 4.38. DC link conditions in a current-source inverter drive: (a) motoring; (b) regeneration.

converter regenerative dc motor drive, which uses an expensive dc motor and in which two inverse-parallel six-thyristor bridges are required for bidirectional control of armature current.

4.8.3. *Control of the CSI Drive*

Induction motor torque and airgap flux are uniquely determined by the stator current and rotor slip frequency, as shown in Chapter 7. Direct control of slip frequency is possible if shaft speed is monitored by means of a tachometer and used to determine stator frequency in a controlled-slip drive. A block diagram of a CSI induction motor drive is shown in Fig. 4.39. The actual speed, n, as measured by the tachometer, is compared with the set speed, n^*. The speed error is amplified and used to define reference values for slip frequency and stator current which, when established in the motor, will maintain a constant airgap flux at all supply frequencies and develop the required torque. The difference between the actual stator current and the set value controls the rectifier firing angle and so regulates the source current I_d and the motor current, as already discussed.

The demanded value of slip frequency is added to the rotational frequency, as measured by the tachometer, to determine the excitation frequency which must be supplied to the motor. If the actual motor speed exceeds the set value, the speed error defines a negative value of slip frequency. The inverter then generates a stator excitation frequency which is less than the motor rotational frequency, and the machine functions as an induction generator, returning energy through the converter to the ac supply network. Controlled-slip operation of the CSI induction motor drive and other control techniques are discussed in Chapter 7 and in the technical literature.[74,75,92-95]

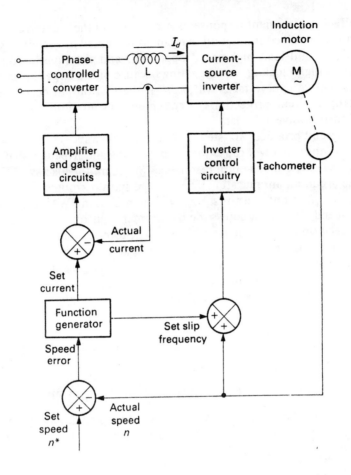

FIG. 4.39. Controlled-slip operation of a CSI induction motor drive.

4.8.4. *Features of the CSI Drive*

As shown in the preceding sections, the CSI drive is a single-motor drive that is inherently capable of regeneration to the ac utility network. Inverter current is under the control of a closed-loop current regulator, and overcurrent transients are effectively suppressed by the regulator and filter inductor. This arrangement gives the drive a reliability and ruggedness which are extremely attractive for industrial applications. The filter inductance is usually several times greater than the total leakage inductance per phase of the induction motor or subtransient reactance per phase of the synchronous motor. This large series inductance limits the rate of rise of fault current in the event of a commutation failure in the inverter or a short-circuit at the output terminals, and the fault can be cleared by suppressing the rectifier gating signals. Consequently, the drive can ride through an inverter misfire or momentary short-circuit at the motor terminals with only a temporary loss of torque and without fuse-blowing or inverter shutdown.

On the other hand, the dc link inductor and commutating capacitors are bulky components that increase the size, weight, and cost of the CSI drive. The large inductance can adversely affect the transient response of the drive, and the maximum frequency of operation is limited by full commutation overlap, as discussed in Section 4.8.1. Stability problems occur for light-load high-speed operation, and the inverter thyristors and diodes must have a high blocking voltage rating because of the large commutating voltage spikes. Destructive overvoltages will occur if the CSI output current is suddenly interrupted, but open-circuit protection can be achieved by providing an alternative current path with a current-divert thyristor.

The basic CSI circuit produces six-step, or quasi-square-wave, currents, and the low-order harmonic components will cause torque pulsations and irregular shaft rotation at low speeds and may excite mechanical resonances in the motor-load system. As shown in Chapter 6, the predominant pulsating torque is the sixth harmonic component due to the fifth and seventh harmonic currents. Harmonic elimination PWM techniques may be used in the low-speed range to suppress the objectionable current harmonics or to minimize the rotor speed ripple resulting from the periodic torque pulsations.[96–98] Alternatively, amplitude modulation of the instantaneous dc link current may be employed to produce a modified output current waveform that minimizes the pulsating torque amplitude.[99,100] In high-power applications, where parallel devices are required, it is more advantageous to operate two CSIs in parallel and combine their phase-shifted outputs to give an improved waveform with reduced harmonic content and smaller pulsating torques.[101,102] Alternatively, the two phase-shifted inverters may feed separate spatially displaced three-phase stator windings in a specially wound machine.[103,104]

4.9. THE LOAD-COMMUTATED INVERTER

The inverters described in the preceding sections of this chapter have relied on the use of power semiconductors that either have a self-turn-off capability (bipolar or field-effect transistors, GTOs, etc.) or require the use of additional capacitors, reactors, and auxiliary thyristors for forced commutation of the main thyristors. The load-commutated inverter achieves commutation of its main thyristors by supplying current to an ac synchronous machine at a leading power factor. This ensures that the ac phase current changes direction before the phase voltage changes polarity. For the time interval after the thyristor current has become zero and before the phase voltage becomes positive, a reverse voltage is applied to the previously conducting thyristor, allowing it to turn off.

Transfer of current from one thyristor to another is initiated by turning on the next thyristor in the gating sequence. The anode voltage of the incoming thyristor is at a higher potential, allowing it to take over the load current. Two variations of the circuit will be described, first, with an ac supply and, second, with a dc input source. Methods of low-speed commutation, current control, and applications are also discussed.

4.9.1. AC Supply/Load-Commutated Inverter

The power circuit for an ac source load-commutated inverter is shown in Fig. 4.40.[105–113] Two power conversion circuits are connected in series. First, a line-commutated, phase-controlled converter rectifies the fixed ac supply voltage and delivers a regulated dc current, as already discussed in Section 4.8. The reversal of ac line voltage at supply frequency allows natural commutation of the thyristors in this circuit.

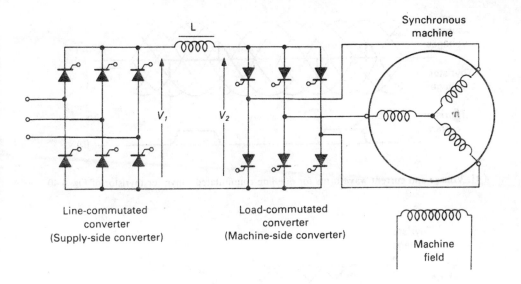

FIG. 4.40. AC supply/load-commutated inverter power circuit.

The second power converter is a load-commutated, phase-controlled converter operating as an inverter. Power flows from the dc side of the inverter to the ac side, which is connected to the synchronous machine. The control system is arranged so that the machine operates at a leading power factor. Polarity reversal of machine voltage at machine frequency provides the commutating voltage for the thyristors. The average voltage, V_1 (Fig. 4.40), is slightly greater than voltage V_2 by the amount needed to force the dc current through the resistance of the dc inductor, L. The inductor filters the harmonic voltages produced by the two converters. The lowest order harmonic voltages are the sixth harmonic of the supply frequency and the sixth harmonic of the motor frequency.

Figure 4.41 shows the voltage and current waveforms for one of the naturally commutated rectifier thyristors, and Fig. 4.42 shows the voltage and current waveforms for one of the load-commutated inverter thyristors. The conduction periods are 120 degrees of the supply frequency and machine frequency, respectively. These thyristor voltage waveforms are derived from the ac line-to-line voltage waveforms, as already discussed in Section 3.3.6 and illustrated in Fig. 3.18. The reverse voltage interval for the rectifier thyristors can approach 240 degrees at the source frequency, thereby ensuring commutation. The reverse voltage interval for the inverter thyristors approaches a minimum value associated with the thyristor turn-off time, and the reverse voltage variation at turn-off is clearly shown in Fig. 4.42. At very low motor frequencies, the conduction time of each inverter thyristor becomes very long, resulting in a reduced current rating for the thyristors. High horsepower drives, at higher voltages, require series-connected thyristors.

During normal load-commutated operation of the drive, motor current is controlled by adjustment of the delay angle of the line-commutated rectifier, and the inverter frequency is controlled by gating of the load-commutated inverter in synchronism with the machine speed. With a fixed field current, the motor-generated emf (or counter emf)

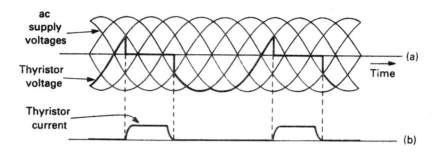

FIG. 4.41. Voltage and current waveforms for one line-commutated converter thyristor of Fig. 4.40.

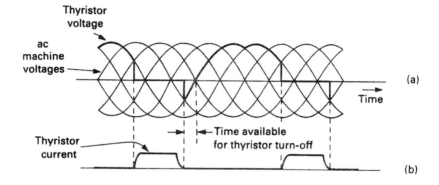

FIG. 4.42. Voltage and current waveforms for one load-commutated inverter thyristor of Fig. 4.40.

increases with increasing machine speed. In order to equalize the two average voltages (V_1 and V_2 in Fig. 4.40), the firing angle of the inverter thyristors must be advanced, giving operation at a reduced leading power factor. If field current is reduced as machine speed increases, then the thyristors can operate at their maximum delay angle and the machine power factor is close to unity. Several strategies for field current control and firing angle adjustment are described in Chapter 8.

At standstill and reduced machine speed, motor-generated emf is insufficient to allow natural commutation of the load-commutated inverter thyristors; a separate means of commutation must be provided. For machines with base frequencies in the range of 60 to 90 Hz, a separate external means of commutation is needed from standstill to 6 to 10 Hz (approximately 10 percent of base speed). The conventional technique for inverter thyristor commutation is to rapidly change the firing angle of the input converter from rectifying to inverting operation. This action quickly forces the current in the dc link to zero. With zero dc current, the two conducting inverter thyristors are starved of current and are able to recover their forward blocking capability. All 12 thyristors in Fig. 4.40 are then nonconducting. After an appropriate time interval to ensure that all thyristors have recovered their voltage blocking capability, the appropriate two rectifier thyristors and the next two inverter thyristors are supplied with gate signals, so that the dc current flows in the loop once more. This "off" interval in dc current occurs every 60 degrees of the machine frequency when one of the inverter thyristors needs to be commutated. This commutation sequence can be implemented at unity power factor to increase the motor torque-per-ampere of dc current. The motor accelerates until the counter emf is sufficient to produce load commutation of the inverter thyristors. In this mode of operation, the rectifier thyristors are gated at the correct firing angle to control the dc current, which is now continuous. The machine power factor decreases from unity to a leading power factor.

Regenerative operation for motor braking is possible by interchanging the roles of the two converters. The machine-side converter now operates as a phase-controlled rectifier while the supply-side converter acts as a line-commutated inverter. This operation allows the synchronous machine to act as a generator and return power to the ac supply. The direction of machine rotation is reversed by changing the gating sequence of the thyristors in the machine-side converter. Thus, a four-quadrant drive is obtained with the 12-thyristor configuration of Fig. 4.40.

The load-commutated inverter (LCI) is used to power high-horsepower (100-10 000 hp) synchronous motors for large fans, blowers, and compressors. This type of drive system has its antecedent in the use of tubes (thyratrons) as controllable electronic valves. A review of the early effort is contained in Reference 114. A second application for this converter is in starting and accelerating to full speed the large synchronous condensers used for utility system transmission line voltage control. After the machine reaches full speed, the load-commutated inverter is disconnected and the synchronous condenser is powered from the ac supply network. The LCI can be rated for a short duty cycle and does not need to be rated for the full machine voltage, as the starting torque is low and the machine can accelerate to full speed with reduced terminal voltage. A third application for the LCI system is starting large pumped storage hydrogenerators. To change from generator operation to pumping operation, the machine direction of rotation must be reversed. The LCI can regeneratively brake the large rotor inertia and return braking power to the utility, reverse its machine-side converter thyristor gating sequence, and then accelerate the rotor of the hydrogenerator in the opposite direction. When the machine reaches full speed, it can be connected to the ac utility network and function either as a hydrogenerator or motor-driven pump to fill the upper reservoir of the pumped storage scheme.

4.9.2. DC Supply/Load-Commutated Inverter

The load-commutated inverter power circuit, supplied from a dc source, is shown in Fig. 4.43.[115,116] The three-phase bridge circuit of main thyristors (TH1 to TH6) is operated in a leading power factor mode similar to that described above for the ac supply system. At standstill and low machine speeds, a forced commutating circuit, with auxiliary thyristors THA and THB and capacitor C, is provided to turn off the main thyristors. When the capacitor is charged with the left-hand plate positive, turning on auxiliary thyristor THA places the capacitor voltage and the machine counter emf in series across thyristors TH1, TH3, and TH5. With sufficient voltage on the capacitor, the conducting thyristor of the three is turned off. The opposite polarity of voltage on the capacitor, in conjunction with gating THB, turns off one of the lower three main thyristors. The auxil-

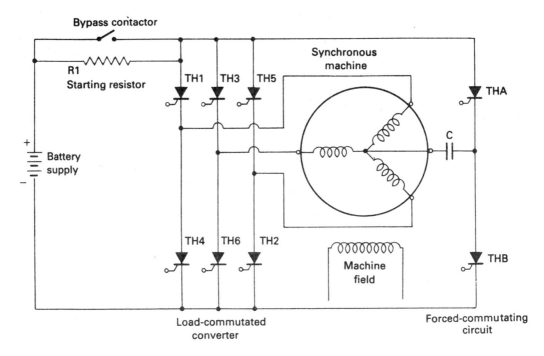

FIG. 4.43. DC supply/load-commutated inverter power circuit.

iary thyristors operate at three times the machine fundamental frequency, and therefore, the circuit has been called the "third harmonic commutated" inverter.[117] At a sufficiently high machine speed and with a suitable control system, the forced commutation function is inhibited by suppressing the gating of auxiliary thyristors THA and THB; the main thyristors are then commutated by the leading power factor operation of the synchronous machine. A starting resistor is provided to limit the dc current below half-speed. Beyond half-speed, the resistor is shorted by a bypass contactor and the motor operates from the dc battery voltage up to full speed. For high-power systems, a dc-dc chopper or an auxiliary low-power ac to dc converter can be provided for current control at reduced motor speeds. Because the battery voltage cannot reverse polarity for regeneration, the dc current must reverse. This is accomplished by an electromechanical reversing contactor or by a separate three-phase bridge converter connected in inverse parallel with the existing converter. This system has been proposed for a high-speed machine with a flywheel energy storage package associated with a battery-powered urban vehicle and a dc overhead-catenary-supplied electric bus.[115]

4.10. THYRISTOR FORCED COMMUTATION METHODS

In general, forced commutation of a thyristor requires that energy be stored in a commutating capacitor that is precharged with the correct polarity. This capacitor is switched in at the desired instant to divert current from the conducting thyristor and to apply reverse voltage until the thyristor has recovered its forward blocking capability.

This section is concerned with forced commutation techniques for thyristor VSI circuits, such as the six-step and PWM inverters. As already discussed, these inverters use feedback diodes to clamp the load voltage to the dc supply voltage and to permit the return of energy from reactive loads to the dc link supply. Because the commutating circuits extinguish a conducting thyristor by applying a brief impulse of reverse voltage, they are termed impulse-commutating circuits. Because these commutation techniques do not affect inverter operation between commutations, it is possible, to a large extent, to separate the functions of the power and commutating circuits, which simplifies commutating circuit analysis and design.[118] This behavior should be contrasted with the autosequentially commutated current-source inverter of Section 4.8.1, where the inverter is load commutated by the delta-connected capacitor banks, and the commutating circuit and load are highly interdependent.

A bewildering variety of forced commutating circuits has been described in the technical literature. This profusion of circuits has led to a number of attempts at classification of the different types.[119–122] Various circuits have also been compared on the basis of cost and efficiency.[123,124] In the earliest thyristor inverter circuits, commutation was achieved by means of a large capacitor in parallel with the load, and feedback diodes were not used.[125,126] This technique was termed parallel-capacitor commutation, and Fig. 4.44 shows a single-phase bridge inverter of this type. Assume TH1 and TH4 are conducting and, as a result, capacitor C is charged with the left-hand plate positive. When TH2 and TH3 are gated, the capacitor discharges around two parallel paths, reducing the currents in TH1 and TH4 to zero and reverse-biasing them. The capacitor then charges with the reverse polarity through thyristors TH2 and TH3 and the series inductor L. A suitable choice of circuit values will ensure that sufficient recovery time is provided for the commutated thyristors. A similar sequence of events occurs 180 degrees later, when TH1 and TH4 are triggered.

The absence of feedback diodes in this circuit means that energy associated with a reactive load cannot be returned to the dc supply but, rather, must be absorbed by the commutating capacitor and inductor. Such operation requires large circuit components and also causes the output voltage waveform to vary with load. The reverse voltage

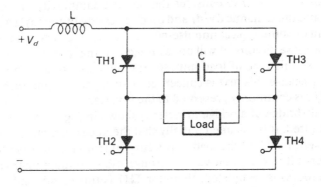

FIG. 4.44. The single-phase bridge inverter with parallel-capacitor commutation.

applied across the commutated thyristor is a function of the load impedance because the commutating capacitor is connected directly across the output terminals. Sufficient recovery time may not be available for commutation under certain loading conditions and, in the case of highly inductive loads, additional commutating capacitance is necessary.

The simple parallel inverter is unsatisfactory, therefore, and modern voltage-source inverters use feedback diodes to allow the transfer of load current to an alternative path at each commutation. This technique avoids a prolonged resonant oscillation following a commutation, permitting a significant reduction in the size of the commutating components and limiting the duration of the commutation interval to a small fraction of a half-cycle. Thus, in the modern impulse-commutated VSI, the commutating circuit only affects inverter operation when required — that is, during the brief commutation interval — and satisfactory inverter operation is achieved for wide variations in output frequency, power factor, and load current. The commutating component values are determined by the peak current to be commutated, the available capacitor voltage, and the circuit turn-off time required.

In general, some of the energy stored in the commutating capacitor prior to commutation is dissipated in the inverter circuit during the commutation process. These commutation losses are often appreciably greater than the on-state conduction losses of the thyristors, particularly at high output frequencies when the number of commutations per second is large. The term "trapped" energy is used to describe energy stored in the magnetic field of a current-carrying inductor at the end of the commutation process.[120,127] The inductor is often in series with the supply or load, and the trapped energy must be returned to the dc supply, if possible, or dissipated in circuit resistance. The amount of trapped energy may be considerable, and the commutating circuit components are often selected on the basis of minimizing this trapped energy.

A high-quality commutating circuit should have good efficiency and a short clearing time to permit successive commutations which are separated by a minimum time interval. These properties are particularly important in PWM inverters, where operation is required at a high switching rate and successive commutations may be very close. In practice, thyristor PWM inverters can have efficiencies of 95 percent, or more, at switching frequencies of 400 to 500 Hz. When selecting a commutating circuit, the inverter designer must adopt a commutation technique that has low energy loss per commutation, but it is also necessary to consider the relative complexity of the different circuit configurations and the dynamic dv/dt and di/dt stressing of the thyristors.

The operation of some typical impulse-commutated inverters will now be considered. For the study of these circuits, it will be assumed that the inverter is supplying an inductive load, because this type of load imposes the most severe duty on the commutating circuit and also because this text is concerned with ac motor applications in which a lagging power factor is commonly presented to the inverter.

One leg or half-bridge of a bridge inverter is shown in Fig. 4.45 with the commutating components omitted. It is assumed initially that the upper thyristor, TH1, is conducting, and therefore, terminal A of the load draws current from the positive terminal of the dc supply and returns it to the negative rail through other conducting thyristors. When the commutation process is completed, thyristor TH1 is turned off, and the inductive load current continues to flow through feedback diode D2.

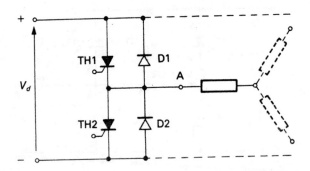

FIG. 4.45. One leg of a three-phase bridge inverter.

The following simplifying assumptions are also made:

1. The load inductance is sufficiently large to maintain the load current constant at a value I_o during the commutation interval. This is justified by the fact that the commutation process is completed in 10 to 100 μs.

2. The brief pulse of reverse thyristor current at turn-off can be ignored.

3. The turn-on time of the thyristor is negligible. This is a reasonable assumption because the turn-on time is only 1 to 4 μs, and the rate of increase of current is determined by external circuit components with much longer time constants.

4. The on-state voltage drop of a thyristor is negligible.

The two thyristors in Fig. 4.45 may be coupled in such a way that each acts as the switch to turn the other off. This is termed complementary commutation and avoids the need for auxiliary switches. The McMurray-Bedford circuit is an example of this technique.

4.10.1. *The McMurray-Bedford Circuit*[118,128]

Figure 4.46 shows one leg of a bridge inverter employing the McMurray-Bedford method of thyristor commutation. The center-tapped inductor is small (of the order of 50 μH), and the two halves are assumed to be perfectly coupled. Figure 4.47 shows the voltage and current waveforms for a lagging power factor load. All voltages are specified with respect to the negative dc rail, and the duration of the commutation interval B is exaggerated.

Interval A. Initially, thyristor TH1 is conducting and drawing load current from the positive side of the dc supply. For a highly inductive load, the rate of change of current is small, and the voltage across the commutating inductor, L1, is negligible compared with the supply voltage. Terminal A of the load is, therefore, effectively connected to the positive supply line, and capacitor C1 has zero voltage, while a similar capacitor C2 is charged to the dc supply voltage V_d.

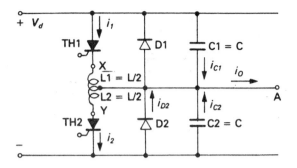

FIG. 4.46. One leg of a bridge inverter employing the McMurray-Bedford commutation method.

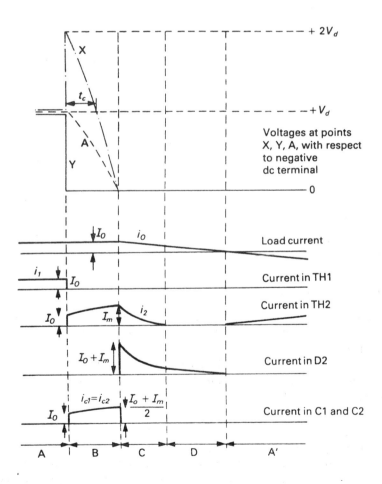

FIG. 4.47. Voltage and current waveforms in the McMurray-Bedford circuit with a lagging power factor load.

Interval B. When TH2 is triggered to initiate commutation of TH1, point Y is connected to the negative side of the supply. The voltages on capacitors C1 and C2 cannot change instantaneously, and hence a voltage V_d appears across winding L2 of the commutating inductor. An equal voltage is induced in winding L1, thus raising the potential of point X to $+2V_d$ with respect to the negative dc line (Fig. 4.47). This places a reverse bias of V_d on thyristor TH1, causing it to turn off. The load current I_o, which was flowing through TH1 and winding L1, will transfer instantaneously to TH2 and winding L2, in order to preserve the ampere-turns balance in the perfectly coupled transformer. The constant current, I_o, demanded by the inductive load is supplied by the charging currents of C1 and C2. These initial conditions determine the subsequent voltage and current variations during the commutation interval.

As capacitor C2 discharges, the voltage applied across L2 decreases, thereby reducing the voltage induced in L1. Simultaneously, capacitor C1 is charging up to the supply voltage, and after a short interval, the forward bias on TH1 due to C1 exceeds the reverse bias due to L1. If L1 = L2 = L/2, and C1 = C2 = C, a rigorous analysis shows that the voltage waveforms are portions of a sine wave of period $2\pi\sqrt{LC}$, and the turn-off time, t_c, provided by the circuit is less than a quarter of a cycle. The thyristor turn-off time, t_q, must be less than t_c for a successful commutation.

During the commutation interval, the inductor L2 supports the voltage on capacitor C2. This causes the current i_2 in L2 to increase from its initial value of I_o to a maximum value I_m, when the voltage on C2 has fallen to zero.

Interval C. Capacitor C1 is now charged to the supply voltage, and the inductive load current is drawn through feedback diode D2, thereby connecting load terminal A to the negative supply line. The inductor L2 has trapped energy of $(L2)\,I_m^2/2$, which is dissipated in resistance losses by a free-wheeling current that circulates in the closed circuit of L2, TH2, and D2. The circuit is usually designed to reduce this trapped energy to a minimum. In small inverters, extra resistance is sometimes added in series with the feedback diode to speed up the decay of circulating current and so permit a reduction in the current ratings of thyristor and diode.

Interval D. When the circulating current has decayed to zero, the commutation process is complete, but diode D2 continues to draw the inductive load current from the negative line.

Interval A'. The load current subsequently decays to zero and then reverses its direction. This current reversal blocks diode D2, and the increasing ac line current in the new direction is carried by TH2, which must be re-gated if the original gating signal (which initiated commutation) has been removed. This completes the transfer of load current from TH1 to TH2.

If the maximum load current to be commutated is I_{oo} and the thyristor turn-off time is t_q, the commutating circuit parameters for minimum trapped energy are[118,128]

$$L = 2.35\,\frac{V_d t_q}{I_{oo}} \qquad\qquad (4.15)$$

and

$$C = 2.35 \frac{I_{oo} t_q}{V_d}.$$ (4.16)

4.10.2. The McMurray Circuit [118,129]

In the McMurray-Bedford circuit, the commutation process is initiated when the complementary thyristor is triggered, but better performance and higher efficiency can usually be achieved when auxiliary thyristors are used to switch in the commutating capacitors. The McMurray circuit uses auxiliary thyristors to switch a high-Q resonant circuit in parallel with the conducting thyristor. Figure 4.48(a) shows one phase of a bridge inverter using this principle, where TH1′ is the auxiliary thyristor used to initiate the commutation of TH1.

Interval A. Suppose TH1 is conducting, thereby connecting terminal A of the load to the positive dc supply line. Capacitor C is already charged to a voltage, V_o, in excess of the supply voltage, V_d, with the right-hand plate positive, from the previous commutation of TH2.

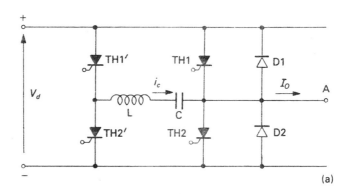

(a)

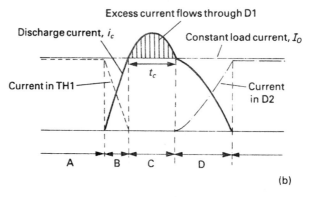

(b)

FIG. 4.48. One leg of a bridge inverter employing the McMurray commutation method: (a) basic circuit; (b) current waveforms with a lagging power factor load.

Interval B. When TH1' is triggered to initiate commutation, capacitor C discharges around the oscillatory circuit consisting of TH1', L, C, and TH1. As the discharge current, i_c, increases from zero on the first half-cycle, it gradually reduces the forward thyristor current to zero, as shown in Fig. 4.48(b).

Interval C. The load current, I_o, is maintained constant by the load inductance, and when the discharge current, i_c, exceeds I_o, the excess current flows through feedback diode D1. Thus, thyristor TH1 has a reverse bias to turn it off equal to the forward voltage drop of the feedback diode (about 1 V). The commutating current, i_c, rises to a peak value when the capacitor voltage is zero, and then decreases as the capacitor is charged in the reverse direction.

Interval D. When i_c falls below the level of I_o, diode D1 ceases to conduct, and diode D2 has to carry the difference between the decreasing current i_c and the constant load current I_o. This clamps load terminal A to the negative supply rail, and the thyristor TH1 must be capable of blocking the applied forward voltage of V_d. At the start of this interval, inductor L carries a current of I_o, and conduction of D2 subjects the LC commutating circuit to a step voltage, V_d, via TH1' and D2. These initial conditions result in a decreasing current i_c, as shown in Fig. 4.48(b), and the magnetic field energy in L is converted to electric field energy in the overcharged capacitor, leaving C with a final voltage, $- V_o$, ready for the next commutation of TH2. All the load current now flows through feedback diode D2, and the commutation transient is complete. The auxiliary thyristor TH1' experiences a reverse voltage and turns off. As usual, the load current subsequently decays to zero and reverses, thereby blocking diode D2 and bringing thyristor TH2 into conduction, provided a gating signal is applied.

In the McMurray circuit, the turn-off time, t_q, of the main thyristor must be less than the circuit turn-off time, t_c, during which discharge current, i_c, exceeds load current I_o. This circuit differs from the McMurray-Bedford circuit in the fact that the reverse voltage applied to the thyristor is limited to a volt or two, and the reverse recovery current of the thyristor is limited to the excess of the commutating current pulse over the load current. In the McMurray-Bedford circuit, the commutating voltage applied across the thyristor has approximately the same magnitude as the supply voltage, and the reverse recovery current is governed by circuit inductance.

Figure 4.48(b) for the McMurray circuit shows that as load current I_o increases, interval C is shortened, and the available circuit turn-off time, t_c, is reduced. This advances the instant at which D2 conducts and enhances the influence of the step voltage, V_d, which is applied to the LC circuit, so that capacitor voltage, V_o, builds up to a higher level over a few commutations. This means that the available commutating current, i_c, is larger, and the commutating capability of the inverter automatically increases with load current. At light load, the commutating current pulse is a minimum, and no-load commutation losses are small. Thus, the McMurray circuit gives a high inverter efficiency, and it can be switched at high frequencies, making it attractive for PWM inverters. For a maximum load current, I_{oo}, and thyristor turn-off time, t_q, the commutating circuit components can be selected on the basis of minimum required energy using the standard expressions[118]

$$L = 0.397 \frac{V_o t_q}{I_{oo}} \tag{4.17}$$

and

$$C = 0.893 \frac{I_{oo} t_q}{V_o} \tag{4.18}$$

where V_o is the steady-state capacitor voltage corresponding to maximum load and minimum supply voltage.

However, as explained above, the McMurray circuit requires several commutations to achieve an increased capacitor voltage, following an increment in load current, and this slow build-up may result in a commutation failure for a sudden increase in load. Consequently, it has been suggested that the capacitor voltage should be stabilized at a high value, independent of load, by advanced firing of main thyristor TH2. If TH2 is gated while excess discharge current, i_c, is still flowing in diode D1, a step voltage V_d, is applied to the LC circuit which permits a controlled build-up of the steady-state capacitor voltage to a value that is adequate for commutation of the maximum load current.[129]

The inherent build-up of capacitor voltage in the McMurray circuit may result in a steady-state voltage that is appreciably higher than the dc supply voltage. In high-voltage inverters, the resulting voltage stresses on the inverter components are unacceptably large, and a modified version of the McMurray circuit is used which sets the capacitor voltage equal to the dc supply voltage, V_d, prior to each commutation.

4.10.3. The Modified McMurray Circuit

The modified McMurray circuit shown in Fig. 4.49 has been widely used in thyristor PWM inverters.[130–132] A resistor, R, and two auxiliary diodes D1′, D2′ are added to the basic McMurray circuit. These components permit the removal of the capacitor overcharge at the end of the commutation process. After commutation of TH1, the excess energy in C, less the energy dissipated in R, is returned to the dc supply by a current flowing through L, R, D1′, the dc source, and D2. The voltage drop across R and D1′

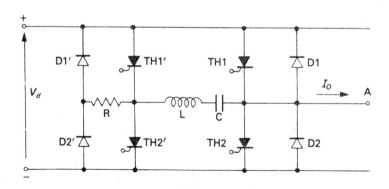

FIG. 4.49. One leg of a bridge inverter employing the modified McMurray commutation method.

commutates TH1'. Resistor R limits the undershoot of capacitor voltage during this discharge interval. If the resistor is large enough to overdamp the series circuit, the capacitor is discharged to the dc supply voltage, V_d, for all values of load current.

Improved operation of the modified McMurray circuit can also be obtained by means of advanced firing of the complementary main thyristor TH2, which must be gated just before D1 stops conducting. [133] Practical circuit design must take into consideration the influence of thyristor snubber circuits and di/dt-limiting inductors on the commutating circuit. [132]

4.10.4. *Input-Circuit Commutation* [134–136]

Instead of commutating each inverter thyristor individually, it is possible to employ a method known as input-circuit commutation, or dc-side commutation, in which the inverter circuit is commutated as a whole. When all conducting thyristors have been turned off, the next set of thyristors in sequence is gated. This commutation technique is also used to extinguish simultaneously all thyristors connected to one rail of the dc supply. Commutation alternates between the positive and negative rails. A typical commutating circuit for this purpose is shown in Fig. 4.50, where it is connected at the input terminals of a three-phase bridge inverter. The commutating circuit consists of four thyristors, THA, THB, THC, and THD, in a bridge connection with a commutating capacitor, C. There are two identical commutating inductors, LP and LN, in the positive and negative supply rails, and these are shunted by free-wheeling diodes DP and DN. It should be noted that the inverter feedback diodes D1 to D6 are connected directly to the input dc terminals.

The operation of the commutating circuit proceeds as follows. Assume that thyristors TH1, TH2, and TH3 are conducting, so that load terminals A and B are connected to the positive dc rail and terminal C is connected to the negative rail. A dc supply current, I_d, is established in the upper and lower commutating inductors, and capacitor C is charged to the dc supply voltage, V_d, with the right-hand plate negative, from previous operation. Suppose that it is desired to commutate thyristor TH1 prior to gating TH4. The gating pulses to the upper thyristors, TH1 and TH3, are removed, but the gating of TH2 continues. Commutation is initiated by simultaneously gating the two auxiliary thyristors, THB and THC, thereby connecting the negative terminal of the charged capacitor to the upper rail of the inverter and diverting the dc current, I_d, through LP, THB, capacitor C, and THC. A voltage of $2V_d$ is impressed across the upper inductor, LP, and the upper thyristors, TH1 and TH3, are subjected to a reverse voltage of V_d because the inductive load currents transfer to the feedback diodes, D4 and D6, thereby clamping terminals A and B to the negative dc supply terminal. Thyristor TH2 continues to conduct the free-wheeling load currents which circulate through TH2, inductor LN, and feedback diodes D4 and D6. Thus, all line-to-line voltages are reduced to zero during this interval.

As explained above, the inverter dc current is diverted into the commutating capacitor, C, when THB and THC are triggered. This current gradually reduces the capacitor voltage to zero and then charges it up with the opposite polarity, leaving it ready for the next commutation of the lower group thyristors. The circuit designer must ensure that the negative voltage on capacitor C is maintained for a time t_c that exceeds the thyristor turn-off time t_q.

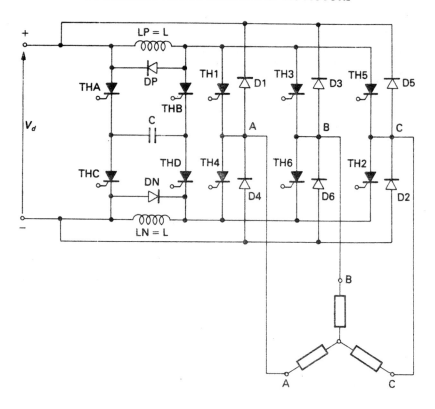

FIG. 4.50. Input-circuit commutation of a three-phase bridge inverter.

During the oscillatory charging of capacitor C, the current in inductor LP increases from its initial value of I_d to a final value of I_m when the capacitor is charged to a voltage V_d with the right-hand plate positive. There is no overshoot of capacitor voltage because the inductor current I_m which was flowing in LP, THB, C, and THC now circulates in inductor LP and free-wheeling diode DP, thereby allowing thyristors THB and THC to turn off. The surplus magnetic field energy in LP, due to its excess current, represents trapped energy that is dissipated in circuit resistance by the circulating current. The commutating circuit is usually designed to reduce this trapped energy to a minimum and so maximize the commutation efficiency. On this basis, the circuit components required to commutate a maximum dc current of I_{do} are[135]

$$L = 1.82 \frac{V_d t_q}{I_{do}}$$

(4.19)

and

$$C = 1.47 \frac{I_{do} t_q}{V_d}.$$

(4.20)

Following successful commutation of the upper group of thyristors, the next inverter conduction sequence is initiated by gating TH3 and TH4, to leave TH2, TH3, and TH4 as the three gated thyristors.

The major advantage of input-circuit commutation is that multiple thyristor turn-off is possible with a reduced number of commutating components, but trapped energy may be appreciable.

4.10.5. *Summary*

The operation of some typical impulse-commutated inverters has been examined in the preceding pages, but many other circuit variations are possible. In practical circuits, voltage and current waveforms will differ somewhat from the idealized waveforms shown above, due to the influence of thyristor and diode reverse recovery currents and the interaction between the commutating circuit and the thyristor snubber circuits. Auxiliary precharging circuits may also be required to prime the commutating capacitors when the inverter is first switched on.

In general, the circuit turn-off time provided by the commutating circuit increases with the size of the commutating capacitor and the voltage to which it is charged, but it also decreases with the current being commutated. If the load current is excessive, sufficient recovery time is not available and a commutation failure occurs, usually resulting in a short-circuit condition. Once the inverter has been constructed with a certain commutating capability, the peak output current must be limited to the design value. A certain amount of overcurrent capacity must be present to prevent a commutation failure on transient overloads, but excessive commutating capability should not be specified, because reduced inverter efficiency under normal operating conditions results. Thus, the commutating capability of an inverter may be increased by redesigning the commutating circuit, but the continuous current rating is unaffected, and the commutation losses at the continuous rating are greater than before.

In many forced-commutated inverters, the commutating capacitors are charged by the inverter supply voltage. In such circuits, a reduction in dc link voltage causes a corresponding reduction in the commutating capability of the inverter. In a six-step inverter drive, the commutating capability is diminished at reduced speeds, when the dc link voltage is low. If the commutating circuit is designed to operate satisfactorily at these reduced link voltages, the commutation losses are excessive at higher voltages and frequencies. For wide-range speed control of the six-step inverter drive, the commutating circuits must be provided with an auxiliary fixed-voltage dc supply. If the inverter is fed by a diode rectifier bridge and series chopper, as in Fig. 4.11(b), the commutating supply can be taken from the input side of the chopper.

4.11. OTHER FORMS OF VOLTAGE-SOURCE INVERTER

The two principal types of VSI are the three-phase six-step and PWM inverters, which have been studied in Sections 4.3 and 4.5, respectively. However, there are other, less common forms of VSI. These forms will now be discussed.

4.11.1. *The Three-Phase 120-Degree-Mode VSI*

As described in Section 4.3, the six-step mode of operation of the three-phase bridge inverter requires a continuous 180-degree firing signal for each of the six main power semiconductor devices. As a result, three devices are turned on at any time. An alternative, but less usual, mode of operation uses the same firing sequence as that of the six-step inverter, but each device is turned on for one-third of a cycle, or 120 degrees, rather than 180 degrees. Consequently, only two devices can conduct simultaneously. The advantage of this mode of operation is that there is a 60-degree period between the turning off of one device and the turning on of the complementary device in the same leg of the inverter. This interval reduces the possibility that a delayed turn-off will result in a shoot-through fault, in which two series devices conduct and short-circuit the dc supply. However, during the 60-degree interval, when neither device in a particular leg of the inverter is conducting, the potential of that output terminal is not uniquely defined. Consequently, the output voltage waveshape of the inverter is load dependent.

The three-phase bridge inverter of Fig. 4.4(b) is repeated in Fig. 4.51(a), where it is shown feeding a balanced resistive load. For 180-degree conduction it is advantageous if the gating of TH4 causes the complementary thyristor, TH1, to turn off. This can be accomplished using the McMurray-Bedford circuit of Fig. 4.46. For the 120-degree-mode of operation, it is desirable that TH1 should be extinguished by gating the adjacent thyristor, TH3, connected to the same dc rail. This can be achieved by means of capacitors connected between the inverter phases, as in the three-phase CSI circuit of Fig. 4.34(a). The six thyristors are gated in the same sequence as before, so that TH6 and TH1 are on for 60 degrees, then TH1 and TH2 for the next 60 degrees, followed by TH2 and TH3, etc. For a balanced three-phase resistive load, the feedback diodes do not conduct, and the pole voltages, v_{AO}, v_{BO} and v_{CO}, are shown in Fig. 4.51(b), (c), and (d), giving the line-to-line voltage, v_{AB}, of Fig. 4.51(e), which has the same waveshape as the line-to-neutral voltage of a wye-connected load fed by a six-step VSI. By Fourier analysis

$$v_{AB} = \frac{3}{\pi} V_d \left(\sin \omega t + \frac{1}{5} \sin 5 \omega t + \frac{1}{7} \sin 7 \omega t \right.$$

$$\left. + \frac{1}{11} \sin 11 \omega t + \frac{1}{13} \sin 13 \omega t + ... \right). \tag{4.21}$$

Hence, the fundamental rms line-to-line voltage is $3 V_d/\sqrt{2}\,\pi$, or $0.675\ V_d$.

If the 120-degree-mode inverter feeds a balanced three-phase RL load, the line voltage waveform is a function of the L/R ratio, or fundamental power factor. When a conducting thyristor is turned off at the end of its 120-degree gating period, the inductive load current transfers to the complementary feedback diode and the pole voltage polarity suddenly reverses. Thus, the commutation of TH1 brings diode D4 into conduction, and the pole voltage changes from $+V_d/2$ to $-V_d/2$, as shown in the v_{AO} pole voltage waveform of Fig. 4.52(a). The feedback diode conducts for an interval whose duration is determined by the L/R ratio of the load. When the diode ceases to conduct, the pole voltage falls suddenly to zero and remains at zero until a thyristor in this inverter leg is gated on. This behavior is evident in the pole voltage waveforms of Fig. 4.52, and the resulting line-to-line voltage waveform for v_{AB} is as indicated in Fig. 4.52(d).

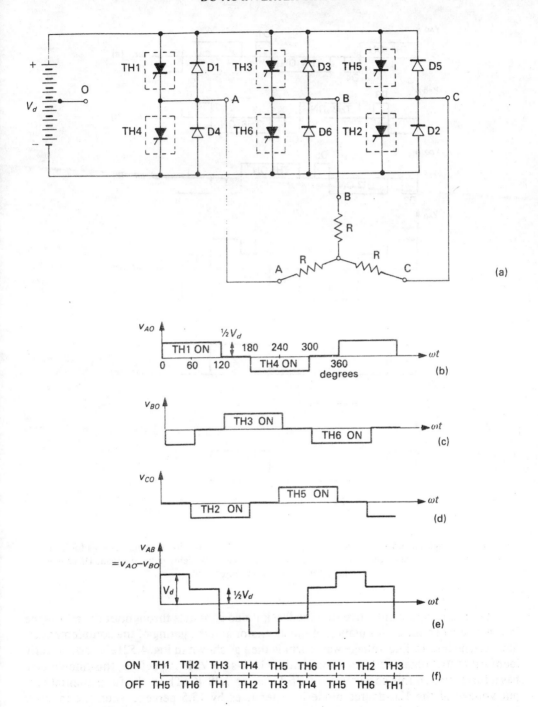

F IG. 4.51. The three-phase bridge voltage-source inverter with 120-degree gating: (a) basic circuit with a balanced resistive load; (b), (c), (d) pole voltages; (e) ac line voltage; (f) gating sequence.

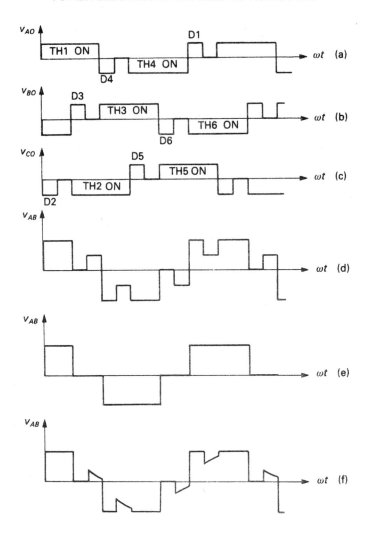

FIG. 4.52. Voltage waveforms for the 120-degree-mode inverter: (a), (b), (c) pole voltages for an RL load; (d) ac line voltage for an RL load; (e) ac line voltage for a highly inductive load; (f) ac line voltage for an induction motor load.

If the load is highly inductive, the feedback diode conducts throughout the 60-degree interval between the commutation of one thyristor and the gating of the complementary device. The line-to-line voltage waveform is then as shown in Fig. 4.52(e), and is clearly identical to that obtained from a conventional six-step VSI. As before, this line voltage has a fundamental rms component of $\sqrt{6}\ V_d/\pi$, or 0.78 V_d. Thus, the fundamental output voltage of the 120-degree-mode inverter rises by 15.5 percent when the inverter loading changes from purely resistive to highly inductive. The fundamental voltage component also undergoes a 30-degree phase shift relative to the gating pulses.

In practice, the line-to-line voltage waveform of Fig. 4.52(e) is obtained when the fundamental power factor of the RL load is 0.55, or less, because the line current is then sustained by the load inductance throughout the 60-degree interval between commutations. At higher power factors, the line current falls to zero at some time within this 60-degree interval. Consequently, the feedback diodes cease to conduct, and the line voltage waveform is as previously shown in Fig. 4.52(d).

When the 120-degree-mode inverter feeds an ac motor, the fundamental voltage applied to the machine varies with shaft loading because of the variation in effective load power factor as seen by the inverter. For an induction motor load, the line-to-line voltage waveform is slightly modified as compared with that for a passive RL load because an open-circuited motor phase has an induced voltage due to its magnetic coupling with the other phase windings of the machine. For discontinuous ac line current, the motor line-to-line voltage waveform is typically like that shown in Fig. 4.52(f). [137–140]

4.11.2. The Stepped-Wave VSI

In the discussion of the three-phase, six-step VSI in Section 4.3, it was pointed out that the low-order output voltage harmonics have undesirable effects in ac motor drive applications. However, the six-step inverter is an elementary form of stepped-wave inverter, and the low-order harmonic content is significantly reduced in more sophisticated stepped waveforms with a larger number of steps. Such waveforms can approximate closely the ideal sinusoidal waveshape, but each change in voltage level corresponds to a switching action in the inverter, and a close approximation to a pure sinusoidal waveshape requires a more complex inverter circuit with additional power semiconductor devices.

In general, assume a sine wave is being simulated by a stepped wave composed of N steps per cycle, each step occupying $2\pi/N$ radians. If the stepped wave is correctly formed, the only harmonics present in the waveform are of order $k = n N \pm 1$, where $n = 1,2,3,4....$ Thus, the principal harmonics present in a six-step wave are the fifth and the seventh. In a 12-step wave, there are no harmonics lower than the eleventh, and an 18-step wave eliminates all harmonics up to the seventeenth.

In order to minimize the total harmonic content of the stepped wave, the voltage level of each step should be equal to the average value in a step interval of the sine wave being simulated. [141] Thus, the height of each step is proportional to

$$v_m = \int_{\theta_{m-1}}^{\theta_m} \sin \theta \, d\theta \tag{4.22}$$

where

$$\theta_{m-1} = (m-1)\frac{2\pi}{N}, \; \theta_m = m\frac{2\pi}{N}, \text{ and } 1 < m < N.$$

This calculation results in a stepped waveform in which the lowest order harmonic is one less than the number of steps per cycle, and the amplitude of the kth harmonic is $1/k^{th}$ of the fundamental amplitude. When these stepped waveforms are used in a three-phase system, a delta connection is possible, because the sum of the instantaneous voltages is always zero, as in the case of pure sine waves.

A 12-step wave has the appearance shown in Fig. 4.53 and has been used in ac motor applications where the harmonic content of the 6-step wave was deemed excessive. It is

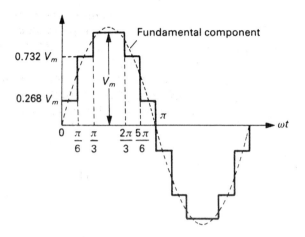

FIG. 4.53. The 12-step voltage waveform.

readily shown by integration that the fundamental amplitude $V_1 = \pi\, V_m/3$, where V_m is the maximum step amplitude (Fig. 4.53). The other step amplitudes are 0.268 V_m and 0.732 V_m. The Fourier series expansion of this 12-step waveform is

$$v = \frac{\pi}{3}\, V_m \left(\sin \omega t + \frac{1}{11} \sin 11\, \omega t + \frac{1}{13} \sin 13\, \omega t \right.$$

$$\left. + \frac{1}{23} \sin 23\, \omega t + \frac{1}{25} \sin 25\, \omega t + \cdots \right). \qquad (4.23)$$

All triplen harmonics are eliminated and the total harmonic content is 15.22 percent of the fundamental, as compared with 31.08 percent for a six-step wave.

Corresponding to the stepped wave defined by Equation 4.22, there is also a second type of stepped wave with a zero voltage or dwell interval of one step width between $-\pi/N$ and $+\pi/N$.[142] The earlier line-to-line and line-to-neutral voltages of Fig. 4.5 and Fig. 4.6 show the two types of six-step waveform, but a corresponding pair of waveforms exists for any value of N. The same harmonics are present in both waveshapes with equal amplitudes, but their relative phase positions are different. This fact is evident from the Fourier series expansions, which show that certain harmonics have opposite phase sense in the two waveforms. Equations 4.5 and 4.8 for the two forms of six-step wave illustrate this point.

A stepped-voltage waveform is generated by adding a number of square-wave or simple pulse-width-modulated voltages so that harmonic neutralization of the objectionable low-order harmonics occurs.[143,144] Figure 4.54 shows a number of single-phase bridge inverters which operate at the same fundamental frequency. These inverters are connected in parallel across the dc source, but the output terminals are series connected as shown. The pulse width of a single-phase inverter output can be adjusted by phase-shifting the gating of one side of the bridge relative to the other, as explained in Section 4.4. If the pulse width of each inverter is suitably adjusted and the output

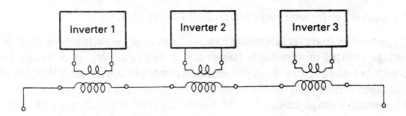

FIG. 4.54. The formation of a stepped voltage waveform by series addition of inverter output voltages.

transformer ratio is correctly selected, the series addition of the secondary voltages gives the desired waveform. Alternatively, all inverters may operate in a square-wave mode with suitable phase displacement between their outputs by means of the firing pulses.

A three-phase, 12-step waveform may be generated by the methods just outlined, but it can also be provided more economically by means of two 6-step inverters which have a phase displacement of 30 degrees and suitable output transformer connections (Fig. 4.55) or external paralleling reactors.[145] Similarly, a three-phase, 24-step output is generated by using four 6-step inverters which are phase displaced by 15 degrees.[146]

Twelve-step waveforms may also be generated by using two 120-degree-mode inverters feeding separate windings on a common transformer core.[147] In motor control applications, the output transformer can be eliminated by combining the inverter outputs in the stator windings of the motor. Two identical isolated three-phase stator winding groups are necessary with a spatial displacement of 30 electrical degrees between them. These winding groups are fed by two 120-degree-mode inverters having a phase displacement of 30 degrees between their gating pulses.[147] Special multilevel inverters have also been suggested as an effective method of generating stepped voltage waveforms.[148,149] These complex schemes are uneconomic for low-power inverters but are more attractive at high power levels and in fixed-frequency uninterruptible power system (UPS) applications.

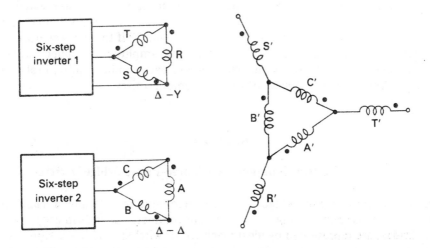

FIG. 4.55. The 3-phase 12-step inverter circuit.

4.11.3. *Stepped-Wave Inverters with Phase-Shift Voltage Control*

The stepped-wave inverters described above require an adjustable-voltage dc link for output voltage control in adjustable-frequency ac motor drives. A constant-voltage dc link has a number of advantages, as discussed in connection with the PWM inverter, and these advantages can be realized if two similar stepped-wave inverters are operated from the same constant-voltage dc supply with their outputs combined in a transformer. Output voltage control is achieved by phase-shifting the output voltage of one inverter relative to the other. This is readily accomplished by introducing a phase displacement between the firing signals of the two inverters, but the harmonic content of the resultant output voltage waveform is high at reduced output voltages.

Two output transformers with their secondary windings connected in series can be used to add the inverter voltages. When the two sets of gating signals are in phase, the fundamental output voltage is twice that of a single inverter. When the gating signals are displaced by an angle, γ, the fundamental output voltage is reduced by a factor $\cos(\gamma/2)$, and a phase displacement of 180 degrees reduces the output voltage to zero. A signal from the inverter reference oscillator can be used to control the phase displacement so that the required output volts/hertz characteristic is obtained.

Figure 4.56 shows the voltage waveforms obtained when the outputs of two six-step inverters are added. With zero phase displacement, the line-to-neutral voltage has a six-step waveshape and the fundamental component has its maximum value. When phase shift is introduced, the voltage waveform changes and the fundamental component is reduced. Waveform quality is inferior to that of the precise stepped waveform described in the preceding section. However, no harmonic which is eliminated from the individual inverter output waveforms is reconstituted when phase-shift control is implemented. Thus, when the two six-step voltage waveforms are combined in a phase-shift inverter, no harmonics of order lower than the fifth will appear in the output voltage. However, the magnitudes of the unneutralized harmonics vary with the phase displacement, γ, and are relatively large compared with the fundamental for large values of γ.

Phase-shift control can be obtained in practice by using the circuit of Fig. 4.57, in which a three-phase output transformer has its primary phases connected between corresponding terminals of two six-step inverters. A six-step output voltage appears at terminals A, B, and C, and also at terminals A', B', and C'. If the two waveforms are 180 degrees out of phase, a six-step voltage of maximum amplitude is applied across the primary windings. When the gating signals of one bridge are phase-shifted relative to the other, a reduced voltage is applied to the transformer primary.

Appendix 4.1

General Voltage Equations for a Three-Phase Bridge Inverter

Consider a three-phase bridge voltage-source inverter feeding a balanced wye-connected load, as shown in Fig. 4.58. The semiconductor devices in each inverter leg, or half-bridge, are represented by ideal mechanical switches, and it is assumed that each half-bridge undergoes the same switching sequence but with a mutual phase displacement of 120 degrees. Consequently, the line voltages relative to the dc mid-point, O,

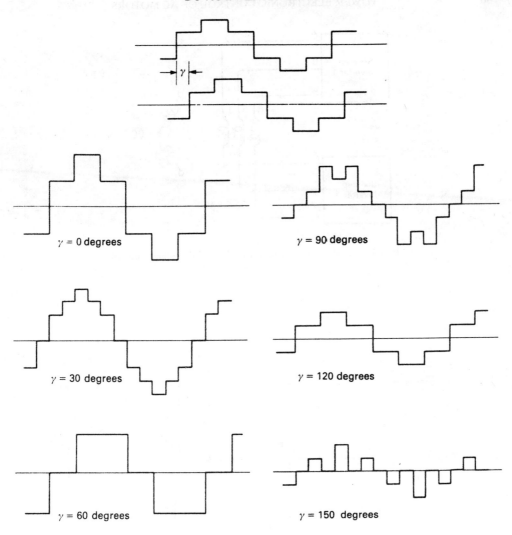

FIG. 4.56. The addition of two six-step voltage waveforms with an adjustable phase displacement, γ.

are mutually displaced in phase by 120 degrees. These are the so-called pole voltages, v_{AO}, v_{BO}, and v_{CO}.

If the neutral point of the load, N, is connected to the dc center-tap, O, then the three half-bridges operate independently, and the line-to-neutral, or phase voltages, of the load are equal to the corresponding pole voltages.

Thus,

$$v_{AN} = v_{AO}$$
$$v_{BN} = v_{BO}$$
$$v_{CN} = v_{CO} .$$

(4.24)

M PEC —G*

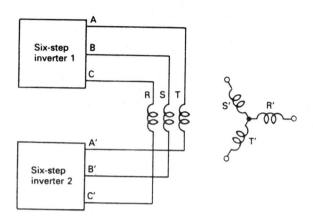

FIG. 4.57. Basic inverter circuit for phase-shift voltage control.

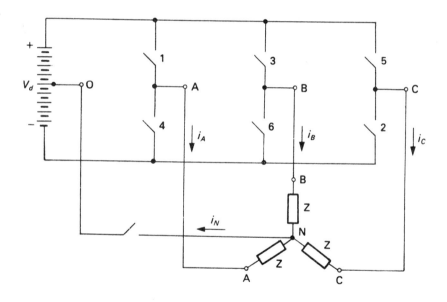

FIG. 4.58. The three-phase bridge voltage-source inverter with a balanced wye-connected load.

In general, these voltages will contain odd and even harmonics, including the triplen harmonics (or zero-sequence harmonics) of order 3,6,9,12..., which are in time phase in the three pole voltages. All other nontriplen voltage harmonics form balanced three-phase systems, with a mutual phase displacement of 120 degrees between the pole voltage components. Each voltage harmonic produces a corresponding current harmonic in the balanced wye-connected load. The instantaneous current in the neutral conductor joining N and O is denoted by i_N, and is the sum of the instantaneous phase currents.

Thus,

$$i_N = i_A + i_B + i_C \qquad (4.25)$$

where the neutral current, i_N, is the arithmetic sum of the in-phase triplen harmonic currents, because all nontriplen currents sum to zero.

In ac motor applications, triplen harmonic currents are undesirable, as they may introduce significant additional I^2R losses. However, these harmonic currents can be suppressed by simply removing the neutral conductor and isolating the neutral point, N. For the resulting three-phase, three-wire load, $i_A + i_B + i_C = 0$, and there is no path for in-phase triplen harmonic currents. The suppression of these harmonic currents in the load means that there are no in-phase triplen voltage harmonics in the line-to-neutral voltages. Hence, the sum of the instantaneous line-to-neutral voltages is zero, so that

$$v_{AN} + v_{BN} + v_{CN} = 0. \qquad (4.26)$$

In general, for the balanced three-phase three-wire wye-connected load, at any instant

$$v_{AO} = v_{AN} + v_{NO}$$
$$v_{BO} = v_{BN} + v_{NO}$$
$$v_{CO} = v_{CN} + v_{NO} \qquad (4.27)$$

where v_{NO} is the instantaneous voltage of the load neutral relative to the dc mid-point.
From Equation 4.27

$$v_{AO} + v_{BO} + v_{CO} = (v_{AN} + v_{BN} + v_{CN}) + 3v_{NO} \qquad (4.28)$$

and substituting Equation 4.26 gives

$$v_{NO} = \frac{1}{3}(v_{AO} + v_{BO} + v_{CO}). \qquad (4.29)$$

In general, therefore

$$v_{AN} = v_{AO} - v_{NO} = \frac{1}{3}(2\,v_{AO} - v_{BO} - v_{CO})$$

$$v_{BN} = \frac{1}{3}(2\,v_{BO} - v_{CO} - v_{AO})$$

$$v_{CN} = \frac{1}{3}(2\,v_{CO} - v_{AO} - v_{BO}). \qquad (4.30)$$

These voltage equations can be used to determine the line-to-neutral voltages for a balanced three-wire wye-connected load or ac motor, when the pole voltages v_{AO}, v_{BO}, and v_{CO} are known.

Each line-to-line voltage is, of course, obtained as the difference of two pole voltages. Thus

$$v_{AB} = v_{AO} - v_{BO}$$
$$v_{BC} = v_{BO} - v_{CO}$$
$$v_{CA} = v_{CO} - v_{AO} \, . \tag{4.31}$$

4.12. REFERENCES

1. KHASAYEV, O.I., Operation of an induction motor from a transistor frequency converter, *Electr. Techn.* U.S.S.R. (Engl. Transl.), 3, 1961, pp. 441-459.

2. MOKRYTZKI, B., Pulse width modulated inverters for ac motor drives, *IEEE Trans. Ind. Gen. Appl.*, IGA-3, 6, Nov./Dec. 1967, pp. 493-503.

3. BELLINI, A., and CIOFFI, G., Induction machine frequency control: Three-phase bridge inverter behavior and performance, *IEEE Trans. Ind. Gen. Appl.*, IGA-7, 4, July/Aug. 1971, pp. 488-499.

4. POLLACK, J.J., Advanced pulse-width modulated inverter techniques, *IEEE Trans. Ind. Appl.*, IA-8, 2, Mar./Apr. 1972, pp. 145-154.

5. ADAMS, R.D., and FOX, R.S., Several modulation techniques for a pulse width modulated inverter, *IEEE Trans. Ind. Appl.*, IA-8, 5, Sept./Oct. 1972, pp. 636-643.

6. SEN, P.C., and PREMCHANDRAN, G., Improved PWM control strategy for inverters and induction motor drives, *Conf. Rec. IEEE Ind. Appl. Soc. Annual Meeting, 1982*, pp. 823-830.

7. SKOGSHOLM, E.A., Efficiency and power factor for a square wave inverter drive, *Conf. Rec. IEEE Ind. Appl. Soc. Annual Meeting, 1978*, pp. 463-469.

8. SCHÖNUNG, A., and STEMMLER, H., Static frequency changers with 'subharmonic' control in conjunction with reversible variable-speed ac drives, *Brown Boveri Rev.*, 51, 8/9, Aug./Sept. 1964, pp. 555-577.

9. BOWES, S.R., New sinusoidal pulse width-modulated inverter, *Proc. IEE*, 122, 11, Nov. 1975, pp. 1279-1285.

10. WILSON, J.W.A., and YEAMANS, J.A., Intrinsic harmonics of idealized inverter PWM systems, *Conf. Rec. IEEE Ind. Appl. Soc. Annual Meeting, 1976*, pp. 967-973.

11. BARTON, T.H., Pulse width modulation waveforms — the Bessel approximation, *Conf. Rec. IEEE Ind. Appl. Soc. Annual Meeting, 1978*, pp. 1125-1130.

12. KING, K.G., A three-phase transistor class-B inverter with sine wave output and high efficiency, IEE Conf. Publ. No. 123, *Power Electronics — Power Semiconductors and Their Applications*, 1974, pp. 204-209.

13. BUJA, G., and INDRI, G., Improvement of pulse width modulation techniques, *Arch. Elektrotech.* (Berlin), 57, 1975, pp. 281-289.

14. KLIMAN, G.B., and PLUNKETT, A.B., Development of a modulation strategy for a PWM inverter drive, *IEEE Trans. Ind. Appl.*, **IA-15**, 1, Jan./Feb. 1979, pp. 72-79.

15. GREEN, R.M., and BOYS, J.T., Implementation of pulsewidth modulated inverter modulation strategies, *IEEE Trans. Ind. Appl.*, **IA-18**, 2, Mar./Apr. 1982, pp. 138-145.

16. ZUBEK, J., ABBONDANTI, A., and NORDBY, C.J., Pulsewidth modulated inverter motor drives with improved modulation, *IEEE Trans. Ind. Appl.*, **IA-11**, 6, Nov./Dec. 1975, pp. 695-703.

17. WILSON, J.W.A., Adaption of pulse-width modulation theory for use in ac motor drive inverters, *IEEE Int. Semicond. Power Converter Conf.*, 1977, pp. 193-199.

18. RAPHAEL, H., Additional losses in PWM inverter-fed squirrel cage motors, *Conf. Rec. IEEE Ind. Appl. Soc. Annual Meeting, 1977*, pp. 932-936.

19. GRANT, D., and SEIDNER, R., Technique for pulse elimination in pulsewidth-modulation inverters with no waveform discontinuity, *IEE Proc.*, **129, Pt. B**, 4, July 1982, pp. 205-210.

20. GREEN, R.M., and BOYS, J.T., PWM sequence selection and optimization: A novel approach, *IEEE Trans. Ind. Appl.*, **IA-18**, 2, Mar./Apr. 1982, pp. 146-151.

21. GRANT, D.A., and HOULDSWORTH, J.A., PWM ac motor drive employing ultrasonic carrier, IEE Conf. Publ. No. 234, *Power Electronics and Variable-Speed Drives*, 1984, pp. 237-240.

22. GRANT, T.L., and BARTON, T.H., Control strategies for PWM drives, *IEEE Trans. Ind. Appl.*, **IA-16**, 2, Mar./Apr. 1980, pp. 211-215.

23. HOULDSWORTH, J.A., and BURGUM, F.J., Induction motor drive using new digital sine-wave PWM system, IEE Conf. Publ. No. 179, *Electrical Variable-Speed Drives*, 1979, pp. 11-14.

24. PLUNKETT, A.B., A current-controlled PWM transistor inverter drive, *Conf. Rec. IEEE Ind. Appl. Soc. Annual Meeting, 1979*, pp. 785-792.

25. PFAFF, G., and WICK, A., Direct current control of ac drives with pulsed frequency converters, *Process Automation*, **2**, 1983, pp. 83-88.

26. HOLTZ, J., and STADTFELD, S., A predictive controller for the stator current vector of ac machines fed from a switched voltage source, *Conf. Rec. Int. Power Electron. Conf.*, Tokyo, 1983, pp. 1665-1675.

27. PFAFF, G., WESCHTA, A., and WICK, A., Design and experimental results of a brushless ac servo-drive, *IEEE Trans. Ind. Appl.*, **IA-20**, 4, July/Aug. 1984, pp. 814-821.

28. BROD, D.M., and NOVOTNY, D.W., Current control of VSI-PWM inverters, *IEEE Trans. Ind. Appl.*, **IA-21**, 4, May/June 1985, pp. 562-570.

29. ANDRIEUX, C., and LAJOIE-MAZENC, M., Analysis of different current control systems for inverter-fed synchronous machine, *Proc. European Power Electronics Conf.*, 1985, pp. 2.159-2.165.

30. TURNBULL, F.G., Selected harmonic reduction in static dc-ac inverters, *IEEE Trans. Commun. and Electron.*, **83**, July 1964, pp. 374-378.

31. PATEL, H.S., and HOFT, R.G., Generalized techniques of harmonic elimination and voltage control in thyristor inverters, Part I — Harmonic elimination, *IEEE Trans. Ind. Appl.*, **IA-9**, 3, May/June 1973, pp. 310-317.

32. PATEL, H.S., and HOFT, R.G., Generalized techniques of harmonic elimination and voltage control in thyristor inverters, Part II — Voltage control techniques, *IEEE Trans. Ind. Appl.*, **IA-10**, 5, Sept./Oct. 1974, pp. 666-673.

33. BUJA, G.S., and INDRI, G.B., Optimal pulsewidth modulation for feeding ac motors, *IEEE Trans. Ind. Appl.*, **IA-13**, 1, Jan./Feb. 1977, pp. 38-44.

34. HALASZ, S., Optimal control of voltage source inverters supplying induction motors, *Proc. 2nd IFAC Symp. on Control in Power Electronics and Electrical Drives*, 1977, pp. 379-385.

35. BELLINI, A., and FIGALLI, G., On the selection of commutation instants in induction motor drives, *IEEE Trans. Ind. Appl.*, **IA-15**, 5, Sept./Oct. 1979, pp. 501-506.

36. BUJA, G.S., Optimum output waveforms in PWM inverters, *IEEE Trans. Ind. Appl.*, **IA-16**, 6, Nov./Dec. 1980, pp. 830-836.

37. PITEL, I.J., TALUKDAR, S.N., and WOOD, P., Characterization of programmed-waveform pulsewidth modulation, *IEEE Trans. Ind. Appl.*, **IA-16**, 5, Sept./Oct. 1980, pp. 707-715.

38. DE CARLI, A., and MAROLA, G., Optimal pulse-width modulation for three-phase voltage supply, *Microelectronics in Power Electronics and Electrical Drives*, VDE Verlag, Darmstadt, 1982, pp. 197-202.

39. MURPHY, J.M.D., and EGAN, M.G., A comparison of PWM strategies for inverter-fed induction motors, *IEEE Trans. Ind. Appl.*, **IA-19**, 3, May/June 1983, pp. 363-369.

40. DE BUCK, F., GISTELINCK, P., and DE BACKER, D., Loss-optimal PWM waveforms for variable-speed induction-motor drives, *IEE Proc.*, **130, Pt. B**, 5, Sept. 1983, pp. 310-320.

41. POLLMANN, A., Comparison of PWM modulation techniques, *Microelectronics in Power Electronics and Electrical Drives*, VDE Verlag, Darmstadt, 1982, pp. 231-236.

42. MENZIES, R.W., and MAHMOUD, A.M.A., Comparison of optimal PWM techniques with a bang-bang reference current controller for voltage source inverter fed induction motor drives, *Conf. Rec. IEEE Ind. Appl. Soc. Annual Meeting, 1983*, pp. 685-692.

43. MAZUR, T., A digital logic PWM speed control for single and polyphase ac motors, *Conf. Rec. IEEE Ind. Appl. Soc. Annual Meeting, 1973*, pp. 1-9.

44. ABBONDANTI, A., A digital modulator circuit for PWM inverters, *Conf. Rec. IEEE Ind. Appl. Soc. Annual Meeting, 1978*, pp. 493-501.

45. EDWARDS, J.D., Three-phase digital PWM inverter, *Proc. IEE*, **122**, 3, Mar. 1975, pp. 302-304.

46. KIRSCHEN, D., MAGGETTO, G., MATHYS, P., and VAN ECK, J.L., Modulateur Digital Pour Un Ondulateur a Thyristors Ultra-Rapids, *Conumel 80*, Lyon, pp. 11-22/11-30.

47. POLLMANN, A., A digital pulse width modulator employing advanced modulation techniques, *IEEE Int. Semicond. Power Converter Conf.*, 1982, pp. 116-121.

48. LE-HUY, H., A microprocessor-controlled pulse-width modulated inverter. *Proc. IEEE IECI, Ind. Appl. Microproc.*, 1978, pp. 223-226.

49. CASTEEL, J.B., and HOFT, R.G., Optimum PWM waveforms of a microprocessor controlled inverter, *IEEE Power Electron. Spec. Conf.*, 1978, pp. 243-250.

50. GRANT, T.L., and BARTON, T.H., A highly flexible controller for a pulse width modulation inverter, *Conf. Rec. IEEE Ind. Appl. Soc. Annual Meeting, 1978*, pp. 486-492.

51. MURPHY, J.M.D., HOWARD, L.S., and HOFT, R.G., Microprocessor control of a PWM inverter induction motor drive, *IEEE Power Electron. Spec. Conf.*, 1979, pp. 344-348.

52. DWYER, E., and OOI, B.T., A lookup table based microprocessor controller for a three-phase PWM inverter, *Proc. IEEE IECI, Ind. and Control Appl. Microproc.*, 1979, pp. 19-22.

53. HUMBLET, L.C.P., DE BUCK, F.G.G., VERBEKE, B., and DE VALCK, P., A realization example of a microprocessor-driven PWM transistor inverter, IEE Conf. Publ. No. 179, *Electrical Variable-Speed Drives*, 1979, pp. 151-156.

54. MURPHY, J.M.D., HOFT, R.G., and HOWARD, L.S., Controlled-slip operation of an induction motor with optimum PWM waveforms, IEE Conf. Publ. No. 179, *Electrical Variable-Speed Drives*, 1979, pp. 157-160.

55. SONE, S., and HORI, Y., Harmonic elimination of microprocessor controlled PWM inverter for electric traction, *Proc. IEEE IECI, Ind. and Control Appl. Microproc.*, 1979, pp. 278-283.

56. POLLMANN, A., and GABRIEL, R., Zündsteurung eines Pulswechselrichters mit Mikrorechner, *Regelungstech. Prax.*, **22**, 5, 1980, pp. 145-150.

57. BRICKWEDDE, A., HEAD, D., and GRAHAM, H., Microprocessor controlled 50 kVA PWM inverter motor drive, *Conf. Rec. IEEE Ind. Appl. Soc. Annual Meeting, 1981*, pp. 666-670.

58. BOWES, S.R., and MOUNT, M.J., Microprocessor control of PWM inverters, *IEE Proc.*, **128, Pt. B**, 6, Nov. 1981, pp. 293-305.

59. RAJASHEKARA, K.S., and VITHAYATHIL, J., Microprocessor-based sinusoidal PWM inverter by DMA transfer, *IEEE Trans. Ind. Electron.*, **IE-29**, 1, Feb. 1982, pp. 46-51.

60. BOSE, B.K., and SUTHERLAND, H.A., A high-performance pulsewidth modulator for an inverter-fed drive system using a microcomputer, *IEEE Trans. Ind. Appl.*, **IA-19**, 2, Mar./Apr. 1983, pp. 235-243.

61. VARNOVITSKY, M., A microcomputer-based control signal generator for a three-phase switching power inverter, *IEEE Trans. Ind. Appl.*, **IA-19**, 2, Mar./Apr. 1983, pp. 228-234.

62. ZACH, F.C., BERTHOLD, R.J., and KAISER, K.H., General purpose microprocessor modulator for a wide range of PWM techniques for ac motor control, *Conf. Rec. IEEE Ind. Appl. Soc. Annual Meeting, 1982*, pp. 446-451.

63. GRANT, D., STEVENS, M., and HOULDSWORTH, J., The effect of word length on the harmonic content of microprocessor-based PWM waveform generators, *Conf. Rec. IEEE Ind. Appl. Soc. Annual Meeting, 1983*, pp. 643-647.

64. THOMAS, G., and LIM, K.M., Recent developments in microprocessor control of variable-speed inverter-fed ac motors, IEE Conf. Publ. No. 234, *Power Electronics and Variable-Speed Drives*, 1984, pp. 241-244.

65. BUJA, G.S., and FIORINI, P., Microcomputer control of PWM inverters, *IEEE Trans. Ind. Electron.*, **IE-29**, 3, Aug. 1982, pp. 212-216.

66. BOLOGNANI, S., BUJA, G.S., and LONGO, D., Hardware and performance-effective microcomputer control of a three-phase PWM inverter, *Trans. IEE of Japan*, **104**, 5/6, May/June 1984, pp. 101-108.

67. FORSYTHE, J.B., and DEWAN, S.B., Output current regulation with PWM inverter-induction motor drives, *IEEE Trans. Ind. Appl.*, **IA-11**, 5, Sept./Oct. 1975, pp. 517-525.

68. DEWAN, S.B., and GALLANT, K., Analysis of a single phase current source inverter, *Conf. Rec. IEEE Ind. Appl. Soc. Annual Meeting, 1974*, pp. 1025-1036.

69. DEWAN, S.B., ROSENBERG, S.A., and NICHOLSON, N.M., Comparison of single phase current source inverter configurations, *Conf. Rec. IEEE Ind. Appl. Soc. Annual Meeting, 1975*, pp. 783-788.

70. PHILLIPS, K.P., Current-source converter for ac motor drives, *IEEE Trans. Ind. Appl.*, **IA-8**, 6, Nov./Dec., 1972, pp. 679-683.

71. MAAG, R.B., Characteristics and application of current source/slip regulated ac induction motor drives, *Conf. Rec. IEEE Ind. Gen. Appl. Group Annual Meeting, 1971*, pp. 411-416.

72. FARRER, W., and MISKIN, J.D., Quasi-sine wave fully regenerative inverter, *Proc. IEE*, **120**, 9, Sept. 1973, pp. 969-976.

73. MANN, S., A current-source converter for multi-motor applications, *Conf. Rec. IEEE Ind. Appl. Soc. Annual Meeting, 1975*, pp. 980-984.

74. WALKER, L.H., and ESPELAGE, P.M., A high-performance controlled-current inverter drive, *IEEE Trans. Ind. Appl.*, IA-16, 2, Mar./Apr., 1980, pp. 193-202.

75. AKAMATSU, M., IKEDA, H., TOMEI, T., and YANO, S., High performance IM drive by co-ordinate control using a controlled inverter, *IEEE Trans. Ind. Appl.*, **IA-18**, 4, July/Aug. 1982, pp. 382-392.

76. SLEMON, G.R., DEWAN, S.B., and WILSON, J.W.A., Synchronous motor drive with current-source inverter, *IEEE Trans. Ind. Appl.*, **IA-10**, 3, May/June 1974, pp. 412-416.

77. HARASHIMA, F., NAITOH, H., and HANEYOSHI, T., Dynamic performance of self controlled synchronous motors fed by current source inverters, *IEEE Trans. Ind. Appl.*, **IA-15**, 1, Jan./Feb. 1979, pp. 36-46.

78. ONG, C.M., and LIPO, T.A., Steady-state analysis of a current source inverter/reluctance motor drive, Part I: Analysis, Part II: Experimental and analytical results, *Conf. Rec. IEEE Ind. Appl. Soc. Annual Meeting, 1975*, pp. 841-851.

79. BRENNEN, M.B., A comparative analysis of two commutation circuits for adjustable current input inverters feeding induction motors, *IEEE Power Electron. Spec. Conf.*, 1973, pp. 201-212.

80. LIENAU, W., MÜLLER-HELLMANN, A., and SKUDELNY, H-C., Power converters for feeding asynchronous traction motors of single-phase ac vehicles, *IEEE Trans. Ind. Appl.*, **IA-16**, 1, Jan./Feb. 1980, pp. 103-110.

81. VAN WYK, J.D., and HOLTZ, H.R., A simple and reliable four quadrant variable frequency ac drive for industrial application up to 50 kW, IEE Conf. Publ. No. 179, *Electrical Variable-Speed Drives*, 1979, pp. 34-37.

82. FEUILLET, R., IVANES, M., and LE-HUY, H., A forced-commutation circuit through smoothing inductor for current source inverters, *Conf. Rec. IEEE Ind. Appl. Soc. Annual Meeting, 1983*, pp. 666-670.

83. OSMAN, R.H., An improved circuit for accelerating commutation in current source inverters, *IEEE Trans. Ind. Appl.*, **IA-20**, 5, Sept./Oct. 1984, pp. 1296-1300.

84. SUBRAHMANYAM, S., YUVARAJAN, S., and RAMASWAMI, B., Analysis and commutation of a current inverter feeding an induction motor load, *IEEE Trans. Ind. Appl.*, **IA-16**, 3, May/June 1980, pp. 332-341.

85. WARD, E.E., Inverter suitable for operation over a range of frequency, *Proc. IEE*, **111**, 8, Aug. 1964, pp. 1423-1434.

86. LIENAU, W., Commutation modes of a current-source inverter, *Proc. 2nd IFAC Symp. on Control in Power Electronics and Electrical Drives*, 1977, pp. 219-229.

87. MOLL, K., and SCHRÖDER, D., Applicable frequency range of current source converters, *Proc. 2nd IFAC Symp. on Control in Power Electronics and Electrical Drives*, 1977, pp. 231-234.

88. PARASURAM, M.K., and RAMASWAMI, B., Analysis and design of a current-fed inverter, *Proc. 2nd IFAC Symp. on Control in Power Electronics and Electrical Drives*, 1977, pp. 235-245.

89. LAZAR, J., The operational modes of a current source inverter induction motor drive system, *Proc. 2nd IFAC Symp. on Control in Power Electronics and Electrical Drives*, 1977, pp. 443-454.

90. REVANKAR, G.N., and BASHIR, A., Effect of circuit and induction motor parameters on current source inverter operation, *IEEE Trans. Ind. Electron. and Control Instrum.*, **IECI-24**, 1, Feb. 1977, pp. 126-132.

91. SHOWLEH, M., MASLOWSKI, W.A., and STEFANOVIC, V., An exact modeling and design of current source inverters, *Conf. Rec. IEEE Ind. Appl. Soc. Annual Meeting, 1979*, pp. 439-459.

92. CORNELL, E.P., and LIPO, T.A., Modeling and design of controlled current induction motor drive systems, *IEEE Trans. Ind. Appl.*, **IA-13**, 4, July/Aug. 1977, pp. 321-330.

93. PLUNKETT, A.B., D'ATRE, J.D., and LIPO, T.A., Synchronous control of a static ac induction motor drive, *IEEE Trans. Ind. Appl.*, **IA-15**, 4, July/Aug. 1979, pp. 430-437.

94. KRISHNAN, R., MASLOWSKI, W.A., and STEFANOVIC, V.R., Control principles in current source induction motor drives, *Conf. Rec. IEEE Ind. Appl. Soc. Annual Meeting, 1980*, pp. 605-617.

95. KRISHNAN, R., LINDSAY, J.F., and STEFANOVIC, V.R., Design of angle-controlled current source inverter-fed induction motor drive, *IEEE Trans. Ind. Appl.*, **IA-19**, 3, May/June 1983, pp. 370-378.

96. LIENAU, W., Torque oscillations in traction drives with current-fed asynchronous machines, IEE Conf. Publ. No. 179, *Electrical Variable-Speed Drives*, 1979, pp. 102-107.

97. ZUBEK, J., Evaluation of techniques for reducing shaft cogging in current fed ac drives, *Conf. Rec. IEEE Ind. Appl. Soc. Annual Meeting, 1978*, pp. 517-524.

98. CHIN, T.H., and TOMITA, H., The principles of eliminating pulsating torque in current source inverter-induction motor systems, *Conf. Rec. IEEE Ind. Appl. Soc. Annual Meeting, 1978*, pp. 910-917.

99. LIPO, T.A., Analysis and control of torque pulsations in current fed induction motor drives, *IEEE Power Electron. Spec. Conf., 1978*, pp. 89-96.

100. CHIN, T.H., A new controlled current type inverter with improved performance, *IEEE Int. Semicond. Power Converter Conf.*, 1977, pp. 185-192.

101. NABAE, A., SHIMAMURA, T., and KUROSAWA, R., A new multiple current-source inverter, *IEEE Int. Semicond. Power Converter Conf.*, 1977, pp. 200-203.

102. PALANIAPPAN, R., and VITHAYATHIL, J., The current-fed twelve-step current wave inverter, *IEEE Trans. Ind. Electron. and Control Instrum.*, **IECI-25**, 4, Nov. 1978, pp. 377-379.

103. WIART, A., Evaluation de la composante pulsatoire du couple des moteurs a courant alternatif alimentes par onduleurs, *Revue Jeumont-Schneider*, **26**, Aug. 1978, pp. 29-40.

104. LIPO, T.A., and WALKER, L.H., Design and control techniques for extending high frequency operation of a CSI induction motor drive, *IEEE Trans. Ind. Appl.*, **IA-19**, 5, Sept./Oct. 1983, pp. 744-753.

105. OWEN, E.L., and WEISS, H.W., Efficient and reliable synchronous motors for large ac adjustable-speed drives, *American Power Conf. 43rd Annual Meeting, Chicago, IL, April 1981*, pp. 1-21.

106. PETERSON, T., and FRANK, K., Starting of large synchronous motor using static frequency converter, *IEEE Trans. Power Appar. Syst.*, **PAS-91**, 1, Jan./Feb. 1972, pp. 172-179.

107. MEYER, H.G., Static frequency changers for starting synchronous machines, *Brown Boveri Rev.*, **51**, Aug./Sept. 1964, pp. 526-530.

108. HOSONO, I., YANO, M., KANDA, M., KATSUKI, K., and HISSATOMI, S., Static converter starting of large synchronous motors, *Conf. Rec. IEEE Ind. Appl. Soc. Annual Meeting, 1976*, pp. 536-543.

109. IMAI, K., New applications of commutatorless motor systems for starting large synchronous motors, *IEEE Int. Semicond. Power Converter Conf., 1977*, pp. 237-246.

110. HABÖCK, A., and KÖLLENSPERGER, D., State of development of converter fed synchronous motors with self control, *Siemens Rev.*, **38**, 9, 1971, pp. 390-395.

111. SHINRYO, Y., HOSONO, I., and SYOJI, K., Commutatorless dc drive for steel rolling mill, *Conf. Rec. IEEE Ind. Appl. Soc. Annual Meeting, 1977*, pp. 263-271.

112. MUELLER, B., SPINANGER, T., and WALLSTEIN, D., Static variable frequency starting and drive system for large synchronous motors *Conf. Rec. IEEE Ind. Appl. Soc. Annual Meeting, 1979*, pp. 429-438.

113. ROSA, J., Utilization and rating of machine commutated inverter-synchronous motor drives, *IEEE Trans. Ind. Appl.*, **IA-15**, 2, Mar./Apr. 1979, pp. 155-164.

114. OWEN, E.L., MORACK, M.M., HERSKIND, C.C., and GRIMES, A.S., AC adjustable-speed drives with electronic power converters — the early days, *Conf. Rec. IEEE Ind. Appl. Soc. Annual Meeting, 1982*, pp. 854-861.

115. PLUNKETT, A.B., and TURNBULL, F.G., Load commutated inverter/synchronous motor drive without a shaft position sensor, *IEEE Trans. Ind. Appl.*, **IA-15**, 1, Jan./Feb. 1979, pp. 63-71.

116. PLUNKETT, A.B., and TURNBULL, F.G., System design method for a load commutated inverter synchronous motor drive, *Conf. Rec. IEEE Ind. Appl. Soc. Annual Meeting, 1978*, pp. 812-819.

117. STEIGERWALD, R.L., and LIPO, T.A., Analysis of a novel forced-commutation starting scheme for a load-commutated synchronous motor drive, *IEEE Trans. Ind. Appl.*, **IA-15**, 1, Jan./Feb. 1979, pp. 14-24.

118. BEDFORD, B.D., and HOFT, R.G., *Principles of Inverter Circuits*, J. Wiley, New York, NY, 1964.

119. GRAFHAM, D.R., and GOLDEN, F.B. (Editors), *SCR Manual*, Sixth Edition, General Electric Co., 1979.

120. HUMPHREY, A.J., Inverter commutation circuits, *IEEE Trans. Ind. Gen. Appl.*, **IGA-4**, 1, Jan./Feb. 1968, pp. 104-110.

121. VERHOEF, A., Basic forced commutated inverters and their characteristics, *IEEE Trans. Ind. Appl.*, **IA-9**, 5, Sept./Oct. 1973, pp. 601-606.

122. DUBEY, G.K., Classification of thyristor commutation methods, *IEEE Trans. Ind. Appl.*, **IA-19**, 4, July/Aug. 1983, pp. 600-606.

123. ABBONDANTI, A., and WOOD, P., A criterion for performance comparison between high power inverter circuits, *IEEE Trans. Ind. Appl.*, **IA-13**, 2, Mar./Apr. 1977, pp. 154-160.

124. DAVIS, R.M., and MELLING, J.R., Quantitative comparison of commutation circuits for bridge inverters, *Proc. IEE*, **124**, 3, Mar. 1977, pp. 237-246.

125. WAGNER, C.F., Parallel inverter with resistive load, *Trans. AIEE*, **54**, Nov. 1935, pp. 1227-1235.

126. WAGNER, C.F., Parallel inverter with inductive load, *Trans. AIEE*, **55**, Sept. 1936, pp. 970-980.

127. NAMUDURI, C., and SEN, P.C., On inverter circuits with least trapped energy, *Conf. Rec. IEEE Ind. Appl. Soc. Annual Meeting, 1983*, pp. 874-883.

128. McMURRAY, W., and SHATTUCK, D.P., A silicon-controlled rectifier inverter with improved commutation, *Trans. AIEE, Pt. 1, Commun. and Electron.*, **80**, 1961, pp. 531-542.

129. McMURRAY, W., SCR inverter commutated by an auxiliary impulse, *IEEE Trans. Commun. and Electron.*, **83**, Nov. 1964, pp. 824-829.

130. PENKOWSKI, L.J., and PRUZINSKY, K.E., Fundamentals of a pulsewidth modulated power circuit, *IEEE Trans. Ind. Appl.*, **IA-8**, 5, Sept./Oct. 1972, pp. 584-592.

131. DIVAN, D.M., and BARTON, T.H., Commutation circuit optimization for the McMurray inverter, *Conf. Rec. IEEE Ind. Appl. Soc. Annual Meeting, 1979*, pp. 394-398.

132. GREEN, R.M., and BOYS, J.T., A reconsideration of the design of the modified McMurray inverter circuit, *IEEE Trans. Ind. Appl.*, **IA-18**, 2, Mar./Apr. 1982, pp. 152-162.

133. BHAGWAT, P.M., and STEFANOVIC, V.R., Some new aspects in the design of PWM inverters, *IEEE Trans. Ind. Appl.*, **IA-20**, 4, July/Aug. 1984, pp. 776-784.

134. BRADLEY, D.A., CLARKE, C.D., DAVIS, R.M., and JONES, D.A., Adjustable-frequency inverters and their application to variable-speed drives, *Proc. IEE*, **111**, 11, Nov. 1964, pp. 1833-1846.

135. DEWAN, S.B., and DUFF, D.L., Optimum design of an input-commutated inverter for ac motor control, *IEEE Trans. Ind. Gen. Appl.*, **IGA-5**, 6, Nov./Dec. 1969, pp. 699-705.

136. DEWAN, S.B., and DUFF, D.L., Practical considerations in the design of commutation circuits for choppers and inverters, *Conf. Rec. IEEE Ind. Gen. Appl. Group Annual Meeting, 1969*, pp. 469-475.

137. JONES, B.L., and CORY, B.J., Polyphase thyristor inverters, IEE Conf. Publ. No. 10, *The Application of Large Industrial Drives*, May 1965, pp. 241-252.

138. SABBAGH, E.M., and SHEWAN, W., Characteristics of an adjustable speed polyphase induction machine, *IEEE Trans. Power Appar. Syst.*, **PAS-87**, 3, Mar. 1968, pp. 613-624.

139. LIPO, T.A., and TURNBULL, F.G., Analysis and comparison of two types of square-wave inverter drives, *IEEE Trans. Ind. Appl.*, **IA-11**, 2, Mar./Apr. 1975, pp. 137-147.

140. RAMAMOORTY, M., Steady state analysis of inverter driven induction motors using harmonic equivalent circuits, *Conf. Rec. IEEE Ind. Appl. Soc. Annual Meeting, 1973*, pp. 437-440.

141. ANDERSON, D.L., WILLIS, A.E., and WINKLER, C.E., Advanced static inverter utilizing digital techniques and harmonic cancellation, *NASA Tech. Note* D-602, 1962.

142. COREY, P.D., Methods for optimizing the waveform of stepped wave static inverters, AIEE Conf. Paper No. CP 62-1147, 1962.

143. KERNICK, A., ROOF, J.L., and HEINRICH, T.M., Static inverter with neutralization of harmonics, *Trans. AIEE, Pt. 2, Appl. and Ind.*, **81**, May 1962, pp. 59-68.

144. SRIRAGHAVEN, S.M., PRADHAN, B.D., and REVANKAR, G.N., Multistage pulse-width-modulated inverter system for generating stepped-voltage waveforms, *Proc. IEE*, **125**, 6, June 1978, pp. 529-530.

145. HUMPHREY, A.J., and MOKRYTZKI, B., Inverter paralleling reactors, *IEEE Int. Semicond. Power Converter Conf.*, May 1972, pp. 2-4-1 to 2-4-6.

146. FLAIRTY, C.W., A 50 kVA adjustable frequency 24 phase controlled rectifier inverter, *Direct Current*, **6**, Dec. 1961, pp. 278-282.

147. WILSON, J.W.A., Double bridge inverters with magnetic coupling — Part 1: Voltage waveforms, *Conf. Rec. IEEE Ind. Appl. Soc. Annual Meeting, 1976*, pp. 1107-1113.

148. BHAGWAT, P.M., and STEFANOVIC, V.R., Generalized structure of a multilevel PWM inverter, *IEEE Trans. Ind. Appl.*, **IA-19**, 6, Nov./Dec. 1983, pp. 1057-1069.

149. OGUCHI, K., and OHTA, M., An improved three-phase multistepped-voltage inverter combined with a single-phase inverter through switching devices, *IEEE Trans. Ind. Appl.*, **IA-20**, 3, May/June 1984, pp. 656-666.

CHAPTER 5

AC-AC Cycloconverter Circuits

5.1. INTRODUCTION

In a cycloconverter, the alternating voltage at supply frequency is converted directly to an alternating voltage at load frequency without any intermediate dc stage. In a line-commutated cycloconverter, the supply frequency is greater than the load frequency. The operating principles were developed in the 1930s, when the grid-controlled mercury-arc rectifier became available. The techniques were applied in Germany, where the three-phase 50 Hz supply was converted to a single-phase ac supply at 16-2/3 Hz for railway traction.[1,2] In the United States, a 400-horsepower scheme in which a synchronous motor was supplied from a cycloconverter comprising 18 thyratrons was in operation for several years as a power station auxiliary drive.[3-5] However, because these early schemes were not sufficiently attractive technically or economically, they were discontinued.

Subsequent invention of the thyristor and the development of reliable transistorized and microprocessor-based control circuitry have led to a revival of interest in the cyclo-conversion principle. Sophisticated control circuits permit the conversion of a fixed input frequency to an adjustable output frequency at an adjustable voltage, and such schemes are attractive for ac motor drives.[6-17] The ruggedness and low weight of the solid-state cycloconverter also make it attractive for aircraft electrical systems, which require the production of a constant output frequency from a variable-speed alternator.[18,19] In addition, cycloconverter systems that drive wound-field- or permanent magnet-excited synchronous motors can use the motor-generated, or back, emf for thyristor commutation and so provide an output frequency greater than the supply frequency.[20,21] Greater flexibility of output frequency and supply power factor can be achieved if the power semiconductors are commutated independently of either the input or output ac voltages. This independence can be achieved by external impulse commutating circuits using capacitors and auxiliary semiconductors,[22] or by using power semiconductors with turn-off capability, such as power transistors, power MOSFETs, or gate turn-off thyristors. Several textbooks have described cycloconverter systems, circuits, and controls in detail.[22-25]

5.2. BASIC PRINCIPLES OF OPERATION

The line-commutated cycloconverter consists of a number of phase-controlled converter circuits connected to an ac supply system that provides the voltages necessary for natural commutation. The individual circuits are controlled so that a low-frequency output voltage waveform is fabricated from segments of the polyphase input voltages.

199

Consider a three-phase, half-wave, phase-controlled converter supplying an inductive load which maintains continuous current flow, and for simplicity, neglect thyristor on-state voltage and commutation overlap. With these simplifications, it is shown in Chapter 3 that the average voltage output is given by Equation 3.3:

$$V_d = V_{do} \cos \alpha$$

where α is the firing angle, or delay angle, and V_{do} is the average output voltage with zero firing delay. This ability to control output voltage by delayed commutation is the basic property on which naturally commutated, phase-controlled cycloconverter operation is based.

Assume that the converter firing angle is slowly varied as shown in Fig. 5.1. At point

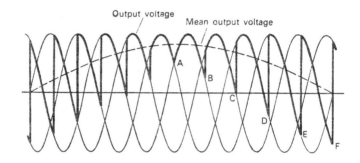

FIG. 5.1. Sinusoidal variation of the average output voltage of a phase-controlled rectifier.

A, there is zero delay, and the average output voltage has its maximum value, V_{do}. At B, the output voltage is slightly reduced by the introduction of a small firing delay. Further reduction is obtained at C, D, and E, while at F the firing angle is 90 degrees and V_d is zero. Thus, if the gating circuitry is suitably designed, a low-frequency sinusoidal variation may be superimposed on the output voltage, V_d. In Fig. 5.1, the rectifier is assumed to conduct during intervals of negative output voltage. This means that the circuit is temporarily inverting, and returning energy from the reactive load to the ac supply. However, for firing angles between zero and 90 degrees, the net power flow is from the ac supply into the load.

The average back emf of a phase-controlled converter operating as an inverter may also be controlled in a sinusoidal manner by suitable variation of the firing angle between 90 and 180 degrees. In Fig. 5.2, the back emf has its maximum value, $-V_{do}$, at point K, where α is 180 degrees. By reducing α, the back emf is also reduced, as shown at points L, M, N, O, P, and Q. For inverter operation, the net power flow is into the ac supply, and a voltage must be available to force current flow against the inverter back emf. In the cycloconverter circuit, this voltage is provided by the induced emf in a reactive load or by regenerative operation of an ac motor.

If the firing angle is varied from zero to 180 degrees and back again to zero, one complete cycle of the low-frequency variation is superimposed on the average output voltage. The superimposed frequency is determined solely by the rate of variation of α and is independent of the supply frequency. Figure 5.3 shows the production of a complete cycle of the low-frequency waveform. This diagram emphasizes the fact that the cyclo-converter is basically a switching arrangement. Each thyristor switch opens and closes at

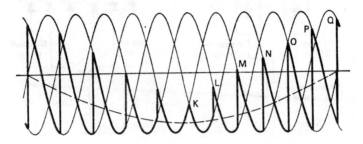

FIG. 5.2. Sinusoidal variation of the average counter emf of a phase-controlled inverter.

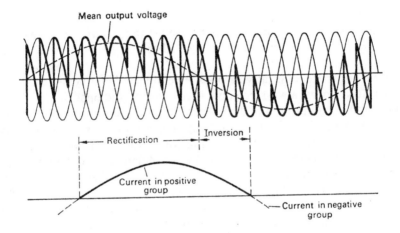

FIG. 5.3. Voltage and current waveforms for the positive group of a phase-controlled cycloconverter when feeding an inductive load at 0.6 power factor.

suitable instants, so that a low-frequency output waveshape is fabricated from segments of the input waveform. The harmonic content of the output voltage waveform decreases as the ratio of output to input frequency is reduced and as the number of supply phases is increased.

The average output voltage of a phase-controlled converter can, therefore, be varied sinusoidally through a complete cycle by suitable variation of the delay angle. However, current can only flow in one direction through the circuit, and, in order to produce a complete cycle of low-frequency output current, two similar circuits must be connected in inverse parallel. The positive converter group permits current flow during the positive half-cycle of the low-frequency output voltage wave, while the negative group allows current flow during the negative voltage half-cycle. The arrangement is shown schemati-cally in Fig. 5.4(a), where the three-phase supply produces a single-phase low-frequency

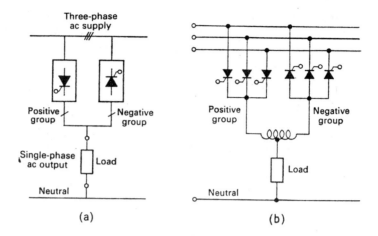

FIG. 5.4. Three-phase to single-phase cycloconverter: (a) schematic diagram; (b) basic circuit.

output. Any converter configuration may be used in the positive and negative group circuits. The simplest arrangement uses two three-phase half-wave circuits, as in Fig. 5.4(b).

Because the positive and negative groups are connected in inverse parallel, their average output voltages must always be equal in magnitude and opposite in sign, in order to avoid large circulating currents at the output frequency. These conditions are achieved when the delay angles of the positive and negative groups, α_P and α_N, are related by the formula $\alpha_P = \pi - \alpha_N$. However, the instantaneous output voltages of the two groups are quite different, and large harmonic currents will circulate around the low-impedance circuit unless they are limited by the use of a center-tapped reactor to combine the outputs, as shown in Fig. 5.4(b). Alternatively, circulating currents may be suppressed by removal of gating signals from each group for the half-cycle during which the group is not delivering load current. The elimination of circulating currents is examined in more detail in Section 5.3.3.

When a three-phase output is required, three single-phase cycloconverters with a phase displacement of 120 degrees between their outputs are connected as shown in Fig. 5.5(a). With a balanced load, the neutral connection is no longer necessary and may be omitted, thereby suppressing all triplen harmonics. The simplest arrangement, using three-phase half-wave circuits, is shown in Fig. 5.5(b). This circuit requires 18 thyristors, but if each group consists of a three-phase bridge or six-phase circuit, then 36 thyristors are required.

It has been assumed that the output frequency of the cycloconverter is less than the supply frequency. In practice, with the basic cycloconverter circuit of Fig. 5.5(b), reasonable power output, efficiency, and harmonic content are obtained in the stepdown region with an output frequency from zero to about one-third of the input frequency. As the cycloconverter output frequency approaches the supply frequency, harmonic distortion in the output voltage increases, because the output voltage waveform is composed of fewer segments of the supply voltage. As a result, losses in the cycloconverter and in the ac motor become excessive, and there is a drop in overall efficiency. By using more

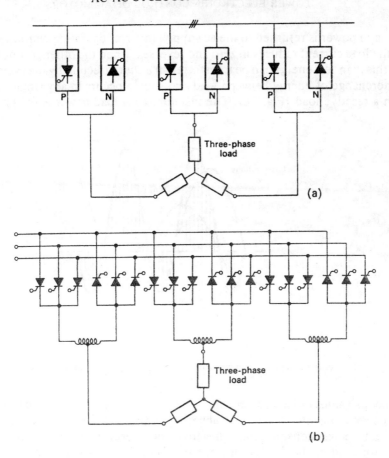

FIG. 5.5. Three-phase to three-phase cycloconverter: (a) schematic diagram; (b) basic circuit.

complex converter circuits with higher pulse numbers, the output voltage waveform is improved and the maximum useful ratio of output to input frequency is increased to about one-half. The ac motor normally presents a high impedance at the ripple frequency, and hence the output current is nearly sinusoidal and no additional filtering is necessary. Systems designed for aircraft power supplies that operate with a fixed output frequency (400 Hz) are provided with an output filter.

5.2.1. *Operation with Reactive Loads*

In a cycloconverter, the positive and negative converter groups must each supply a half-cycle of the low-frequency output current. When a resistive load is being supplied, there is no need for inverter operation, because the positive group supplies load current during the positive half-cycle of output voltage, and the negative group conducts for the negative half-cycle. When an inductive load is being supplied, the low-frequency current lags the output voltage. This means that each converter group must continue to conduct after its output voltage changes polarity. During this period, the group functions as an inverter, and power is returned to the ac supply. Inverter operation continues until such time as the load current reduces to zero and reverses, when the other group starts to conduct. In this manner, energy can be transferred in either direction by the cycloconverter.

An interchange of energy between load and source occurs in any single-phase ac circuit with a reactive load (Fig. 5.6). The instantaneous load power is the product of the

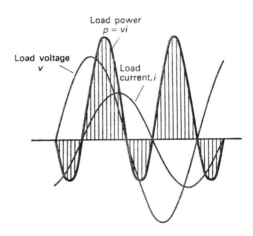

FIG. 5.6. Instantaneous voltage, current, and power in a single-phase ac circuit.

instantaneous values of load current and voltage. During part of the ac cycle, the instantaneous power is positive and the load absorbs energy from the supply. Later in the cycle, instantaneous power consumption is negative and energy is returned to the supply from the magnetic field of the load inductance (or from the electric field of the load capacitance). The difference between the two energies is the net energy delivered to the load. Real power and energy are associated with the component of current that is in phase with the applied voltage, and reactive power is associated with the current component that is 90 degrees out of phase with the applied voltage. Thus, reactive power represents power that oscillates between source and load and has an average value of zero.

The flow of reactive power associated with an inductive load occurs directly in an ac circuit, but when a cycloconverter separates the load from the supply, reactive power transfer must take place through the cycloconverter. At certain parts of the low-frequency output cycle, this power transfer demands inverter operation, which occurs automatically whenever the voltage and current of a converter group have opposite polarities. Thus, the cycloconverter is able to deliver low-frequency ac power to any type of reactive load, and the phase displacement between the half-cycles of current conduction and output voltage determines the intervals of rectifier and inverter operation. In Fig. 5.3, a sinusoidal output current is assumed, with a lagging power factor of 0.6, and the periods of rectification and inversion are indicated for the positive group. The flow of reactive power is considered further in Section 5.3.2.

5.3. CYCLOCONVERTER CIRCUIT RELATIONSHIPS

The behavior of the line-commutated phase-controlled cycloconverter circuit will now be examined in more detail.

5.3.1. *Output Voltage Equation*

The average direct voltage output of a half-wave, phase-controlled converter with zero firing delay is given by Equation 3.2:

$$V_{do} = \sqrt{2}\, V_{ph} \left(\frac{m}{\pi} \right) \sin \left(\frac{\pi}{m} \right)$$

where V_{ph} is the rms phase voltage and m is the number of secondary phases. As already explained, this result assumes instantaneous commutation and negligible thyristor on-state voltage drop.

If the cycloconverter delay angle is slowly varied, the output phase voltage at any point of the low-frequency cycle may be calculated as the average output voltage for the appropriate delay angle. This calculation ignores the rapid fluctuations superimposed on the average low-frequency waveform. Assuming continuous current conduction, the average output voltage is given by Equation 3.3:

$$V_d = V_{do} \cos \alpha \; .$$

If V_o is the fundamental rms output voltage per phase of the cycloconverter, then the peak output voltage corresponding to zero delay is

$$\sqrt{2}\, V_o = V_{do} = \sqrt{2}\, V_{ph} \left(\frac{m}{\pi} \right) \sin \left(\frac{\pi}{m} \right) \qquad (5.1)$$

and hence

$$V_o = V_{ph} \left(\frac{m}{\pi} \right) \sin \left(\frac{\pi}{m} \right) . \qquad (5.2)$$

However, the firing angle of the positive group cannot be reduced to zero, for this value corresponds to a firing angle of 180 degrees in the negative group ($\alpha_P = \pi - \alpha_N$). In practice, inverter firing cannot be delayed by 180 degrees because sufficient margin must be allowed for commutation overlap and thyristor turn-off time. Consequently, the delay angle of the positive group cannot be reduced below a certain finite value, α_{min}. The maximum output voltage per phase is, therefore,

$$V_{dmax} = V_{do} \cos \alpha_{min} = r V_{do} \qquad (5.3)$$

where $r = \cos \alpha_{min}$ and is called the "voltage reduction factor." The expression for the fundamental rms phase voltage delivered by the cycloconverter is therefore modified to

$$V_o = r \left[V_{ph} \left(\frac{m}{\pi} \right) \sin \left(\frac{\pi}{m} \right) \right] . \qquad (5.4)$$

Since α_{min} is necessarily greater than zero, the voltage reduction factor, r, is always less than unity. By deliberately increasing α_{min}, and thereby reducing the range of variation of α about the 90 degree value, the output voltage, V_o, can be reduced, and a static method of voltage control is obtained. In practice, the output voltage is less than the theoretical value given by Equation 5.4, due to the influence of commutation overlap and the circulating currents between positive and negative groups.

5.3.2. Displacement Factor

The cycloconverter consists, essentially, of a number of phase-controlled converters operating with a variable firing angle. As already explained in Section 3.3.4, the ac supply currents are nonsinusoidal in a converter circuit and one must distinguish between power factor, displacement factor, and distortion factor. The power factor, λ, in such cases, is defined as the ratio of real input power in watts to total apparent power in volt-amperes. The displacement factor, $\cos\phi$, is the fundamental power factor, because ϕ is the phase displacement between the fundamental phase current and the fundamental phase voltage. The distortion factor, μ, is the ratio of the fundamental rms current to the total rms current. As shown in Section 3.3.4, these quantities are related by Equation 3.24: $\lambda = \mu\cos\phi$. If the commutation overlap between phases is negligible, the displacement factor, $\cos\phi$, equals $\cos\alpha$, where α is the firing angle. Thus, when α is 90 degrees, the displacement factor is zero. In practice, the presence of overlap tends to increase ϕ and so further reduce the displacement factor and power factor of the system.

In a cycloconverter, voltage reduction is obtained by reducing the variation of the delay angle around the 90-degree value. At low output voltages, the average phase displacement between input current and voltage is large, and the cycloconverter has a low displacement factor. The input current always lags the supply voltage, because phase delay is always present, irrespective of the nature of the load. The cycloconverter cannot transmit leading reactive power, and the lagging reactive power drawn from the supply is always greater than that delivered to the load. A capacitive load consumes leading reactive power, but this consumption appears as a demand for lagging reactive power on the input side of the cycloconverter. Thus, the displacement factor has its greatest value when the load is purely resistive, and a capacitive load with a given leading power factor will reduce the input displacement factor by exactly the same amount as an inductive load with the same lagging power factor.

In the classical analysis by von Issendorff[26], the cycloconverter is assumed to have an infinite number of input phases and negligible commutating reactance. When the output voltage has its maximum value, the following expressions are obtained for the reactive power consumption and the input displacement factor:

$$Q_i = Q_o \left[\frac{2}{\pi} \cdot \frac{\cos\phi_o + \phi_o \sin\phi_o}{\sin\phi_o} \right] \tag{5.5}$$

and

$$\cos\phi_i = \frac{1}{\left[1 + 4 \left(\dfrac{\cos\phi_o + \phi_o \sin\phi_o}{\pi\cos\phi_o} \right)^2 \right]^{1/2}} \cdot \tag{5.6}$$

where $\cos\phi_i$ is the input displacement factor of the cycloconverter and $\cos\phi_o$ is the output displacement factor, or fundamental load power factor; Q_i is the reactive power supplied to the cycloconverter, and Q_o is the reactive power consumption of the load. It has been shown that Equations 5.5 and 5.6 are also valid for any finite number of input phases.[19]

When voltage control is implemented in the cycloconverter, the displacement factor is reduced, and in Fig. 5.7, input displacement factor, $\cos \phi_i$, is plotted against load phase

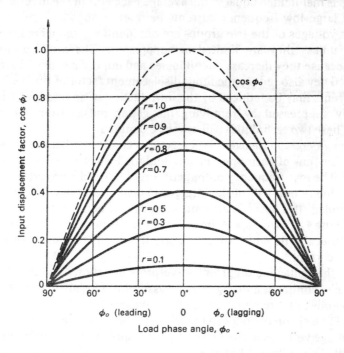

FIG. 5.7. Variation of input displacement factor for a phase-controlled cycloconverter.

angle, ϕ_o, for different values of r, the voltage reduction factor. The displacement factor has a maximum value of 0.843 with a unity power factor load and maximum output voltage. In practice, as already explained, the voltage reduction factor, r, is always less than unity, and the maximum displacement factor is somewhat less than the theoretical value. Figure 5.7 also confirms that lagging and leading loads of the same power factor cause similar reductions in the input displacement factor.

The classical analysis can also be used to prove that the input power factor is given by

$$\lambda = \frac{r}{\sqrt{2}} \cos \phi_o \qquad (5.7)$$

and the distortion factor is, therefore

$$\mu = \frac{\lambda}{\cos \phi_i} = \frac{r}{\sqrt{2}} \frac{\cos \phi_o}{\cos \phi_i}. \qquad (5.8)$$

5.3.3. Circulating Currents

As explained in Section 5.2, the cycloconverter delivers a low-frequency alternating current to each phase of the load through two antiparallel thyristor groups. The positive group has a delay angle of α_P and delivers positive current to the load. The negative group has a delay angle α_N and permits current flow in the opposite or negative direction. If the on-state voltage of a thyristor is negligible, the delay angles must be controlled so that $\alpha_P = \pi - \alpha_N$. In this manner, as already explained, the average output voltage of the rectifier group is maintained equal to the average back emf of the inverter group, and the circulation of large low-frequency currents between groups is avoided. However, the instantaneous voltages of the two groups are not identical, and large harmonic currents will circulate unless they are limited or suppressed. These intergroup currents are undesirable because they increase circuit losses and impose a heavier current loading on the thyristors. They also reduce the input displacement factor of the system. The flow of harmonic currents may be reduced by the introduction of an intergroup reactor, or may be completely suppressed by removing the gating pulses from the nonconducting group.[27,28] These two techniques will now be examined.

Limitation by Intergroup Reactor. The reactor is connected between the two groups in order to limit the flow of harmonic current, and the load is connected to a center-tap, as in Fig. 5.4(b). The low-frequency output current of the cycloconverter is opposed by the reactance, X, due to one-half of the intergroup reactor. If the two halves of the reactor are tightly coupled, the flow of harmonic current between groups is opposed by a reactance approaching $4kX$, where k is the order of the harmonic. A suitable choice of inductance will restrict the flow of harmonic current without seriously affecting the fundamental output current.

In Fig. 5.8, the voltage and current waveforms are shown for the three-phase half-wave antiparallel connection of Fig. 5.4(b), assuming a highly inductive load. The positive group operates as a rectifier with a delay angle, α_P, which is less than 90 degrees, while the negative group operates in the inverter region with a delay angle $\alpha_N = \pi - \alpha_P$. Thus, the average voltages of the two groups are equal, but the instantaneous values are quite different. The voltage difference appears across the intergroup reactor and has the waveform shown in Fig. 5.8(c). The circulating current flows through the two series-connected thyristor circuits and is limited only by the inductance of the intergroup reactor, assuming negligible circuit resistance. The circulating current waveform is therefore determined by the volt-second integral of the intergroup voltage and has the variation shown in Fig. 5.8(d). In the three-phase half-wave system under consideration, the maximum circulating current occurs when the firing angle α_P is 60 degrees and α_N is 120 degrees. During the intervals when the positive and negative thyristor groups have different instantaneous voltages, the intergroup reactor behaves as a potential divider, and the output voltage at the center-tap is the average of the two group voltages, as shown in Fig. 5.8(e).

In a cycloconverter application, the delay angles α_P and α_N are continuously modulated, and the output voltage waveform passes smoothly through each of the stages shown in Fig. 5.8(e). The output waveform is considerably improved if six-pulse converters are used. Intergroup voltages and circulating currents are then significantly reduced, and maximum circulating current occurs with a delay angle of 90 degrees, corresponding to zero output voltage.[29,30]

Suppression by Intergroup Blanking. In modern versions of the cycloconverter, the circulating current is usually suppressed by blocking all thyristors in the rectifier group that is not delivering load current. Blockage is achieved by removing the gating pulses for the appropriate periods, and the intergroup reactors of Fig. 5.5(b) can then be reduced in

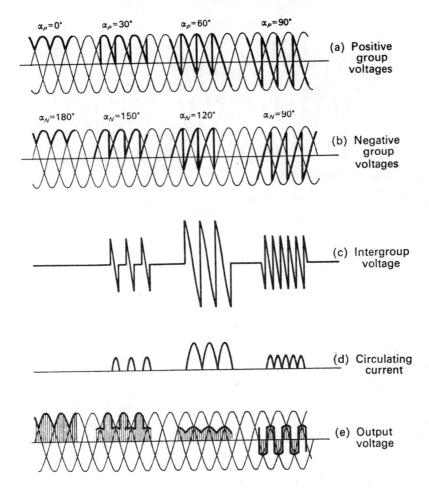

$\alpha_P = 0°$ $\alpha_P = 30°$ $\alpha_P = 60°$ $\alpha_P = 90°$

(a) Positive group voltages

$\alpha_N = 180°$ $\alpha_N = 150°$ $\alpha_N = 120°$ $\alpha_N = 90°$

(b) Negative group voltages

(c) Intergroup voltage

(d) Circulating current

(e) Output voltage

FIG. 5.8. Voltage and current waveforms for the three-phase, half-wave, antiparallel connection of Fig. 5.4(b).

size or completely eliminated. A current-sensing device is incorporated in each output phase of the cycloconverter. This sensor detects the direction of the output current and feeds a signal to the control circuitry to inhibit, or "blank," the gating of thyristors in the nonconducting group. If an overload or fault current flows in the system, all gating pulses are removed in order to protect the thyristors.

Suppression of the circulating current by intergroup blanking improves the efficiency and displacement factor of the cycloconverter and also increases the maximum usable output frequency. With resistive loads, the cycloconverter output voltage has the discontinuous waveform shown in Fig. 5.9, and the effect of the intergroup blanking is evident by the intervals of zero output voltage. The positive group delivers voltage and current to the load for the positive half-cycle, and the negative group determines the voltage and current during the negative half-cycle. Because only one rectifier group is gated at a time, the load voltage is the same as the output voltage of the conducting group, and the potential-divider effect of Fig. 5.8 is not present. If the load is sufficiently inductive, the

Output voltage

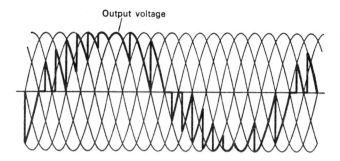

FIG. 5.9. Output voltage waveform for a phase-controlled cycloconverter with intergroup blanking and a resistive load.

output current is continuous and lags the voltage. As the load current passes through zero, it transfers from one converter group to the other. Since $\alpha_P + \alpha_N = \pi$, the incoming group delivers the same voltage as the outgoing group, and the load voltage also transfers smoothly from one group to the other.

5.3.4. *Cycloconverter Analysis*

The asymmetrical nature of the thyristor results in nonsinusoidal current and voltage waveforms in phase-controlled cycloconverter circuits. In order to simplify the analysis, the classical derivation of the circuit relations by von Issendorff[26] makes a number of simplifying approximations. An infinite number of input phases is assumed, and a highly inductive load maintains continuous current conditions. Because circulating currents between converter groups are limited by intergroup reactors, intergroup blanking is not employed. In present-day practical cycloconverters,[31] these circuit conditions and approximations are often invalid, and an exact analysis becomes extremely complicated. The difficulty may be avoided by making other appropriate circuit approximations in order to obtain equivalent circuits for the purposes of analysis.[32,33] Alternatively, the operation of the thyristors, sources, and loads can be analyzed with a digital computer.[34,35] The output can be in the form of waveforms as a function of time or calculated values such as rms currents. The analog computer has been used to simulate the operation of the cycloconverter during transient changes in operating conditions.[36]

5.4. FREQUENCY AND VOLTAGE CONTROL

An ideal phase-controlled converter, in which current flow is continuous, delivers an output voltage waveform with an average value of zero when the firing delay, α, is 90 degrees. If the delay angle is biased to the 90-degree position and is slowly varied by ± 90 degrees on either side of this point, a low-frequency output voltage is obtained with peak values of $\pm V_{do}$. When the range of oscillation of α on either side of the 90-degree value is reduced, the magnitude of the output voltage is also reduced and voltage control is obtained. The cycloconverter output frequency is determined by the rate at which α is varied above and below the 90-degree point; hence output voltage and frequency are governed by the timing of the gating pulses from the control circuitry.

The repetition frequency and sequence of the gating pulses can be determined by the cycloconverter ac supply voltages and a reference or control voltage signal which is generated in a low-power master oscillator. This oscillator has the same number of phases as the cycloconverter output, generating reference voltages with the waveshape, frequency, and phase sequence required in the output. The supply voltages and reference voltages are fed to comparator circuits, and the output signals from the comparators control the gating circuits. Gating signals are designed to duplicate the reference conditions in the cycloconverter output voltages. Normally, the reference voltages are sinusoidal, and the gating pulses modulate the various delay angles so that the average output voltage has a superimposed sinusoidal variation. A positive reference voltage advances firing of the positive group and delays firing of the negative group by the same amount in order to satisfy the relationship $\alpha_P + \alpha_N = \pi$. Variation of the reference oscillator frequency produces a corresponding variation in output frequency giving stability, accuracy, and fast transient response. The direction of rotation of the drive motor is reversed by simply changing the phase sequence of the reference oscillator; no switching is necessary in the power circuits. The amplitude of the fundamental component of cycloconverter output voltage is proportional to the reference voltage amplitude, and hence the cycloconverter may be regarded as a high-gain power amplifier. Independent control of voltage and frequency is possible, and maximum output voltage may be obtained at any operating frequency. Alternatively, the required volts/hertz variation for a motor load may be programmed into the control circuitry.

The microprocessor, together with digital logic circuits, has taken over many of the control tasks in modern cycloconverter equipment, replacing the low-level reference waveform generator and analog signal comparator circuits.[17,37-41] The reference waveforms can be stored in a read only memory and accessed under the control of the stored program and microprocessor clock oscillator. In a similar manner, analog supply voltage signals can be converted to digital signals in analog-to-digital converters. The waveform comparisons can then be accomplished with the comparison features of the microprocessor software. The addition of time delays for thyristor turn-off time requirements and intergroup blanking can also be accomplished with digital techniques and computer software. Recent microprocessors have sufficient speed of operation to be able to perform the necessary calculations and supply gate drive signals in the time interval between successive thyristor firings.

The foregoing discussion has assumed that a sinusoidal output voltage waveform is required, but the control circuitry can be modified to develop square or trapezoidal waveforms.[42] A trapezoidal waveform has reduced phase delay, improved power factor, and a slightly greater fundamental content than a sine wave with the same peak amplitude. It is generally advantageous to provide trapezoidal waveforms for large ac motor drive systems. The cycloconverter can also be operated under closed-loop current control; it is then provided with a current reference signal and functions as a low-frequency current source.

5.5. LOAD-COMMUTATED CYCLOCONVERTER

The load-commutated cycloconverter differs from the previously described line-commutated cycloconverter in that the thyristors can be commutated by the reversal of the load voltage. In this case, the load must possess a generated, or back, emf that is independent of the source voltage. The most common example of such a load is a wound-field or permanent magnet synchronous machine. In such cases, the load frequency can be equal to or greater than the source frequency and still allow natural thyristor commutation. The thyristor gating is based on two control signals: first, with respect to the source voltage, to control the load voltage; and second, with respect to the synchronous machine-generated emf, to insure that current will flow in the correct phase of the load machine at the correct time. For optimum torque per ampere in a synchronous machine, commutation should take place with 180 degrees of firing delay with respect to the machine voltage. This, however, leaves no margin for commutation overlap and thyristor turn-off. As in other types of thyristor-based synchronous motor drive, the commutation must be advanced from the optimum position. It is the function of the shaft position sensor and the control electronics to insure that the commutation is sufficiently in advance of the line-to-line voltage crossing. This condition results in a phase current that leads the phase voltage.

In aircraft applications, load-commutated cycloconverters are used to start and accelerate a gas turbine through its high-frequency synchronous ac generator operating as a motor. After the turbine has started and reached its operating speed, the generator and cycloconverter deliver 400 Hz power to the aircraft.[20,21]

5.6. FORCED-COMMUTATED CYCLOCONVERTER

The line-commutated and the load-commutated cycloconverters described in the previous sections have relied on the reversal of the source or load voltage to achieve natural commutation of the thyristors. Another approach to commutation is to provide additional circuit components to force-commutate the thyristors independent of the source and load voltages. This independent turn-off technique allows the output frequency to be lower than, equal to, or higher than the input frequency. Two forms of forced, or impulse, commutation are described in the literature.[22] These are reverse voltage commutation and reverse current commutation. Each of the numerous load current-carrying thyristors in the cycloconverter needs to be connectable to the forced commutating circuit or circuits, and there are many different locations for the commutating components. As discussed in Chapter 4, these components may include auxiliary thyristors, diodes, capacitors, reactors, transformers, and auxiliary power supplies.[22] The added cost of these additional components has tended to limit the application of forced-commutated cycloconverters, but the advent of devices with self-turn-off capability may allow these types of cycloconverter to be constructed without auxiliary components for commutation.

5.7. COMPARISON OF THE CYCLOCONVERTER AND DC LINK CONVERTER

Both the cycloconverter and dc link converter deliver an alternating voltage at a frequency determined by the reference oscillator. This technique provides an extremely stable and accurate output frequency that is independent of fluctuations in the ac network frequency and voltage. Speed reversal of machines is readily obtained by reversing the thyristor firing sequence.

The cycloconverter has some advantages and some disadvantages compared with the dc link converter. The following advantages can be listed:

1. In a cycloconverter, ac power at one frequency is converted directly to ac at another frequency in a single conversion stage. The dc link converter has two power controllers in cascade, and the full output power is converted twice, resulting in higher losses.

2. Naturally commutated cycloconverter operation is based on traditional line-commutated-converter technology, as used in dc drives. Because natural commutation is employed, standard converter-grade thyristors are used, and no auxiliary forced commutating circuits are necessary. Natural commutation results in a more compact power circuit and also eliminates the circuit losses associated with forced commutation.

3. The cycloconverter is inherently capable of power transfer in either direction between source and load, and can therefore supply ac power to loads of any power factor. It is also capable of regenerative operation at full torque over the complete speed range down to standstill. This feature is also available in a current-source dc link converter, but regenerative operation is not readily incorporated in a voltage-source dc link converter; hence, the cycloconverter is often preferable for large reversing drives requiring rapid acceleration and deceleration. This type of application occurs principally in the metal rolling industry.[9,12,17]

4. In the cycloconverter, a commutation failure may cause a short-circuit of the ac supply, but if an individual thyristor fuse ruptures, a complete shutdown is not necessary because the cycloconverter continues to function with a somewhat distorted output waveform. A balanced load is presented to the ac supply, even with unbalanced output conditions.

5. The cycloconverter delivers a high-quality sinusoidal waveform at low output frequencies because the low-frequency wave is fabricated from a large number of segments of the supply waveform. On the other hand, many voltage- and current-source dc link converters generate a stepped-wave output voltage or current that may cause nonuniform rotation of an ac motor at frequencies below 10 Hz. For this reason, the cycloconverter is superior for very low speed applications.

The cycloconverter also has the following disadvantages:

1. Load-commutated and forced-commutated cycloconverter systems permit load frequencies greater or less than source frequency, but in a line-commutated cycloconverter the maximum output frequency must be less than about one-third or one-half of the input frequency for reasonable power output and efficiency. This upper frequency limit is a serious constraint when operation is on 50 or 60 Hz ac supplies, because the synchronous speed of the motor is restricted to a maximum of approximately 1800 rev/min, even with a two-pole machine. This disadvantage can be overcome in applications where ac power for the cycloconverter is generated by an engine-driven alternator. By selection of a high alternator frequency, high maximum speeds can be obtained with compact high-frequency ac motors.

2. The cycloconverter requires a large number of thyristors, and its control circuitry is more complex than that employed in many dc link converters. This expensive circuitry is not justified in small installations, but the cycloconverter is economical for units of 100 kVA or more. It may be advantageous to use two-phase motors in order to reduce the number of thyristors and improve thyristor utilization.

3. The line-commutated cycloconverter has a low input power factor, particularly at reduced output voltages. In a dc link converter, a high input power factor may be obtained for all operating conditions with the use of a simple uncontrolled rectifier input.

To summarize, this comparison of the cycloconverter and dc link converter indicates that the dc link system is more suitable for industrial applications at high output frequencies, but the cycloconverter is extremely attractive for low-speed, high-horsepower reversible drives.

5.8. REFERENCES

1. SCHENKEL, M., Eine unmittelbare asynchrone Umrichtung für niederfrequente Bahnnetze, *Elektr. Bahnen.*, **8**, 1932, pp. 69-73.

2. BOSCH, M., and KASPEROWSKI, O., Die Steuerumrichter-Versuchsanlage der Siemens-Schuckertwerke A.-G. im Reichsbahn-Saalachkraftwerk, *Elektr. Bahnen.*, **11**, 1935, pp. 235-249.

3. ALEXANDERSON, E.F.W., and MITTAG, A.H., The thyratron motor, *Electr. Eng.* (New York), **53**, Nov. 1934, pp. 1517-1523.

4. BEILER, A.H., The thyratron motor at the Logan plant, *Electr. Eng.* (New York), **57**, Jan. 1938, pp. 19-24.

5. OWEN, E.L., MORACK, M.M., HERSKIND, C.C., and GRIMES, A.S., AC adjustable-speed drives with electronic power converters — the early days, *Conf. Rec. IEEE Ind. Appl. Soc. Annual Meeting*, 1982, pp. 854-861.

6. GRIFFITH, D.C., and ULMER, R.M., A semiconductor variable-speed ac motor drive, *Electr. Eng.* (New York), **80**, May 1961, pp. 350-353.

7. HECK, R., and MEYER, M., A static-frequency-changer-fed squirrel-cage motor drive for variable speed and reversing, *Siemens Rev.*, **30**, Nov. 1963, pp. 401-405.

8. ANNIES, B., Steuerumrichter für Käfigläufermotoren, *AEG Mitt.*, **54**, 1/2, 1964, pp. 123-125.

9. GUYESKA, J.C., and JORDAN, H.E., Static ac variable frequency drive, *Proc. IEEE Nat. Electron. Conf.*, **20**, 1964, pp. 358-365.

10. BOWLER, P., The application of a cycloconverter to the control of induction motors, IEE Conf. Publ. No. 17, *Power Applications of Controllable Semiconductor Devices*, 1965, pp. 137-145.

11. SLABIAK, W., and LAWSON, L.J., Precise control of a three-phase squirrel-cage induction motor using a practical cycloconverter, *IEEE Trans. Ind. Gen. Appl.*, **IGA-2**, 4, July/Aug. 1966, pp. 274-280.

12. HAMILTON, R.A., and LEZAN, G.R., Thyristor adjustable frequency power supplies for hot strip mill run-out tables, *IEEE Trans. Ind. Appl.*, **IGA-3**, 2, Mar./Apr. 1967, pp. 168-175.

13. ALLAN, J.A., WYETH, W.A., HERZOG, G.W., and YOUNG, J.A.I., Electrical aspects of the 8750 hp gearless ball mill drive at the St. Lawrence Cement Company, *IEEE Trans. Ind. Appl.*, **IA-11**, 6, Nov./Dec. 1975, pp. 681-687.

14. STEMMLER, H., Drive systems and electronic control equipment of the gearless tube mill, *Brown Boveri Rev.*, **57**, 3, Mar. 1970, pp. 121-129.

15. MAENO, T., and KOBATA, M., AC commutatorless and brushless motor, *IEEE Trans. Power Appar. Syst.*, **PAS-91**, 4, July/Aug. 1972, pp. 1476-1484.

16. TERENS, L., BOMMELI, J., and PETERS, K., The cycloconverter-fed synchronous motor, *Brown Boveri Rev.*, **69**, 4/5, Apr./May 1982, pp. 122-132.

17. SUGI, K., NAITO, Y., KUROSAWA, R., KANO, Y., KATAYAMA, S., and YOSHIDA, T., A microcomputer-based high capacity cycloconverter drive for main rolling mill, *Int. Power Electron. Conf., Tokyo*, 1983, pp. 744-755.

18. CHIRGWIN, K.M., STRATTON, L.J., and TOTH, J.R., Precise frequency power generation from an unregulated shaft, *Trans. AIEE, Pt. 2, Appl. and Ind.*, **79**, Jan. 1961, pp. 442-451.

19. PLETTE, D.L., and CARLSON, H.G., Performance of a variable speed constant frequency electrical system, *IEEE Trans. Aerosp.*, **AS-2**, 2, Apr. 1964, pp. 957-970.

20. LAFUZE, D.L., VSCF Starter Generator, *IEEE Power Electron. Spec. Conf.*, 1974, pp. 327-333.

21. LAFUZE, D.L., and RICHTER, E., A high power rare earth permanent magnet generator VSCF starter generator system, *IEEE Nat. Aerosp. Electron. Conf.*, 1976, pp. 971-977.

22. GYUGYI, L., and PELLY, B.R., *Static Power Frequency Changers*, Wiley-Interscience, New York, NY, 1976.

23. McMURRAY, W., *The Theory and Design of Cycloconverters*, Massachusetts Institute of Technology Press, Cambridge, MA, 1972.

24. MÖLTGEN, G., *Line Commutated Thyristor Converters*, Pitman, London, 1972.

25. PELLY, B.R., *Thyristor Phase-Controlled Converters and Cycloconverters*, Wiley-Interscience, New York, NY, 1971.

26. VON ISSENDORFF, J., Der gesteuerte Umrichter, *Wiss. Veröff, Siemens-Werke*, **14**, 1935, pp. 1-31.

27. TAMURA, Y., TANAKA, S., and TADAKUMA, S., Control method and upper limit of output frequency in circulating-current type cycloconverter, *IEEE Int. Semicond. Power Converter Conf.*, 1982, pp. 313-323.

28. UEDA, R., SONODA, T., MOCHIZUKI, T., and TAKATA, S., Stabilization of bank selection in no circulating cycloconverter by means of reliable current-zero and current-polarity detection, *Conf. Rec. IEEE Ind. Appl. Soc. Annual Meeting*, 1982, pp. 651-656.

29. MEYER, M., and MÖLTGEN, G., Kreisströme bei Umkehrstromrichtern, *Siemens-Z.*, **37**, 5, May 1963, pp. 375-379.

30. VAN ECK, R.A., Frequency changer systems using the cycloconverter principle, *IEEE Trans. Appl. and Ind.*, **82**, May 1963, pp. 163-168.

31. LAWSON, L.J., The practical cycloconverter, *IEEE Trans. Ind. Gen. Appl.*, **IGA-4**, 2, Mar./Apr. 1968, pp. 141-144.

32. AMATO, C.J., An ac equivalent circuit for a cycloconverter, *IEEE Trans. Ind. Gen. Appl.*, **IGA-2**, 5, Sept./Oct. 1966, pp. 358-362.

33. AMATO, C.J., Sub-ripple distortion components in practical cycloconverters, *IEEE Trans. Aerosp. (Suppl.)*, **AS-3**, June 1965, pp. 98-106.

34. TADAKUMA, S., and TAMURA, Y., Current response simulation in six-phase and twelve-phase cycloconverter, *Conf. Rec. IEEE Ind. Appl. Soc. Annual Meeting*, 1978, pp. 423-431.

35. CHATTOPADHYAY, A.K., and RAO, T.J., Digital simulation of a phase-controlled cycloconverter induction motor drive system, *Conf. Rec. IEEE Ind. Appl. Soc. Annual Meeting*, 1978, pp. 853-860.

36. AMATO, C.J., Analog computer simulation of an SCR as applied to a cycloconverter, *IEEE Trans. Ind. Gen. Appl.*, IGA-2, 2, Mar./Apr. 1966, pp. 137-140.

37. CHEN, H.H., A microprocessor control of a three-pulse cycloconverter, *IEEE Trans. Ind. Electron. and Control Instrum.*, **IECI-24**, 3, Aug. 1977, pp. 226-230.

38. SINGH, D., and HOFT, R.G., Microcomputer-controlled single-phase cycloconverter, *IEEE Trans. Ind. Electron. and Control Instrum.*, **IECI-25**, 3, Aug. 1978, pp. 233-238.

39. FUKAO, T., AKAGI, H., and MIYAIRI, S., A direct digital control of a three-phase six-pulse cycloconverter using a microprocessor, *Proc. IEEE IECI, Ind. and Control Appl. Microproc.*, 1979, pp. 2-7.

40. TSO, S.K., SPOONER, E.D., and COSGROVE, J., Efficient microprocessor-based cycloconverter control, *IEE Proc.*, **127**, **Pt. B**, 3, May 1980, pp. 190-196.

41. TSO, S.K., and LEUNG, C.C., Microprocessor control of triac cycloconverter, *IEE Proc.*, **130**, **Pt. B**, 3, May 1983, pp. 193-200.

42. BLAND, R.J., Factors affecting the operation of a phase-controlled cycloconverter, *Proc. IEE*, **114**, 12, Dec. 1967, pp. 1908-1916.

CHAPTER 6

AC Motor Operation with Nonsinusoidal Supply Waveforms

6.1. INTRODUCTION

Most static frequency converters generate an output voltage or current waveform with a significant harmonic content. In this chapter, motor performance with nonsinusoidal supplies is compared with normal sine wave operation. Harmonic effects are studied for six-step voltage and current supplies, and for pulse-width-modulated (PWM) waveforms.

Initially, the conventional equivalent circuit theory of the polyphase induction motor with a sinusoidal supply voltage is reviewed. This theory is also relevant to the harmonic behavior of synchronous machines and synchronous reluctance motors, because harmonic airgap fields rotate asynchronously with respect to rotor cage windings or damper bars.

6.2. INDUCTION MOTOR EQUIVALENT CIRCUIT

The equivalent circuit of the polyphase induction motor is very similar to the usual transformer equivalent circuit, because the induction motor is essentially a transformer with a rotating secondary winding. As in a static transformer, the primary or stator current establishes a mutual flux that links the secondary or rotor winding, and also a leakage flux that links only the primary winding. This leakage flux induces a primary emf which is proportional to the rate of change of primary current. Its effect may be represented, in the usual manner, by a series leakage reactance, X_1, in each stator phase (Fig. 6.1). R_1 is the stator resistance per phase and $(R_1 + jX_1)$ is termed the stator leakage impedance. The mutual flux in the airgap induces slip frequency emfs in the rotor and supply frequency emfs in the stator. The voltage drop across the stator leakage impedance causes the airgap emf per phase, E_1, and the mutual flux per pole, Φ_1, to decrease slightly as load is applied to the motor. The resultant stator current, I_1, is composed of the magnetizing current, I_m, and the load component of stator current, I_2, which cancels the magnetomotive force (mmf) due to rotor current. Core losses and saturation effects are neglected.

In deriving the rotor equivalent circuit, the actual cage or phase-wound rotor winding is considered to be replaced by an equivalent short-circuited rotor winding having the same number of turns and the same winding arrangement as the stator. This is equivalent to the usual transformer procedure of referring secondary quantities to the primary. At standstill, the induced emf per phase in the equivalent rotor is equal to the stator emf, E_1, and the rotor frequency equals the supply frequency, f_1. If the rotor slip

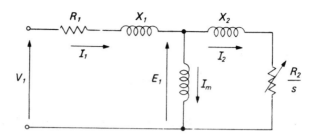

FIG. 6.1. Single-phase equivalent circuit of the polyphase induction motor.

with respect to the fundamental rotating field is denoted by s, the rotor emf, E_2, equals sE_1, and the rotor frequency, f_2, equals sf_1. If R_2 is the equivalent rotor resistance per phase and X_2 is the equivalent rotor leakage reactance per phase at standstill, then the rotor current is given by

$$I_2 = \frac{E_2}{R_2 + jsX_2} = \frac{sE_1}{R_2 + jsX_2}$$ (6.1)

and hence

$$I_2 = \frac{E_1}{(R_2/s) + jX_2}.$$ (6.2)

In Equation 6.1, all rotor quantities are at slip frequency, but in Equation 6.2 they are at supply frequency. These equations show that the rotor current, I_2, is unaltered in magnitude if the rotor is brought to standstill and the resistance increased from R_2 to R_2/s. The rotor equivalent circuit may therefore be joined directly to the stator circuit, as in Fig. 6.1, to give the complete equivalent circuit for one phase of the motor.

6.2.1. Torque Equation

At a slip, s, the rotor power loss in the equivalent circuit is $I_2^2 R_2/s$ watts per phase, whereas, in the actual machine, the rotor copper loss is $I_2^2 R_2$ watts per phase. The additional power loss in the equivalent circuit is the electrical equivalent of the mechanical power output of the motor. If P_{mech} denotes the gross mechanical power output, including windage and friction losses, then

$$P_{mech} = m_1 \left[(I_2^2 R_2/s) - (I_2^2 R_2) \right]$$

$$= m_1 I_2^2 R_2 (\frac{1-s}{s})$$ (6.3)

where m_1 is the number of stator phases.

If ω_m is the mechanical angular velocity of the rotor and T is the electromagnetic torque, the gross mechanical power output can be written as

$$T\omega_m = m_1 I_2^2 R_2 (\frac{1-s}{s})$$ (6.4)

and

$$T = \frac{m_1 I_2^2 R_2}{\omega_m} \left(\frac{1-s}{s}\right).$$ (6.5)

This is the internal motor torque, which is greater than the useful shaft torque by the amount required to overcome windage and friction torques.

Since the synchronous angular velocity in mechanical radians per second is given by $\omega_s = \omega_m/(1-s) = 2\pi f_1/p$, where p is the number of pole-pairs, the torque equation can be rewritten as

$$T = \frac{m_1 I_2^2 R_2}{s\omega_s}$$ (6.6)

or

$$T = \frac{pm_1}{2\pi f_1} (I_2)^2 \frac{R_2}{s}.$$ (6.7)

6.2.2. Power Division in the Rotor

It can be seen from the equivalent circuit that the total electrical power input to the rotor across the airgap from the stator is

$$P_{ag} = m_1 I_2^2 R_2/s.$$ (6.8)

This power is divided between mechanical power output, P_{mech}, and rotor copper loss, P_2. Thus,

$$P_{ag} = P_{mech} + P_2$$ (6.9)

where

$$P_{mech} = T\omega_m$$ (6.10)

and

$$P_2 = m_1 I_2^2 R_2.$$ (6.11)

Combining Equations 6.6 and 6.8 gives

$$P_{ag} = T\omega_s$$ (6.12)

and hence the total electrical power input to the rotor is equal to the internal mechanical torque multiplied by the synchronous angular velocity.

6.2.3. Phasor Diagram

The phasor diagram corresponding to the equivalent circuit of Fig. 6.1 is shown in Fig. 6.2. The magnetizing current, I_m, is taken as the reference phasor, and the stator

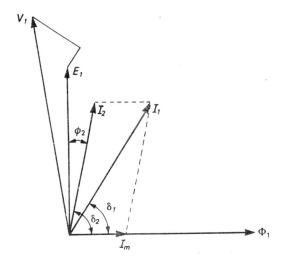

FIG. 6.2. Phasor diagram of the polyphase induction motor.

emf, E_1, leads I_m by 90 degrees. The load component of stator current is I_2. It neutralizes the rotor mmf and lags E_1 by the rotor power factor angle, ϕ_2. The net stator current, I_1, is the phasor sum of I_m and I_2. Fundamental airgap flux, Φ_1, is in phase with the magnetizing current, I_m, and their amplitudes are proportional when saturation is negligible.

The flux and current phasors can also be regarded as space vectors of flux and mmf, which represent the combined effect of all three phases and rotate anticlockwise at the synchronous angular velocity of the machine. Thus, I_1, I_2, and I_m represent stator, rotor, and mutual mmf waves, respectively. Similarly, Φ_1 represents the airgap mutual flux wave. In general, torque is developed by the tendency for alignment of mutual airgap flux and winding mmf. The magnitude of the motor torque is proportional to the product of the amplitudes of the mutual flux and winding mmf waves and the sine of the electrical angular displacement between them. From Fig. 6.2,

$$T = K\,\Phi_1\,I_2 \sin \delta_2 = K\,\Phi_1\,I_1 \sin \delta_1 . \tag{6.13}$$

Since the torque angles, δ_1 and δ_2, are constant, a steady positive, or anticlockwise, motoring torque is developed. Obviously, $\delta_2 = \pi/2 - \phi_2$, and substituting in Equation 6.13 gives the earlier torque equation, Equation 1.1,

$$T = K\,\Phi_1\,I_2 \cos \phi_2 . \tag{6.14}$$

6.3. AIRGAP MMF HARMONICS

The elementary two-pole, three-phase winding of a synchronous or induction motor has three stator coils displaced 120 degrees apart in space, which are excited by a three-phase system of currents that are 120 degrees apart in time. In normal operation, each coil carries a sinusoidal current which establishes a pulsating magnetic field, or mmf. It may be assumed that each phase establishes its mmf wave in the airgap independently of the other two phases. Since the phase winding is usually distributed in a number of slots in an iron surface, the spatial distribution of mmf is nonsinusoidal. However, the actual mmf distribution may be resolved into a fundamental component and a series of higher, odd space harmonics. For the present, only the fundamental component is considered, and consequently, the airgap of the machine has three sinusoidally distributed mmf waves that are 120 degrees apart in space. As each phase current varies, the corresponding mmf wave pulsates in magnitude but retains its sinusoidal space distribution.

Let F_1 denote the amplitude of the mmf wave from coil 1 at some instant. The spatial mmf distribution is therefore given by

$$f_1 = F_1 \cos \theta \qquad (6.15)$$

where θ represents the angular displacement on the armature surface in electrical radians, with the origin chosen on the axis of coil 1. As the coil current varies sinusoidally, with an angular frequency ω, the amplitude F_1 varies proportionately, and a standing mmf wave is produced. The instantaneous amplitude, F_1, is given by

$$F_1 = \hat{F}_1 \sin \omega t \qquad (6.16)$$

where $\hat{F}_1$ is the maximum value of F_1 corresponding to the peak coil current, and a time origin is taken at the instant when the current in coil 1 is zero.

Combining Equations 6.15 and 6.16 gives the spatial distribution of mmf from coil 1 at time t as

$$f_1 = \hat{F}_1 \cos \theta \sin \omega t . \qquad (6.17)$$

The mmf wave from coil 2 is displaced by 120 degrees in space and time relative to f_1, and hence is given by

$$f_2 = \hat{F}_1 \cos(\theta - 2\pi/3) \sin (\omega t - 2\pi/3) . \qquad (6.18)$$

Similarly, for coil 3,

$$f_3 = \hat{F}_1 \cos(\theta - 4\pi/3) \sin (\omega t - 4\pi/3) . \qquad (6.19)$$

The resultant airgap mmf due to the entire winding is obtained by adding the contributions of the three phases. Thus,

$$\begin{aligned} f &= f_1 + f_2 + f_3 \\ &= \hat{F}_1 \left[\cos \theta \sin \omega t + \cos (\theta - 2\pi/3) \sin (\omega t - 2\pi/3) \right. \\ &\quad \left. + \cos (\theta - 4\pi/3) \sin (\omega t - 4\pi/3) \right] . \end{aligned} \qquad (6.20)$$

From the trigonometric relation that $\cos A \sin B = \frac{1}{2} \sin (A+B) - \frac{1}{2} \sin (A-B)$, the above expression reduces to

$$f = \frac{3}{2} \hat{F}_1 \sin (\omega t - \theta) . \qquad (6.21)$$

This expression represents a sinusoidal distribution of mmf of constant amplitude, which moves with a uniform angular velocity, ω, in the direction of increasing θ. This movement may be understood by noting that the sine wave travels through a distance, ωt, in a time, t, and hence the displacement increases uniformly with time. Alternatively, a point P may be considered which moves with the wave so that the mmf at P is constant at all times. For such a point, $\sin(\omega t - \theta)$ is constant, and hence $(\omega t - \theta) = K$, a constant. On differentiating with respect to t, $\omega - d\theta/dt = 0$, or $d\theta/dt = \omega$. This result verifies the familiar fact that a three-phase winding excited by balanced sinusoidal currents produces a fundamental mmf wave rotating at synchronous speed.

For a machine with p pole-pairs, the electrical angular displacement, θ, is p times the mechanical angular displacement, θ_m. Thus, $\theta = p\theta_m$, and the synchronous angular velocity in mechanical radians/second is $d\theta_m/dt = (1/p)\,d\theta/dt = \omega/p = 2\pi f_1/p = \omega_s$, as before.

If the phase currents have a nonsinusoidal waveform, additional harmonic mmf waves are present in the airgap. The existence of a particular harmonic may be confirmed by an analysis similar to that presented above for the fundamental mmf.

6.3.1. Time Harmonic MMF Waves

Time harmonic waves of mmf are produced by current harmonics in the phase windings. Assume, for example, that the phase current has a fifth harmonic component, and consequently, each phase establishes a standing mmf wave having the same spatial distribution as the fundamental but which pulsates at five times the supply frequency. The fifth harmonic mmf from coil 1 is therefore given by

$$f_1 = \hat{F}_{1,5} \cos \theta \sin 5\omega t \tag{6.22}$$

where $\hat{F}_{1,5}$ is the maximum amplitude of the fundamental space mmf wave due to the fifth harmonic of current. Similarly,

$$f_2 = \hat{F}_{1,5} \cos(\theta - 2\pi/3) \sin 5(\omega t - 2\pi/3) \tag{6.23}$$

and

$$f_3 = \hat{F}_{1,5} \cos(\theta - 4\pi/3) \sin 5(\omega t - 4\pi/3). \tag{6.24}$$

The resultant mmf is obtained, as before, by adding the three mmf contributions. Thus,

$$f = \frac{3}{2} \hat{F}_{1,5} \sin(5\omega t + \theta). \tag{6.25}$$

This result confirms that a rotating mmf is produced by the fifth harmonic currents. The speed of rotation is given by $d\theta/dt = -5\omega$, which means that the wave is moving at five times synchronous speed in the opposite direction to the fundamental mmf.

Similarly, it can be shown that the seventh harmonic currents produce a rotating mmf wave that moves at seven times synchronous speed in the same direction as the fundamental. In general, current harmonics of order $k = (3n+1)$, where $n = 1,2,3...$, produce forward-rotating mmf waves, while harmonics of order $k = (3n-1)$ produce backward-rotating waves.[1] The speed of rotation of a time harmonic field is always k times the synchronous speed.

6.3.2. Space Harmonic MMF Waves

In the analysis above, it has been assumed that each phase current establishes a fundamental space mmf wave and that the presence of higher odd space harmonics can be ignored. In fact, even when the polyphase winding is excited by purely sinusoidal currents, the space harmonic distributions of mmf due to different phases combine to produce rotating harmonic mmf waves. Thus, the fifth space harmonic mmf due to fundamental current in coil 1 may be written

$$f_1 = \hat{F}_{5,1} \cos 5\theta \sin \omega t . \tag{6.26}$$

Corresponding expressions are written for f_2 and f_3 and, when the three mmfs are combined, the result is

$$f = \frac{3}{2} \hat{F}_{5,1} \sin (\omega t + 5\theta) . \tag{6.27}$$

This result confirms the existence of a fifth space harmonic mmf rotating backward at one-fifth of synchronous speed. Similarly, the seventh space harmonic can be shown to rotate forward at one-seventh of synchronous speed. These are the well-known space harmonic waves of mmf produced by a symmetrical integral-slot winding excited by balanced sinusoidal currents.

When harmonic currents are present in the phases of the integral-slot winding, the time and space harmonic mmf waves predicted above are present simultaneously. Additional rotating mmf waves are produced by the space harmonic distributions of mmf resulting from time harmonics in the phase currents. The existence of a particular harmonic is again confirmed by summing the contributions of the three phases. For example, the fifth harmonic currents in each of the three phases produce the following seventh harmonic space mmfs:

$$f_1 = \hat{F}_{7,5} \cos 7\theta \sin 5\omega t \tag{6.28}$$

$$f_2 = \hat{F}_{7,5} \cos 7(\theta - 2\pi/3) \sin 5(\omega t - 2\pi/3) \tag{6.29}$$

$$f_3 = \hat{F}_{7,5} \cos 7 (\theta - 4\pi/3) \sin 5(\omega t - 4\pi/3) . \tag{6.30}$$

On combining these three components

$$f = \frac{3}{2} \hat{F}_{7,5} \sin (5\omega t + 7\theta) . \tag{6.31}$$

The existence of a seventh space harmonic mmf rotating backward at five-sevenths of synchronous speed is, therefore, confirmed.

The complete results are summarized in Table 6.1, where all speeds are expressed as multiples of the synchronous speed, which is denoted by +1. A positive sign denotes a wave rotating in the same direction as the fundamental, and a negative sign indicates a backward-rotating harmonic. The first row of the table shows the time harmonic waves of mmf already analyzed. These are the fundamental space mmf waves resulting from harmonics in the phase currents. The first column shows the spatial harmonic waves of mmf due to fundamental phase currents. The remainder of the table shows space harmonics up to the fifteenth due to time harmonics up to the thirteenth. Because most inverter circuits do not generate even-numbered time harmonics, these are omitted from the table. A similar mmf analysis can be performed for a two-phase ac winding. [2]

TABLE 6.1. MMF COMPONENTS OF A THREE-PHASE ARMATURE WINDING

Order of space harmonic h	Order of time harmonic, k						
	1	3	5	7	9	11	13
1	$+1$	$-$	-5	$+7$	$-$	-11	$+13$
3	$-$	± 1	$-$	$-$	± 3	$-$	$-$
5	$-1/5$	$-$	$+1$	$-7/5$	$-$	$+11/5$	$-13/5$
7	$+1/7$	$-$	$-5/7$	$+1$	$-$	$-11/7$	$+13/7$
9	$-$	$\pm 1/3$	$-$	$-$	± 1	$-$	$-$
11	$-1/11$	$-$	$+5/11$	$-7/11$	$-$	$+1$	$-13/11$
13	$+1/13$	$-$	$-5/13$	$+7/13$	$-$	$-11/13$	$+1$
15	$-$	$\pm 1/5$	$-$	$-$	$\pm 3/5$	$-$	$-$

6.3.3. Amplitude of MMF Harmonics

For a particular stator current waveform, the harmonic mmf amplitudes relative to the fundamental are determined by the winding arrangement used. For a three-phase winding, the amplitude of the hth space harmonic rotating mmf due to the kth time harmonic of current is readily shown to be

$$F_{h,k} = 1.35 \, k_{wh} \, \frac{1}{h} \, \frac{N_{ph}}{p} \, I_k \text{ ampere-turns per pole} \qquad (6.32)$$

where
k_{wh} is the hth space harmonic winding factor,
N_{ph} is the number of series turns per phase,
p is the number of pole-pairs, and
I_k is the rms value of the kth harmonic phase current.
For the fundamental rotating mmf wave, $h = k = 1$ and the amplitude is

$$F_{1,1} = 1.35 \, k_w \, \frac{N_{ph}}{p} \, I_1 \text{ ampere-turns per pole} \qquad (6.33)$$

where
k_w is the fundamental winding factor, and
I_1 is the fundamental rms phase current.

The harmonic mmf amplitude expressed as a fraction of the fundamental amplitude is

$$\frac{F_{h,k}}{F_{1,1}} = \frac{k_{wh}}{k_w} \frac{1}{h} \frac{I_k}{I_1}. \tag{6.34}$$

For normal, well-designed three-phase motors, the harmonic winding factor k_{wh} is much less than k_w and the space harmonic mmf waves have negligible amplitudes. This is also true for two-phase motors. In the remainder of this chapter, attention is therefore confined to the time harmonic mmf waves which have a fundamental space distribution.

6.3.4. *Positive-, Negative-, and Zero-Sequence Harmonics*

As shown in the first row of Table 6.1, the time harmonic waves of mmf due to harmonic currents of order $k = (6n+1)$, where n is an integer, rotate in the same direction as the main field, while harmonic currents of order $(6n-1)$ generate backward-rotating mmf waves. In general, the direction of rotation of the harmonic field is determined by the phase sequence of the harmonic currents. Harmonics of order $(6n+1)$ have the same phase sequence as the fundamental currents, and hence produce mmf waves that rotate in the same direction as the main field. Harmonics of order $(6n+1)$ are, therefore, called positive-sequence harmonics. Conversely, harmonic currents of order $(6n-1)$ are called negative-sequence harmonics, because they have opposite phase sequence to the fundamental and generate backward-rotating mmf waves. Harmonic currents of order $(6n-3)$ do not produce any fundamental airgap mmf, because they are in exact time phase in each of the three windings (see Table 6.1). These are termed the zero-sequence currents, and since the harmonic order is always a multiple of three, they are also termed triplen harmonics.

It has been assumed in Table 6.1 that only odd time harmonics of mmf are present in the airgap, and this is usually the case. In general, if both odd and even time harmonics are present, positive-sequence, negative-sequence, and zero-sequence harmonics are of order $k = (3n+1)$, $k = (3n-1)$, and $k = (3n)$, respectively, where n is an integer.

6.4. HARMONIC BEHAVIOR OF AC MOTORS

When an ac motor is operated on a nonsinusoidal supply, the stator voltage or current can be analyzed into a fundamental component and a series of harmonics. If magnetic saturation is neglected, the motor may be regarded as a linear device, and the principle of superposition can be applied. This means that motor behavior can be analyzed independently for the fundamental component and for each harmonic term. The overall response to the nonsinusoidal supply is then obtained as a summation of the responses to the individual components. Thus, if the motor is fed with a nonsinusoidal voltage, the net motor current and torque are given by the sum of the current and torque contributions of each voltage component in the supply waveform.

6.4.1. *Harmonic Equivalent Circuits*

The conventional induction motor equivalent circuit of Fig. 6.1 is repeated in Fig. 6.3(a). As before, X_1 and X_2 are the stator and rotor leakage reactances at the supply

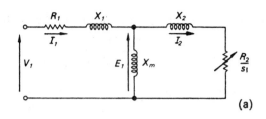

(a)

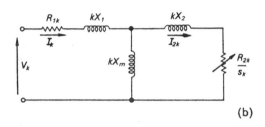

(b)

FIG. 6.3. Induction motor equivalent circuit diagrams: (a) fundamental-frequency equivalent circuit; (b) equivalent circuit for the kth time harmonic (positive or negative sequence).

frequency, and X_m is the corresponding magnetizing reactance. The rotor slip with respect to the fundamental rotating field, which has been denoted by s, is henceforth denoted by s_1, to distinguish it clearly from the harmonic slip. Hence,

$$s_1 = \frac{n_1 - n}{n_1} \tag{6.35}$$

where n_1 is the synchronous speed of the fundamental rotating field, and n is the actual rotor speed.

The kth harmonic components in the phase currents produce a time harmonic mmf wave rotating forward or backward at a speed kn_1. The rotor slip in a forward-rotating harmonic field is

$$s_k = \frac{kn_1 - n}{kn_1} \tag{6.36}$$

and for a backward-rotating field

$$s_k = \frac{kn_1 + n}{kn_1} . \tag{6.37}$$

In general, therefore,

$$s_k = \frac{kn_1 \mp n}{kn_1} \tag{6.38}$$

where the negative sign is valid for positive-sequence harmonics and the positive sign applies to negative-sequence harmonics.

The harmonic slip, s_k, is expressed in terms of s_1 by substituting for n from Equation 6.35, to give

$$s_k = \frac{(k \mp 1) \pm s_1}{k}. \tag{6.39}$$

The fundamental equivalent circuit of Fig. 6.3(a) may be adapted for the kth harmonic voltage and current as shown in Fig. 6.3(b). The harmonic slip, s_k, is substituted for the fundamental slip, s_1, and all inductive reactances are increased by a factor k. The stator and rotor resistances are also larger due to skin effect at the harmonic frequency. Strictly speaking, the rotor leakage inductance is also modified by the skin effect, and this difference must be taken into consideration in precise calculations.[3]

It may be verified by means of Equation 6.39 that there is very little variation in s_k for normal motor operation. If motor speed varies from synchronous speed to standstill, the fundamental slip, s_1, varies from 0 to 1, but the fifth harmonic slip, s_5, varies only from 1.2 to 1. The corresponding variation in s_7 is 0.857 to 1, and for higher harmonics, s_k is even closer to unity. The harmonic equivalent circuit of Fig. 6.3(b) can be simplified as shown in Fig. 6.4(a) by removing the resistances. This is justified by the fact that inductive reactances increase linearly with frequency, while the increase in rotor resistance with frequency due to skin effect is much less than linear. Since s_k is approximately unity, circuit resistance can be neglected in comparison with reactance at the harmonic frequencies. Further simplification is possible as in Fig. 6.4(b), because shunt magnetizing reactance is much greater than rotor leakage reactance and may be omitted. The motor impedance presented to harmonic currents is, therefore, approximately $k(X_1 + X_2)$ where X_1 and X_2 are the stator and rotor leakage reactances at fundamental supply frequency.

The zero-sequence stator current harmonics are in time phase, and consequently do not produce a space fundamental rotating mmf wave. However, zero-sequence harmon-

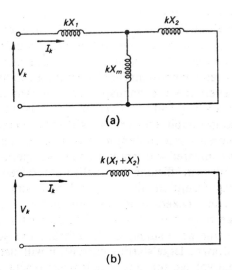

(a)

(b)

FIG. 6.4. Approximate equivalent circuits for harmonic current calculations.

ics may establish small pulsating space harmonic mmf waves in the airgap, and each pulsating wave can be resolved into a forward and backward traveling wave, as shown in Table 6.1. These flux waves induce unequal harmonic currents in the moving rotor, and hence the presence of zero-sequence stator currents can have some effect on motor torque.

The reactance presented to the flow of the kth zero-sequence harmonic is kX_0, where X_0 is the stator zero-sequence reactance at fundamental frequency. If X_0 is small and the applied voltage has a large zero-sequence component, the resulting zero-sequence currents may cause a significant stator copper loss, seriously reducing motor efficiency. However, zero-sequence currents can flow only in a wye-connected (star-connected) motor having a neutral connection between the source and load; otherwise there is no return path for the in-phase zero-sequence currents. In practice, most inverter circuits do not generate zero-sequence voltages, but if these components are present, they are offered an infinite zero-sequence impedance by eliminating the neutral connection.

The fundamental equivalent circuit of Fig. 6.3 (a) applies only to the polyphase induction motor, but the harmonic equivalent circuits of Fig. 6.4 are also valid for the harmonic behavior of synchronous motors of the field-excited, permanent magnet-excited or synchronous reluctance type, because these machines have cage windings or damper bars, and operate asynchronously with respect to time harmonic mmf waves. In the case of a salient-pole synchronous motor, the effective reactance in Fig. 6.4 (b) is the mean of the direct and quadrature axis subtransient reactances.

For analyzing motor operation at very low supply frequencies, the approximate harmonic equivalent circuits of Fig. 6.4 may not be valid, because winding resistance can be a significant factor at low harmonic frequencies. The simplifications are usually justified, however, if the fundamental frequency exceeds about 10 Hz.

6.4.2. Harmonic Currents

Since s_k is nearly unity at all motor speeds from standstill to synchronism, the harmonic equivalent circuit of Fig. 6.3 (b) is practically independent of motor speed, and this is emphasized by the approximate circuits derived from it. When the ac motor is fed by a voltage-source inverter with a specific output waveform at a particular frequency, the harmonic currents remain constant for all operating conditions of the motor from no load to full load and even down to standstill. The fundamental stator current is determined by the motor loading and, as a result, the relative harmonic content of the machine current is considerably greater for light-load operation than for the full-load condition. This greater harmonic content may cause a significant increase in the no-load losses of the machine compared with normal sine wave operation. However, as shown in the next section, the full-load efficiency is usually not reduced excessively.

The approximate equivalent circuit of Fig. 6.4 (b) is similar to that used for normal sine wave calculations on a locked-rotor induction motor, when the motor current is also limited by the leakage reactance $(X_1 + X_2)$. The standstill or starting behavior of the induction motor on a sine wave supply is, therefore, a measure of its harmonic performance. If the motor draws a large starting current, it will also draw large harmonic currents on nonsinusoidal voltage supplies. The leakage reactance of a synchronous reluctance motor also determines its harmonic current flow. If the motor has a very low leakage reactance, it should be used with caution on nonsinusoidal voltages, because excessive harmonic currents may flow and overheat the motor.

If V_k denotes the kth harmonic component of the supply voltage, the corresponding stator current harmonic is $I_k = V_k / Z_k$, where Z_k is the kth harmonic input impedance. For positive- and negative-sequence harmonics, the approximate equivalent circuit of Fig. 6.4(b) is valid, and $Z_k = k(X_1 + X_2)$. Thus,

$$I_k = \frac{V_k}{k(X_1 + X_2)} . \qquad (6.40)$$

For zero-sequence harmonics, $Z_k = kX_0$, and

$$I_k = \frac{V_k}{kX_0} . \qquad (6.41)$$

These formulae permit rapid evaluation of the harmonic currents due to a nonsinusoidal voltage waveform whose harmonic content is known. Usually, there are no zero-sequence harmonics and no even-numbered harmonics, and hence the total rms harmonic current is given by

$$I_{har} = \left[I_5^2 + I_7^2 + I_{11}^2 + I_{13}^2 + \cdots + I_k^2 + \cdots \right]^{1/2}$$

$$= \left[\sum_{k \neq 1} I_k^2 \right]^{1/2} . \qquad (6.42)$$

If I_1 is the fundamental rms current of the motor, the total rms stator current, including the fundamental, is

$$I_{rms} = \left[I_1^2 + I_5^2 + I_7^2 + I_{11}^2 + I_{13}^2 + \cdots + I_k^2 + \cdots \right]^{1/2}$$

$$= \left[I_1^2 + I_{har}^2 \right]^{1/2} . \qquad (6.43)$$

For a given voltage waveform, the relative harmonic content of the stator current is closely related to the leakage reactance of the motor. It is convenient to express motor current and leakage reactance in normalized or per-unit (pu) form: that is, the actual values of current and reactance are expressed as fractions of certain base values. Base current is rated full-load sine wave current, I_{FL}, and base reactance, X_{base}, equals V_R / I_{FL}, where V_R is rated sine wave phase voltage. Thus, the per-unit leakage reactance of the motor at the rated fundamental frequency, f_R, is

$$X_{pu} = \frac{(X_1 + X_2)}{X_{base}} = (X_1 + X_2) \frac{I_{FL}}{V_R} . \qquad (6.44)$$

Also,

$$X_{pu} = \frac{I_{FL}}{I_s} \sin \phi_s \qquad (6.45)$$

where I_s is the fundamental standstill current of the motor for a direct-on-line start at rated voltage and frequency, and ϕ_s is the corresponding power factor angle.

Combining Equations 6.40 and 6.44, the kth harmonic current can be expressed in per-unit form as a fraction of the rated full-load current. Thus,

$$I_k = \frac{V_k}{kX_{pu}}$$

(6.46)

where V_k is now the per-unit k^{th} harmonic voltage based on the rated sine wave voltage of the motor.

For operation at rated frequency, X_{pu} is the normal per-unit leakage reactance parameter of the motor, but this reactance obviously varies linearly with frequency. It is convenient to retain X_{pu} as the per-unit reactance at rated or base frequency, f_R, and to take account of the frequency dependence of leakage reactance by means of a multiplying factor, f_1, which is the per-unit fundamental frequency, and is unity at rated motor frequency. The kth harmonic per-unit current at a per-unit fundamental frequency, f_1, is then given by

$$I_k = \frac{V_k}{k f_1 X_{pu}}.$$

(6.47)

For the six-step voltage waveform, the magnitude of each harmonic voltage is inversely proportional to the order of the harmonic. Thus,

$$V_k = V_1/k$$

(6.48)

and Equation 6.47 gives the per-unit harmonic current as

$$I_k = \frac{V_1}{k^2 f_1 X_{pu}}.$$

(6.49)

Fundamental airgap flux is directly proportional to stator induced emf, E_1, and inversely proportional to frequency. To a good approximation, except at low frequency, airgap flux is therefore proportional to V_1/f_1. If the base value of airgap flux is that corresponding to rated terminal voltage, V_R, at rated frequency, f_R, then the per-unit fundamental airgap flux is

$$\Phi_1 = V_1/f_1$$

(6.50)

where V_1 and f_1 are also expressed in per-unit form.

Substituting in Equation 6.49 gives

$$I_k = \frac{\Phi_1}{k^2 X_{pu}}.$$

(6.51)

For normal constant volts/hertz operation below base frequency, Φ_1 is unity and

$$I_k = \frac{1}{k^2 X_{pu}}.$$

(6.52)

Since X_{pu} is the per-unit reactance at base frequency, it is evident that harmonic current amplitudes are independent of supply frequency and motor load if the volts/hertz ratio is constant.

Using Equations 6.42 and 6.52, the total rms per-unit harmonic current with a six-step voltage supply is evaluated as $0.046/X_{pu}$. Similar calculations on a 12-step supply give a value of $0.0105/X_{pu}$. Thus, the harmonic rms current is inversely proportional to the per-unit reactance. The total rms stator current at full load on a per-unit basis is $[1 + (0.046/X_{pu})^2]^{1/2}$ for a six-step voltage waveform and $[1 + (0.0105/X_{pu})^2]^{1/2}$ with a 12-step supply. In Fig. 6.5, the total rms stator current is plotted as a function of the per-unit reactance. The increase in rms current is almost negligible with a 12-step supply, but the six-step voltage waveform can produce a significant increase, particularly when the per-unit reactance is small. This large increase in current occurs principally with synchronous reluctance motors, which may have a per-unit reactance as low as 0.05, resulting in a 35 percent increase in the rms full-load current. The polyphase induction motor usually has a per-unit reactance in the range of 0.1 to 0.2, and the total rms current at full load on a six-step voltage supply is from 2 to 10 percent greater than the fundamental current.

Figure 6.6 shows a typical stator current waveform for an ac motor operating on a six-step voltage supply. This waveform was calculated for a per-unit reactance of 0.1, assuming that fundamental current lags fundamental voltage by 60 degrees. This fundamental phase angle is determined by the loading conditions and, in the present case, it corresponds to a fundamental power factor of 0.5. The corresponding current waveform with a 12-step voltage supply is shown in Fig. 6.7. The harmonic distortion not only increases the rms value of the stator current but also produces large current peaks, which increase the required current rating of the inverter transistors (Chapter 12), or increase the commutating duty imposed on the inverter thyristors (Chapter 13). In Fig. 6.8, the

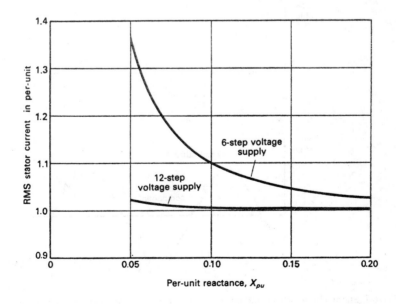

FIG. 6.5. RMS stator current as a function of the per-unit leakage reactance of the motor.

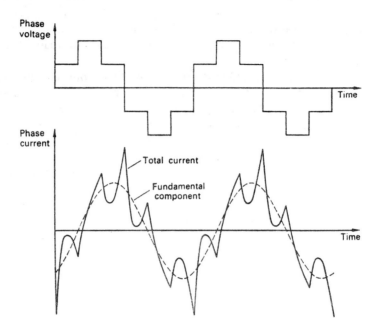

FIG. 6.6. Stator phase voltage and current waveforms for a wye-connected ac motor with a six-step voltage supply.

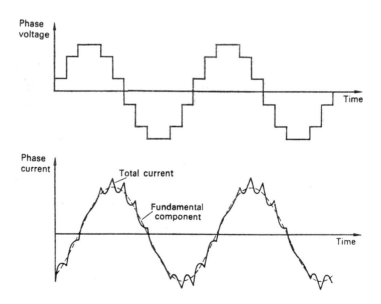

FIG. 6.7. Stator phase voltage and current waveforms for a wye-connected ac motor with a 12-step voltage supply.

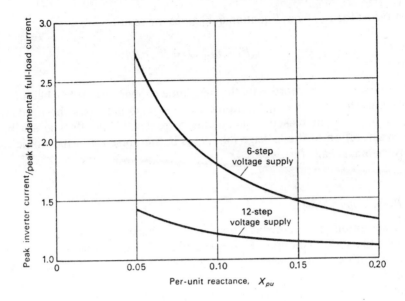

FIG. 6.8. Peak inverter current as a function of the per-unit leakage reactance of the motor.

ratio of peak inverter current to peak fundamental full-load current is plotted as a function of per-unit reactance. These characteristics were derived theoretically, assuming a fundamental power factor or displacement factor of 0.5. If the displacement factor is greater than this, the peak inverter currents are lower than the values indicated in Fig. 6.8.

6.5. MOTOR LOSSES ON NONSINUSOIDAL SUPPLIES

The additional losses that are present in an ac motor with nonsinusoidal excitation are now considered. For comparison with normal sine wave operation at rated frequency of a 50 or 60 Hz ac motor, typical numerical values are quoted for operation with a six-step voltage supply, having a fundamental frequency of 50 or 60 Hz and a fundamental voltage component of the same amplitude as the rated sine wave voltage of the motor. Appropriate PWM strategies can reduce harmonic motor losses below the values for six-step voltage operation, but a careless choice of modulation strategy can result in significantly higher losses. Harmonic motor losses for PWM voltage waveforms and six-step current waveforms are considered in more detail later in this chapter.

6.5.1. *Stator Copper Loss*

The presence of harmonic currents in the stator winding causes an increased I^2R loss. When skin effect is negligible, the stator copper loss on a nonsinusoidal supply is proportional to the square of the total rms current. If m_1 is the number of stator phases and R_1 is the stator resistance per phase, the total stator copper loss is

$$P_1 = m_1 I_{rms}^2 R_1. \tag{6.53}$$

Substituting for I_{rms} from Equation 6.43 gives the result

$$P_1 = m_1 (I_1^2 + I_{har}^2) R_1 \tag{6.54}$$

where the second term represents the harmonic copper loss. It has been found experimentally that the presence of harmonic currents also increases the fundamental term slightly, due to an increased magnetizing current.[4] This effect is attributed to the increased saturation of the leakage flux paths in the presence of harmonic currents and their associated leakage fluxes.

6.5.2. Rotor Copper Loss

The assumption of a constant resistance at harmonic frequencies is reasonably justified for the stator windings of wire-wound machines. For large ac motors there is an increase in stator resistance with frequency which depends on the shape, size, and disposition of the conductors in the stator slot. However, the skin effect is much more pronounced in the cage rotor, which exhibits a significant increase in resistance at harmonic frequencies, particularly in the case of deep-bar rotors. In a synchronous machine operating on a nonsinusoidal supply, time harmonic mmf waves induce rotor-cage harmonic currents, as in an induction motor operating near its fundamental synchronous speed. In both synchronous and asynchronous motors, the fifth harmonic mmf rotating backward and the seventh harmonic rotating forward will induce rotor currents of six times fundamental frequency — that is, 300 Hz in the case of a 50 Hz supply. Similarly, the eleventh and thirteenth harmonics induce rotor currents of 12 times fundamental frequency, or 600 Hz. At these harmonic frequencies, the rotor resistance is much greater than the dc value. The actual increase depends on the geometrical shape of the conductor cross section and of the rotor slot in which it is placed. Alger has published curves giving the factor by which resistance is increased.[3] For a rectangular copper conductor of 2 cm depth in a rectangular rotor slot, the ratio of the 50 Hz ac resistance to the dc resistance is approximately 1.3; at 300 Hz, the ratio is 3.3; and at 600 Hz, it is 4.7. At higher frequencies, the ratio increases as the square root of the frequency.

Since the rotor resistance is a function of harmonic frequency, the rotor copper loss is calculated independently for each harmonic. In general, for the kth harmonic, the rotor copper loss is

$$P_{2k} = m_1 (I_{2k})^2 R_{2k} \tag{6.55}$$

where I_{2k} is the kth harmonic rotor current, and R_{2k} is the corresponding rotor resistance, corrected for skin effect. It may also be appropriate to use a reduced value of per-unit reactance because rotor leakage inductance is reduced significantly as a result of skin effect.[3] The total leakage inductance of the motor at high harmonic frequencies is typically 80 or 90 percent of its value at rated frequency.[5,6] The total harmonic copper loss is obtained as a summation of the harmonic contributions. In many induction motors, the additional rotor copper loss due to harmonic currents is the principal cause of reduced efficiency on nonsinusoidal supplies.

6.5.3. Harmonic Core Loss

The core loss in the machine is also increased by the presence of harmonics in the supply voltage and current. As already explained, a time harmonic mmf wave is established in the airgap by each stator current harmonic. These time harmonic waves of mmf have the same number of poles as the fundamental field, but rotate forward or backward at a multiple of the fundamental speed. However, the resultant time harmonic airgap fluxes are small. In the case of a three-phase, six-step voltage supply, for example, the fundamental flux is normally about 1 pu. At harmonic frequencies, stator and rotor voltage drops are approximately equal, and the fifth harmonic applied voltage of 0.2 pu produces an airgap emf of 0.1 pu. Since the harmonic frequency is five times fundamental frequency, the fifth harmonic airgap flux is only 0.02 pu. Similarly, the seventh harmonic airgap flux is about 0.01 pu. These small time harmonic main fluxes cause negligible increase in the core loss of the motor, a fact that has been confirmed by experiment.[7]

Core loss due to space harmonic airgap fluxes is also negligible, but the end-leakage and skew-leakage fluxes, which normally contribute to the stray load loss, may produce an appreciable core loss at harmonic frequencies. Consequently, these effects must be taken into consideration for motor operation on nonsinusoidal supplies.[7-9] The end-leakage loss is the eddy-current loss in the core laminations due to end-winding leakage flux, which enters the laminations in an axial direction. The end-leakage effect is present in both stator and rotor windings, and this loss is present in the stator and rotor laminations and in associated structural components.

As shown above, time harmonic airgap fluxes are normally negligible, indicating that stator and rotor harmonic mmfs almost completely neutralize one another. However, there is a skew-leakage effect in cage-rotor induction motors in which rotor slots are skewed with respect to stator slots. This construction results in an angular phase difference along the core length between the peak values of stator and rotor mmf. If the airgap mmfs of stator and rotor conductors balance one another midway along the core length, there is a resultant radial airgap mmf as one moves axially in either direction. This skew mmf, which is greatest at the core ends, establishes a skew-leakage flux in the airgap that produces a core loss in the stator and rotor laminations. For fundamental currents, the core loss is small, because the skew flux changes at fundamental frequency in the stator core and at slip frequency in the rotor core. For harmonic currents, however, the skew-leakage flux changes at harmonic frequency in both stator and rotor cores, and the associated skew-leakage loss can be substantial. Chalmers and Sarkar have confirmed the importance of time harmonic end-leakage and skew-leakage losses in a detailed analytical and experimental study of harmonic losses in inverter-fed induction motors.[7] They have shown that these time harmonic core losses may approach or even exceed the harmonic copper losses in the motor.

6.5.4. Motor Efficiency

The magnitude of the harmonic losses obviously depends on the harmonic content of the motor voltage and current. Large harmonic voltages at low harmonic frequencies cause significantly increased machine losses and reduced efficiency. However, most inverters do not generate harmonics of lower order than the fifth, and high-order harmonic currents usually have small amplitudes. For such waveforms, the reduction in full-load motor efficiency is not excessive.[10-12]

Typical harmonic loss values are now quoted for a 50 or 60 Hz three-phase induction motor operating at rated frequency on a six-step voltage supply. The motor receives a fundamental voltage component equal to the rated sine wave voltage of the machine. For a typical leakage reactance of 0.15 pu, Fig. 6.5 shows that the full-load rms current is 1.045 pu, or 4.5 percent greater than the fundamental value. If skin effect is neglected, motor copper loss is proportional to the square of the total rms current, and harmonic copper loss is approximately 9 percent of the fundamental copper loss. Allowing an average three-fold increase in motor resistance due to skin effect, the harmonic copper loss can be estimated at 27 percent of the fundamental loss. If the fundamental copper loss constitutes 50 percent of total machine losses, an increase of 13.5 percent in overall machine losses results.

The increase in core loss is less predictable because it is influenced by the machine construction and magnetic materials used. If the high-order harmonic content of the stator voltage waveform is relatively low, as in the six-step wave, the total harmonic core loss due to skew-leakage and end-leakage effects should not exceed 25 percent of the fundamental core loss. Assuming that the fundamental core loss constitutes approximately 40 percent of the machine losses, the harmonic core loss contribution to total machine losses is about 10 percent. Friction and windage losses are obviously unaffected, and consequently, the overall increase in motor losses is about 23.5 percent. If the motor has a normal full-load efficiency of 90 percent on a sinusoidal supply, the presence of harmonics increases the machine losses from 10 percent to 12.2 percent, and motor efficiency falls by about 2 percent. This is a small reduction in efficiency, but the increased motor losses may necessitate some derating of the motor to avoid overheating. This aspect is discussed in Chapter 13.

If the harmonic content of the applied voltage waveform is significantly greater than that of the six-step wave, the harmonic motor losses can be increased considerably, and may be greater than the fundamental losses.[4] Even with a six-step voltage supply, a synchronous reluctance motor of low leakage reactance may draw very large harmonic currents, causing a drop in motor efficiency of 5 percent or more. In the induction motor or synchronous motor, the harmonic currents and losses are practically independent of load, and the no-load losses of an inverter-fed motor may be appreciably greater than those of a sinusoidally excited machine. However, as shown above, full-load efficiency is usually not seriously affected. The magnitude of the total time harmonic loss is readily determined by comparing no-load losses on sinusoidal and nonsinusoidal supplies.

6.6. HARMONIC TORQUES

The presence of time harmonic mmf waves in the airgap results in additional harmonic torques on the rotor. These torques are of two types: (a) steady harmonic torques and (b) pulsating harmonic torques.

6.6.1. Steady Harmonic Torques

Constant or steady torques are developed by the reaction of harmonic airgap fluxes with harmonic rotor mmfs, or currents, of the same order. However, these steady harmonic torques, which are a very small fraction of rated torque, have negligible effect on motor operation. This may be verified by calculating the torque contribution from the harmonic equivalent circuit, just as fundamental torque is derived from the fundamental equivalent circuit. Thus, the fundamental torque is given by Equation 6.7 as

$$T_1 = \frac{pm_1}{2\pi f_1} (I_2)^2 \frac{R_2}{s_1} .$$

Similarly, the kth harmonic torque contribution, T_k, can be calculated from the equivalent circuit of Fig. 6.3(b), giving

$$T_k = \pm \frac{pm_1}{2\pi k f_1} (I_{2k})^2 \frac{R_{2k}}{s_k} \qquad (6.56)$$

where forward torque due to a positive-sequence harmonic is positive and backward torque due to a negative-sequence harmonic is negative.

The fundamental slip, s_1, is small for normal full-load operation of the induction motor, and consequently, Equation 6.39 for the harmonic slip becomes

$$s_k = \frac{k \mp 1}{k} \qquad (6.57)$$

Substituting in Equation 6.56 gives

$$T_k = \pm \frac{pm_1}{2\pi f_1} (I_{2k})^2 \frac{R_{2k}}{k \mp 1} \qquad (6.58)$$

The kth harmonic torque, expressed as a fraction of the fundamental torque, is therefore

$$\frac{T_k}{T_1} = \pm \left(\frac{I_{2k}}{I_2} \right)^2 \left(\frac{R_{2k}}{R_2} \right) \left(\frac{s_1}{k \mp 1} \right) . \qquad (6.59)$$

If the motor operates at rated load, then I_2 is near rated current, and (I_{2k}/I_2) is approximately equal to the per-unit k^{th} harmonic current I_k, as given by Equation 6.47. Substituting this expression in Equation 6.59 gives the general equation

$$\frac{T_k}{T_1} = \pm \left(\frac{V_k}{k f_1 X_{pu}} \right)^2 \left(\frac{R_{2k}}{R_2} \right) \left(\frac{s_1}{k \mp 1} \right) . \qquad (6.60)$$

In the case of a six-step voltage waveform, the per-unit value of I_k is given by Equation 6.51, and substituting for I_{2k}/I_2 in Equation 6.59 gives the corresponding harmonic torque equation as

$$\frac{T_k}{T_1} = \pm \left(\frac{\Phi_1}{k^2 X_{pu}} \right)^2 \left(\frac{R_{2k}}{R_2} \right) \left(\frac{s_1}{k \mp 1} \right) \qquad (6.61)$$

where Φ_1 is the per-unit airgap flux as defined by Equation 6.50.

Now consider the fifth harmonic torque in the case of an induction motor operating on a six-step voltage supply. Rated fundamental voltage and frequency are applied so that Φ_1 is unity. A three-fold increase in rotor resistance due to skin effect is assumed, so that

$R_{2k}/R_2 = 3$. If the fundamental full-load slip is 0.03, substituting in Equation 6.61 gives $T_5/T_1 = -0.24 \times 10^{-4}/X_{pu}^2$. Thus, for a leakage reactance of 0.1 pu, $T_5/T_1 = -0.0024$, or T_5 is 0.24 percent of the fundamental torque. For a leakage reactance of 0.2 pu, T_5 is only 0.06 percent of T_1. This small counter-torque due to the negative-sequence fifth harmonic is opposed by a somewhat smaller forward torque due to the positive-sequence seventh harmonic. The combined effect of the fifth and seventh harmonics is, therefore, to produce a very small negative torque opposing the fundamental motor torque. This also applies to the eleventh and thirteenth harmonics, and the overall effect of the supply harmonics is a negligible reduction in steady torque of a fraction of 1 percent.

6.6.2. Pulsating Harmonic Torques

As shown above, a steady component of torque is produced by the reaction of each rotor harmonic mmf with a harmonic airgap flux of the same order. Pulsating torque components are produced by the reaction of harmonic rotor mmfs with harmonic rotating fluxes of a different order. As shown earlier, the harmonic airgap fluxes are small, and the dominant pulsating torques arise from the interaction between harmonic rotor currents, or mmfs, and fundamental rotating flux. The fifth harmonic stator currents, for example, form a negative-sequence system and produce a space fundamental mmf wave that rotates at five times fundamental synchronous speed in the opposite direction to the fundamental field. The rotor currents induced by this time harmonic field react with the fundamental rotating field to produce a pulsating torque at six times fundamental frequency, because the relative speed of the rotor mmf wave and the fundamental airgap field is six times synchronous speed.

The seventh harmonic stator currents also produce a pulsating torque at six times fundamental frequency. The seventh harmonic has positive phase sequence and therefore produces a time harmonic field rotating at seven times synchronous speed in the same direction as the fundamental field. The relative speed of the main airgap field and the rotor harmonic mmf wave is again six times synchronous speed, and the two pulsating torques at six times fundamental frequency combine to produce a fluctuation in the electromagnetic torque developed by the motor. Similarly, the eleventh and thirteenth harmonics produce a twelfth harmonic pulsating torque, but the sixth harmonic component predominates in the case of a six-step supply.

The pulsating torques have zero average value, but their presence causes the angular velocity of the rotor to vary during a revolution. At very low speeds, motor rotation takes place in a series of jerks or steps, and this irregular cogging motion sets a lower limit to the useful speed range of the motor. The point at which the speed pulsation becomes objectionable depends on the inertia of the rotating system. In certain applications, such as machine tool drives, the speed fluctuation is intolerable. Abnormal wearing of gear teeth can also occur, particularly if the torque pulsation coincides with a shaft mechanical resonance. This resonant frequency is usually below 100 Hz, however, and a sixth harmonic torque pulsation is outside the range of shaft resonance over most of the speed range. Sustained operation in the low-speed region requires an improved inverter output waveform in which low-order harmonics are suppressed.

In Section 6.2, fundamental torque has been expressed by Equation 6.14 as

$$T = K \, \Phi_1 \, I_2 \cos \phi_2 \, .$$

This equation can also be written in per-unit form. As usual, fundamental airgap flux has a base value corresponding to rated stator voltage at rated frequency, and base current is rated motor current. Base torque is then defined as the torque corresponding to a rotor current of 1 per unit, with an airgap flux of 1 per unit and a rotor power factor of unity. This base torque is somewhat larger than the rated torque of the motor. In per-unit form, the fundamental torque equation becomes

$$T = \Phi_1 I_2 \cos \phi_2 . \tag{6.62}$$

Similarly, the per-unit form of Equation 6.13 is

$$T = \Phi_1 I_2 \sin \delta_2 = \Phi_1 I_1 \sin \delta_1 \tag{6.63}$$

where a forward or positive motoring torque is produced when currents I_1 and I_2 lead airgap flux Φ_1 by torque angles δ_1 and δ_2, respectively, as shown in Fig. 6.2.

These equations can also be used to determine pulsating torque amplitudes.[13,14] Initially, the dominant airgap flux and stator current phasors are superimposed in a single diagram. Consider an arbitrary phase voltage waveform with fundamental, fifth, and seventh harmonic voltages. Figure 6.9 shows each voltage phasor rotating at its own synchronous speed. The positive-sequence seventh harmonic rotates at 7ω in the positive, or anticlockwise, direction, while the negative-sequence fifth harmonic rotates at 5ω in the negative, or clockwise, direction. Selecting the phasor rotation in this manner ensures that each rotor mmf space vector has the correct sense of rotation. At the instant chosen for the phasor diagram, the instantaneous fundamental voltage is zero and about to become positive. The fifth harmonic voltage, V_5, is negative and increasing, while the seventh harmonic voltage, V_7, is positive and increasing.

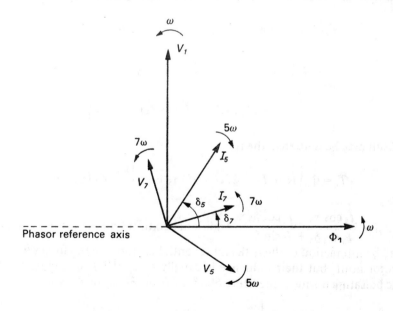

FIG. 6.9. Combined phasor diagram for fundamental, fifth, and seventh harmonics.

Assuming that the fundamental stator voltage, V_1, is in phase with the stator induced emf, E_1, then the fundamental flux, Φ_1, lags V_1 by 90 degrees, as shown in Fig. 6.9. Neglecting the effects of machine resistance at the harmonic frequencies and assuming that harmonic magnetizing currents are negligible, the harmonic phasor diagrams are readily completed. Since the harmonic equivalent circuit is purely reactive, the harmonic stator current lags the corresponding terminal voltage by 90 degrees. If the influence of machine resistance is significant, the modifications to the phasor diagram are straightforward.

At the instant chosen for Fig. 6.9, harmonic current, or mmf, I_5, leads fundamental flux, Φ_1, by a torque angle, δ_5. By analogy with the fundamental phasor diagram, these waves interact to produce a positive motoring torque given by

$$T = \Phi_1 I_5 \sin \delta_5 \qquad (6.64)$$

where all quantities are in per-unit form.

However, angle δ_5 is decreasing at a rate of 6ω and, after time, t, the initial angle, δ_5, becomes $(\delta_5 - 6\omega t)$, giving

$$T = \Phi_1 I_5 \sin (\delta_5 - 6\omega t) . \qquad (6.65)$$

Similarly, at the instant chosen for Fig. 6.9, the seventh harmonic current, I_7, leads fundamental flux, Φ_1, by a torque angle, δ_7, which is increasing at a rate of 6ω. Thus, there is an additional positive torque component of

$$T = \Phi_1 I_7 \sin (\delta_7 + 6\omega t) . \qquad (6.66)$$

Obviously, each torque component is pulsating with an angular frequency of 6ω, and the net sixth harmonic pulsating torque is obtained by addition. Hence,

$$T_6 = \Phi_1 \left[I_5 \sin (\delta_5 - 6\omega t) + I_7 \sin (\delta_7 + 6\omega t) \right]$$

$$= \Phi_1 \left[(I_5 \sin \delta_5 + I_7 \sin \delta_7) \cos 6\omega t \right.$$

$$\left. - (I_5 \cos \delta_5 - I_7 \cos \delta_7) \sin 6\omega t \right] . \qquad (6.67)$$

This result may be written in the form

$$T_6 = \Phi_1 \left[I_5^2 + I_7^2 - 2 I_5 I_7 \cos (\delta_5 + \delta_7) \right]^{1/2} \cos (6\omega t + \beta) \qquad (6.68)$$

where $\tan \beta = \dfrac{I_5 \cos \delta_5 - I_7 \cos \delta_7}{I_5 \sin \delta_5 + I_7 \sin \delta_7}$. Additional sixth harmonic torque components are developed by interaction of the fifth and seventh harmonic airgap fluxes with the fundamental rotor mmf, but their influence is usually small.[13] The amplitude of the sixth harmonic pulsating torque is therefore obtained from Equation 6.68 as

$$T_6 = \Phi_1 \left[I_5^2 + I_7^2 - 2 I_5 I_7 \cos (\delta_5 + \delta_7) \right]^{1/2} . \qquad (6.69)$$

A similar analysis gives the per-unit amplitude of the twelfth harmonic torque due to the eleventh and thirteenth harmonic currents as

$$T_{12} = \Phi_1 \left[I_{11}^2 + I_{13}^2 - 2 I_{11} I_{13} \cos (\delta_{11} + \delta_{13}) \right]^{1/2}. \tag{6.70}$$

In general, harmonic currents of order $k = (6n-1)$ and $(6n+1)$, where n is an integer, develop a pulsating torque of order $6n$, with a per-unit amplitude given by

$$T_{6n} = \Phi_1 \left[I_{6n-1}^2 + I_{6n+1}^2 - 2 I_{6n-1} I_{6n+1} \cos(\delta_{6n-1} + \delta_{6n+1}) \right]^{1/2}. \tag{6.71}$$

This result is also applicable to a synchronous motor with a pole-face cage or damper winding.

Equation 6.71 can be used to evaluate pulsating torque amplitudes for an ac motor with a six-step voltage supply. In this case, the phase voltage waveform is given by Equation 4.8 as

$$v_{AN} = \frac{2}{\pi} V_d \left[\sin \omega t + \frac{1}{5} \sin 5\omega t + \frac{1}{7} \sin 7\omega t \right.$$

$$\left. + \frac{1}{11} \sin 11\omega t + \frac{1}{13} \sin 13\omega t + \cdots \right].$$

The combined phasor diagram for the fundamental, fifth, and seventh harmonics is shown in Fig. 6.10, and it is evident that the initial torque angles δ_5 and δ_7 are both zero.

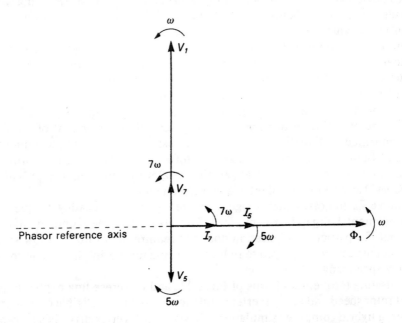

FIG. 6.10. Combined phasor diagram for an ac motor with a six-step voltage supply.

M–PEC—I

This conclusion is valid for all higher order harmonics so that, in general, δ_{6n-1} and δ_{6n+1} are zero. Also, with a six-step voltage supply, the kth harmonic current is given by Equation 6.51 as

$$I_k = \frac{\Phi_1}{k^2 X_{pu}}$$

where X_{pu} is the per-unit leakage reactance of the induction motor or per-unit mean subtransient reactance for a synchronous motor.

Substituting in Equation 6.71 gives the pulsating torque amplitude as

$$T_{6n} = \Phi_1\,[I_{6n-1} - I_{6n+1}] = \frac{\Phi_1^2}{X_{pu}}\left[\frac{1}{(6n-1)^2} - \frac{1}{(6n+1)^2}\right]. \tag{6.72}$$

The negative sign in this equation indicates that the individual pulsating torque components due to I_{6n-1} and I_{6n+1} are in direct phase opposition, thereby reducing the resultant torque amplitude. The dominant sixth harmonic pulsating torque is obtained when $n = 1$. Thus,

$$T_6 = \frac{\Phi_1^2}{X_{pu}}\left[\frac{1}{5^2} - \frac{1}{7^2}\right] = 0.02\,\frac{\Phi_1^2}{X_{pu}}. \tag{6.73}$$

For constant volts/hertz operation, the airgap flux is nearly constant at 1 pu, and typical values of leakage reactance lie between 0.1 and 0.2 pu. From Equation 6.73, the corresponding sixth harmonic pulsating torque amplitude is in the range from 0.2 to 0.1 pu — that is, from 20 percent to 10 percent of base torque — with lower reactance machines obviously developing larger torque pulsations. The corresponding twelfth harmonic pulsating torque is between 2.3 percent and 1.2 percent of base torque.

The instantaneous electromagnetic torque waveform can be readily obtained from an analog or digital computer simulation in which the six-step voltage waveform is applied to the ac motor. Figure 6.11 shows the torque variation during one cycle of the supply frequency for a 2 hp induction motor operating with nominal volts/hertz at a stator frequency of 50 Hz. These waveforms were obtained by means of a digital simulation. As expected, the sixth harmonic torque component is dominant. The oscillograms also show that the pulsating torque amplitude is practically independent of motor loading. This is confirmed by Equation 6.72, which expresses harmonic torque amplitude solely in terms of fundamental airgap flux and motor reactance. In fact, for normal constant volts/hertz operation where the airgap flux is almost constant, the pulsating torque amplitude will also be independent of supply frequency.

Equation 6.72 indicates that a significant reduction in pulsating torque is possible when fundamental airgap flux is reduced. This also diminishes fundamental torque, of course, but if a particular application does not require high torque at low speeds, then field weakening can be employed to reduce pulsating torque amplitude and so extend the useful low-speed range of operation.

The pulsating torque waveforms of Fig. 6.11, and the preceding expressions, assume constant rotor speed and zero inverter regulation — that is, negligible output impedance. However, a hybrid computer simulation of a six-step inverter drive has shown that pulsating torque amplitude may be considerably accentuated by the voltage fluctuations

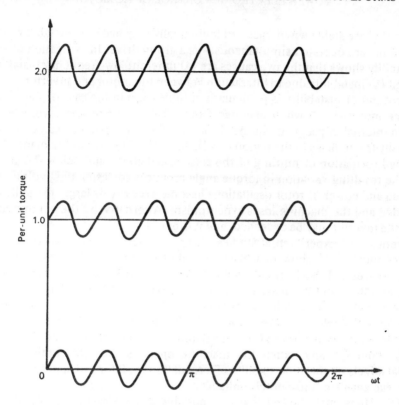

FIG. 6.11. Induction motor torque pulsations at no load, at rated torque, and at twice rated torque, with a six-step voltage supply.

present in a practical system with finite inverter regulation.[15] The harmonic currents delivered by the inverter produce a sixth harmonic voltage ripple on the dc link capacitor, which amplifies the harmonic voltage components in the inverter output voltage. As a result, harmonic current flow to the motor increases and the sixth harmonic pulsating torque is amplified. It has been found that the low-speed torque pulsations may be two or three times the theoretical values predicted by an analysis which assumes constant terminal voltage and speed. A large dc link capacitor will limit this effect.

Low-order torque pulsations can be avoided by supplying the motor with an improved voltage or current waveform, such as a PWM waveform employing a sinusoidal modulation strategy. However, it is a characteristic of PWM techniques that large-amplitude torque pulsations are produced at high switching frequencies. This aspect will be considered further in Section 6.9 of this chapter. Increasing the number of stator phases has also been suggested as a means of reducing the amplitude of the torque pulsation.[16,17]

6.7. DRIVE STABILITY

When ac motors are operated on adjustable-frequency supplies, system instability may occur for certain critical frequency ranges and loading conditions. Machines that are perfectly stable on an ac utility network may become unstable with an inverter supply, and machines that are stable when operated individually may become unstable when several of the motors are operated simultaneously as a group drive. Investigation of sources of this instability shows that the two causes are (a) inherent low-frequency instability in the motor and (b) instability due to interaction between the motor and inverter.

Inherent motor instability is particularly troublesome in the case of the synchronous reluctance motor. Even when operated from a balanced sinusoidal voltage source of negligible internal impedance, the synchronous reluctance machine can exhibit instability at supply frequencies in the region of 5 Hz to 20 Hz. This instability manifests itself as a sustained oscillation or hunting of the rotor speed above and below the synchronous value. The resulting variation in torque angle produces corresponding pulsations in output torque and power. If rotor oscillations become excessively large, the pull-out torque is exceeded and the machine loses synchronism. Alternatively, the commutating capability of the inverter may be exceeded, and it trips out.

Theoretical and experimental studies of the inherent instability of the synchronous reluctance motor[18,19] show that machine stability is improved by reducing stator and rotor resistances and also by reducing the direct-axis/quadrature-axis reactance ratio. An increase in stator and rotor leakage reactances also reduces the tendency of the machine to oscillate, and machine stability generally improves with an increase in pole number. Lawrenson and Bowes[19] have shown that the direct-axis/quadrature-axis resistance ratio of the rotor circuits has a pronounced influence on machine stability. The optimum value is about 0.5 and, when the machine operates near this optimum ratio, the beneficial effects on machine stability of a reduction in stator resistance or an increase in leakage reactance are considerably enhanced.

In 1972, Honsinger showed that it is possible to construct synchronous reluctance motors that are inherently stable over a wide range of operating frequencies.[20,21] These motors have magnetically saturable bridges in the rotor quadrature axis — that is, between the rotor poles — and the degree of saturation is highly variable with respect to motor load. By designing the machine so that each bridge is unsaturated at no load and is completely saturated at pull-out, the motor can be inherently stable and yet display higher pull-out torque than earlier designs, together with improved pull-in torque and lower starting current.

Inherent instability in the dc-excited synchronous motor and induction motor has also been investigated.[22–25] A well-designed dc-excited synchronous motor should not exhibit instability at low speeds, but it may become unstable in certain cases. The transient response of the induction motor becomes more oscillatory as the supply frequency is reduced, but the normal machine does not usually become unstable on an infinite system. However, small motors with a low inertia constant may be unstable. Reducing the magnetizing reactance and increasing the stator and rotor resistances may improve induction motor stability.

Instability due to interaction between motor and converter can occur with synchronous, synchronous reluctance, or induction motors. It arises when the converter which supplies the adjustable-speed motor has a finite source impedance. The impedance may be introduced by a transformer or by the filter which smooths the dc link supply in a static inverter drive. System instability usually occurs at frequencies below 25 Hz when an interchange of energy takes place between motor inertia and filter inductance and capacitance.[26]

In an open-loop adjustable-speed drive, the unstable region of operation is normally confined to a certain torque and frequency range. A torque-frequency diagram can be prepared, as in Fig. 6.12, in which the unstable zone is enclosed by the contour. The critical contour can be predicted by use of the D-decomposition method,[19] or by the application of the root-locus technique or the Nyquist stability criterion to a linearized set of machine equations which are valid for small frequency excursions above and below a fixed base frequency. It is found that the size of the unstable zone is affected by the system inertia, the load damping, and the electrical parameters of the motor and supply source. The harmonic content of the stator voltage waveform may slightly affect system stability, particularly if the inertia is small. However, open-loop system stability can usually be assured over the desired range of torque and speed by proper coordination of motor and inverter. Stability of the inverter-fed induction motor drive is generally enhanced (a) by increasing load torque and inertia, (b) by reducing stator voltage, and (c) by increasing filter capacitance and reducing filter inductance and resistance.

The PWM inverter has a stepdown transformer action that reduces its effective output impedance at low frequencies and so improves drive stability. At low speeds, the fundamental output voltage is appreciably less than the dc link voltage, whereas output current is significantly greater than dc link current. If the dc link filter impedance is referred to the inverter output terminals, as in the case of a stepdown transformer, the effective shunt capacitance is magnified and the series inductance is diminished. Both of these parameter changes are in the correct sense for an enhancement of drive stability, and consequently, the PWM inverter can give stable open-loop operation at low speeds. Typically, the filter capacitance is in the range from 5000 to 10 000 μF. The six-step inverter does not exhibit this stepdown transformer action, and it requires a large and costly filter capacitor in the range from 50 000 to 200 000 μF.

Appropriate modification of machine parameters can also help to eliminate instability in an open-loop adjustable-frequency drive, but the provision of special machine designs may be uneconomical, and steady-state performance and efficiency may not be as good

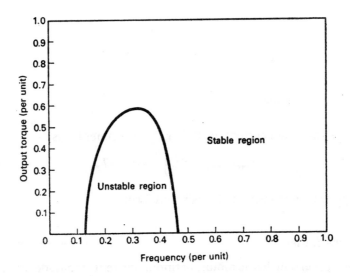

FIG. 6.12. Stability boundary for a converter-fed ac motor drive.

as in a normal motor. Various closed-loop feedback methods have therefore been developed to suppress rotor oscillations. A synchronous reluctance motor can be stabilized by altering the amplitude of the stator voltage in accordance with the fluctuations in rotor speed.[27] An inverter-fed induction motor has been stabilized by controlling inverter frequency with the motor emf[28] or a derivative of dc link current.[29] Alternative feedback methods have also given satisfactory stabilization of adjustable-frequency drives.[30-32]

6.8. OPERATION ON A SIX-STEP CURRENT SUPPLY

The ac motor may receive its supply from a six-step current-source inverter, as described in Chapter 4. Motor performance on a six-step current supply is similar to that on a six-step voltage supply. In each case, the drive control is usually designed to maintain constant airgap flux over a constant-torque range of operation below base frequency.

The six-step current waveform contains all odd harmonic components except those that are multiples of three. Assuming an ideal six-step waveshape, each harmonic current of order k has a defined magnitude that is inversely proportional to the harmonic order. Thus,

$$I_k = \frac{I_1}{k} \tag{6.74}$$

where I_1 is the fundamental rms current of the motor.

The total rms harmonic current is given by

$$I_{har} = \left(\sum_{k \neq 1} I_k^2 \right)^{1/2}$$

$$= \left(I_5^2 + I_7^2 + I_{11}^2 + I_{13}^2 + \cdots \right)^{1/2}$$

$$= I_1 \left(\frac{1}{5^2} + \frac{1}{7^2} + \frac{1}{11^2} + \frac{1}{13^2} + \cdots \right)^{1/2}$$

$$= 0.31 \, I_1. \tag{6.75}$$

Hence, the total rms stator current, including the fundamental, is

$$I_{rms} = \left(I_1^2 + I_{har}^2 \right)^{1/2} = 1.047 \, I_1. \tag{6.76}$$

These equations are obviously valid in per-unit form.

6.8.1. Harmonic Losses

In the six-step, current-fed ac motor, harmonic current amplitudes are defined and are practically independent of motor parameters. In the six-step, voltage-fed motor, on the other hand, harmonic voltage amplitudes are defined, and harmonic current amplitudes are inversely proportional to per-unit leakage reactance. If a very low leakage reactance machine has a six-step voltage supply, it will obviously have larger harmonic currents and higher harmonic copper losses than a current-fed motor. However, for a typical motor with a leakage reactance of 0.15 pu, which is operating at rated fundamental current, the total harmonic rms current and harmonic copper losses are approximately the same on both six-step voltage and current supplies. Consequently, when airgap flux is held constant, the full-load efficiencies are approximately equal at all speeds for the two supply modes.

Harmonic current amplitudes decrease proportionally with fundamental current in a six-step current-fed machine, thereby reducing harmonic losses at light loads. In a voltage-fed motor, however, harmonic current flow is undiminished at partial loads, and harmonic losses are high. Assuming comparable full-load efficiencies, the voltage-fed motor has a lower part-load efficiency, as compared with a lightly loaded current-fed motor. Conversely, for load torques greater than rated torque, operation from a voltage-source inverter is more efficient.[33]

6.8.2. Pulsating Harmonic Torques

With a six-step current supply, steady harmonic torques are negligible, as in the case of a six-step voltage supply. However, pulsating harmonic torques are again significant, and attention is confined to these.

When a three-phase stator winding is supplied with quasi-square-wave currents, the current distribution does not vary during the 60-degree interval when a particular pair of phases is conducting. Consequently, the winding mmf is stationary until a phase commutation occurs, causing the current distribution and mmf to jump forward by 60 electrical degrees to a new stationary position. Harmonic airgap fluxes are small, however, and the airgap flux is predominantly a fundamental component of constant amplitude, rotating uniformly at synchronous speed. Interaction of this uniformly rotating flux wave with the step-wise rotating mmf results in a torque waveform that has a predominant sixth harmonic variation superimposed on the constant shaft torque.

From another point of view, it is evident that the power delivered to the motor at any instant is the product of the instantaneous values of current and terminal voltage. Figure 6.13 is a simplified diagram of the current-source inverter (CSI) induction motor drive which omits the auxiliary forced commutating circuitry. During the 60-degree interval when thyristors TH1 and TH2 conduct, the constant source current, I_d, is directed through phases A and C of the motor, and the instantaneous power input is $v_{AC}I_d$, where v_{AC} is the instantaneous line voltage. When TH1 is turned off, thyristors TH2 and TH3 deliver current I_d to phases B and C, giving an instantaneous power input of $v_{BC}I_d$ during this 60-degree interval. It has been shown in Chapter 4 that the terminal voltage of the motor is virtually sinusoidal, and since I_d is constant, the instantaneous power input waveform is a series of 60-degree sine wave segments, as shown in Fig. 6.14. The selected segments are determined by the angle ϕ. This is the phase angle by which fundamental motor current lags fundamental voltage. The instantaneous power waveform has a dominant ripple component at six times the inverter output frequency, and obviously resembles the output voltage waveform of a conventional three-phase bridge rectifier with a firing angle ϕ.

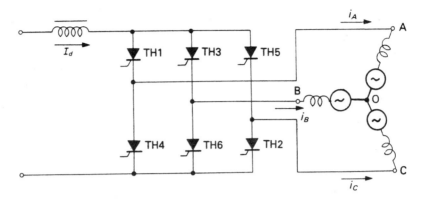

FIG. 6.13. Simplified current-source inverter drive.

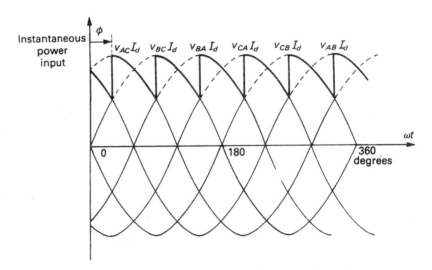

FIG. 6.14. Idealized instantaneous power (and torque) waveform.

Neglecting motor losses, the instantaneous input power to the motor is also the instantaneous output power which, when divided by the constant mechanical angular velocity, gives the instantaneous torque developed. Consequently, the instantaneous power variation of Fig. 6.14 is also an instantaneous torque variation. Analytical expressions for pulsating torque may be determined from this approach. The Fourier series expression for six-step phase current is multiplied by the corresponding sinusoidal phase voltage to give instantaneous phase power. The product of fundamental current and sinusoidal voltage for all three phases represents a constant power input and results in the main developed torque. Fifth and seventh harmonic currents, in association with the sinusoidal phase voltage yield a pulsating sixth harmonic power input, resulting in a sixth harmonic pulsating torque. Similarly, a twelfth harmonic pulsating torque is developed by the eleventh and thirteenth harmonic currents.

It is probably easier to quantify the pulsating torque contribution of individual current harmonics by adopting the combined phasor diagram approach, as used previously in Section 6.6.2. The six-step or quasi-square-wave current has the series expansion of Equation 4.13.

$$i_A = \frac{2\sqrt{3}}{\pi} I_d \left[\sin \omega t - \frac{1}{5} \sin 5\omega t - \frac{1}{7} \sin 7\omega t \right.$$

$$\left. + \frac{1}{11} \sin 11\omega t + \frac{1}{13} \sin 13\omega t - \cdots \right].$$

The fundamental, fifth, and seventh harmonic currents can be represented in a combined phasor diagram, as shown in Fig. 6.15(a). As before, each phasor has a sense of rotation appropriate to its phase sequence. Each current is zero at the instant chosen, with the fundamental current, I_1, about to become positive, and I_5 and I_7 about to become negative, in accordance with the Fourier series expression for i_A.

When an induction motor is fed with a six-step current waveform, the fundamental stator current lags the near-sinusoidal phase voltage by the phase angle, ϕ. A combined phasor diagram may be drawn for this operating condition, as shown in Fig. 6.15(b).

The earlier equations (6.69 and 6.71) for pulsating torque amplitudes are again valid. Thus, the sixth harmonic torque is given by

$$T_6 = \Phi_1 \left[I_5^2 + I_7^2 - 2I_5 I_7 \cos (\delta_5 + \delta_7) \right]^{1/2}$$

and, in general,

$$T_{6n} = \Phi_1 \left[I_{6n-1}^2 + I_{6n+1}^2 - 2I_{6n-1} I_{6n+1} \cos (\delta_{6n-1} + \delta_{6n+1}) \right]^{1/2}$$

where all quantities are in per-unit form.

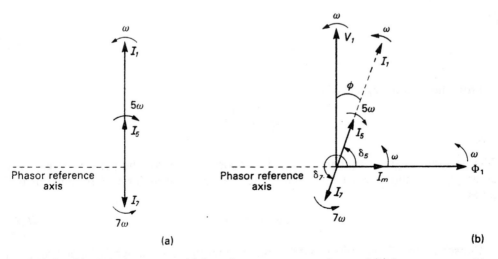

FIG. 6.15. Combined phasor diagrams (a) for a six-step current waveform and (b) for an ac motor with a six-step current supply.

If the motor is developing rated torque at rated fundamental current, then $I_5 = \frac{1}{5}$, $I_7 = \frac{1}{7}$, and the airgap flux is near its base value of unity. From Fig. 6.15(b), $\delta_5 = \frac{\pi}{2} - \phi$ and $\delta_7 = \frac{3\pi}{2} - \phi$. Substituting these values gives the sixth harmonic torque amplitude as

$$T_6 = \left[\frac{1}{25} + \frac{1}{49} - \frac{2}{35} \cos 2\phi \right]^{1/2} . \tag{6.77}$$

The $\cos 2\phi$ term in this equation indicates that the pulsating torque contributions of I_5 and I_7 are no longer in antiphase, as in the earlier equation (6.73) for the six-step voltage-source inverter (VSI) drive, but are now 2ϕ away from the antiphase condition. For an induction motor operating at rated load, ϕ is typically about 30 degrees, corresponding to a fundamental power factor of 0.87. Consequently, T_6 is 0.178 pu or 17.8 percent of base torque. The corresponding value of the twelfth harmonic pulsating torque is 8.5 percent of base torque. If the motor has a large magnetizing reactance, resulting in a small magnetizing current and reduced phase angle, ϕ, torque pulsations are reduced in amplitude. However, a larger filter inductance is required in the dc link to maintain the same ripple current, and it has been shown that this current ripple has an influence on the torque pulsations.[34]

At light load, the six-step current is small, and harmonic current amplitudes are reduced. However, the fundamental phase angle, ϕ, is larger than at full load and, as a result, pulsating torque amplitudes are undiminished. For six-step current operation, the earlier sixth harmonic torque equation (6.67) can be written as

$$T_6 = \Phi_1 I_1 \left[\left(\frac{2}{35} \cos \phi \right) \cos 6\omega t - \left(\frac{12}{35} \sin \phi \right) \sin 6\omega t \right] . \tag{6.78}$$

For a lightly loaded motor, angle ϕ is large and the term in $\cos \phi$ may be neglected, giving

$$T_6 \simeq - \Phi_1 I_1 \left(\frac{12}{35} \sin \phi \right) \sin 6\omega t . \tag{6.79}$$

From the phasor diagram of Fig. 6.15(b),

$$\sin \phi \simeq \frac{I_m}{I_1} . \tag{6.80}$$

The per-unit magnetizing current, I_m, is proportional to Φ_1, the per-unit airgap flux, assuming saturation is negligible, and, therefore, substituting for $\sin\phi$ in Equation 6.79 gives

$$T_6 \propto \Phi_1^2 \sin 6\omega t . \tag{6.81}$$

Consequently, T_6 is approximately proportional to Φ_1^2 for a lightly loaded motor in which the stator current has a large magnetizing component. This result indicates that field weakening can effect a significant reduction in pulsating torque and so enhance the low-speed operation of the drive, but fundamental torque is also reduced. A similar conclusion was reached for the six-step voltage-fed motor.

If high torque is required at standstill and low speeds, torque pulsations in CSI induction motor drives may be reduced by imposing an appropriate modulation on instantaneous dc link current. This causes a corresponding variation in the output phase current waveform so that pulsating torque amplitudes are minimized.[35–38] Pulse-width modulation of inverter output current can also be employed to eliminate low-order current harmonics and associated pulsating torque components. However, the PWM strategies are more complex than the simple sinusoidal PWM technique which is used in voltage-source inverters.[37–39] A multiple inverter configuration, in which CSI output waveforms are combined, has also been suggested as a method of reducing harmonic content and torque ripple.[40]

6.9. OPERATION ON A PWM INVERTER SUPPLY

In earlier sections of this chapter, attention has been focused on the harmonic effects associated with six-step VSI and CSI supplies. PWM voltage-source inverters are widely used in industry, but it is difficult to draw general conclusions regarding the harmonic effects of PWM waveforms, because of the wide variety of modulation strategies available. The harmonic motor losses associated with a six-step supply may be significantly reduced by adopting an appropriate PWM approach, but an ill-advised modulation technique can result in substantially higher motor losses. On poor PWM waveforms, Klingshirn and Jordan have shown that total harmonic motor losses can even exceed fundamental losses.[4] This emphasizes the importance of a careful choice of modulation strategy.

In a PWM VSI drive, as motor speed increases, the modulation strategy is usually altered and the number of inverter switchings per cycle is reduced. This practice is designed to give a gradual transition to six-step voltage operation at base speed and to minimize inverter switching losses. The PWM strategies used immediately prior to the change to six-step operation have few commutations per cycle and must be carefully selected, because a poor choice of transition strategy can result in very high harmonic losses and rapid overheating of the motor.[41–43]

For optimum drive performance, an appropriate modulation strategy is required for each portion of the speed range below base speed. In order to compare different PWM strategies, it is convenient to develop loss criteria which permit a general comparison of waveform quality with respect to harmonic motor losses, without requiring detailed loss calculations for a particular machine design.[43]

6.9.1. Harmonic Loss Factors

An optimum PWM technique should minimize additional harmonic losses in the motor. These losses are primarily harmonic copper losses. For a VSI supply, where leakage reactance determines harmonic current flow, the per-unit k^{th} harmonic current is given, as before, by Equation 6.47:

$$I_k = \frac{V_k}{k\,f_1\,X_{pu}}$$

where X_{pu} is the per-unit leakage reactance at base frequency, and f_1 is the per-unit fundamental frequency.

The kth harmonic copper loss is $I_k^2 R_k$, where R_k is the resistance of the motor to the kth harmonic. The total harmonic copper loss per phase is therefore

$$P_{loss} = \sum_{k \neq 1} I_k^2 R_k = \frac{1}{X_{pu}^2} \sum_{k \neq 1} \left(\frac{V_k}{kf_1} \right)^2 R_k . \qquad (6.82)$$

If R_k can be assumed constant and unaffected by frequency, the harmonic copper loss is proportional to the quantity

$$\sigma_1 = \sum_{k \neq 1} \left(\frac{V_k}{kf_1} \right)^2 . \qquad (6.83)$$

This is a loss factor which ideally has a value of zero and can be used to compare the harmonic copper losses due to different PWM techniques.

In practice, as shown in Section 6.5, skin effect can have a significant influence on harmonic copper losses, particularly if the rotor has a deep-bar construction. The slot leakage component of rotor inductance decreases with frequency, but the overall reduction in the leakage inductance of the motor is less significant than the appreciable increase in rotor resistance which occurs. Because the loss factor, σ_1, as defined in Equation 6.83, ignores skin effect, it may not be a reliable criterion for comparing PWM waveforms. Thus, a fifth harmonic voltage component of 0.2 pu makes the same contribution to the loss factor as a 25th harmonic of 1 pu, whereas, in practice, the motor offers a significantly higher resistance to the 25th harmonic, resulting in greater copper losses. The loss factor, σ_1, is therefore unduly favorable to waveforms with pronounced high-order harmonics.

Stator and rotor resistance increase with frequency due to skin effect, but the additional harmonic copper losses are primarily in the rotor. If f_{2k} is the rotor frequency corresponding to the kth harmonic, the rotor resistance, R_{2k}, taking skin effect into account, is proportional to the square root of f_{2k} at high rotor frequencies.[3] Thus,

$$R_{2k} \propto \sqrt{f_{2k}} . \qquad (6.84)$$

In general, the harmonic slip, s_k, is $f_{2k}/(kf_1)$, and assuming that the motor operates near its fundamental synchronous speed, then from Equation 6.39,

$$f_{2k} = (k \mp 1)f_1 \simeq kf_1 \qquad (6.85)$$

and hence

$$R_{2k} \propto (kf_1)^{1/2} . \qquad (6.86)$$

The harmonic rotor copper loss per phase is given by

$$P_{2loss} = \frac{1}{X_{pu}^2} \sum_{k \neq 1} \left(\frac{V_k}{kf_1} \right)^2 R_{2k} . \qquad (6.87)$$

Substituting for R_{2k} from Equation 6.86 gives a modified loss factor

$$\sigma_2 = \sum_{k \neq 1} \frac{V_k^2}{(kf_1)^{3/2}} . \tag{6.88}$$

As explained in Section 6.5, harmonic iron losses are difficult to predict accurately because they are strongly influenced by machine construction and magnetic materials. The increase in core loss due to time-harmonic main fluxes is negligible, but the end-leakage and skew-leakage fluxes, which normally contribute to the stray load loss, may cause a significant core loss at harmonic frequencies. If an unskewed rotor construction is employed, end-leakage losses are the dominant component and may be calculated with the equation of Alger, Angst, and Davies,[44] which indicates that these losses are proportional to frequency times current squared. Hence, the stator and rotor end losses associated with the kth harmonic are nearly proportional to $I_k^2(kf_1)$, and the total harmonic end loss is given by

$$P_{endloss} \propto \sum_{k \neq 1} (I_k)^2 kf_1 . \tag{6.89}$$

An end loss factor can be defined for these dominant harmonic iron losses and is

$$\sigma_3 = \sum_{k \neq 1} \frac{V_k^2}{kf_1} . \tag{6.90}$$

The total stray load (SL) losses are given more generally by

$$P_{SLloss} \propto \sum_{k \neq 1} (I_k)^x (kf_1)^y \tag{6.91}$$

where the x and y coefficients depend on the machine construction. It has been determined experimentally that the total stray load losses due to harmonics are obtained with reasonable accuracy by putting $x = 2$ and $y = 1.5$.[5] This gives a loss factor

$$\sigma_4 = \sum_{k \neq 1} \frac{V_k^2}{(kf_1)^{0.5}} . \tag{6.92}$$

6.9.2. Comparison of PWM Strategies

The general loss factors derived above can be used to compare the different PWM strategies introduced in Chapter 4. Overall drive efficiency is the product of inverter efficiency and motor efficiency. Inverter losses are a function of the number of commutations per second, and in order to compare drive performance with different PWM strategies, it is desirable that the number of switchings per cycle should be the same in each case. For this reason, the following PWM techniques were selected: (a) harmonic elimination PWM with the fifth, seventh, eleventh, and thirteenth harmonics suppressed (control of the fundamental voltage and elimination of these four harmonics requires 22 switchings per phase per cycle, including the switchings at 0 degrees and 180 degrees); (b) distortion minimization PWM with five switching angles per quarter-cycle, also requiring 22 commutations per phase per cycle; (c) sinusoidal PWM with a carrier ratio of 9, requiring 18 switchings per phase per cycle; (d) sinusoidal PWM with a carrier ratio of 12, requiring 24 switchings per phase per cycle. Waveforms (a) and (b) are the so-called optimum PWM waveforms discussed in Section 4.5.4. The sinusoidal PWM waveform (c) has fewer commutations per cycle than waveforms (a) and (b), whereas waveform (d) has too many commutations. Exact correspondence is not possible because sinusoidal PWM must have a carrier ratio which is a multiple of three. The study of sinusoidal PWM strategies is confined to the region where the modulation index is less than unity and pulse dropping does not occur.

Comparison of waveforms (a) through (d) is performed over the constant volts/hertz range of operation below base frequency. In practice, of course, each modulation strategy would have an increased number of switchings per cycle at low fundamental frequencies to minimize low-order harmonic currents and torque pulsations. As base frequency is approached, the number of switchings per cycle is reduced to minimize inverter switching losses and allow a gradual transition to six-step operation. However, it is possible to draw general conclusions regarding the relative merits of the PWM strategies under consideration by confining the comparison to the particular number of switchings per cycle specified above.

Figure 6.16 compares the four PWM waveforms over the fundamental voltage and frequency range, using the basic harmonic copper loss factor, σ_1, which neglects skin effect. Evidently, the harmonic elimination PWM technique (a) is superior to sinusoidal PWM above 0.6 pu voltage. Despite having fewer commutations per cycle, curve (a) shows harmonic losses of less than one-third of those for curve (d) in the region of 0.9 pu voltage. At low fundamental voltages, however, harmonic elimination PWM has large losses.

The distortion minimization curve (b) is a composite curve consisting of a number of segments and gives the absolute minimum value of loss factor which is possible with five switching angles per quarter-cycle. Harmonic losses in the region of 0.9 pu voltage are now less than one-sixth of those for sinusoidal PWM with a carrier ratio of 12.

It is instructive to introduce the six-step voltage waveform into the comparison. For six-step operation, σ_1 is a horizontal straight line at a value of 2.15×10^{-3}. The constant value of σ_1 over the constant volts/hertz range is explained by the fact that the six-step waveshape is retained at all frequencies, and so the relative harmonic content and harmonic losses do not vary. The results indicate that sinusoidal PWM with a carrier ratio of 9, or less, is always inferior to the six-step wave.

Figure 6.17 plots the modified copper loss factor, σ_2, for the previous PWM waveforms. The percentage loss reduction obtained by the use of optimum PWM techniques is slightly less than in Fig. 6.16, but their superiority over the sinusoidal PWM strategies is again quite evident. The distortion minimization curve is calculated for the same switching angles as used previously, although a slightly better solution is possible.

Figure 6.18 plots the end loss factor, σ_3. The distortion minimization strategy is again seen to be the optimum, despite the fact that the switching angles were chosen to minimize harmonic copper losses. The stray load loss factor, σ_4, is plotted in Fig. 6.19, which confirms the conclusions reached in Fig. 6.18 but shows that the superiority of distortion minimization PWM is somewhat reduced.

Peak current is an important factor in inverter design, and the various modulation strategies can also be compared on this basis. It is assumed that the inverter delivers rated fundamental current to the ac motor at 0.85 fundamental power factor over the full

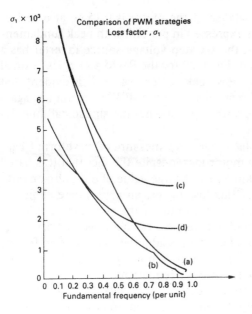

FIG. 6.16. Copper loss factor as a function of per-unit fundamental frequency.

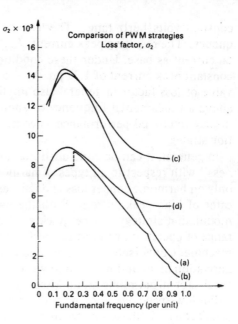

FIG. 6.17. Modified copper loss factor as a function of per-unit fundamental frequency.

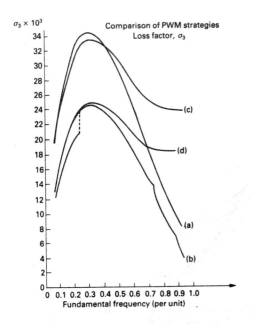

FIG. 6.18. End loss factor as a function of per-unit fundamental frequency.

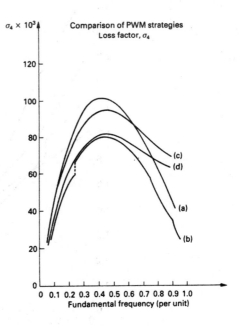

FIG. 6.19. Stray load loss factor as a function of per-unit fundamental frequency.

constant volts/hertz range. The leakage reactance of the motor is 0.15 pu at base frequency. The resulting peak current, I_{max}, is expressed in per unit with peak fundamental current as base. Under these conditions, the six-step voltage-source inverter has a constant peak current of 1.32 pu. As shown in Fig. 6.20 for the PWM strategies, a small value of loss factor in general also implies a low peak current value. It is evident that above a fundamental frequency of about 0.6 pu, the optimum PWM techniques again display improved performance as compared with the conventional sinusoidal modulation strategy.

In general, it can be concluded that loss factor, σ_1, is a measure of waveform "badness" with respect to all types of harmonic motor loss, despite the fact that it is based only on harmonic copper loss and ignores skin effect. A large value of σ_1 is also an indicator of high peak current. With the use of this loss factor, an appropriate choice of modulation strategy can be quickly made for each portion of the constant volts/hertz range of operation, without the performance of detailed loss calculations for a particular machine. Conversely, if minimization of loss factor σ_1 is adopted as a criterion for the derivation of an optimum PWM waveform, the resulting solution gives nearly optimum results for all harmonic motor losses and also for peak current amplitude.

For low-speed operation with a high switching frequency, sinusoidal PWM is perfectly satisfactory. At these low frequencies, computation of the numerous switching angles

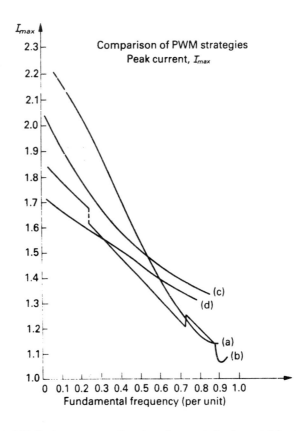

FIG. 6.20. Peak current as a function of per-unit fundamental frequency.

for the optimum PWM techniques is very tedious, and subsequent implementation does not yield a significant improvement in efficiency. As motor speed increases, the number of switching angles per cycle must be reduced to avoid an excessive number of commutations per second, and to allow a gradual transition to six-step operation at about base frequency. At these higher speeds, the optimum PWM strategies are superior to sinusoidal PWM with respect to harmonic motor losses and peak current amplitude. Their superiority with respect to harmonic losses has also been confirmed by other authors.[45-47]

6.9.3. Pulsating Harmonic Torques

As already mentioned, it is characteristic of a PWM strategy that low-frequency cogging torques which are detrimental to smooth, low-speed rotation can be eliminated at the expense of large-amplitude, high-frequency harmonic torques. This is advantageous if the pulsating torque frequencies lie above the shaft mechanical resonances.[42]

As shown earlier, harmonic currents of order $k = (6n + 1)$ and $(6n - 1)$ have positive and negative phase sequence, respectively, and combine to produce a pulsating torque component of order $6n$, with a per-unit amplitude as given by the earlier equation (6.71). Thus,

$$T_{6n} = \Phi_1 \left[I_{6n-1}^2 + I_{6n+1}^2 - 2I_{6n-1} I_{6n+1} \cos(\delta_{6n-1} + \delta_{6n+1}) \right]^{1/2}.$$

Sinusoidal PWM with a carrier ratio p is characterized by large-amplitude voltage harmonics at $(p \pm 2)$ and $(2p \pm 1)$ times the fundamental frequency. The $(2p \pm 1)th$ harmonics have opposite phase sequence and combine to develop a large, pulsating torque component at $2p$ times the fundamental frequency. Following a Fourier analysis of the PWM waveform, a combined phasor diagram may be drawn. This shows that the term $(\delta_{6n-1} + \delta_{6n+1})$ in the torque equation is 180 degrees and the equation can be written as

$$T_{2p} = \Phi_1 \left(I_{2p-1} + I_{2p+1} \right). \tag{6.93}$$

This result indicates that pulsating torque components due to I_{2p-1} and I_{2p+1} are in phase, and therefore reinforce one another. Direct addition of pulsating torques is a feature of most PWM waveforms.[13,48] On the other hand, Equation 6.72 indicates that for six-step voltage operation, harmonic torque components are in phase opposition.

For a PWM voltage-source inverter, where harmonic voltage amplitudes are defined, the dominant pulsating torque components may be calculated by using the above equation for T_{6n}, but harmonic current amplitudes are first determined with Equation 6.47:

$$I_k = \frac{V_k}{kf_1 X_{pu}}.$$

Again, harmonic current and pulsating torque amplitudes are determined by motor leakage reactance and are independent of load.

The four PWM strategies for which a loss comparison was performed in the previous section can now be compared on the basis of dominant pulsating torques. These torques are a measure of the low-speed capability of the various PWM techniques. Induction motor operation at 0.2 pu fundamental frequency and voltage is assumed, with a typical motor leakage reactance of 0.15 pu (at base frequency). Possible amplification of harmonic torques due to rotor speed fluctuations and dc link voltage variations is again neglected.

In the case of sinusoidal PWM, Fourier analysis shows that each of the $(2p \pm 1)th$ voltage harmonics has an amplitude of 0.184 pu and is independent of p for p greater than 9. For a carrier ratio of 12 and a fundamental flux of 1 pu, the dominant pulsating torque is therefore of order 24 and has an amplitude of 0.513 pu. For a carrier ratio of 9, the torque amplitude is 0.684 pu and the harmonic order is 18. The harmonics of order $(p \pm 2)$ cause lower order harmonic torques. In the case of $p = 12$, there are additional ninth and fifteenth harmonic torques of amplitude 0.065 pu and 0.047 pu, respectively. For $p = 9$, there are sixth and twelfth harmonic torques with amplitudes of 0.094 pu and 0.060 pu, respectively.

Harmonic elimination PWM seeks to suppress the specific lower order torque harmonics that cause speed fluctuation at reduced speeds. Elimination of the fifth, seventh, eleventh, and thirteenth harmonic voltages removes the sixth and twelfth harmonic pulsating torques, but higher order torques may be significant. For 0.2 pu fundamental voltage, harmonic analysis shows that the seventeenth and nineteenth harmonic voltages have amplitudes of 0.157 pu and 0.218 pu, respectively. Each of the resulting current harmonics reacts with the fundamental airgap flux to produce an eighteenth harmonic pulsating torque. The two torque components are additive, giving a resultant torque amplitude of 0.69 pu, which is approximately the same as that for sinusoidal PWM with $p = 9$. Lower order torques are absent so that low-speed capability may be somewhat improved as compared with sinusoidal PWM.

Distortion minimization PWM cannot be seriously considered for low-frequency operation. The overall harmonic distortion is minimized, but no specific attention is paid to the lower order harmonics, so that large low-frequency pulsating torques are developed.

It can be concluded that for low-speed operation with a high switching frequency, sinusoidal PWM gives near-optimum torque smoothness with a simpler modulation strategy.

6.10. REFERENCES

1. JAIN, G.C., The effect of voltage waveshape on the performance of a three-phase induction motor, *IEEE Trans. Power Appar. Syst.*, **83**, 6, June 1964, pp. 561-566.

2. LANGSDORF, A.S., *Theory of Alternating Current Machinery*, McGraw-Hill, New York, NY, 1955.

3. ALGER, P.L., *Induction Machines*, Gordon and Breach, New York, NY, 1970.

4. KLINGSHIRN, E.A., and JORDAN, H.E., Polyphase induction motor performance and losses on nonsinusoidal voltage sources, *IEEE Trans. Power Appar. Syst.*, **PAS-87**, 3, Mar. 1968, pp. 624-631.

5. LARGIADER, H., Design aspects of induction motors for traction applications with supply through static frequency changers, *Brown Boveri Rev.*, **57**, Apr. 1970, pp. 152-167.

6. RAPHAEL, H., Additional losses in PWM inverter-fed squirrel cage motors, *Conf. Rec. IEEE Ind. Appl. Soc. Annual Meeting*, 1977, pp. 932-936.

7. CHALMERS, B.J., and SARKAR, B.R., Induction motor losses due to non-sinusoidal supply waveforms, *Proc. IEE*, **115**, 12, Dec. 1968, pp. 1777-1782.

8. HONSINGER, V.B., Induction motors operating from inverters, *Conf. Rec. IEEE Ind. Appl. Soc. Annual Meeting*, 1980, pp. 1276-1285.

9. MURPHY, J.M.D., and HONSINGER, V.B., Efficiency optimization of inverter-fed induction motor drives, *Conf. Rec. IEEE Ind. Appl. Soc. Annual Meeting*, 1982, pp. 544-552.

10. HEUMANN, K., and JORDAN, K.G., Einfluss von Spannungs- und Stromober-schwingungen auf den Betrieb von Asynchronmaschinen, *AEG Mitt.*, **54**, 1/2, 1964, pp. 117-122.

11. VON ZWEYGBERGK, S., and SOKOLOV, E., Verlustermittlung im stromrichter-gespeisten Asynchronmotor, *Elektrotech. Z. Ausg. A.*, **90**, 23, 1969, pp. 612-616.

12. BYSTRON, K., Einflüsse von Strom- und Spannungsoberschwingungen eines Zwischenkreisumrichters auf Asynchronmaschinen, *Siemens-Z.*, **41**, 3, 1967, pp. 244-247.

13. ROBERTSON, S.D.T., and HEBBAR, K.M., Torque pulsations in induction motors with inverter drives, *IEEE Trans. Ind. Gen. Appl.*, **IGA-7**, 2, Mar./Apr. 1971, pp. 318-323.

14. WILLIAMSON, A.C., The effects of system harmonics upon machines, *Int. J. Electr. Eng. Educ.*, **19**, 2, Apr. 1982, pp. 145-155.

15. LIPO, T.A., KRAUSE, P.C., and JORDAN, H.E., Harmonic torque and speed pulsations in a rectifier-inverter induction motor drive, *IEEE Trans. Power Appar. Syst.*, **PAS-88**, 5, May 1969, pp. 579-587.

16. WARD, E.E., and HÄRER, H., Preliminary investigation of an inverter-fed 5-phase induction motor, *Proc. IEE*, **116**, 6, June 1969, pp. 980-984.

17. McLEAN, G.W., NIX, G.F., and ALWASH, S.R., Performance and design of induction motors with square-wave excitation, *Proc. IEE*, **116**, 8, Aug. 1969, pp. 1405-1411.

18. LIPO, T.A., and KRAUSE, P.C., Stability analysis of a reluctance-synchronous machine, *IEEE Trans. Power Appar. Syst.*, **PAS-86**, 7, July 1967, pp. 825-834.

19. LAWRENSON, P.J., and BOWES, S.R., Stability of reluctance machines, *Proc. IEE*, **118**, 2, Feb. 1971, pp. 356-369.

20. HONSINGER, V.B., Stability of reluctance motors, *IEEE Trans. Power Appar. Syst.*, **PAS-91**, 4, July/Aug. 1972, pp. 1536-1543.

21. HONSINGER, V.B., Inherently stable reluctance motors having improved performance, *IEEE Trans. Power Appar. Syst.*, **PAS-91**, 4, July/Aug. 1972, pp. 1544-1554.

22. LIPO, T.A., and KRAUSE, P.C., Stability analysis for variable frequency operation of synchronous machines, *IEEE Trans. Power Appar. Syst.*, **PAS-87**, 1, Jan. 1968, pp. 227-234.

23. NELSON, R.H., LIPO, T.A., and KRAUSE, P.C., Stability analysis of a symmetrical induction machine, *IEEE Trans. Power Appar. Syst.*, **PAS-88**, 11, Nov. 1969, pp. 1710-1717.

24. RAMESH, N., and ROBERTSON, S.D.T., Induction machine instability predictions — based on equivalent circuits, *IEEE Trans. Power Appar. Syst.*, **PAS-92**, 2, Mar./Apr. 1973, pp. 801-807.

25. BOWLER, P., and NIR, B., Steady-state stability criterion for induction motors, *Proc. IEE*, **121**, 7, July 1974, pp. 663-667.

26. LIPO, T.A., and KRAUSE, P.C., Stability analysis of a rectifier-inverter induction motor drive, *IEEE Trans. Power Appar. Syst.*, **PAS-88**, 1, Jan. 1969, pp. 55-66.

27. KRAUSE, P.C., Methods of stabilizing a reluctance synchronous machine, *IEEE Trans. Power Appar. Syst.*, **PAS-87**, 3, Mar. 1968, pp. 641-649.

28. RISBERG, R.L., A wide speed range inverter fed induction motor drive, *Conf. Rec. IEEE Ind. Gen. Appl. Group Annual Meeting, 1969*, pp. 629-633.

29. FALLSIDE, F., and WORTLEY, A.T., Steady-state oscillation and stabilization of variable-frequency inverter-fed induction-motor drives, *Proc. IEE*, **116**, 6, June 1969, pp. 991-999.

30. BEJACH, B., and PETERSON, C.V., Stability of static power inverters and synchronous reluctance multimotor systems, *IEEE Trans. Ind. Appl.*, **IA-12**, 3, May/June 1976, pp. 275-283.

31. STEFANOVIC, V.R., Closed loop performance of induction motors with constant volts/hertz control, *Elect. Mach. and Electromechan.*, **1**, 1977, pp. 255-266.

32. WOLFINGER, J.F., and LIPO, T.A., Stability improvement of inverter driven induction motors by use of feedback, *IFAC Symp. on Control in Power Electronics and Electrical Drives*, 1974, pp. 237-251.

33. VENKATESAN, K., and LINDSAY, J.F., Comparative study of the losses in voltage and current source inverter fed induction motors, *IEEE Trans. Ind. Appl.*, **IA-18**, 3, May/June 1982, pp. 240-246.

34. CREIGHTON, G.K., Current-source inverter-fed induction motor torque pulsations, *Proc. IEE*, **127**, 7, July 1980, pp. 231-239.

35. LIPO, T.A., Analysis and control of torque pulsations in current fed induction motor drives, *IEEE Power Electron. Spec. Conf.*, 1978, pp. 89-96.

36. CHIN, T.H., A new controlled current type inverter with improved performance, *IEEE Int. Semicond. Power Converter Conf.*, 1977, pp. 185-192.

37. CHIN, T.H., and TOMITA, H., The principles of eliminating pulsating torque in current source inverter induction motor systems, *Conf. Rec. IEEE Ind. Appl. Soc. Annual Meeting,* 1978, pp. 910-917.

38. ZUBEK, J., Evaluation of techniques for reducing shaft cogging in current fed ac drives, *Conf. Rec. IEEE Ind. Appl. Soc. Annual Meeting,* 1978, pp. 517-524.

39. LIENAU, W., Torque oscillations in traction drives with current-fed asynchronous machines, IEE Conf. Publ. No. 179, *Electrical Variable-Speed Drives,* 1979, pp. 102-107.

40. NABAE, A., SHIMAMURA, T., and KUROSAWA, R., A new multiple current-source inverter, *IEEE Int. Semicond. Power Converter Conf.,* 1977, pp. 200-203.

41. KLIMAN, G.B., and PLUNKETT, A.B., Development of a modulation strategy for a PWM inverter drive, *IEEE Trans. Ind. Appl.,* **IA-15,** 1, Jan./Feb. 1979, pp. 72-79.

42. ANDRESEN, E.C., and BIENIEK, K., On the torques and losses of voltage and current source inverter drives, *IEEE Int. Semicond. Power Converter Conf.,* 1982, pp. 428-437.

43. MURPHY, J.M.D., and EGAN, M.G., A comparison of PWM strategies for inverter-fed induction motors, *IEEE Trans. Ind. Appl.,* **IA-19,** 3, May/June 1983, pp. 363-369.

44. ALGER, P.L., ANGST, G., and DAVIES, E.J., Stray-load losses in polyphase induction machines, *AIEE Trans. Power Appar. Syst.,* **78, Pt. III-A,** June 1959, pp. 349-357.

45. BUJA, G.S., and INDRI, G.B., Optimal pulsewidth modulation for feeding ac motors, *IEEE Trans. Ind. Appl.,* **IA-13,** 1, Jan./Feb. 1977, pp. 38-44.

46. DE BUCK, F., GISTELINCK, P., and DE BACKER, D., Loss-optimal PWM waveforms for variable-speed induction motor drives, *Proc. IEE,* **130, Pt. B,** 5, Sept. 1983, pp. 310-320.

47. WILLIAMSON, S., and CANN, R.G., A comparison of PWM switching strategies on the basis of drive system efficiency, *IEEE Trans. Ind. Appl.,* **IA-20,** 6, Nov./Dec. 1984, pp. 1460-1472.

48. WIART, A., Evaluation de la composante pulsatoire des moteurs á courant alternatif alimentes par onduleurs, *Jeumont-Schneider Rev.,* **26,** Aug. 1978, pp. 29-40.

CHAPTER 7

Control Systems for Adjustable-Frequency Induction Motor Drives

7.1. INTRODUCTION

Modern methods of static frequency conversion have liberated the induction motor from its historical role as a fixed-speed machine, but the inherent advantages of adjustable-frequency operation cannot be fully realized unless a suitable control technique is employed; the choice of control strategy is vital in determining the overall characteristics and performance of the drive system. Also, the power converter has little excess current capability; during normal operation, the control strategy must ensure that motor operation is restricted to regions of high torque per ampere, thereby matching the motor and inverter ratings and minimizing system losses. Overload or fault conditions must be handled by sophisticated control rather than overdesign.

Open-loop speed control of an induction motor with an adjustable-frequency supply provides a satisfactory adjustable-speed drive when the transient performance characteristics are undemanding and when the motor operates at steady speeds for long periods. However, feedback control is necessary for precise steady-state operation in the presence of supply voltage fluctuations and load disturbances. Also, when the drive requirements include rapid acceleration and deceleration, an open-loop system is unsatisfactory because the supply frequency cannot be varied quickly without exceeding the rotor breakdown frequency. Beyond the breakdown point, the motor currents are large, but the power factor, output torque, and efficiency are low. When fast dynamic response is important, closed-loop control methods are essential, but precise feedback signals are required, and stability problems must be avoided by proper system design.

Figure 7.1(a) is a block diagram of a drive system with closed-loop torque control. This torque loop is the essential element in a traction drive and is also the basic building block in high-performance speed-controlled or position-controlled drives. In Fig. 7.1(a), a dc reference voltage representing the commanded or set torque, T^*, is compared with the actual torque signal, T, as determined from measured electrical quantities such as current and flux. The error, $T^* - T$, is fed to a torque controller or regulator that amplifies the error and applies a compensating transfer function. As usual, in a high-gain feedback system, the control acts to nullify the error, but for good transient and steady-state performance, the controller must be properly designed so that the closed-loop transfer function of the drive system has the desired structure.

When satisfactory torque control has been achieved, an outer speed loop can be readily added to give an adjustable-speed drive, as shown in Fig.7.1(b). Again, the reference signal is an analog voltage whose magnitude and polarity represent the desired motor

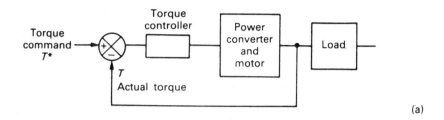

(a)

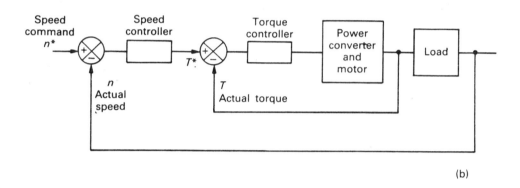

(b)

FIG. 7.1. Closed-loop control of a drive system: (a) torque-controlled drive; (b) speed-controlled drive.

speed and direction. This commanded or set speed, n^*, is compared with the actual shaft speed, n, as measured by a tachometer. The resulting speed error, $n^* - n$, is passed through the speed controller; the compensated error signal becomes the torque command signal for the inner torque loop.

In a position-controlled drive system, an outer position loop is superimposed on the speed loop, as shown in Fig. 7.1(c). In general, for such a cascaded or hierarchical control structure, the output of each control loop serves as the command signal for the next inner loop; the outermost loop has the slowest response, with the control action becoming progressively faster in the nested inner loops. Thus, if the torque control loop is properly designed, the system seen by the speed control loop can usually be approximated by a simple second-order system, and the design of the speed controller (and later the position controller) follows classical control system principles. Proportional-integral (PI) controllers are commonly used because of their easy tuning and zero steady-state error; the controller output is amplitude limited to limit the excursion of the control variable. For the fastest dynamic response, the inner control loops can be activated by additional feedforward signals.

In a dc motor drive, the torque control loop of Fig. 7.1(a) is readily implemented because torque is proportional to armature current when airgap flux is constant. Thus, a fast-acting armature current loop gives effective torque control and also protects the power converter and motor from excessive currents during fast transients and steady-state overloads. The induction motor, on the other hand, is a complex nonlinear mul-

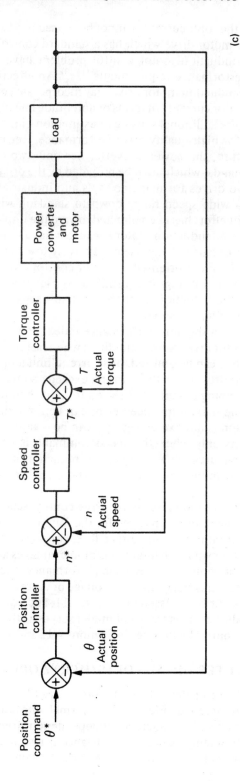

FIG. 7.1 (continued). Closed-loop control of a drive system: (c) position-controlled drive.

tivariable control plant, and the rotor currents cannot be sensed in a cage-rotor machine. Consequently, unlike the dc motor drive which has a standard control structure, a wide variety of solutions to the induction motor control problem have emerged to meet different applications and performance requirements.[1-20] An effective torque control loop for a high-performance induction motor drive may incorporate control loops for air-gap flux, slip frequency, or stator current (in terms of amplitude and phase), as discussed later. These inner loops require additional sensing or estimation of motor quantities, but their presence is necessary for high-quality drive performance. Regardless of the complexity of the control structure, the power converter has only two control inputs, the voltage and frequency commands, which must be provided by the drive control system.

In high-performance servo drives for machine tools and industrial robots, the drive system must operate over a wide speed range down to standstill with a fast transient response. Much development effort has been directed toward the implementation of servo control of the mass-produced induction motor because it has a rugged brushless construction and is inexpensive because of the use of relatively low cost materials and because of the economy of scale. Field-oriented or vector control provides the key to the operation of the induction motor in a high-performance four-quadrant servo drive. This control technique results in the induction motor acting like a separately excited dc machine with decoupled control of torque and flux.

Good dynamic response is not the only performance index, however, and the control structure of the induction motor drive can be simplified when fast response is not a priority. Thus, the inner torque loop can be omitted, or a current limit loop can function as an intermittent torque limit control when the current exceeds some preset value. In a speed-controlled drive, a ramping circuit can be used to limit the rate of change of the speed reference signal, or the maximum speed error can be limited. These techniques ensure that motor acceleration is not excessive; they can be regarded as an intermittent torque control that functions only when the speed command is changed too rapidly. These simplified control structures allow a significant reduction in drive cost, and are perfectly satisfactory for many industrial applications where very precise and fast control is not required.

In this chapter, various open- and closed-loop drive control schemes are described. Initially, steady-state motor operation is analyzed for the following operating conditions: constant terminal volts/hertz ratio, constant airgap flux, constant terminal voltage, and controlled stator current. As shown in Chapter 6, harmonic voltages and currents usually have a minor influence on the motor's steady-state performance; the present treatment therefore assumes a balanced sinusoidal supply of voltage or current. The transient analysis of induction motor behavior is discussed in Section 7.11, and a general mathematical model of the motor is introduced. This transient model serves as the basis for a detailed discussion of field-oriented control techniques in Sections 7.12 and 7.13.

7.2. CONSTANT TERMINAL VOLTS/HERTZ OPERATION

The steady-state performance of the induction motor is readily analyzed by means of the fundamental equivalent circuit of Fig. 6.1. For normal sine wave operation, skin effect is usually neglected; hence, resistances are independent of frequency, while reactances are proportional to frequency. Core loss resistance is also neglected; hence, the magnetizing branch is composed of only the magnetizing reactance (X_m). The rotating flux wave in the airgap induces an emf, E_1, in the stator winding. This emf is less than the applied voltage, V_1, because of the voltage drop, $(R_1 + jX_1)I_1$, across the stator leakage impedance. Because the presence of space harmonic mmf waves is ignored, the rotating flux wave has a sinusoidal space distribution, and the flux linking each stator turn has a sinusoidal time variation. If Φ_1 denotes the fundamental flux per pole of the rotating field, the instantaneous flux linking a full-span stator turn is

$$\phi = \Phi_1 \sin \omega_1 t \tag{7.1}$$

where $\omega_1 = 2\pi f_1$, the angular frequency of the supply voltage. The induced emf per turn is, therefore,

$$e_1 = d\phi/dt = \omega_1 \Phi_1 \cos \omega_1 t \tag{7.2}$$

and the rms stator emf is given by

$$E_1 = \omega_1 \Phi_1 k_w N_1 / \sqrt{2} = 4.44 \, k_w f_1 N_1 \Phi_1 \tag{7.3}$$

where N_1 is the number of series turns per phase and k_w is the winding factor. If the winding factor is unity, the usual transformer emf equation is obtained; hence, for a motor or transformer, Φ_1 is proportional to E_1/ω_1 or E_1/f_1.

For effective utilization, the airgap flux of the induction motor must be sustained at all frequencies. A constant airgap flux is obtained when the ratio E_1/f_1 is constant, but if the voltage drop across the stator leakage impedance is small, then V_1 and E_1 are approximately equal. Consequently, airgap flux is nearly constant when the ratio V_1/f_1 has a fixed value. This is the constant terminal volts/hertz mode of operation that is commonly used in simple open-loop systems. The linear output voltage-frequency characteristic is provided by the inverter or cycloconverter, with the use of the voltage- and frequency-control techniques already described. Unfortunately, the motor performance deteriorates at low frequencies when the airgap flux decreases, because of the voltage drop across the stator leakage impedance.

7.2.1. Torque Characteristics

From the equivalent circuit of Fig. 6.1, the following phasor equations may be obtained:

$$V_1 = (R_1 + jX_1)I_1 + \left[\frac{R_2}{s} + jX_2\right]I_2 \tag{7.4}$$

$$jX_m(I_1 - I_2) = \left[\frac{R_2}{s} + jX_2\right]I_2 \ . \tag{7.5}$$

Equation 6.7 can be written in the form

$$T = \frac{pm_1}{\omega_1}(I_2)^2 \frac{R_2}{s} \tag{7.6}$$

where m_1 is the number of stator phases and p is the number of pole pairs.

By definition, the fractional slip is

$$S = \frac{\omega_1 - \omega_m}{\omega_1} \qquad (7.7)$$

where the synchronous angular velocity, ω_1, and the shaft angular velocity, ω_m, are specified in the same units, either electrical or mechanical radians per second. When using electrical radians per second, $\omega_1 = 2\pi f_1$, the angular frequency of the supply.

If f_2 is the rotor frequency corresponding to a stator frequency, f_1, then

$$S = \frac{f_2}{f_1} = \frac{\omega_2}{\omega_1} \qquad (7.8)$$

where $\omega_2 = 2\pi f_2$.

The speed difference, $\omega_1 - \omega_m$, is termed the slip speed. Combining Equations 7.7 and 7.8 gives

$$\omega_1 - \omega_m = \omega_2 = S\omega_1 . \qquad (7.9)$$

Thus, the slip speed in electrical radians per second equals the rotor angular frequency, ω_2.

On combining Equations 7.4, 7.5, 7.6, and 7.8, the motor torque can be expressed in terms of the applied voltage, V_1, and the angular frequencies, ω_1 and ω_2. The resulting expression is

$$T = pm_1 \left[\frac{V_1}{\omega_1} \right]^2 \frac{\omega_2 X_m^2 / R_2}{\left[R_1 - \frac{\omega_2}{\omega_1 R_2} \left(X_{11} X_{22} - X_m^2 \right) \right]^2 + \left[X_{11} + \frac{\omega_2 R_1 X_{22}}{\omega_1 R_2} \right]^2} \qquad (7.10)$$

where $X_{11} = X_1 + X_m$, the total stator reactance at the supply frequency ω_1, and $X_{22} = X_2 + X_m$, the total rotor reactance at the same frequency.

For a fixed volts/hertz ratio at the motor terminals, the quantity V_1/ω_1 in Equation 7.10 is constant, and the torque characteristics can be evaluated when the machine parameters are known. Torque can be plotted as a function of rotor frequency for different values of supply frequency using Equation 7.10;[21,22] alternatively, ω_2 can be replaced by $(\omega_1 - \omega_m)$ in accordance with Equation 7.9 to allow torque-speed characteristics to be plotted. Figure 7.2 reproduces the calculated torque-speed characteristics published in Reference 23 for a 5-horsepower, 60-Hz, induction motor. Torque is expressed in per-unit form with the rated torque of the motor as a base value; the characteristics are extended into the induction generator region where the slip is negative.

As shown in Fig. 7.2, the peak torque (breakdown torque) decreases rapidly in the motoring region when the stator frequency is reduced below about 10 Hz. This torque decrease is caused by the reduction in airgap flux at low frequency resulting from the increased influence of the stator resistance. The stator resistance voltage drop at rated current has the same magnitude at all frequencies; it therefore constitutes a considerably higher fraction of the supply voltage at low frequency than at rated frequency, when it may often be neglected.

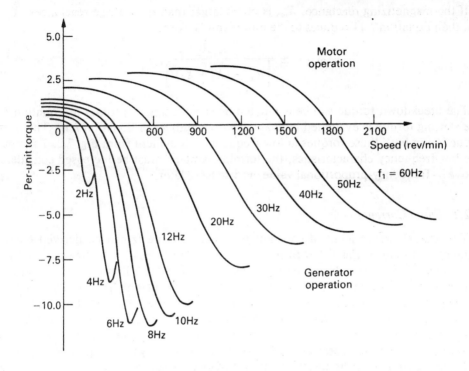

FIG. 7.2. Torque-speed characteristics for adjustable-frequency operation of the induction motor with a constant volts/hertz ratio. (Motor: 5-horsepower, 4-pole, 220 V, 60 Hz)

In the induction generator region, the machine operates with a reversed power flow and reversed stator voltage drop, resulting in an increased emf, E_1, and increased airgap flux. Consequently, large generator torques are produced, particularly at low frequencies. However, the characteristics of Fig. 7.2 are based on the linear equivalent circuit of Fig. 6.1, in which magnetic saturation is neglected. In practice, saturation effects will cause the braking torques in the generator region to be significantly less than the theoretical values indicated in Fig. 7.2; a nonlinear analysis is essential if realistic quantitative results are required.

In the motoring region, saturation effects are not severe and Equation 7.10 can be used to determine motor torque. The breakdown torque occurs at a particular rotor frequency known as the rotor breakdown frequency. By differentiating Equation 7.10 with respect to ω_2, and equating the derivative to zero, the rotor breakdown frequency, ω_{2b}, is obtained as

$$\omega_{2b} = \pm \, \omega_1 R_2 \left[\frac{R_1^2 + X_{11}^2}{(X_{11}X_{22} - X_m^2)^2 + R_1^2 X_{22}^2} \right]^{1/2} \tag{7.11}$$

where the positive and negative values apply to motor and generator operation, respectively.

If the magnetizing reactance, X_m, is much larger than the leakage reactances, X_1 and X_2, then Equation 7.11 reduces to the more familiar form

$$\omega_{2b} = \pm \frac{\omega_1 R_2}{\left[R_1^2 + (X_1 + X_2)^2 \right]^{1/2}} \, . \tag{7.12}$$

The breakdown torque can be evaluated by substituting ω_{2b} for ω_2 in Equation 7.10; the starting torque is evaluated by making ω_1 equal to ω_2 in Equation 7.10. The serious reduction in both these torques at low frequencies is evident from Fig. 7.2. To improve the low-frequency characteristics, the terminal voltage must be increased considerably above its frequency-proportional value, as discussed later.

7.2.2. Stator Current Locus

The stator current locus at constant terminal volts/hertz can be derived from the equivalent circuit of Fig. 6.1 with the use of complex inversion, as treated in detail by Nürnberg.[24] The procedure can be summarized briefly as follows: the stator input impedance, Z_1, is a function of the rotor slip, s. The stator current $I_1 = V_1 / Z_1$, where V_1 is the applied phase voltage. Consequently, if the locus of Z_1 is determined and then inverted with respect to the origin, the current locus is obtained. In order to place the eventual current diagram in its usual position, the positive resistance axis is drawn vertically upward, and the positive reactance axis is drawn horizontally to the right.

From the equivalent circuit, the input impedance is expressed by

$$Z_1 = R_1 + jX_1 + \frac{jX_m[(R_2/s) + jX_2]}{(R_2/s) + j(X_m + X_2)} = R_1 + jX_{11} + \frac{X_m^2}{(R_2/s) + jX_{22}} \, . \tag{7.13}$$

As the slip varies from minus to plus infinity, the input impedance traces a circular locus. Two points on the circle are readily determined: at synchronism, $s = 0$ and $Z_1 = R_1 + jX_{11}$, while at $s = \infty$, $Z_1 = R_1 + jX_{11} - jX_m^2/X_{22}$. These two values define the end points A and B of the horizontal diameter of the impedance circle in Fig. 7.3. The diametrical length is X_m^2/X_{22} and is independent of rotor resistance, R_2. The center point, M, of the circle is at $R_1 + jX_{11} - \frac{1}{2}jX_m^2/X_{22}$, and its position is also independent of R_2. A circle transforms into another circle after inversion with respect to a point outside it. Consequently, the impedance circle transforms into an admittance circle after inversion with respect to the origin. If the scales are suitably chosen, the admittance circle coincides with the impedance circle. The stator current is proportional to the admittance; hence, the admittance circle also represents the current locus. The current scale is obtained by multiplying the admittance scale by the phase voltage.

When the induction motor operates on an adjustable-frequency supply, inductive reactances vary linearly with frequency, but winding resistances remain constant. Under constant terminal volts/hertz operation, the impedance locus at different supply frequencies can be obtained by the assumption that the supply voltage and frequency remain constant while the resistances R_1 and R_2 vary inversely with frequency. As the supply frequency is reduced, the size of the impedance circle is unaffected, but the equivalent increase in R_1 causes the impedance circle to rise vertically between the tangents T_A and T_B shown in Fig. 7.3. After inversion, the tangents transform into the

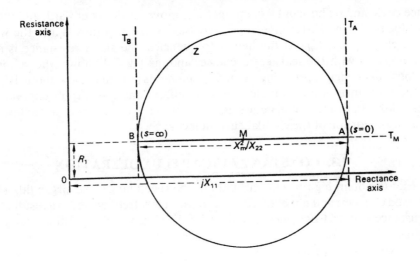

FIG. 7.3. Input impedance locus for the induction motor.

circles C_A and C_B of Fig. 7.4. The horizontal diameter, T_M, transforms into the circle C_M, intersecting C_A and C_B at A' and B'. These interstices are the inverse points corresponding to A and B on the impedance locus. The impedance circle, Z, transforms into the current circle, C, touching C_A and C_B at the points A' and B'.

As the supply frequency is reduced at constant terminal volts/hertz, the circles C_A and C_B are unchanged because the tangents T_A and T_B are unaffected. However, the im-

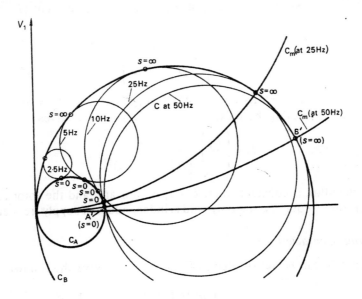

FIG. 7.4. Derivation of induction motor current locus for adjustable-frequency operation with a constant volts/hertz ratio.

pedance circle and its horizontal diameter, T_M, move vertically upward, and the circular current locus obtained after inversion decreases in size as it moves into the wedge between circles C_A and C_B. At the higher frequencies, the stator resistance is relatively small compared with the leakage reactance, and, as Fig. 7.4 shows, the stator current locus does not change appreciably. As the frequency is reduced below about 10 Hz, however, the resistance becomes increasingly significant and the current circle rapidly decreases in size. In practice, magnetic saturation, which is often significant at low frequencies, causes distortion of the circular current locus.[25]

7.3. CONSTANT AIRGAP FLUX OPERATION

In order to obtain high torque throughout the speed range, the airgap flux should be maintained constant and not allowed to decrease at low frequencies as a result of increasing resistance effects. Constancy is achieved if the airgap emf, E_1, rather than the terminal voltage, V_1, is adjusted linearly with frequency.

From the equivalent circuit of Fig. 6.1,

$$E_1 = jX_m I_m = j\omega_1 L_m I_m \tag{7.14}$$

where L_m is the magnetizing inductance and I_m is the magnetizing current.

The airgap flux, Φ_1, is proportional to E_1/ω_1 and is therefore proportional to the product $L_m I_m$, from Equation 7.14. If L_m is constant and unaffected by saturation, the airgap flux is proportional to I_m; a constant airgap flux requires a constant magnetizing current at all speeds and loads. In practice, I_m may be held constant at the rated (or nominal) value corresponding to normal full-load operation at rated voltage and frequency; alternatively, I_m may be held at the slightly larger value corresponding to no-load operation at rated voltage and frequency.

Again, from the equivalent circuit of Fig. 6.1, the rotor current, I_2, is given by the equation

$$I_2 = \frac{E_1}{\left[(R_2/s)^2 + X_2^2\right]^{1/2}}. \tag{7.15}$$

Now, $X_2 = \omega_1 L_2$, where L_2 is the rotor leakage inductance. Also, $s = \omega_2/\omega_1$, from Equation 7.8. Substituting these expressions in Equation 7.15 gives

$$I_2 = \left(\frac{E_1}{\omega_1}\right) \frac{\omega_2}{\left[R_2^2 + (\omega_2 L_2)^2\right]^{1/2}}. \tag{7.16}$$

If the airgap flux is constant, then E_1/ω_1 is constant, and the rotor current, I_2, is a function of rotor frequency, ω_2, only, and is independent of supply frequency, ω_1.

7.3.1. Torque Characteristics

Equation 7.16 can be substituted in Equation 7.6 to give the following torque equation:

$$T = pm_1 \left[\frac{E_1}{\omega_1}\right]^2 \left[\frac{\omega_2 R_2}{R_2^2 + (\omega_2 L_2)^2}\right]. \tag{7.17}$$

Since the airgap flux is proportional to E_1/ω_1, the electromagnetic torque is proportional to the square of the airgap flux at a given rotor frequency ω_2. Consequently, if the airgap flux is maintained constant under all operating conditions, the induction motor torque, like the rotor current, is determined solely by ω_2 and is independent of ω_1. Equation 7.9 shows that ω_2 equals the slip speed $(\omega_1 - \omega_m)$. It follows that motor torque is uniquely determined by the slip speed; the shape of the torque-speed characteristic therefore remains the same at different supply frequencies, but the curve is shifted laterally along the speed axis by changes in supply frequency, as shown in Fig. 7.5. These characteristics were computed for the same 5-horsepower, 60 Hz induction motor whose constant volts/hertz characteristics were plotted in Fig. 7.2.[26] The airgap flux is now maintained constant at the value corresponding to no-load operation at rated voltage and frequency. Obviously, the breakdown torque is the same at all supply frequencies and the motor has a genuine constant-torque capability over the full speed range, thus avoiding any deterioration in low-speed performance.

Differentiating Equation 7.17 with respect to rotor frequency, ω_2, and equating to zero, yields the rotor breakdown frequency as

$$\omega_{2b} = \pm \frac{R_2}{L_2} . \tag{7.18}$$

On substituting this value in Equation 7.17, the breakdown torque is obtained as

$$T_b = \pm \, pm_1 \left[\frac{E_1}{\omega_1} \right]^2 \frac{1}{2L_2} . \tag{7.19}$$

Thus, the breakdown torque is proportional to the square of the airgap flux and inversely proportional to the rotor leakage inductance. The rotor resistance does not affect

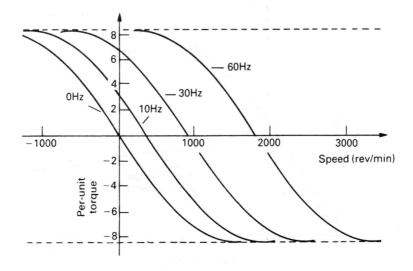

FIG. 7.5. Torque-speed characteristics for adjustable-frequency operation of the induction motor with constant airgap flux. (Motor: 5-horsepower, 4-pole, 220 V, 60 Hz)

the breakdown torque but influences the rotor frequency at which breakdown occurs. On combining Equations 7.17, 7.18, and 7.19, the following torque equation is obtained for constant-flux operation:

$$\frac{T}{T_b} = \frac{2}{(\omega_2/\omega_{2b}) + (\omega_{2b}/\omega_2)} . \tag{7.20}$$

This universal torque equation is valid at all stator frequencies and for both motor and generator operation. In Fig. 7.6, torque is plotted as a function of rotor frequency to give a universal torque characteristic for constant-flux operation of the 5-horsepower motor. The torque characteristic for normal fixed-frequency operation at rated voltage and frequency is included in Fig. 7.6 for comparison. Clearly, the available torque is much greater when the airgap flux level is maintained constant.

7.3.2. Stator Current Locus

From the equivalent circuit, the stator current, I_1, is given by the phasor sum of I_m and I_2. Thus,

$$I_1 = \frac{E_1}{jX_m} + \frac{E_1}{(R_2/s) + jX_2} = \left[\frac{E_1}{\omega_1}\right]\left[\frac{R_2 + j\omega_2 L_{22}}{-\omega_2 L_2 L_m + jR_2 L_m}\right] \tag{7.21}$$

where $L_{22} = L_2 + L_m$, the total rotor inductance.

This equation shows that the stator current, I_1, is independent of the supply frequency when the airgap flux is constant. If the stator current locus is plotted according to Equation 7.21, a general circle diagram is obtained that is valid at all stator frequencies. The

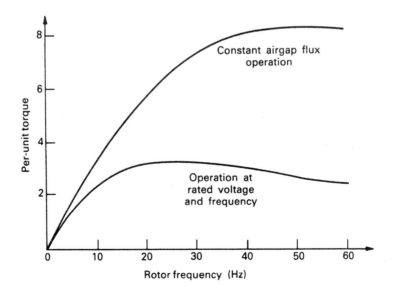

FIG. 7.6. The universal torque characteristic for constant airgap flux operation compared with the normal torque characteristic at rated voltage and frequency. (Motor: 5-horsepower, 4-pole, 220 V, 60 Hz)

circular locus has a larger diameter than the normal circle diagram at rated voltage and frequency, indicating that the machine currents are greater.[22] Sustained operation in the high-torque region may not be possible unless special cooling methods are provided.

7.3.3. Stator Voltage

The terminal voltage, V_1, is obtained by phasor addition of the airgap emf E_1, and stator voltage drop, $(R_1 + jX_1)I_1$. Thus,

$$V_1 = E_1 + (R_1 + jX_1)I_1 . \tag{7.22}$$

Under constant-flux conditions, the airgap emf, E_1, is varied linearly with stator frequency, but the stator voltage drop is determined by the supply current, I_1, and supply frequency, ω_1. The terminal voltage necessary to maintain a constant flux density in the machine is, therefore, a function of stator frequency and motor load. The load torque determines the rotor frequency, ω_2, and hence the stator current, I_1, as expressed by Equation 7.21. Curves of terminal voltage as a function of stator frequency can be plotted with the rotor frequency or stator current as a parameter.[22,27,28] These characteristics were computed for a 2-horsepower, 60 Hz, laboratory machine and are plotted in Fig. 7.7. The load-dependent increase in terminal voltage in the motor region is clearly seen. For generator operation, the terminal voltage is significantly less than that for motoring operation, except at low frequencies.

The large breakdown torque which can be developed for constant airgap flux conditions cannot be realized without a substantial increase in terminal voltage above the constant volts/hertz value. At low speeds, the power converter can readily deliver the increased stator voltage required, but at rated supply frequency, the terminal voltage needed to establish the rated airgap flux at the breakdown point may be almost double the rated voltage of the motor.[26]

Airgap flux control can be achieved by means of a feedback control system that employs a flux sensor in the airgap, as discussed later (Section 7.9).

7.4. CONSTANT VOLTAGE OPERATION

In the preceding sections, the objective was to maintain a constant-torque capability throughout the speed range by increasing both stator voltage and frequency so that a constant (or nearly constant) airgap flux was established. The constant-torque speed range extends up to the base speed of the drive at which the power converter delivers its maximum output voltage. In a PWM voltage-source inverter drive, the modulation index is varied with frequency to control the output volts/hertz ratio in the constant-torque range; maximum output voltage occurs when the modulation index has been increased to its upper limit, or the inverter has transitioned to the six-step mode of operation. In a six-step voltage-source inverter drive, maximum output voltage occurs when the front end thyristor rectifier circuit is phased fully on and the dc link voltage has its maximum value.

The maximum voltage delivered by the power converter is often approximately equal to the rated motor voltage; hence the base speed of the drive is the normal operating speed at rated frequency and voltage. However, the converter can deliver a constant out-

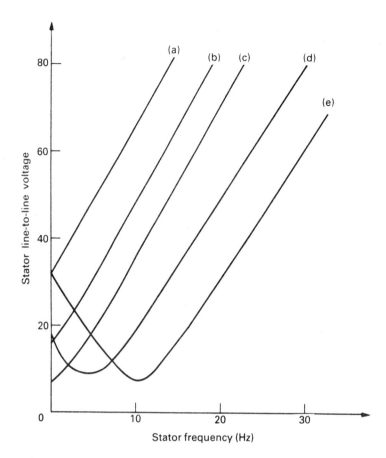

Fig. 7.7. Stator voltage required for adjustable-frequency constant-flux operation of the induction motor: (a) $I_1 = 2.0$ pu, motoring; (b) $I_1 = 1.0$ pu, motoring; (c) $I_1 = $ nominal magnetizing current; (d) $I_1 = 1.0$ pu, generating; (e) $I_1 = 2.0$ pu, generating. (Motor: 2-horsepower, 4-pole, 208 V, 60 Hz)

put voltage at higher frequencies. By increasing the stator frequency with constant stator voltage, the induction motor operates above base speed with a reduced volts/hertz ratio and a reduced torque capability.

7.4.1. Torque Characteristics

In the high-frequency, field-weakening region above base speed, the magnetizing current is small, and the leakage reactances are relatively large compared with the stator resistance. Consequently, there is little error if the magnetizing reactance, X_m, and the stator resistance, R_1, are eliminated from the equivalent circuit of Fig. 6.1. The current I_2 is therefore given by

$$I_2 = \frac{V_1}{\left[(R_2/s)^2 + (X_1 + X_2)^2 \right]^{1/2}} \, . \qquad (7.23)$$

This expression can be substituted in Equation 7.6 to give the motor torque as

$$T = \frac{pm_1}{\omega_1} \cdot \frac{V_1^2}{[(R_2/s)^2 + (X_1 + X_2)^2]} \cdot \frac{R_2}{s} \cdot \qquad (7.24)$$

Substituting for s from Equation 7.8 and simplifying gives

$$T = \frac{pm_1}{\omega_1^2} \cdot \frac{V_1^2 \omega_2 R_2}{[R_2^2 + \omega_2^2(L_1 + L_2)^2]} \qquad (7.25)$$

where $(L_1 + L_2)$ is the total leakage inductance of the motor.

As usual, by partial differentiation of Equation 7.25 with respect to ω_2, and by equating to zero, the rotor breakdown frequency, ω_{2b}, at a given stator frequency, ω_1, can be determined. Thus,

$$\omega_{2b} = \pm \frac{R_2}{L_1 + L_2} , \qquad (7.26)$$

and substituting in Equation 7.25 gives the breakdown torque as

$$T_b = \pm \frac{pm_1}{\omega_1^2} \cdot V_1^2 \cdot \frac{1}{2(L_1 + L_2)} \cdot \qquad (7.27)$$

Thus, ω_{2b} is independent of ω_1, but T_b is inversely proportional to ω_1^2. The family of torque-speed characteristics shown in Fig. 7.8 demonstrates the rapid reduction in torque capability when the supply frequency is increased at constant voltage. In this diagram, shaft speed is expressed in per-unit form with the base speed as 1.0 pu.

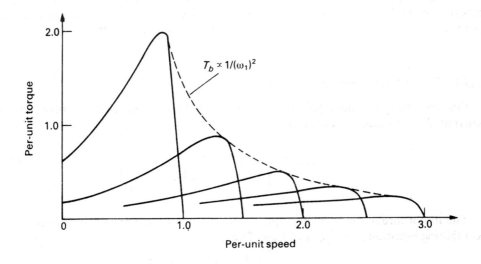

FIG. 7.8. Typical torque-speed characteristics for adjustable-frequency operation of the induction motor at constant voltage.

7.5. CONTROLLED-CURRENT OPERATION

The airgap flux of the machine can be indirectly determined by controlling the stator current, I_1, and rotor slip frequency, ω_2. The practical implementation of this control technique is discussed later, but, for the moment, it can be noted that, since the motor torque is determined by the airgap flux and machine current, an improved dynamic performance is achieved by directly controlling the stator current rather than the terminal voltage. There are also benefits arising from the operation of the converter as a controlled-current source. Inherent short-circuit protection is provided, and there are no transient current surges, so that a more economical inverter design is possible.

As discussed earlier, the airgap flux is proportional to the magnetizing current, I_m, with the assumption that the magnetizing inductance, L_m, is unaffected by saturation. In order to preserve constant airgap flux in the motor, the magnetizing current obviously must be held constant. The stator current, I_1, required to achieve this condition is a function of the rotor frequency, ω_2, but is independent of the stator frequency, ω_1, as is readily shown by analyzing the equivalent circuit of Fig. 6.1. The exact relationship between I_m and I_1 is

$$I_m = I_1 \left[\frac{R_2^2 + (\omega_2 L_2)^2}{R_2^2 + (\omega_2 L_{22})^2} \right]^{1/2}. \tag{7.28}$$

If constant-flux operation is required, the prescribed values of stator current and rotor frequency must always be suitably related so that the right-hand side of Equation 7.28 gives the desired constant value of I_m. Typically, I_m is held constant at the value corresponding to no-load operation at rated voltage and frequency. The precise relationship between I_1 and ω_2 for a particular machine can be implemented by an analog or digital function generator. It is clear from Equation 7.28 that the functional relationship is independent of the sign of ω_2, indicating that the stator current is the same for motoring or regenerating; when ω_2 is zero, I_1 equals I_m. Figure 7.9 shows the calculated characteristic for the 2-horsepower, 60 Hz motor when the airgap flux is held constant at the nominal no-load value corresponding to the nameplate voltage and frequency.

7.5.1. Torque Characteristics

Further analysis of the equivalent circuit shows that the relationship between rotor current, I_2, and stator current, I_1, is

$$I_2 = I_1 \frac{\omega_2 L_m}{\left[R_2^2 + (\omega_2 L_{22})^2 \right]^{1/2}}. \tag{7.29}$$

The motor torque can be expressed in terms of stator current and rotor frequency by combining Equations 7.6, 7.8, and 7.29. Thus,

$$T = pm_1 (I_1)^2 \left[\frac{(\omega_2 L_m)^2 R_2/\omega_2}{R_2^2 + (\omega_2 L_{22})^2} \right]. \tag{7.30}$$

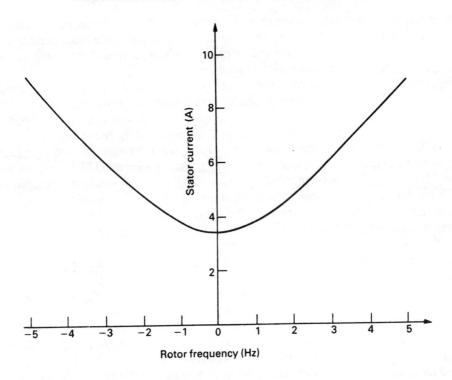

FIG. 7.9. Relationship between stator current and slip frequency for constant airgap flux. (Motor: 2-horsepower, 4-pole, 208 V, 60 Hz)

This equation gives the gross output torque, assuming negligible saturation. Clearly, torque is a function of ω_2 only, and is independent of stator frequency. It is readily shown that, for a given stator current, the torque expression has a maximum value at the rotor breakdown frequency, ω_{2b}, where

$$\omega_{2b} = \pm \frac{R_2}{L_{22}} . \qquad (7.31)$$

The corresponding value of breakdown torque is given by

$$T_b = \pm pm_1 (I_1)^2 \frac{L_m^2}{2L_{22}} . \qquad (7.32)$$

Equations 7.28, 7.29, and 7.30 show that when the stator current and slip frequency are defined, the motor condition is completely characterized; for given values of I_1 and ω_2, there is a unique value of magnetizing current, I_m (Equation 7.28), of rotor current, I_2 (Equation 7.29), and of torque, T (Equation 7.30).

If the values of I_1 and ω_2 are not properly related, the current-fed machine may be subjected to high levels of magnetic saturation resulting in a substantial variation in machine parameters. In particular, magnetic saturation causes a significant reduction in the magnetizing inductance, L_m, and therefore, at high current levels, the actual breakdown torque may be considerably lower than that predicted when the unsaturated value of L_m is used in Equation 7.32. Also, as the stator current increases, the rotor breakdown frequency, ω_{2b}, which is theoretically independent of stator current (Equation 7.31), is displaced to a higher value because of the reduction in L_{22}.[22,29-31]

Figure 7.10 shows calculated curves of per-unit torque and per-unit airgap voltage (or flux) as a function of per-unit speed for a 25-horsepower, 60 Hz induction motor.[30] The magnetizing inductance of the motor is assumed constant, and the motor is current fed with rated stator current at rated frequency. The developed torque builds up from a very low value at standstill to a breakdown value of 2.0 pu at 0.98 pu speed and then falls rapidly to zero at synchronous speed. This characteristic is explained by the corresponding airgap voltage curve which indicates that the airgap flux is very small near standstill when the low value of referred rotor impedance at high slip limits the airgap voltage developed by the rated stator current. As motor speed rises, the rotor impedance and airgap voltage increase, allowing the development of greater flux and torque. Near synchronous speed, the motor is highly saturated and the assumption of a constant magnetizing reactance is not justified. Consequently, the breakdown torque of 2.0 pu will not be realized in practice.

In Fig. 7.10, the normal torque-speed curve of the motor for constant-voltage excitation at rated voltage and frequency is included for comparison. Conventional voltage-fed induction motor operation takes place on the negatively sloped portion of this characteristic, resulting in rated airgap flux. Thus, the intersection point A defines the operating point of the current-fed machine at which it has rated current, rated airgap flux and rated terminal voltage. This operating point clearly lies on the positively sloped or statically unstable part of the torque-speed curve for the current-fed motor.

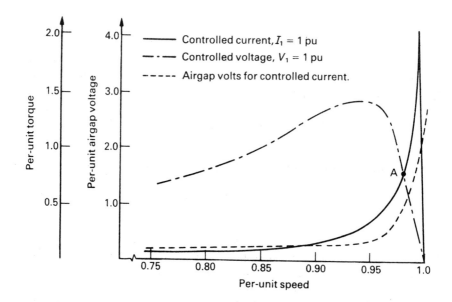

FIG. 7.10. Theoretical torque-speed characteristic for controlled-current operation at rated current and frequency compared with the normal torque characteristic at rated voltage and frequency. (Motor: 25-horsepower, 4-pole, 230 V, 60 Hz)

The torque-speed characteristics for the current-fed machine are replotted in Fig. 7.11, with the reduction in magnetizing reactance that results from saturation taken into account. The modified torque characteristic for 1 pu stator current shows that the theoretical breakdown torque of 2.0 pu has been reduced to 0.8 pu. In addition, the breakdown point is at a somewhat lower speed, indicating a higher value of rotor breakdown frequency. Several other torque-speed curves are plotted in Fig. 7.11 for different values of stator current; the torque curve at rated voltage is also included. As before, the intersection point A indicates the operating point for rated conditions. The same torque can be developed at point B on the stable side of the torque-speed characteristic, but the airgap flux and core losses are larger and the terminal voltage is higher. Consequently, controlled-current operation is usually implemented at point A, but this mode of operation precludes open-loop control; a feedback loop is essential for stable operation. As discussed earlier in this section, the motor torque is varied by adjusting the stator current and rotor frequency so that rated airgap flux is maintained. The motor operating points therefore fall on the torque-speed curve defined for rated voltage operation.

Practical control schemes for controlled-current operation are discussed in Section 7.10.

7.6. CONTROLLED-SLIP OPERATION

In general, the induction motor operates at high power factor and high efficiency provided the rotor frequency does not exceed the breakdown value corresponding to maximum torque. Beyond the breakdown point, the motor power factor and torque per am-

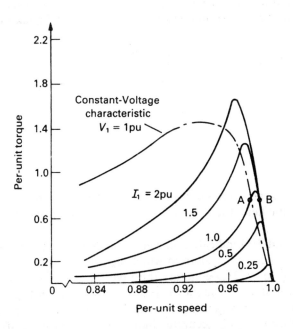

FIG. 7.11. Torque-speed characteristics for controlled-current operation at rated frequency when saturation effects are included. (Motor: 25-horsepower, 4-pole, 230 V, 60 Hz)

pere are low; when a typical cage-rotor induction motor is started from standstill at rated voltage and frequency on the ac utility supply, the sudden application of rated voltage results in a standstill current of five or six times rated current, but the starting torque may not significantly exceed the rated torque. In an adjustable-frequency drive, the control strategy must ensure that motor operation is restricted to low slip frequencies, resulting in stable operation with a high power factor and a high torque per stator ampere, thereby minimizing the inverter current rating.

In simple open-loop drive systems, induction motor operation is restricted to the low-slip region by a ramping circuit that limits the rate of change of inverter frequency when a sudden speed change is requested; motor speed can then track the slow variation in stator frequency without exceeding the rotor breakdown frequency. In more sophisticated drives with an improved dynamic performance, the rotor slip frequency is controlled either directly or indirectly. Indirect control of slip frequency can be achieved if the stator current and airgap flux are both directly controlled. However, as discussed later, precise control of airgap flux usually requires the fitting of flux sensors and precludes the use of a standard commercial induction motor. This disadvantage can be overcome by direct control of slip frequency, but shaft speed must be accurately monitored by means of a tachometer. This controlled-slip technique has been widely used in adjustable-frequency drive systems. [1,3,5,32–42]

The fractional slip has already been defined in Equation 7.7 as

$$s = \frac{\omega_1 - \omega_m}{\omega_1}$$

where ω_1 and ω_m are the synchronous and actual shaft speeds, respectively, in electrical radians per second. The corresponding slip speed is given by Equation 7.9 as

$$\omega_1 - \omega_m = \omega_2 = s\omega_1 \ .$$

Thus,

$$\omega_1 = \omega_2 + \omega_m \ . \tag{7.33}$$

Alternatively,

$$f_1 = f_2 + pn \tag{7.34}$$

where n is the shaft speed in revolutions per second and p is the number of pole-pairs in the rotating magnetic field; the product pn is sometimes termed the rotational frequency.

Equations 7.33 and 7.34 indicate that a commanded rotor frequency, ω_2^* (or f_2^*), can be implemented by a control system in which ω_m (or pn) is measured by a tachometer and is added to ω_2^* (or f_2^*) to generate the inverter frequency command ω_1^* (or f_1^*). Thus, direct control of slip frequency is possible, as shown in Fig. 7.12. The rotational frequency is usually much larger than the slip frequency and, for precise control, the tachometer signal must be highly accurate; digital techniques can satisfy this requirement.

At starting, the drive control system reduces both stator voltage and frequency, thereby maintaining the airgap flux level; the low rotor frequency ensures a high starting

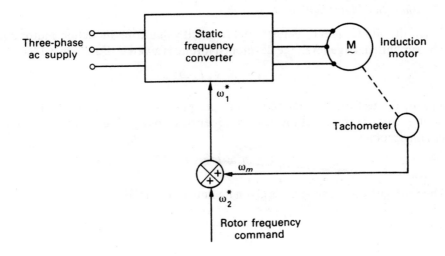

FIG. 7.12. Induction motor drive with direct control of rotor frequency.

torque per ampere and rapid acceleration from standstill. Because the induction motor is not subjected to a direct-on-line start with the step application of rated voltage and frequency, there is no need for a high-resistance or double-cage rotor winding to limit starting current and increase starting torque. The use of a low-slip motor will ensure reduced rotor losses and improved efficiency under normal operating conditions.

If ω_2^* is negative, the stator excitation frequency, ω_1^*, is less than ω_m by an amount ω_2^*. Under these conditions, the machine has a negative slip and operates as an induction generator, returning energy to the power converter which also supplies the lagging kVA required by the induction generator. As explained in earlier chapters, a regenerative power converter (such as a current-source converter or cycloconverter) can produce rapid deceleration by returning the kinetic energy of the rotating masses to the ac utility network. A nonregenerative dc link converter must dissipate the energy in a dynamic braking resistor in the dc link circuit. A slip limit is introduced in the control system so that the demanded rotor frequency can never exceed the breakdown value in the motoring or generating regions. Thus, stable operation is possible close to the breakdown point, giving a high torque per ampere and good dynamic performance for sudden changes in demanded speed.

The direction of motor rotation is determined by the phase sequence of the stator supply. Reversal of the phase sequence is obtained statically by means of the converter firing circuitry; no switching of power leads is necessary. Independent control of the direction of rotation and the slip polarity means that four-quadrant operation is possible, giving motor or generator operation for either direction of rotation.

The use of a shaft-coupled tachometer detracts somewhat from the mechanical simplicity and ruggedness of the cage-rotor induction motor; tachometer expense is also undesirable in small low-cost drive systems. Electromechanical transducers can be eliminated by using indirect methods to determine the slip information. Thus, the precise value of slip frequency at any instant can be deduced from the stator voltages and currents by appropriate signal-processing techniques.[38]

7.6.1. *Simplified Motor Equations*

When the slip frequency is restricted to small values, the induction motor equations can be simplified. In general, the developed torque is given by Equation 6.14 as

$$T = K \, \Phi_1 \, I_2 \cos \phi_2$$

where Φ_1 is the fundamental airgap flux per pole, and $I_2 \cos \phi_2$ is the in-phase component of rotor current. At low rotor frequencies, $\cos \phi_2$ is nearly unity and the torque equation becomes

$$T = K \, \Phi_1 \, I_2 \ . \tag{7.35}$$

The rotor current, in general, is given by Equation 7.15 as

$$I_2 = \frac{E_1}{\left[(R_2/s)^2 + X_2^2 \right]^{1/2}} \ .$$

When s is small, $X_2 \ll R_2/s$, and

$$I_2 = \frac{sE_1}{R_2} = \frac{\omega_2}{\omega_1} \cdot \frac{E_1}{R_2} \ . \tag{7.36}$$

Because the airgap flux, Φ_1, is proportional to E_1/ω_1, the rotor current is proportional to the product of airgap flux and rotor slip frequency. Hence,

$$I_2 = K' \Phi_1 \omega_2 \ . \tag{7.37}$$

On combining Equations 7.35 and 7.37

$$T = K'' \Phi_1^2 \omega_2 \ . \tag{7.38}$$

Thus, with constant airgap flux and small rotor slip, torque is directly proportional to rotor frequency. The shaft speed has no effect on the available torque, apart from a slight reduction with increasing speed because of windage and friction. The torque equation is valid for positive or negative slip frequencies corresponding to motor or generator operation.

7.7. CONSTANT-HORSEPOWER OPERATION

In certain applications, the mechanical load has a constant-horsepower requirement, and the adjustable-speed drive is designed to have a constant-horsepower capability over a specified speed range. In a traction drive, for example, it is often necessary to constrain the operation of the drive so that it conforms to a constant-horsepower characteristic. This constraint arises because of the limited power capability of the power distribution system or on-board supply, and because of the need to achieve an economical propulsion unit; a constant-horsepower characteristic is therefore imposed on the drive system to allow operation over a wide speed range with a limit on the maximum power demand.

General-purpose drives are often required to have a constant-torque capability below a certain base speed and a constant-horsepower capability above base speed. In the constant-torque region, airgap flux is maintained constant by increasing both terminal voltage and frequency, as already discussed. However, the stator voltage can only be increased up to a certain level, either because of source voltage limitations or maximum voltage restrictions imposed by power semiconductor devices or stator insulation. This limiting voltage and frequency condition defines the base speed of the drive, which often corresponds to the normal operating speed of the motor at its rated utility voltage and frequency. A limited constant-horsepower range of operation can be achieved above base speed by supplying the induction motor with a constant stator voltage at higher frequencies. For these operating conditions, the airgap flux is reduced, and the breakdown torque is inversely proportional to frequency squared, as shown in Section 7.4, whereas constant-horsepower operation requires that torque vary inversely with frequency. By suitable control of rotor frequency, ω_2, in a controlled-slip drive, the motor torque can be made to vary inversely with ω_1, as required. This constant-horsepower mode of operation corresponds to the field weakening operation of a dc motor.

In general, the induction motor torque at small slip is given by Equation 7.38:

$$T = K'' \Phi_1^2 \omega_2 .$$

At high frequencies, the airgap flux can be assumed to be proportional to the terminal volts/hertz ratio. Thus,

$$T \simeq K''' \left[\frac{V_1}{\omega_1} \right]^2 \omega_2 . \tag{7.39}$$

This torque equation for low-slip operation is an approximate version of Equation 7.25 and confirms that for a constant stator voltage, V_1, and rotor frequency, ω_2, the torque decreases as the square of the supply frequency, ω_1. However, if ω_2 is increased linearly with ω_1, the output torque varies as the inverse of ω_1, and approximately as the inverse of speed, giving a constant-horsepower characteristic. The upper speed limit is determined by the maximum working value of rotor frequency, as discussed below.

In general, it may be required to operate the motor at speeds higher than can be achieved in the constant-horsepower mode. In this case, motor voltage and slip frequency are maintained constant at their maximum values as the stator frequency is further increased; hence the output torque varies inversely as speed squared. This high-speed motoring characteristic corresponds to the torque-speed characteristic of a dc series motor.

Figure 7.13 shows a family of motor torque-speed characteristics at different stator frequencies in the constant-torque, constant-horsepower, and high-speed motoring regions. The maximum torque, or breakdown torque, which is indicated by a dashed line, is constant below base speed and decreases inversely with speed squared above base speed. The lower solid line in Fig. 7.13 is an imposed operating characteristic for the drive; slip frequency is held constant below base speed but is then increased with supply frequency to give a constant-horsepower characteristic up to twice base speed. In a traction application, the motor can be constrained to accelerate along this composite characteristic by appropriate control of rotor frequency, but steady-state operation must also be possible at any operating point below this characteristic. A corresponding operating profile can be imposed for braking operation in the fourth quadrant of the torque-speed diagram.

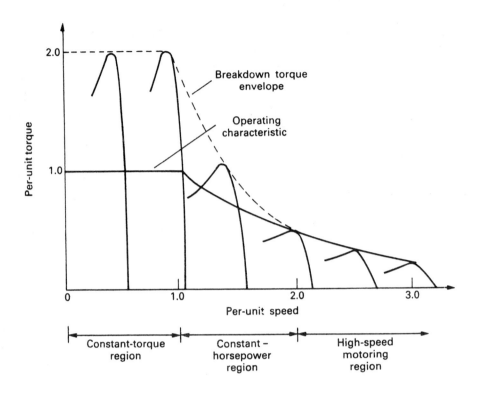

FIG. 7.13. Typical torque-speed characteristics for an adjustable-frequency induction motor drive.

Figure 7.14 shows the variation in motor voltage, current, slip frequency, and torque, as a function of speed for the imposed operating characteristic of Fig. 7.13. At base speed, the motor is supplied with rated voltage and frequency, and draws rated current; the developed torque is approximately half the breakdown value (Fig. 7.13). At higher speeds in the constant-horsepower range, the stator current stays nearly constant at the rated value, but rotor frequency is increased and the operating point moves closer to the breakdown point. As shown in Fig. 7.13, an upper speed limit is reached at constant horsepower when the rotor frequency is increased to the breakdown value, and the required torque is equal to the breakdown torque. In the diagram, this condition occurs at twice base speed, but the speed range at constant horsepower can be extended by oversizing the motor; for a larger frame size, the operating torque at base speed is a smaller fraction of the breakdown torque, and hence the constant-horsepower mode can be maintained up to a higher speed. In practice, it may be desirable to limit the rotor frequency to about half the breakdown frequency at the upper speed limit, because operation nearer the breakdown point produces increased currents and increased copper losses without a significant increase in torque output.[37]

A constant-horsepower characteristic may also be obtained by operating with a constant slip frequency and controlling the stator voltage and frequency together, so that the required torque is developed.

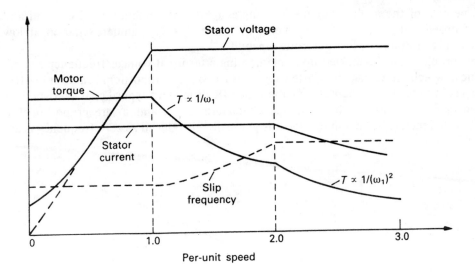

FIG. 7.14. Induction motor voltage, current, slip frequency, and torque, as a function of speed for the torque-speed operating characteristic of Fig. 7.13.

Again, from Equation 7.38

$$T = K'' \Phi_1^2 \omega_2$$

and, for a fixed value of ω_2,

$$T \propto \Phi_1^2 \propto (V_1/\omega_1)^2 . \tag{7.40}$$

If V_1^2 is proportional to ω_1, the torque is inversely proportional to ω_1, and a constant-horsepower output is obtained. Thus, the range of voltage variation is equal to the square root of the frequency range. For a given speed range, the motor has a smaller frame size than before, but the power converter and motor must have a higher voltage capability than required for base speed operation. For traction applications that require constant-horsepower operation over a very wide speed range, this method may be used for part of the range and the constant-voltage method for the remainder of the range.

7.8. TERMINAL VOLTS/HERTZ CONTROL

As shown in Section 7.2, the application of a constant volts/hertz supply at the motor terminals gives constant airgap flux if the stator voltage drop, $(R_1 + jX_1)I_1$, is negligible. This condition is reasonably well satisfied near rated motor frequency, but the stator IR voltage drop that is developed by the rated motor current remains constant as the frequency is reduced, so that at low frequencies, this IR drop is a large proportion of the terminal voltage. Thus, if the stator IR drop at rated frequency and full load is 4 percent of the phase voltage, the effect on airgap flux is relatively insignificant. However, at one-tenth of rated frequency, with a constant volts/hertz supply, the IR drop at rated current is 40 percent of the applied voltage, causing a significant reduction in airgap emf and flux. Consequently, for motoring operation, there is severe underexcitation at low speeds and an intolerable loss of torque capability.

This problem can be tackled by implementing a terminal voltage/frequency characteristic in which the voltage is boosted above its frequency-proportional value at low frequencies in order to compensate for the stator IR drop. Two typical characteristics are shown in Fig. 7.15(a). In the nonlinear characteristic, V_1 and ω_1 are proportional at higher frequencies, but a voltage boost is gradually introduced as zero frequency is ap-

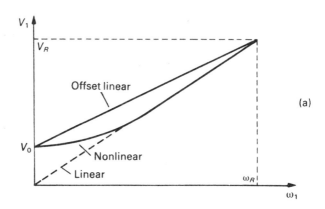

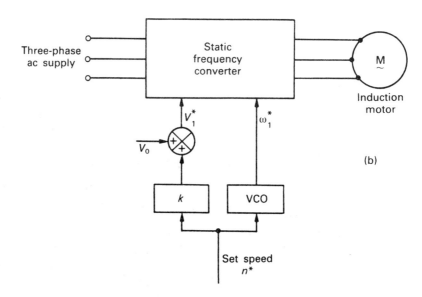

FIG. 7.15. Open-loop adjustable-speed induction motor drive with terminal volts/hertz control: (a) typical voltage/frequency characteristics; (b) basic control scheme.

proached. In the offset linear characteristic, a constant voltage component, V_o, is added to a frequency-proportional component, $k\omega_1$, to define the stator voltage as

$$V_1 = V_o + k\omega_1 . \tag{7.41}$$

V_o and k are chosen so that the required voltage boost is applied at zero frequency, and the rated voltage, V_R, is applied at the rated (50 or 60 Hz) frequency, ω_R. If the load requires a high starting torque, V_o can be adjusted so that a large motor current flows at zero frequency. However, this large boost may cause overheating if the motor operates for long periods at low speeds when the effectiveness of the internal cooling fan is significantly reduced. If the motor drives a fan load, the required torque is very small at low speeds, and the volts/hertz ratio can be reduced to minimize motor heating. In some commercial drives, the low-frequency voltage boost can be adjusted by the user to suit different load requirements.

In general, a preprogrammed voltage/frequency characteristic cannot maintain constant airgap flux under varying load conditions, because the stator voltage drop is a function of stator current, I_1. As load is applied to the motor shaft, I_1 increases and E_1 is reduced. To overcome this difficulty, the motor current may be monitored and used to determine a compensating increase in terminal voltage that is load dependent. The voltage boost can be made proportional to the magnitude of the stator current, or the in-phase component of stator current, but neither technique is totally satisfactory.[28] Commercial ac drive systems usually have a preprogrammed voltage/frequency characteristic appropriate to a particular motor or application; hence, the low-frequency voltage boost does not take any account of varying torque levels.

7.8.1. Open-Loop Speed Control

A static frequency converter can independently control motor voltage, V_1, and frequency, ω_1, but a programmed voltage/frequency characteristic reduces the number of command variables to one, allowing simple volts/hertz control. Figure 7.15(b) shows a block diagram of an open-loop adjustable-speed drive with terminal volts/hertz control; the offset linear voltage/frequency characteristic of Fig. 7.15(a) is implemented. The setpoint or reference signal is the speed command, n^*, which generates the inverter frequency command, ω_1^*, via a voltage-controlled oscillator (VCO). The voltage command, V_1^*, is also determined directly from the set speed signal, as shown.

The static frequency converter in Fig. 7.15(b) may be either a cycloconverter or a dc link converter employing a six-step or PWM voltage-source inverter. Open-loop speed control of the six-step inverter drive was discussed briefly in Chapter 4. In this drive configuration, direct control of inverter frequency is possible as shown in Fig. 4.12, or the dc link voltage can be directly controlled, with inverter frequency tracking voltage, as shown in Fig. 4.13. These control techniques are perfectly satisfactory for single or multiple induction motor drives where high dynamic performance is not required. Open-loop speed control of the PWM inverter drive is also widely used for general-purpose applications. The absence of motor speed or flux sensing gives a low-cost adjustable-speed drive that can employ standard commercial induction motors and is readily implemented in retrofit applications. Because rapid changes in the inverter frequency command may cause excessive current, the speed reference must be passed through a ramping circuit that gives a timed rate of change of frequency. Alternatively, a current limit control can be introduced, as described below.

7.8.2. Closed-Loop Speed Control

Open-loop speed control has the disadvantage that the rotor slip increases and the induction motor slows down slightly when load torque is applied. Improved speed regulation can be achieved when the motor's natural droop in speed with load is compensated by means of a slip compensation technique in which the inverter frequency is boosted by a signal proportional to motor current. However, open-loop speed control has a poor dynamic performance, and a closed-loop system with tachometer feedback is preferred.[43,44] Motor speed is then adjusted to the commanded value, giving improved speed regulation and reduced speed sensitivity to shaft load fluctuations. Speed feedback also ensures uniform drive performance over the entire frequency range, thereby eliminating the decrease in stability that has traditionally been a problem with open-loop adjustable-frequency drive systems at lower supply frequencies.

In a six-step inverter drive with simple volts/hertz control, the motor voltage can be adjusted by closed-loop regulation of the dc link voltage using the front-end phase-controlled rectifier (Fig. 4.12). This technique stabilizes the dc link voltage against changes in the ac utility voltage, but it is still an open-loop control system as far as speed is concerned, and hence the advantages of tachometer feedback are not available.

A drive with terminal volts/hertz control and a speed feedback loop is shown in Fig. 7.16. The set speed, n^*, is compared with the actual speed, n, to determine the speed error which is then passed through the speed controller and defines the inverter frequency and voltage. As usual, proper controller design is essential if the benefits of closed-loop speed control are to be fully realized. The control also features a current-limit signal that only comes into effect when the motor current rises to a preset maximum level; this signal then controls the rate at which the inverter frequency and voltage are ramped. Thus, if the speed command is suddenly increased, the motor current quickly rises to the preset limit; the demanded output frequency and voltage are then gradually increased so that the motor speed tracks the inverter frequency and the rotor breakdown frequency is not exceeded. The machine accelerates at constant torque under the influence of the current-limit control until the set speed is reached. The current then falls below the limit, and steady-state operation is achieved. Above base speed, the power converter cannot deliver an increasing voltage, but operation at constant voltage and increasing frequency causes the motor to operate in the field-weakening region.

Speed Control with Slip Regulation. Figure 7.17 shows an alternative volts/hertz control scheme using slip frequency control rather than current limit control. The speed error is compensated in the speed controller and is used to generate the slip frequency command, ω_2^*. This signal is added to the tachometer signal, in the usual manner, to determine the inverter frequency command, ω_1^*. The latter signal is also supplied to a function generator that develops the voltage command, V_1^*. The function generator implements a voltage boost at low frequencies so that the airgap flux is approximately constant. The induction motor torque is therefore proportional to the slip frequency, as

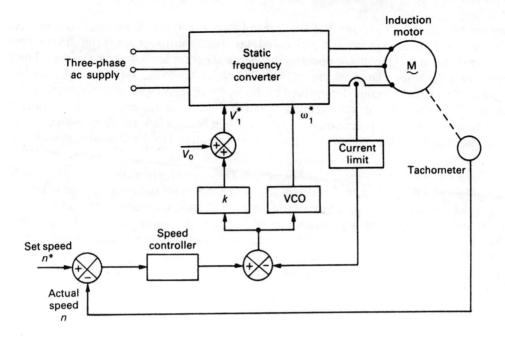

FIG. 7.16. Closed-loop adjustable-speed induction motor drive with terminal volts/hertz control and current limiting.

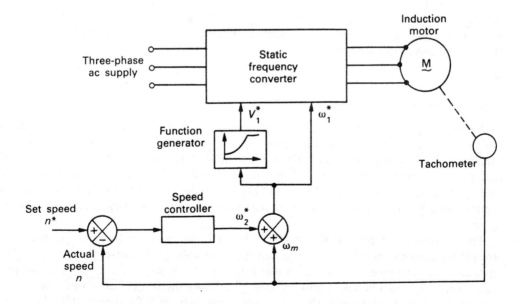

FIG. 7.17. Closed-loop adjustable-speed induction motor drive with terminal volts/hertz control and slip regulation.

shown by Equation 7.38, and the slip control loop functions as an inner torque loop. When a sudden speed increase is demanded, the speed controller limits the slip frequency command so that the motor operates just below the rotor breakdown frequency. Thus, the motor develops a large torque and rapidly accelerates to the set speed. The slip frequency then falls to a value determined by the load torque. A rapid reduction in demanded speed results in the generation of a negative slip command, causing the machine to function as an induction generator. As discussed in Section 7.6, the regenerated energy must be dissipated in a dynamic braking resistor or returned to the ac utility network.

Speed Control with Constant Slip Frequency. The control structure in Fig. 7.17 can be simplified by operating with a constant slip frequency. This constant slip-speed control has been popular in the past because of its ease of implementation. The inverter frequency is defined by adding a constant preset component, representing the fixed slip frequency, to the tachometer signal. The speed error signal now controls the motor voltage only. Thus, there is no longer a fixed relationship between motor voltage and frequency and, as a result, the airgap flux varies widely. Equation 7.38 gives the steady-state torque at small slip frequencies as

$$T = K''\Phi_1^2\omega_2$$

and for constant ω_2, torque is clearly proportional to Φ_1^2. The speed loop now controls the motor voltage, the airgap flux, and the developed torque. At light load, the motor voltage is automatically reduced, and the airgap flux falls to a low value. Consequently, the core losses in the motor are reduced, and the overall efficiency of the motor is improved. However, a change in motor operating conditions involves a change in airgap flux level, and hence the dynamic performance is very poor. Motor speed is highly sensitive to load fluctuations, and the response to a sudden change in set speed is very sluggish, making this control unsuitable for high-performance drives.

7.8.3. *Low-Frequency Performance with Increased Volts/Hertz*

As already noted, commercial ac drive systems usually have a preprogrammed voltage/frequency relationship with a significant voltage boost at low frequencies. Because this built-in relationship is independent of load, stator current variations are not taken into account. Consequently, if the voltage boost is adjusted to give approximately constant airgap flux for the full-load condition, then the volts/hertz ratio is excessive at light loads and low frequencies, causing magnetic saturation and large magnetizing currents.

The conventional equivalent circuit of Fig. 6.1 can be used to calculate the low-frequency performance with an increased terminal voltage. Because of saturation, the magnetizing reactance, X_m, is a nonlinear function of the magnetizing current, and this variation in X_m must be taken into account.[29,45,46] As the frequency of operation is reduced, the leakage reactances decrease proportionately with frequency, but the resistances remain constant. At low frequencies, the voltage drop across the stator resistance is, therefore, a relatively large fraction of the terminal voltage, and the approximate equivalent circuit in which X_m is moved to the input terminals is not valid.

Some unusual effects are noted when the machine is operated at low frequency with an increased terminal voltage. The stator resistance voltage drop is large compared with the leakage reactance drop; the phasor diagrams of Fig. 7.18(a) and Fig. 7.18(b) show the no-load and full-load conditions, respectively. The magnetizing current, I_m, lags the air-gap emf, E_1, by 90 degrees, and at no load the magnitude of emf E_1 is almost equal to the applied voltage, V_1. Thus, the machine is highly saturated, and a large magnetizing current flows. This no-load current may equal or even exceed the full-load current.

On load, the rotor current I_2 is more in phase with E_1, and the airgap emf is reduced, as shown in Fig. 7.18(b). Consequently, there is a large drop in magnetizing current as the load current increases. The reduction in magnetizing current more than offsets the increase in referred rotor current, and hence the total supply current decreases as load is applied. This process is illustrated by the characteristics of Fig. 7.19, which were measured at 5 Hz on a 2-horsepower, 50 Hz motor. The relationship between torque and stator current is plotted for the rated terminal volts/hertz condition and also for a terminal voltage which is 40 percent greater than the frequency-proportional value. The unusual characteristic obtained at the higher voltage is obviously caused by magnetic saturation. This characteristic shows that appreciable torque can be developed at low frequencies provided the terminal voltage is increased sufficiently.

7.8.4. *Optimum Efficiency Operation*

In preceding sections, the need to sustain the airgap flux by appropriate volts/hertz control has been emphasized. Operation at, or near, rated airgap flux gives good utilization of the motor iron and a high torque per stator ampere; rated motor torque can be developed at all supply frequencies. For these reasons, the constant flux and constant volts/hertz modes of control have often been regarded as optimum control techniques.

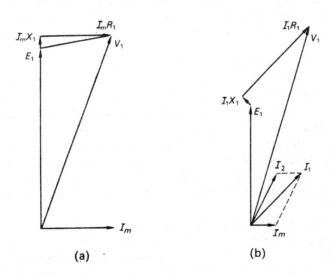

FIG. 7.18. The influence of magnetic saturation on induction motor performance at low frequency: (a) no-load phasor diagram; (b) full-load phasor diagram.

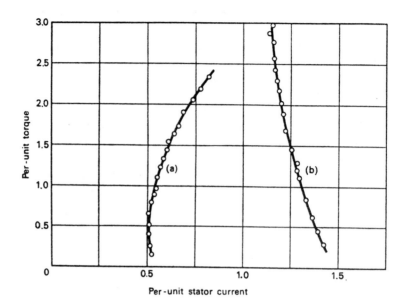

FIG. 7.19. Relationship between induction motor torque and stator current for a supply frequency of 5 Hz: (a) at rated volts/hertz; (b) at 140 percent of rated volts/hertz. (Motor: 2-horsepower, 6-pole, 230 V, 50 Hz)

However, at light load, the airgap flux may be greater than necessary for the development of the required torque, and losses are high, resulting in a motor efficiency that is less than optimum. This section discusses the adjustment of the applied volts/hertz ratio as a function of motor load in order to operate the motor at its highest efficiency. It is shown that in a lightly loaded motor, maximum efficiency is achieved with a reduced voltage and increased slip, as compared with the usual constant volts/hertz mode of operation.

In an adjustable-frequency induction motor drive, the voltage and frequency applied to the motor are independently variable, and a specific torque-speed operating point can be achieved with a variety of different voltage-frequency combinations. Each voltage-frequency pairing defines a particular motor torque-speed characteristic passing through the specified operating point, P, as in Fig. 7.20, but motor efficiency may vary widely. If the volts/hertz ratio is too high, then magnetizing current and core losses are excessive. If the volts/hertz ratio is too low, the rotor frequency rises unduly so that rotor currents and copper losses are excessive. Consequently, there is an optimum voltage-frequency pairing that gives the required steady-state torque at the specified speed with maximum efficiency.[47]

Consideration of the phasor diagrams in Fig. 7.21, which refer to the conventional equivalent circuit of Fig. 6.1, will clarify the influence of stator voltage reduction on motor losses. For a given supply frequency, motor torque is proportional to the power dissipated in rotor resistance, R_2/s, and consequently, is proportional to the product $E_1 I_2$, if the influence of rotor leakage reactance is assumed to be small. Figure 7.21(a) shows the phasor diagram for light-load operation at normal voltage and emphasizes the large magnetizing current and low power factor inherently associated with this operating condi-

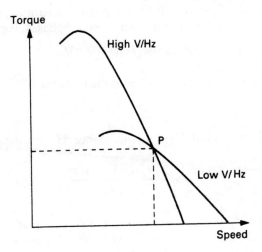

FIG. 7.20. Induction motor torque-speed characteristics for operation at point P.

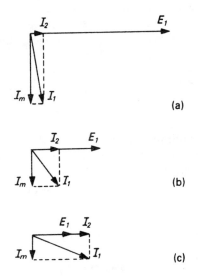

FIG. 7.21. Phasor diagrams for reduced voltage operation.

tion. If the voltage E_1 is halved, as in Fig. 7.21 (b), rotor current I_2 must double in order to develop the same motor torque as before. Airgap flux and magnetizing current, I_m, are also halved, and the total stator current, I_1, is reduced. The resulting reduction in stator copper loss and stator core loss (associated with voltage E_1) more than offsets the increase in rotor copper loss, so that overall motor losses are reduced. However, a further halving of voltage E_1, as in Fig. 7.21 (c), results in an increased stator current and increased motor losses. Evidently, there is an optimum value of E_1, and optimum value of rotor slip, s, which develop the required torque with maximum motor efficiency.

From the equivalent circuit (Fig. 6.1), the gross mechanical power output, including windage and friction losses, for a motor with m_1 stator phases, is given by Equation 6.3 as

$$P_{mech} = m_1 I_2^2 R_2 (1 - s)/s .$$

The corresponding electrical power input to the motor is

$$P_{in} = m_1 I_1^2 R_{in} \tag{7.42}$$

where R_{in} is the resistive part of the input impedance. Hence, the efficiency is given by

$$\eta = \frac{I_2^2 R_2 (1 - s)/s}{I_1^2 R_{in}} . \tag{7.43}$$

If the rotor current, I_2, is expressed in terms of I_1, the current terms in Equation 7.43 cancel, and motor efficiency is expressed exclusively in terms of equivalent circuit parameters, with fractional slip, s, as the only variable. By differentiating with respect to s, and equating to zero, an expression is obtained for the slip which yields maximum efficiency. However, core loss effects must be included, and a core loss resistance, R_m, is placed in parallel with the magnetizing reactance, X_m, of Fig. 6.1. This modified equivalent circuit is used to derive a more precise current relationship for substitution in Equation 7.43. After some approximation, analysis yields the optimum slip as[48]

$$S_{op} = \frac{R_2}{X_m + X_2} \left\{ \frac{1 + A}{1 + (R_2/R_1)} \right\}^{1/2} \tag{7.44}$$

where $A = X_m^2/(R_1 R_m)$.

Motor operation at this optimum slip implies that the equivalent circuit has no variable impedances, and hence motor input impedance and power factor are constant. Motor terminal voltage, V_1, is controlled to deliver the current required for a particular torque; because the equivalent circuit is linear, torque is proportional to V_1^2. Circuit linearity also implies that P_{mech}/P_{in} is constant, and efficiency is independent of motor voltage and torque.

The optimum rotor frequency, ω_{2op}, is given by

$$\omega_{2op} = \omega_1 S_{op} \tag{7.45}$$

and hence from Equation 7.44,

$$\omega_{2op} = \frac{R_2}{L_m + L_2} \left\{ \frac{1 + A}{1 + (R_2/R_1)} \right\}^{1/2} . \tag{7.46}$$

Since A is proportional to X_m^2, ω_{2op} is a function of the supply frequency.

Assuming a magnetically linear model of the induction motor, Reference 49 predicts the improvement in motor efficiency with a six-step voltage-source inverter supply; Reference 50 extends the analysis and shows that, in the presence of saturation, the optimum slip frequency is load dependent as well as frequency dependent.

At low flux levels, the peak torque capability of the motor is diminished, thereby impairing the drive response to a large torque demand. In general, attention must be paid to the dynamic response of the system when excitation control is employed. Several approaches can be identified for the practical realization of an efficiency-optimized motor drive. A slip control method is the most obvious solution, as slip regulation techniques have been successfully implemented for many years, although not specifically to reduce losses. The implementation of the control is aided by the relative insensitivity of the drive efficiency to wide variations in slip frequency about the optimum value.[51,52] In fact, the constant slip frequency drive described in Section 7.8.2 is a crude implementation of the optimum efficiency control technique.

Sophisticated optimizing controllers may also be employed in which an accurate loss model of the drive is used for on-line loss computation.[53] At any operating point, the controller adjusts one or more variables in the model, until the optimum values are found. These optimized values then become the commanded values for the drive controller. The effectiveness of this approach obviously depends on the accuracy of the loss model. Efficiency optimization may also be accomplished by measuring the power input to the drive and perturbing one or more variables while seeking the minimum input power at the particular operating point.[54] The minimization of total power input may not be a very sensitive procedure for minimization of losses, but accurate loss modeling and precise information regarding motor parameter values are not required. If the power input is measured on the source side of the power converter, a direct measurement of power input is easier to obtain because of the lower harmonic content of the source voltage and current waveforms, as compared with the corresponding motor waveforms. Moreover, the loss minimization is not restricted to motor losses but covers total drive system losses.

7.9. AIRGAP FLUX CONTROL

Below base speed, the torque-producing capability of a given motor frame size can be fully utilized by holding the airgap flux constant at the nominal value defined by the motor nameplate voltage and frequency. As shown in Section 7.3, the induction motor torque capability is then independent of the supply frequency, giving maximum torque per stator ampere at all speeds, and permitting a fast transient response of the drive system.

For constant airgap flux, the inverter output voltage, V_1, and frequency, ω_1, must be appropriately controlled for each operating condition. The inverter frequency is closely related to the shaft speed and must be adjusted to a value that is dictated by the speed command. Consequently, the only quantity available for flux regulation is the motor voltage, V_1.

Precise control of airgap flux is possible with a closed-loop feedback system in which the actual airgap flux is sensed and compared with a fixed reference voltage representing the desired flux value, as shown in Fig. 7.22. The flux controller amplifies the error voltage and generates a terminal voltage command, V_1^*, so that the commanded flux level is maintained for all shaft loads and stator frequencies. However, direct measurement of airgap flux is difficult. Flux-sensitive devices such as Hall-effect sensors can be introduced into the airgap, but Hall elements are sensitive to heat and mechanical vibration,

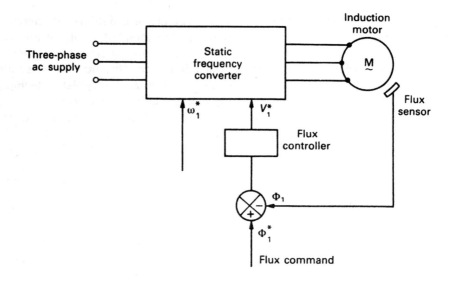

FIG. 7.22. Induction motor drive with closed-loop control of airgap flux.

and the flux signal is distorted by large slot harmonics that cannot be filtered effectively because their frequency varies with motor speed. Another disadvantage is that a special machine is required with built-in Hall sensors.

Search coils can be placed in the wedges closing the stator slots to sense the rate of change of airgap flux;[10,55] slot harmonic effects are suppressed when a full-pitch search coil is used. The induced voltage in the search coil is proportional to the rate of change of flux; integration of this voltage in a high-quality integrator gives an airgap flux signal. However, the use of search coils also precludes the use of a standard induction motor.

Indirect sensing of airgap flux gives a more versatile drive system that can be used with standard commercial motors, but this approach usually results in a relatively complex control system. Thus, the fundamental airgap emf, E_1, can be determined from the measured fundamental terminal voltage and line current of the motor by computing the phasor difference between the phase voltage, V_1, and the stator leakage impedance drop, $(R_1 + jX_1)I_1$. This method requires sensing of V_1 and I_1, but, for an inverter-fed machine, these signals may have significant harmonic ripple. Constant airgap flux operation is achieved by regulating the synthesized airgap voltage signal, E_1, so that it is proportional to the stator frequency, ω_1. However, the computation of E_1 also requires a knowledge of the stator resistance and leakage inductance parameters. The substantial variation in stator resistance with motor temperature causes a loss of accuracy, particularly at low frequencies; changes in stator leakage reactance resulting from saturation are also present.

Information on the airgap flux level can also be derived by using the terminal voltage and current signals to compute the motor reactive power.[28] The advantage of this technique is that the signal processing circuits do not require a knowledge of the stator resistance parameter. Of course, indirect control of airgap flux is possible by directly controlling the stator current and slip frequency, as already discussed in Section 7.5. Practical current-controlled drive systems of this type are discussed in the next section.

7.10. STATOR CURRENT CONTROL

In an induction motor drive, there are advantages in controlling stator current rather than stator voltage. Direct control of the stator phase currents gives fast, effective control of the amplitude and spatial phase angle of the stator mmf wave, thereby facilitating high-quality torque control and rapid dynamic response. This approach is analogous to torque control of a dc motor by closed-loop control of armature current. The controlled stator current can be delivered by a current-controlled PWM inverter or by a current-source inverter (CSI), as described in Chapter 4. The PWM approach is widely used in high-performance servo drives at power levels up to 10 kW or more. The CSI drive is used at higher power levels in applications where its low-speed torque ripple can be tolerated.

7.10.1. The Current-Source Inverter Drive

The CSI induction motor drive was introduced in Chapter 4, where its control was also briefly discussed. As subsequently explained in Section 7.5, the current-fed motor is normally operated in the magnetically unsaturated state, requiring stable operation in the statically unstable region of the induction motor torque-speed characteristic. Consequently, unlike the VSI drive, the CSI drive cannot operate in an open-loop configuration, and feedback control is essential. Control techniques often involve some form of slip frequency control with stator current regulation. [5,29,30,56–59]

In a conventional dc motor drive with its double-loop control system, the speed error signal becomes the set current and hence the set torque in the dc machine. As shown in Section 7.5, induction motor torque is determined by stator current and rotor frequency; the speed error is therefore used to furnish a torque command signal by generating appropriate stator current and slip frequency commands. Figure 7.23 is a block diagram of a typical adjustable-speed drive system. The motor current is determined by the dc link current, I_d, which is regulated by the front-end phase-controlled rectifier, in the usual manner. The commanded current, I_d^*, is determined by the compensated speed error signal, as shown. The measured dc link current, I_d, is fed to a function generator that provides the appropriate slip frequency command, ω_2^*. As usual, ω_2^* is added to the rotational frequency of the motor to determine the inverter frequency command, ω_1^*. Thus, the airgap flux is indirectly controlled at its rated value by varying the slip frequency as a function of the dc link current. This functional relationship involves the motor parameters and must be accurately precomputed for a particular machine. The function generator can be based on analog hardware, a look-up table algorithm, or an off-line microprocessor calculation. For no-load operation of the motor, ω_2^* is near zero, but the the dc link current has a minimum value corresponding to the magnetizing current of the motor. This minimum current is also necessary to maintain satisfactory commutation of the inverter.

In order to take advantage of the inherent regenerative capability of the CSI drive, a negative slip frequency command must be defined when the actual motor speed exceeds the set value. This requirement is satisfied by the polarity sensor in Fig. 7.23, which detects a change in sign of the speed error and reverses the polarity of ω_2^*; of course, I_d^* is positive for positive or negative speed errors. Because the negative slip frequency is clamped below the breakdown value, full regenerative braking torque is maintained

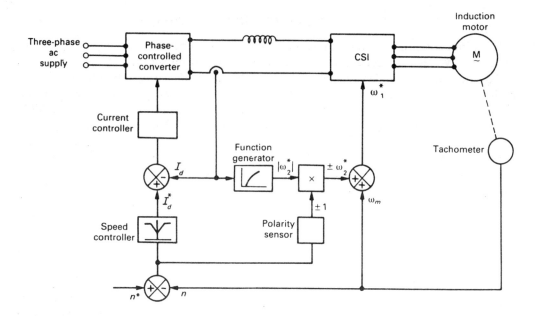

Fig. 7.23. The adjustable-speed current-source inverter drive with slip frequency control.

throughout deceleration. By reversal of the inverter firing sequence, the direction of motor rotation can be reversed, resulting in a full four-quadrant capability.

In Fig. 7.23, the current command, I_d^*, is set by the speed loop error, but it is also possible for the speed controller to define the slip frequency command, ω_2^*, from which the current command is then derived by a function generator. It has been shown that there is little difference in performance between these two schemes.[59]

As the speed of the induction motor rises, its airgap emf increases, and a larger dc link voltage is required to force the commanded motor current. At high speeds, the firing angle of the phase-controlled rectifier is fully advanced, and the rectifier produces its maximum output voltage. The drive cannot operate in a current-fed mode when the rectifier is phased fully on, but voltage-fed operation is possible; a constant-horsepower characteristic can be imposed by boosting the slip frequency command with an additional speed-dependent component.

For normal current-fed operation in the constant-torque region, the control scheme functions as described above, with a preprogrammed relationship between I_d and ω_2^* so that the magnetizing current is constant for all speeds and loads. However, as already noted, this relationship is parameter dependent; in particular, it involves the rotor resistance and inductance parameters (Equation 7.28). Rotor resistance may vary considerably with temperature, while leakage inductance is affected by current level and local saturation resulting from leakage flux. In practice, therefore, it is difficult to achieve constant airgap flux with a precomputed relationship based on fixed parameter values.

It should also be noted that the control scheme of Fig. 7.23 is based on a *steady-state* relationship between motor current and slip frequency, rather than a general transient relationship. Despite this fundamental limitation, the transient performance is satisfactory for many applications, but improved control structures have been developed that give better dynamic performance. These fast-response strategies control the instantaneous torque angle between the airgap flux phasor and the stator current phasor. Angle control schemes may incorporate an additional control loop that quickly alters the inverter output frequency during motor transients so that the torque angle is rapidly corrected and the torque response is improved[13,60,61]; direct sensing of airgap flux may be introduced to reduce the parameter sensitivity of the control.[11,62] Field-oriented control is now widely used in high-performance CSI induction motor drives; this control technique is studied in detail in later sections of this chapter.

The CSI Drive with Stator Voltage Feedback. The CSI induction motor drive with tachometer feedback is a rugged single-motor drive with inherent regenerative capability, but speed feedback is unsuitable for a multimotor drive. An alternative control technique using stator voltage feedback has been successfully applied in industry.[63-65] This approach retains the advantages of current-source operation and is suitable for multimotor applications; it is also appropriate for general-purpose single-motor drives when the presence of a tachometer is undesirable.

Figure 7.24 is a block diagram of a drive control scheme using stator voltage feedback. The speed command, n^*, is rate limited by a ramping circuit before defining the inverter frequency reference, ω_1^*, in a simple open-loop control. The frequency command also specifies a proportional stator voltage command, V_1^*, thereby demanding a constant volts/hertz ratio. The terminal voltage of the motor is rectified by a diode rectifier bridge and filtered before being compared with the commanded voltage. The error signal from the voltage controller defines the set current, I_d^*, for the inner current loop. Thus, the motor current is adjusted to a level that develops the commanded terminal voltage, and the airgap flux level is approximately constant if stator resistance effects are small. A low-frequency voltage boost can be readily introduced by adding a voltage offset to the frequency-proportional voltage reference.

The CSI drive with a voltage feedback loop can operate stably in the region corresponding to the rated airgap flux condition. The resulting steady-state torque-speed characteristics are identical to those of a voltage-fed machine with terminal volts/hertz control; parallel motors can be supplied from a single inverter. However, the voltage-regulated drive has a slower response than the drive with tachometer feedback, and is also more difficult to stabilize.

7.10.2. *The Current-Controlled PWM Inverter Drive*

When the PWM inverter is operated as a controlled-current source, using conventional current regulating loops, the stator phase currents track the sine wave reference values, as explained in Chapter 4. The current controller can be based on a simple hysteresis comparator for each phase; the current error signal, with some hysteresis, controls the switching of an inverter leg. Alternatively, a PWM controller can be employed in which the current error for each phase is compared with a high-frequency triangular wave; the resulting PWM signal controls the inverter switching, as before.

As long as the maximum or ceiling voltage of the inverter is adequate, the stator phases are effectively fed by sinusoidal current sources, thereby eliminating the

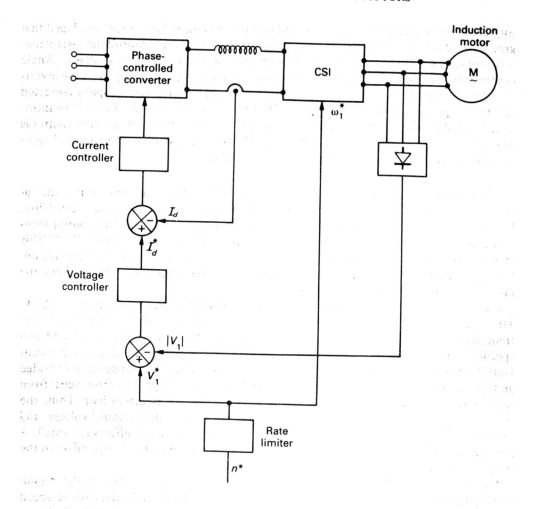

FIG. 7.24. The adjustable-speed current-source inverter drive with stator voltage feedback.

influence of the stator resistance and leakage inductance, and simplifying the induction motor control. Accurate current tracking is achieved by adopting a high switching frequency, which has the additional advantage of giving a wide bandwidth for control. At low speeds, the sinusoidal phase currents ensure smooth rotation. At high speeds, the output current is unable to track the reference current because of insufficient dc link voltage; as a result, the PWM inverter operation will transition to the six-step voltage-source mode.

Figure 7.25 is a block diagram of an adjustable-speed drive in which the slip-frequency control strategy is very similar to that described previously for the CSI drive. The speed controller generates a slip frequency command, ω_2^*, that is added to the rotational frequency of the motor, in the usual manner. The rotor frequency command is also fed to a function generator that develops the appropriate current amplitude command, I_1^*. The two command signals, I_1^* and ω_1^*, control the amplitude and frequency of the three-

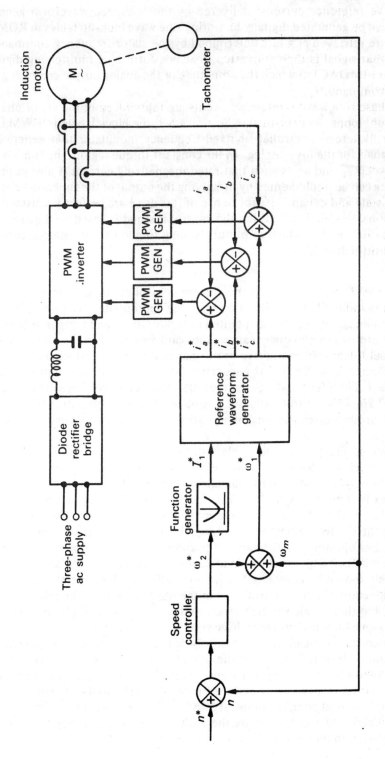

Fig. 7.25. The adjustable-speed current-controlled PWM inverter drive with slip frequency control.

phase sine wave reference currents delivered by the reference waveform generator. These signals can be generated digitally by storing sine wave look-up tables in ROM. The stored values are retrieved at a rate determined by the stator frequency command, ω_1^*. The digital output signal is then converted to analog form by a multiplying digital-to-analog converter (MDAC) in which the amplitude of the analog output can be regulated by the current command, I_1^*.

The three-phase sine wave reference currents are reproduced in the stator phases by the PWM current loops. As explained in Section 4.5.3, the block labeled "PWM GEN" represents the hysteresis controller or fixed-frequency modulator that generates the PWM firing signals for the inverter leg. In the constant-torque region, the functional relationship between I_1^* and ω_2^* is such that rated magnetizing current is always present. Field weakening can be implemented by modifying the output of the function generator.

Both steady-state and dynamic performance of this drive are perfectly satisfactory for many applications, despite the fact that the control method is based on a quasi-steady-state representation of the induction motor, leading to good steady-state accuracy but only average control dynamics.

Airgap Flux and Torque Control. In the current-controlled drive system of Fig. 7.25, the airgap flux is indirectly controlled by specifying appropriate stator current and slip frequency commands; but, as already pointed out, saturation- and temperature-induced variations in motor parameters may cause significant variations in the airgap flux level with a consequent degradation in drive performance. In a high-performance drive system, precise control of airgap flux is necessary; fast-response torque control is also an essential element. Therefore, independent torque and flux control loops are desirable, as shown in Fig. 7.26. This control structure can be used in a traction drive for an electric vehicle, or the torque loop can be used as an inner loop in a speed- or position-controlled drive system.

Direct sensing of airgap flux is desirable, but there are some practical difficulties and objections, as discussed in Section 7.9. In Fig. 7.26, the terminal voltages and currents of the motor are monitored and used to calculate motor torque and airgap flux. The calculated airgap flux is compared with the commanded value; the error determines the current amplitude command, I_1^*, for the reference sine wave generator. The torque error is used to specify the slip frequency command, ω_2^*, that is added to the tachometer signal to determine ω_1^*. For optimum performance, the computation of motor torque and airgap flux should be unaffected by variations in machine parameters; otherwise, the measured feedback signals should be compensated for such effects. The general transient equations of the induction motor, as introduced in the next section, can be used to calculate motor torque, but this simple slip frequency control method cannot be regarded as the optimum basis for a high-performance drive system.

For best dynamic performance, it is obviously desirable that the control philosophy be based on a genuine transient model of the induction motor. As in the case of the CSI drive, instantaneous torque angle control and direct sensing of airgap flux can be utilized to give a fast torque response with reduced parameter sensitivity.[12] However, field-oriented control, as studied in Sections 7.12 and 7.13, is the best of the improved fast-response strategies and has a rigorous theoretical basis in the general mathematical model of the induction motor.

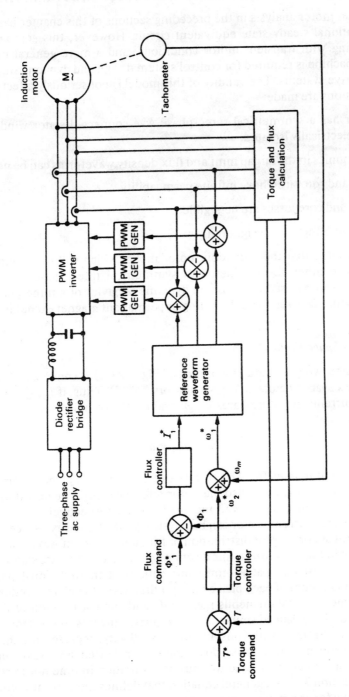

Fig. 7.26. Induction motor drive with independent torque and flux control loops.

7.11. TRANSIENT ANALYSIS OF INDUCTION MOTOR BEHAVIOR

The induction motor analysis in the preceding sections of this chapter has been based on the conventional steady-state equivalent circuit. However, this approach is inadequate for dealing with transient motor conditions, and a more general mathematical model of the machine is required for control system design and dynamic studies in high-performance drive systems. The validity of this model is not seriously affected if the following assumptions are made:

1. The motor has a symmetrical three-phase wye-connected stator winding and the neutral is electrically isolated.

2. Space harmonics in the airgap mmf and flux density waveforms can be neglected.

3. The stator and rotor iron have infinite permeability.

4. Skin effect and core losses are negligible.

5. Slot- and end-effects can be ignored.

It should be noted that the induction motor model which will be developed can be readily adapted to other ac machines with symmetrical three-phase stator windings. Thus, the model can be used to study the dynamic behavior of a three-phase synchronous machine with a dc-excited field winding or permanent magnet excitation.

7.11.1. *Complex Space Vectors*

The three-phase two-pole induction motor of Fig. 7.27(a) is conveniently analyzed in terms of complex space vectors or "space phasors."[66-68] Thus, if i_{as}, i_{bs}, and i_{cs} are the instantaneous currents in stator phases as, bs and cs, the complex stator current vector, $\bar{i}_s$, is defined by

$$\bar{i}_s = i_{as} + ai_{bs} + a^2 i_{cs} \tag{7.47}$$

where $a = e^{j2\pi/3}$ and $a^2 = e^{j4\pi/3}$. The real axis of the complex plane is coincident with the axis of the as phase, which is also the stator reference axis. The operators, a and a^2, represent a phase shift in space of $2\pi/3$ and $4\pi/3$ radians, respectively.

The complex stator current vector can be physically interpreted as a space vector related to the resultant stator current distribution or fundamental mmf wave due to the three phases, as shown in Fig. 7.27(b). In this diagram, i_{as}, ai_{bs}, and $a^2 i_{cs}$ are stationary vectors representing the sinusoidal spatial mmf contributions of the individual stator phases, with the spatial maximum of each phase mmf on the axis of the phase winding. The vector $\bar{i}_s$, then, defines the instantaneous magnitude and angular position of the resultant stator mmf wave. For balanced sinusoidal three-phase currents, the vector $\bar{i}_s$ has a constant amplitude and rotates with constant angular velocity, representing the uniformly rotating field for normal three-phase operation on a balanced sine-wave supply. However, space vectors are not restricted to sinusoidal variation in time nor to constant frequency, and Equation 7.47 is a general equation that defines the current vector, $\bar{i}_s$, at any instant and is valid for any stator currents, provided

$$i_{as} + i_{bs} + i_{cs} = 0, \tag{7.48}$$

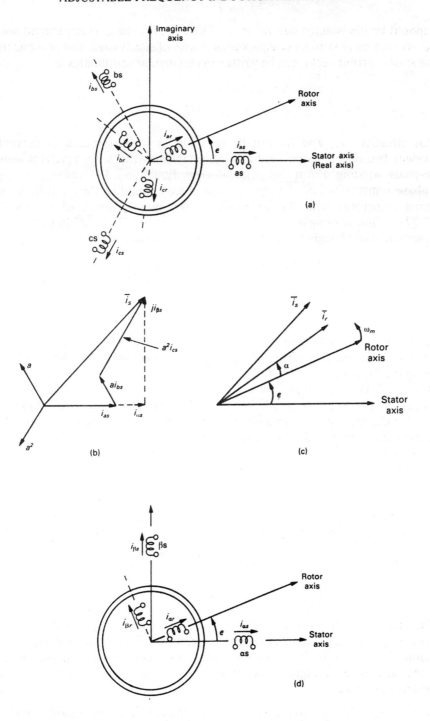

F<small>IG</small>. 7.27. Modeling of a symmetrical induction motor: (a) basic three-phase two-pole machine; (b) construction of the stator current vector; (c) angular relationship between current vectors; (d) equivalent two-phase two-pole machine.

as imposed by the isolated neutral point. The complex space vector should not be confused with the more familiar complex time phasor of steady-state sine-wave ac theory.

The stator current vector can be written in rectangular coordinates as

$$\bar{i}_s = i_{\alpha s} + j i_{\beta s}$$

$$= i_{as} + a i_{bs} + a^2 i_{cs} . \tag{7.49}$$

In this equation, $i_{\alpha s}$ and $i_{\beta s}$ can be regarded as the instantaneous currents in an equivalent two-phase stator winding that establishes the same resultant mmf as the three-phase winding. From this point of view, Equation 7.49 implies a three-phase to two-phase transformation of variables with the αs phase of the equivalent two-phase machine coincident with the as phase of the three-phase machine, as shown in Fig. 7.27 (d). Substituting $a = -1/2 + j\sqrt{3}/2$ and $a^2 = -1/2 - j\sqrt{3}/2$ in Equation 7.49 and separating real and imaginary terms, gives the matrix equation

$$\begin{bmatrix} i_{\alpha s} \\ i_{\beta s} \end{bmatrix} = \begin{bmatrix} 1 & -\dfrac{1}{2} & -\dfrac{1}{2} \\ 0 & \dfrac{\sqrt{3}}{2} & -\dfrac{\sqrt{3}}{2} \end{bmatrix} \begin{bmatrix} i_{as} \\ i_{bs} \\ i_{cs} \end{bmatrix} . \tag{7.50}$$

This transformation equation is also used for stator voltages so that a three-phase machine can be defined in terms of an equivalent two-phase machine with a simpler mathematical model.

In Fig. 7.27 (a), the three-phase wound rotor has instantaneous phase currents, i_{ar}, i_{br}, i_{cr}. A cage rotor can also be represented by an equivalent three-phase wound rotor, and consequently, one can always define a rotor current vector, $\bar{i}_r$, representing the resultant rotor mmf distribution. Thus,

$$\bar{i}_r = i_{ar} + a i_{br} + a^2 i_{cr}$$

$$= i_{\alpha r} + j i_{\beta r} \tag{7.51}$$

where $\bar{i}_r$ is obviously defined with respect to a rotor reference axis that is coincident with the axis of the ar phase. The rotor winding also has an isolated neutral point, and hence

$$i_{ar} + i_{br} + i_{cr} = 0. \tag{7.52}$$

7.11.2. Vector Rotation

The stator current vector, $\bar{i}_s$ of Equation 7.47, is clearly specified in a stator-based coordinate system, or stationary reference frame, whereas the rotor current vector, $\bar{i}_r$ of Equation 7.51, is expressed in moving rotor-based coordinates. As shown in Fig. 7.27 (c), the instantaneous angular velocity of the rotor is $\omega_m = d\epsilon/dt$, where ϵ is the angle of rotation of the rotor.

In a reference frame moving with the rotor, the rotor current vector, $\bar{i}_r$ in Fig. 7.27 (c), can be expressed as $i_r e^{j\alpha}$, where i_r and α are the polar coordinates with respect to the rotor axis. If this vector is defined with respect to the stationary stator axis, it has a value $i_r e^{j(\alpha+\epsilon)} = i_r e^{j\alpha} e^{j\epsilon} = \bar{i}_r e^{j\epsilon}$, which indicates that $\bar{i}_r$ is rotated through an angle ϵ. Thus, the rotor-based current vector, $\bar{i}_r$, is expressed in a stationary reference frame as $\bar{i}_r e^{j\epsilon}$. Conversely, the equivalent stator current vector in rotor-based coordinates is $\bar{i}_s e^{-j\epsilon}$. It is often necessary to express stator and rotor quantities in a common reference frame using the transformation factor, $e^{j\epsilon}$ or $e^{-j\epsilon}$, which is termed a vector rotator or coordinate transformation.

Let $\bar{i}_r'$ denote the equivalent rotor current vector in a stationary reference frame, and assume its two-phase rectangular coordinates are $i_{\alpha r}'$ and $i_{\beta r}'$. Thus,

$$\bar{i}_r' = i_{\alpha r}' + j i_{\beta r}' . \tag{7.53}$$

But, as above

$$\bar{i}_r' = \bar{i}_r e^{j\epsilon} = \left(i_{\alpha r} + j i_{\beta r} \right) \left(\cos \epsilon + j \sin \epsilon \right)$$

$$= \left(i_{\alpha r} \cos \epsilon - i_{\beta r} \sin \epsilon \right) + j \left(i_{\alpha r} \sin \epsilon + i_{\beta r} \cos \epsilon \right) . \tag{7.54}$$

Comparing real and imaginary terms in Equations 7.53 and 7.54 gives the matrix equation

$$\begin{bmatrix} i_{\alpha r}' \\ i_{\beta r}' \end{bmatrix} = \begin{bmatrix} \cos \epsilon & -\sin \epsilon \\ \sin \epsilon & \cos \epsilon \end{bmatrix} \begin{bmatrix} i_{\alpha r} \\ i_{\beta r} \end{bmatrix} . \tag{7.55}$$

Equation 7.55 is the familiar two-phase rotor-to-stator transformation in which the vector rotator takes the form of a 2×2 transformation matrix involving $\sin \epsilon$ and $\cos \epsilon$.

In general, it is possible to apply a vector rotation that transforms stator and rotor quantities to a common reference frame rotating at an arbitrarily chosen speed. A synchronously rotating reference frame is often an advantageous choice.

7.11.3. *Voltage and Torque Equations*

By analogy with Equation 7.47, if λ_{as}, λ_{bs}, and λ_{cs} are the instantaneous flux linkages of stator phases as, bs, and cs, the complex stator flux linkage vector can be defined as

$$\bar{\lambda}_s = \lambda_{as} + a \lambda_{bs} + a^2 \lambda_{cs} . \tag{7.56}$$

If one regards the instantaneous flux linkage terms on the right-hand side of Equation 7.56 as representing the airgap flux contributions of the three phases, then the flux linkage vector $\bar{\lambda}_s$ can be physically interpreted as a space vector representing the magnitude and direction of the resultant sinusoidal flux distribution in the airgap of the machine. In practice, however, $\bar{\lambda}_s$ will also include a component due to leakage flux.

Likewise for the rotor

$$\bar{\lambda}_r = \lambda_{ar} + a \lambda_{br} + a^2 \lambda_{cr} \tag{7.57}$$

and is defined in rotor-based coordinates.

If v_{as}, v_{bs}, and v_{cs} are the instantaneous stator line-to-neutral voltages, then

$$v_{as} = R_s i_{as} + \frac{d\lambda_{as}}{dt}$$

$$v_{bs} = R_s i_{bs} + \frac{d\lambda_{bs}}{dt}$$

$$v_{cs} = R_s i_{cs} + \frac{d\lambda_{cs}}{dt} \tag{7.58}$$

where R_s is the stator resistance per phase (previously denoted by R_1).

A complex stator voltage vector, $\bar{v}_s$, can be defined according to the usual procedure. Thus,

$$\bar{v}_s = v_{as} + a v_{bs} + a^2 v_{cs} \tag{7.59}$$

and the three voltage equations (Equation 7.58) can be combined in a single vectorial equation:

$$\bar{v}_s = R_s \bar{i}_s + \frac{d\bar{\lambda}_s}{dt} \ . \tag{7.60}$$

However, it is not possible to attach a physical significance to this stator voltage vector.

Similarly, the voltage equation for the short-circuited rotor is

$$0 = R_r \bar{i}_r + \frac{d\bar{\lambda}_r}{dt} \tag{7.61}$$

where R_r now denotes the rotor resistance per phase.

The flux linkages can be expressed in terms of L_s, L_r, and M — the stator, rotor and mutual inductances per phase.[68] Thus,

$$\bar{\lambda}_s = L_s \bar{i}_s + M \bar{i}_r e^{j\epsilon} \tag{7.62}$$

$$\bar{\lambda}_r = L_r \bar{i}_r + M \bar{i}_s e^{-j\epsilon} \ . \tag{7.63}$$

These flux linkage relationships are readily derived for the equivalent two-phase machine of Fig. 7.27(d), but they can also be derived directly for the original three-phase machine of Fig. 7.27(a).[68] The exponential term $e^{j\epsilon}$ attached to $\bar{i}_r$ in Equation 7.62 indicates that the rotor current vector must be transformed to a stationary reference frame before its effect is added to that of the stator current vector $\bar{i}_s$. Similarly, the vector rotator, $e^{-j\epsilon}$ in Equation 7.63 transforms the stator current vector to rotor-based coordinates before its effect is added to that of $\bar{i}_r$.

If the stator and rotor windings have the same number of turns, the usual stator and rotor leakage factors, σ_s and σ_r, can be defined by the equations

$$L_s = (1 + \sigma_s) M \tag{7.64}$$

$$L_r = (1 + \sigma_r) M \ . \tag{7.65}$$

The total leakage factor is given by

$$\sigma = 1 - \frac{1}{(1+\sigma_s)(1+\sigma_r)} . \tag{7.66}$$

On substituting Equations 7.62 and 7.63 in Equations 7.60 and 7.61, the following voltage equations are obtained:

$$\bar{v}_s = R_s \bar{i}_s + L_s \frac{d\bar{i}_s}{dt} + M \frac{d}{dt}\left(\bar{i}_r e^{j\epsilon}\right) \tag{7.67}$$

$$0 = R_r \bar{i}_r + L_r \frac{d\bar{i}_r}{dt} + M \frac{d}{dt}\left(\bar{i}_s e^{-j\epsilon}\right) . \tag{7.68}$$

These general voltage equations must be supplemented by an equation for the electromagnetic torque. The torque equation is derived from the tangential force exerted on the rotor windings by the radial magnetic field produced by the stator mmf.[68] This calculation shows that the torque is proportional to the vector cross product of $\bar{i}_s$ and $\bar{i}_r$ in any common frame of reference, and is therefore proportional to the sine of the angle between the stator and rotor current vectors. Again, considering the two-pole machine of Fig. 7.27, the torque equation in stator-based coordinates is usually written in the form

$$T = \frac{2}{3} M \operatorname{Im}\left[\bar{i}_s\left(\bar{i}_r e^{j\epsilon}\right)^*\right] \tag{7.69}$$

where Im indicates the imaginary part of the term in brackets and $\bar{i}^*$ is the complex conjugate of $\bar{i}$. As before, the rotor current vector has an associated exponential term $e^{j\epsilon}$, indicating that it is referred to the fixed stator reference frame.

If T_L is the load torque and J is the combined inertia of motor and load, then

$$J \frac{d\omega_m}{dt} = T - T_L = \frac{2}{3} M \operatorname{Im}\left[\bar{i}_s\left(\bar{i}_r e^{j\epsilon}\right)^*\right] - T_L . \tag{7.70}$$

Also, the instantaneous mechanical angular velocity of the rotor is

$$\omega_m = d\epsilon/dt . \tag{7.71}$$

The general set of nonlinear differential equations (Equations 7.67, 7.68, 7.70, and 7.71) describe the dynamic behavior of the symmetrical induction machine with arbitrary voltage and current waveforms. As a special case, they describe steady-state operation with sinusoidal voltages and currents, and can be used to derive the steady-state equivalent circuit of Fig. 6.1.[68] This derivation shows that the inductances L_s, L_r, and M are the inductances L_{11}, L_{22}, and L_m, used in the earlier sections of this chapter.

The stator voltage equation (Equation 7.67) and the rotor voltage equation (Equation 7.68) are in stator- and rotor-based coordinates, respectively. Using the vector rotator, the two equations can be referred to a common reference frame, either stator- or rotor-based, or a general reference frame rotating at an arbitrary speed. The voltage and current vectors can then be expressed in rectangular coordinates with respect to orthogonal direct and quadrature axes in the common reference frame. This procedure defines the familiar dq-axis components of voltage and current. If the resulting pair of complex equations are separated into real and imaginary parts, four dq-axis voltage equations are obtained, thus defining the usual two-axis model of the induction motor.[69–72] When the four voltage equations are combined with Equations 7.70 and 7.71, a general set of six real nonlinear differential equations is obtained. These equations have been widely used to study the dynamic performance of the induction motor. The equations are difficult to solve analytically but can be readily programmed on an analog or digital computer.[71–75]

7.11.4. *Dynamic Performance*

Lawrenson and Stephenson[75] have used a digital computer to study the transient performance of a particular voltage-fed induction motor with an adjustable supply frequency. They have shown that the motor speed can be changed rapidly and efficiently by varying the supply frequency uniformly as a function of time at constant volts/hertz. Under these conditions, an acceleration from standstill can be achieved more quickly than by the usual direct-on-line start, or step application of power at rated voltage and frequency; the machine currents are also smaller, but the optimum rate of change of frequency is dependent on the load inertia. A complete speed reversal from full speed in one direction to full speed in the opposite direction also takes a shorter time and has lower losses when the voltage and frequency are varied uniformly with time. When the usual plugging method of reversal is used, the phase sequence of the supply voltage is suddenly reversed, thereby changing the direction of the field rotation. As the motor is brought to rest, the kinetic energy in the rotating masses is dissipated as losses in the rotor circuit, with the result that heating problems may arise. When the voltage and frequency are reduced uniformly, the motor is brought to rest by regenerative braking, because the supply frequency is reduced faster than the rotor speed, and the machine operates with negative slip. The kinetic energy of the rotating masses is therefore returned to the supply and is not dissipated in the rotor. Chapter 13 includes further discussion on the digital simulation of an adjustable-frequency induction motor drive.

The induction motor also exhibits oscillatory behavior that may prove objectionable under adjustable-frequency conditions of operation. If a lightly loaded induction motor is subjected to a normal direct-on-line start, it is found that the speed may overshoot the synchronous value and oscillate above and below synchronism for a few cycles before steady speed conditions are achieved. Similar speed oscillations are obtained when the load torque is suddenly altered or the supply voltage is changed rapidly. It has been shown that the damping coefficient for these speed oscillations is a function of the supply frequency. As the stator frequency is reduced below its rated value, the speed oscillations become more persistent, and the damping coefficient has a minimum value at some reduced frequency that is dependent on the system inertia. For a normal 50 or 60 Hz motor, the damping is usually least when the supply frequency lies between 10 and 20 percent of the rated frequency. The lightly damped oscillations occurring at these reduced frequencies impair the system performance at low speeds. As mentioned in Section 6.7, small induction motors may even become unstable and oscillate in speed around a given speed set-point; these instability problems can be exacerbated in converter-fed drives with simple open-loop speed control by terminal volts/hertz regulation.

In order to facilitate the transient analysis of the induction motor, the general dq-axis equations may be simplified with the use of linearization about a steady-state operating point. A set of linear differential equations which are valid for small deviations from the steady-state condition results.[76] The stability of the machine can then be investigated by the conventional methods of linear feedback control theory. Rogers[77] has used the root locus method to analyze induction motor transients due to small disturbances in stator voltage and load torque. Pfaff[78] and De Carli *et al.*[79] have also made approximations to the machine equations in order to derive simple models of the induction motor that are valid for small frequency oscillations above and below a fixed base frequency. An idealized two-coil rotor model of the induction motor has been used to study the transient performance on a constant-current supply.[80,81] A parity simulation technique has been described that combines digital and analog simulation principles.[82] In this approach, individual components — that is, power semiconductors, capacitors, inductors, or machines — maintain their terminal equivalence: a thyristor simulation, for example, has voltage, current, and time properties, like a real thyristor, except that the voltages and currents are scaled in magnitude. Like modular building blocks, machines, power conductors, and other components can be connected together to achieve the desired drive configuration. The parity simulation of the drive system can be driven by breadboard control hardware and software. This simulation allows the control system to be debugged without subjecting the power devices and machines to overvoltages, fault currents, and damaging machine torques.

Microprocessor-based control systems, where the microprocessor performs complex calculations and decision-making functions, can be emulated with a digital computer or with the digital computer portion of a hybrid (analog/digital) computer. Care must be exercized to ensure that the time required for the microprocessor calculations is sufficiently short that the microprocessor can be used at the speed associated with the real-time hardware rather than the time-scaled simulation. Other approximate methods of analysis have been developed, but in the case of complex adjustable-speed drives, it appears that a computer simulation of the system is necessary at the design stage in order to predict the transient performance and to ensure that system instability does not occur in the operating range.

7.12. FIELD-ORIENTED CONTROL

The induction motor is superior to the dc motor with respect to size, weight, rotor inertia, maximum speed capability, efficiency, and cost. However, the ease of control of the dc motor cannot be equaled, because the induction motor has a nonlinear and highly interacting multivariable control structure, whereas the separately excited dc motor has a decoupled control structure with independent control of flux and torque. This section describes the technique of field-oriented control, or vector control, which can be used with both induction and synchronous machines and essentially transforms the dynamic structure of the ac machine into that of a separately excited compensated dc motor. As a result, the induction motor drive can achieve four-quadrant operation with a fast torque response and good performance down to zero speed.

In a dc motor, a number of coils are distributed around the armature surface and interconnected to form a closed winding. Stationary poles with dc-excited field windings, or permanent magnets, establish a magnetic field in which the armature rotates. Current is supplied to the armature through the commutator brushes so that the armature mmf axis is established at 90 degrees electrical to the main field axis. This orthogonal or perpendicular relationship between the flux and mmf axes is independent of the speed of rotation and so the electromagnetic torque of the dc motor is proportional to the product of the field flux and armature current. Assuming negligible magnetic saturation, field flux is proportional to field current and is unaffected by armature current because of the orthogonal orientation of the stator and rotor fields. Thus, in a separately excited dc motor with a constant value of field flux, torque is directly proportional to armature current. Direct control of armature current gives direct control of motor torque and fast response, because motor torque can be altered as rapidly as armature current can be altered.

In the induction motor, the space angle between the rotating stator and rotor fields varies with load, giving rise to complex interactions and oscillatory dynamic response. Ideally, the space angle should be controlled so that the stator input current can be decoupled into flux-producing and torque-producing components. This objective is achieved by using the principle of field orientation, which implements a 90-degree space angle between specific field components to impart dc motor control characteristics to the induction motor. As a result, the induction motor dynamics lose most of their complexity and a high-performance drive can be realized.

The implementation of field orientation, or vector control, requires information regarding the magnitude and position of the rotor flux vector. The control action takes place in a field-coordinate system using the rotating rotor flux vector as a frame of reference for the stator currents and voltages. The transformed stator currents are resolved into direct and quadrature axis components that correspond to the field and armature currents of a compensated dc motor.

Field-oriented control can be employed in a voltage-source or current-source inverter drive, but controlled-current operation of the motor results in a much simpler implementation of the control. In high-performance servo drives, the current-controlled PWM inverter is commonly used, but the current-source inverter (CSI) is employed in larger drives for applications that do not require very smooth operation down to zero speed. The voltage-source PWM inverter is also suitable for the implementation of field-oriented control, but the six-step voltage-source inverter is not very appropriate because it does not allow accurate control of the magnitude and phase of the stator currents.

The principle of field orientation originated in West Germany in the work of Hasse[6] and Blaschke[7,83] at the Technical Universities of Darmstadt and Braunschweig, and in the laboratories of Siemens AG. A variety of implementation methods has now been developed but these techniques can be broadly classified into two groups: direct control and indirect control.[84] The classification is based on the method used to determine the rotor flux vector. Indirect field-oriented control, as proposed by Hasse, requires a high-resolution rotor position sensor, such as an encoder or resolver, to determine the rotor flux position. Direct field-oriented control, as originally suggested by Blaschke, determines the magnitude and position of the rotor flux vector by direct flux measurement or by a computation based on terminal conditions. The development effort in direct control methods at the University of Braunschweig has been described by Professor Leonhard and his students in a series of classic papers and, more recently, in a textbook.[68,85-91] The treatment of field-oriented control in the next section follows that of Leonhard.

Field-oriented control has been criticized as being an unduly complicated approach requiring sophisticated signal processing and complex coordinate transformation. This criticism was justified in the past when the implementation was attempted using analog control electronics. However, digital control with microprocessors is making the criticism increasingly irrelevant due to the spectacular advances in the speed and processing power of modern digital processors and the significant reductions in processor size and cost. Today, all the signal processing required in a high-performance ac servo drive can be executed by a single microprocessor.[91]

7.12.1. *Principles of Field Orientation*

The instantaneous electromagnetic torque of the induction motor has been expressed in Equation 7.69 as

$$T = \frac{2}{3} M \operatorname{Im}\left[\bar{i}_s\left(\bar{i}_r e^{j\epsilon}\right)^*\right].$$

This general torque equation emphasizes the difficulty of controlling the cage-rotor induction motor: because there is no direct access to the rotor current vector, it must be indirectly controlled through the stator voltages and currents. It is logical, therefore, to replace the term $\bar{i}_r e^{j\epsilon}$ by an equivalent quantity that can be measured with stator-based sensing equipment. The mutual airgap flux is a measure of the magnetizing current and can be detected by stator search coils or Hall-effect sensors. This measurement defines the magnetizing current vector, $\bar{i}_m$, which is the sum of the stator and rotor current vectors in a common reference frame. Thus, in stator coordinates

$$\bar{i}_m = \bar{i}_s + \bar{i}_r e^{j\epsilon}. \tag{7.72}$$

This expression can be used to eliminate the rotor current term, $\bar{i}_r e^{j\epsilon}$, from the torque equation. However, some alternative choices are available. Thus, the stator flux vector, including stator leakage flux, may be used to define a modified magnetizing current vector, but the best choice, as will appear in due course, is provided by the modified magnetizing current vector representing rotor flux, including rotor leakage flux. This current vector is denoted by $\bar{i}_{mr}$ and is defined in stator coordinates as

$$\bar{i}_{mr} = \bar{i}_s + \left(1+\sigma_r\right)\bar{i}_r e^{j\epsilon}. \tag{7.73}$$

Introducing $\bar{i}_{mr}$ into the torque equation gives

$$T = \frac{2}{3} \frac{M}{1+\sigma_r} \operatorname{Im}\left[\bar{i}_s\left(\bar{i}_{mr} - \bar{i}_s\right)^*\right]$$

$$= \frac{2}{3} \frac{M}{1+\sigma_r} \operatorname{Im}\left[\bar{i}_s\, \bar{i}_{mr}^*\right]. \tag{7.74}$$

The magnetizing current vector $\bar{i}_{mr}$ can be expressed as $i_{mr} e^{j\rho}$, where i_{mr} and ρ are the polar coordinates with respect to the stator reference axis, as shown in Fig. 7.28. Substituting for $\bar{i}_{mr}$ in Equation 7.74 gives

$$T = \frac{2}{3} \frac{M}{1+\sigma_r} i_{mr} \operatorname{Im}\left[\bar{i}_s e^{-j\rho}\right]. \tag{7.75}$$

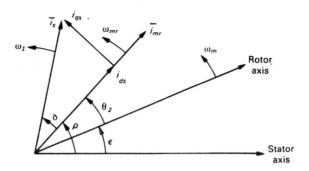

FIG. 7.28. Vector diagram for induction motor currents.

The vector rotator, $e^{-j\rho}$, implements a coordinate transformation from stator-based coordinates to a moving frame of reference which is defined by the magnetizing current vector $\bar{i}_{mr}$, representing rotor flux. Consequently, the vector $\bar{i}_s e^{-j\rho}$ in Equation 7.75 is the stator current vector as seen from the position of the rotor flux and is termed the stator current vector in field coordinates. As indicated in Fig. 7.28, the rotor of the two-pole machine has an angular velocity ω_m, and $\bar{i}_s$ rotates at the synchronous angular velocity, ω_1, whereas $\bar{i}_{mr}$ has an instantaneous angular velocity, ω_{mr}.

The stator current vector in field coordinates has orthogonal direct and quadrature components, i_{ds} and i_{qs}, parallel and perpendicular to the $\bar{i}_{mr}$ vector, as shown in Fig. 7.28. Obviously

$$i_{ds} = \mathrm{Re}\left[\bar{i}_s e^{-j\rho}\right] = i_s \cos\delta \tag{7.76}$$

$$i_{qs} = \mathrm{Im}\left[\bar{i}_s e^{-j\rho}\right] = i_s \sin\delta . \tag{7.77}$$

If Equation 7.77 is substituted in Equation 7.75, the electromagnetic torque is

$$T = \frac{2}{3}\frac{M}{1+\sigma_r} i_{mr} i_{qs} = k\, i_{mr}\, i_{qs} \tag{7.78}$$

where $k = \dfrac{2}{3}\dfrac{M}{1+\sigma_r}$.

Thus, the induction motor torque is proportional to the product of i_{mr}, the magnitude of the magnetizing current vector, and i_{qs}, the quadrature component of the stator current vector. This result is similar to that for an unsaturated, separately excited dc machine where the torque is proportional to the product of the field and armature currents. The analogy can be developed further if Equation 7.73 is used to eliminate $\bar{i}_r$ in the rotor voltage equation (Equation 7.68), yielding the following differential equation for $\bar{i}_{mr}$,

$$\tau_r \frac{d\bar{i}_{mr}}{dt} + \left(1 - j\omega_m \tau_r\right)\bar{i}_{mr} = \bar{i}_s \tag{7.79}$$

where $\tau_r = L_r/R_r$, the rotor time constant.

Equation 7.79 is expressed in stator coordinates, but it can be transformed to field coordinates by multiplying each term of the equation by the vector rotator, $e^{-j\rho}$. If $\bar{i}_{mr}$ is written in the usual polar form, $i_{mr}e^{j\rho}$, the resulting equation is

$$\tau_r \frac{d i_{mr}}{dt} + j\tau_r i_{mr} \frac{d\rho}{dt} + \left(1 - j\omega_m \tau_r\right) i_{mr} = \bar{i}_s e^{-j\rho} \,. \qquad (7.80)$$

This single vectorial equation can be separated into real and imaginary parts, giving the following pair of real differential equations in field coordinates:

$$\tau_r \frac{d i_{mr}}{dt} + i_{mr} = i_{ds} \qquad (7.81)$$

$$\frac{d\rho}{dt} = \omega_{mr} = \omega_m + \frac{i_{qs}}{\tau_r i_{mr}} \,. \qquad (7.82)$$

In the induction motor, i_{mr} is analogous to the main field flux of the dc machine and is controlled by i_{ds}, the direct component of the stator current vector, as shown by Equation 7.81. However, it is clear that the rotor time constant, τ_r, introduces a significant time lag in the response of i_{mr} to a variation in i_{ds}. This time constant may be up to a second in a large machine and is analogous to the time lag in the response of the field flux of a dc machine to a variation in field voltage.

The quadrature current component, i_{qs}, is analogous to the armature current of the dc machine, and it can be rapidly varied by an appropriate change in stator current to give a fast response to a sudden torque demand. Equation 7.82 shows that i_{qs} controls ω_{mr}, the instantaneous angular velocity of the rotor flux vector which produces slip and torque. From Fig. 7.28, the synchronous angular velocity of the stator current vector is given by

$$\omega_1 = \omega_{mr} + d\delta/dt \qquad (7.83)$$

where δ is a torque angle that is zero on no load. For steady-state operation with sinusoidal currents, δ is constant and ω_{mr} equals ω_1, so that the current vectors $\bar{i}_s$ and $\bar{i}_{mr}$ rotate in synchronism. Hence, i_{ds} and i_{qs} are constant dc quantities, and a steady motor torque is developed. In general, if i_{ds} and i_{qs} can be independently controlled, the induction motor will behave like a dc motor with decoupled control of flux and torque. This is the basic principle of field orientation, or field-oriented control.

Equations 7.70, 7.78, 7.81, and 7.82 define a model of the induction motor in field coordinates which is represented by the block diagram of Fig. 7.29. The transition from stator to field coordinates is shown in two stages, with a three-phase to two-phase current transformation followed by a vector rotation from two-phase to two-axis field coordinates. The three-phase to two-phase transformation yields the orthogonal two-phase currents $i_{\alpha s}$ and $i_{\beta s}$, as already described in Section 7.11.1. The two-phase to dq-axis transformation is now based on the flux angle ρ, to give the stator current vector in field coordinates, $\bar{i}_s e^{-j\rho}$. Thus,

$$\begin{aligned}
\bar{i}_s e^{-j\rho} &= \left(i_{\alpha s} + j\, i_{\beta s}\right)\left(\cos\rho - j\sin\rho\right) \\
&= \left(i_{\alpha s}\cos\rho + i_{\beta s}\sin\rho\right) + j\left(i_{\beta s}\cos\rho - i_{\alpha s}\sin\rho\right) \\
&= i_{ds} + j\, i_{qs}
\end{aligned}$$

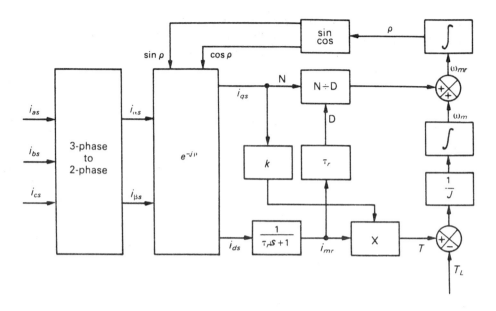

FIG. 7.29. Block diagram of induction motor in field coordinates.

or in matrix form:

$$\begin{bmatrix} i_{ds} \\ i_{qs} \end{bmatrix} = \begin{bmatrix} \cos\rho & \sin\rho \\ -\sin\rho & \cos\rho \end{bmatrix} \begin{bmatrix} i_{\alpha s} \\ i_{\beta s} \end{bmatrix}.$$

(7.84)

7.12.2. Acquisition of the Flux Vector

The implementation of direct field-oriented control requires the measurement or calculation of the rotor flux vector, as expressed by the magnitude and angle of the rotor magnetizing current vector, $\bar{i}_{mr}$. Sensing devices in the airgap of the machine will determine the airgap flux vector as expressed by the magnetizing current vector, $\bar{i}_m$, of Equation 7.72. Thus,

$$\bar{i}_m = \bar{i}_s + \bar{i}_r e^{j\epsilon}$$

but this signal can be combined with a stator current signal to generate the rotor magnetizing current vector, $\bar{i}_{mr}$, of Equation 7.73. Thus,

$$\bar{i}_{mr} = \bar{i}_s + (1+\sigma_r)\,\bar{i}_r e^{j\epsilon}$$

$$= (1+\sigma_r)\bar{i}_m - \sigma_r \bar{i}_s$$

(7.85)

and hence $\bar{i}_{mr}$ can be measured by stator-based sensors.

As described in Section 7.9, the airgap flux can be sensed using Hall-effect devices or stator search coils, but both techniques have the disadvantage that a specially modified induction motor is required. In addition, the Hall element is fragile and temperature-sensitive, while the search coil method is not usable below about 1 Hz because of the inherent drift and offset problems in the analog integrator. A standard induction motor can be used if the stator winding itself is employed as the flux sensing coil. As explained earlier, the airgap emf signal is obtained by sensing the terminal voltage and deducting the stator voltage drop, but accurate IR compensation is difficult because of the temperature dependence of stator resistance; hence the lower frequency limit for useful operation is about 3 Hz.

In general, sensing coil methods measure flux changes which are accurately integrated to give a flux signal. Consequently, operation at zero frequency is precluded, and these methods are unsuitable for position control drives. The difficulties of analog integration can be avoided by using Equation 7.79, the rotor voltage equation in terms of $\overline{i_{mr}}$, as a flux model that is solved continuously for the current magnitude i_{mr} and angle ρ in stator coordinates.[87] Alternatively, the calculation can be performed more efficiently in field coordinates by using Equations 7.81 and 7.82.[88] In both cases, three input variables are required, the shaft speed and two of the three stator phase currents; the third phase current can be deduced. It has been shown that the differential equations are readily solved in real time with a microprocessor.[87-91] This approach also has the advantage that zero-speed operation is feasible because the equations are valid when ω_m is zero. However, the calculation requires a value for the rotor time constant, τ_r, which involves the temperature-dependent rotor resistance, R_r, and saturation-dependent inductance, L_r. A thermal model of the motor may be used to track temperature-induced variations in rotor resistance, but more sophisticated adaptive techniques have been developed, where parameter changes are continually sensed and appropriate corrections made. In one technique, a small random disturbance is injected in the presumed direct-axis channel and is correlated with the signal in the actual quadrature-axis channel. If there is noticeable correlation, then the axes are not orthogonal and the parameter τ_r in the flux model is incorrect and must be modified until the autocorrelation function is brought to zero.[87] This method can be used to track slow changes in τ_r due to temperature-induced variation of the rotor resistance.

In servo drives, the operating conditions change very rapidly and the resulting variation in the degree of magnetic saturation causes sudden variations in τ_r. The correlation technique is too slow for such applications, but very fast adaptation of the control has been achieved by correcting the flux model on the basis of the stator voltages.[90]

The indirect method of field orientation eliminates measurement or computation of the rotor flux vector but determines the instantaneous position of the rotor flux by summing a rotor position signal and a commanded slip position signal. The latter signal is calculated from a model of the induction motor that requires the value of the rotor time constant, τ_r; consequently, all indirect methods are sensitive to variations in machine parameters. If the value of τ_r used in the calculation is not equal to the actual value, the desired decoupling of flux and torque is not achieved, and there is a deterioration in both the steady-state and dynamic behavior of the drive. Parameter adaptation is essential to overcome the undesirable effects of parameter variation due to changes in temperature and magnetic saturation level. However, indirect field orientation can be implemented down to zero speed and is therefore suitable for a servo drive.

7.13. IMPLEMENTATION OF FIELD-ORIENTED CONTROL

As stated in Section 7.11, the practical implementation of field-oriented control is considerably simplified if the induction motor is fed with impressed stator currents by a current-controlled PWM inverter with fast current control loops. The stator currents then accurately track the reference currents, and the time lags due to stator leakage inductance are effectively suppressed by the rapid action of the current loops. Consequently, the stator voltage equation can be omitted from the machine model. At lower power levels, a transistor PWM inverter switching at 10 kHz, or more, provides a fast-acting power converter for effective closed-loop current control. As noted earlier, this form of power converter is used in high-performance induction motor servo drives for machine tool applications and industrial robots where the continuous power rating is usually less than 10 kW.

In general, a servo drive should have the following attributes:

- High torque/inertia ratio
- Four-quadrant operation
- Fast transient response
- Large short-term overload capability
- High power density
- Smooth torque at low speeds
- Controlled torque at standstill

These demanding characteristics can be realized by field-oriented control of a cage-rotor induction motor powered by a current-controlled PWM inverter. Braking energy is normally dissipated in a dynamic braking resistor that is switched across the dc link when required. A special low-inertia rotor or a disk rotor can be used for enhanced dynamic response.

In a servo drive, flux-sensing coils cannot be used because the flux acquisition technique must function at all speeds down to standstill. However, the flux model described in Section 7.12.2 calculates the magnetizing current vector on the basis of terminal currents and speed, and remains effective at zero speed. Consequently, this technique can be used to implement direct field-oriented control of a servo drive. Indirect field orientation methods dispense with the measurement or calculation of flux and are widely used for servo control.

The control simplification offered by controlled-current operation of the induction motor is not possible with a PWM voltage-source thyristor inverter switching at several hundred hertz, because the inverter time delays are excessive. Nevertheless, the principle of field orientation can be successfully applied, as discussed later.

Controlled-current operation is also feasible in large low-speed drives employing a cycloconverter to supply the induction motor with low-frequency sinusoidal currents. If each phase current is controlled by a current loop, the motor is again effectively fed with impressed stator currents and the implementation of field-oriented control is facilitated.

The current-source inverter (CSI) is used in field orientation control systems to give a four-quadrant drive with excellent performance, except at very low speeds. When operation down to zero speed is not required, it may be advantageous to implement direct field orientation with flux acquisition by means of sensing coils, because an actual measurement of flux eliminates the parameter sensitivity of the indirect method of field orientation and of the flux model approach to direct field orientation.

In a field-oriented control system, as already explained, the induction motor behaves like a dc machine under both steady-state and transient conditions. Consequently, similar drive control strategies can be employed, and below base speed, the magnetizing current, i_{mr}, of the induction motor, representing rotor flux magnitude, is maintained at its maximum possible value but is limited by magnetic saturation. Above base speed, the flux is reduced because of the voltage limitations of the converter, giving the usual field-weakening region of operation, as in a dc drive.

A number of practical drive implementations are now discussed in more detail.

7.13.1. *Direct Field Orientation Methods*

Figure 7.30 shows a simplified block diagram of a field-oriented control scheme using a current-controlled PWM inverter. The two-axis reference currents, i_{qs}^* and i_{ds}^*, are the demanded torque and flux components of stator current, respectively, and are generated by the outer control loops. As shown in Fig. 7.30, i_{ds}^* and i_{qs}^* undergo a coordinate

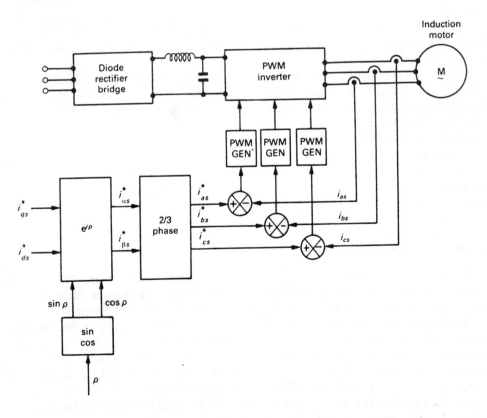

FIG. 7.30. Basic field orientation system for an induction motor with a current-controlled PWM inverter.

transformation to two-phase stator-based quantities, followed by a two-phase to three-phase transformation which generates the stator reference currents i_{as}^*, i_{bs}^*, and i_{cs}^*. These reference currents are reproduced in the stator phases by the current-controlled PWM inverter. The internal action of the motor, as shown in Fig. 7.29, is to transform the impressed three-phase stator currents to equivalent two-axis currents, i_{ds} and i_{qs}. Thus, the external reference currents, i_{ds}^* and i_{qs}^*, are reproduced within the induction motor, and control is executed in terms of these direct and quadrature axis current components to give decoupled control of flux and torque, as in a dc machine.

The vector rotation in Fig. 7.30 is based on the rotor flux angle, ρ, and hence precise information on rotor flux position is vital. The block diagram of Fig. 7.31 shows a complete field-oriented control scheme and incorporates a flux model for acquisition of the flux vector. As explained in Section 7.12.2, two stator current values are used in conjunction with shaft speed to continuously calculate the values of ρ, i_{mr} and i_{qs} in the motor. The values of i_{mr} and i_{qs} are also multiplied to determine actual motor torque, T, in accordance with Equation 7.78.

The outer control structure of Fig. 7.31 is similar to that of a dc drive. A speed control system is shown, but an outer position loop can be superimposed, if required. The speed error is fed to the speed controller to generate the torque command, T^*, which is compared with the calculated torque, T, for precise torque control. The torque error generates the quadrature axis reference current, i_{qs}^*. The direct axis reference current, i_{ds}^*, is produced by a magnetizing current control loop in which the reference value, i_{mr}^*, is compared with the actual value, i_{mr}. Below base speed, i_{mr}^* is held constant, but field weakening is implemented above base speed by making i_{mr}^* speed dependent using a function generator, as shown in Fig. 7.31.

Induction motor behavior under field-oriented control is similar to that of a dc motor, and control system design is also similar. The induction motor now loses its characteristic nonlinear features, and the dynamic interactions of the drive are greatly simplified. The control scheme of Fig. 7.31 does not employ a reference oscillator, but the stator frequency is determined by the flux signal derived from the actual machine or its model. Consequently, the motor is self-controlled and behaves like a dc motor with its mechanical commutator. There is no pull-out effect, and if the motor is overloaded, or if the speed reference is changed too rapidly, the speed error signal saturates so that the torque cannot exceed the prescribed maximum value.

7.13.2. The Current-Source Inverter Drive

Field orientation requires fast control of stator current in both magnitude and phase. In the current-source inverter (CSI) drive, the magnitude and phase of the stator current are independently controlled by the supply-side converter and CSI respectively.

As explained in Chapter 4, the CSI has six main power devices in the usual three-phase bridge configuration, and, in each 60-degree interval of the output current waveform, current flows through two phases of the wye-connected stator winding. Consequently, a stationary mmf vector is established whose magnitude is directly proportional to the dc link current. Six different conduction sequences are implemented in each cycle of output current; hence the stator mmf vector has a resolution of 60 degrees electrical. A particular vector angle is chosen when the demanded angle comes within 30 de-

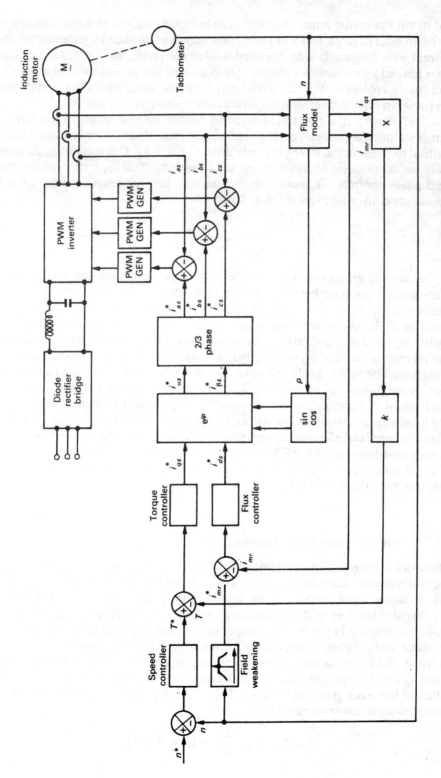

FIG. 7.31. Induction motor speed control by direct field orientation with a current-controlled PWM inverter.

grees of it, but the spatial mmf orientation can be rapidly altered by gating the appropriate pair of inverter devices. Stator current magnitude is controlled by regulation of the dc link current with the supply-side phase-controlled converter, but the rate of change of current is limited by the dc link inductor. Despite these limitations, field orientation can be successfully applied to the CSI drive, but very low speed operation is not smooth enough for servo drive applications requiring continuous position control.

Figure 7.32 illustrates a direct field-oriented control scheme in which the direct and quadrature stator reference currents, i_{ds}^* and i_{qs}^*, are generated by outer control loops that are identical to those in the PWM inverter drive of Fig. 7.31. Consequently, these outer loops will not be discussed further. Because of the functional separation of the magnitude and phase controls for stator current, the reference currents in rectangular field coordinates are converted to polar form. Thus,

$$i_{ds}^* + j\, i_{qs}^* = i_s^* e^{j\delta^*} \qquad (7.86)$$

where i_s^* is directly proportional to the reference value for dc link current, I_d^*, and δ^* is the demanded torque angle between the stator current and rotor flux vectors.

As shown in Fig. 7.32, I_d^* determines the dc link current by means of the usual closed-loop control of the converter firing angle, and δ^* controls the inverter gating circuits. The switching state of the inverter is determined by a six-step ring counter whose state may be represented by a complex number, $e^{j\xi}$, where ξ changes in discrete increments of 60 degrees. The torque angle reference, δ^*, is converted to stator coordinates by the addition of the flux angle, ρ, as in Fig. 7.32. This reference angle is compared with the feedback signal, ξ, representing the status of the ring counter (and inverter) in an angle control loop which determines the gating of the CSI. At low speeds, this angle control loop has the beneficial effect of automatically generating forward and reverse switching commands that implement a PWM operation of the inverter. The resulting improvement in the stator current waveforms reduces the low-speed cogging effect that is normally associated with the CSI drive.

7.13.3. The PWM Voltage-Source Inverter Drive

In field orientation systems employing direct control of stator current, the stator voltage equation has no effect on the drive dynamics and can be omitted from the machine model. At higher power levels, the induction motor is often fed from a PWM voltage-source thyristor inverter, and the switching frequency is usually less than 1 kHz. Consequently, the time lag in the inverter response does not allow effective closed-loop current control, and the stator voltage equation must be taken into account.

As usual, field orientation requires independent control of the stator current components, i_{ds} and i_{qs}, but these currents must now be indirectly controlled by appropriate variation of the stator terminal voltages. The earlier Equation 7.67 gives the stator voltage relationship in stator coordinates as

$$\bar{v}_s = R_s \bar{i}_s + L_s \frac{d\bar{i}_s}{dt} + M \frac{d}{dt}\left(\bar{i}_r e^{j\epsilon}\right).$$

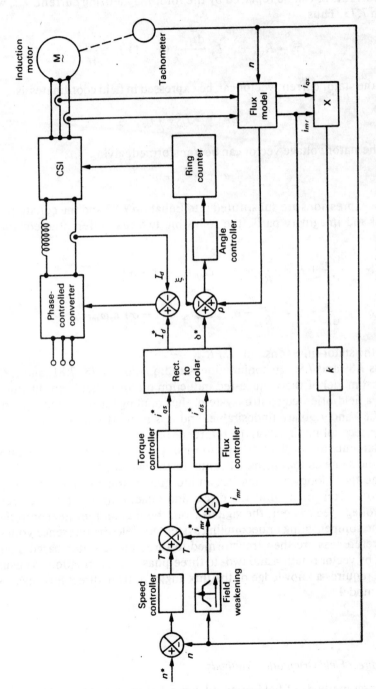

FIG. 7.32. Induction motor speed control by direct field orientation with a current-source inverter.

The rotor current, $\bar{i}_r$, can be replaced by the rotor magnetizing current, $\bar{i}_{mr}$, with the use of Equation 7.73. Thus,

$$\bar{v}_s = R_s \bar{i}_s + \sigma L_s \frac{d\bar{i}_s}{dt} - (\sigma - 1) L_s \frac{d\bar{i}_{mr}}{dt}. \tag{7.87}$$

As before, the stator current vector can be expressed in field coordinates as

$$\bar{i}_s e^{-j\rho} = i_{ds} + j\,i_{qs}.$$

Similarly, the stator voltage vector can be transformed, giving

$$\bar{v}_s e^{-j\rho} = v_{ds} + jv_{qs}. \tag{7.88}$$

When these expressions are substituted into Equation 7.87 and the equation is separated into its real and imaginary parts, the following two real differential equations are obtained:

$$\sigma \tau_s \frac{di_{ds}}{dt} + i_{ds} = \frac{v_{ds}}{R_s} + \sigma \tau_s \omega_{mr} i_{qs} - (1 - \sigma) \tau_s \frac{di_{mr}}{dt} \tag{7.89}$$

$$\sigma \tau_s \frac{di_{qs}}{dt} + i_{qs} = \frac{v_{qs}}{R_s} - \sigma \tau_s \omega_{mr} i_{ds} - (1 - \sigma) \tau_s \omega_{mr} i_{mr} \tag{7.90}$$

where τ_s is the stator time constant, L_s/R_s.

Equations 7.89 and 7.90, in conjunction with Equations 7.78, 7.81, and 7.82, form the mathematical model of the voltage-fed induction motor in field coordinates. When implementing a field-oriented control system, the coupling terms on the right-hand side of Equations 7.89 and 7.90 are undesirable, and it is necessary to introduce suitable compensating or decoupling signals at the reference inputs.

The implementation of field orientation using voltages as the controlled variables is essentially the same as that using currents. In the current-fed systems of Fig. 7.31 and Fig. 7.32, the speed loop generates flux and torque commands that are translated by the flux and torque controllers into the direct and quadrature reference currents, i_{ds}^* and i_{qs}^*. In the voltage-fed system, the signals from the flux and torque controllers are processed by the compensating or decoupling system to yield the reference voltages, v_{ds}^* and v_{qs}^*. These references are then transformed to three-phase stator-based quantities, v_{as}^*, v_{bs}^*, and v_{cs}^*, by vector rotation and two- to three-phase transformation. As usual, the vector rotation requires a knowledge of the flux angle, ρ, from direct flux measurements or from a flux model.

7.13.4. Indirect Field Orientation Methods

The indirect methods of field-oriented control eliminate the need for a flux sensor or flux model but require an accurate measurement of shaft position in order to determine the precise location of the rotor flux vector.[92-94]

As shown in Section 7.12.1, induction motor behavior in field coordinates is described by Equations 7.78, 7.81, and 7.82. Thus,

$$T = k\, i_{mr} i_{qs}$$

$$\tau_r \frac{d i_{mr}}{dt} + i_{mr} = i_{ds}$$

and

$$\omega_{mr} = \omega_m + \frac{i_{qs}}{\tau_r i_{mr}}.$$

The latter equation states that the rotor flux vector has an instantaneous angular velocity, ω_{mr}, which is the sum of the instantaneous shaft angular velocity, ω_m, and the instantaneous slip angular velocity of the rotor flux relative to the rotor. If this slip angular velocity is denoted by ω_2, then

$$\omega_2 = \frac{i_{qs}}{\tau_r i_{mr}} \tag{7.91}$$

and

$$\omega_2 = \omega_{mr} - \omega_m = s\omega_{mr} \tag{7.92}$$

where s is the fractional slip of the rotor with respect to the rotor flux vector. Equations 7.91 and 7.92 show that ω_2 and s are determined by i_{qs} and i_{mr}.

Direct field orientation methods, as already described, measure or compute the rotor flux vector and synchronize the stator current vector with the rotor flux vector. This synchronization of $\bar{i}_s$ and $\bar{i}_{mr}$ ensures that the slip relationship of Equation 7.82 is always satisfied. However, the slip equation can be implemented in the field-oriented controller so that direct measurement of rotor flux position is unnecessary. This approach is the basis of the indirect methods of field orientation, which are often termed slip frequency control methods.

Electromagnetic torque and rotor flux are independently controlled by appropriately regulating i_{ds}, i_{qs}, and ω_2. The field-coordinate equations, Equations 7.78, 7.81, and 7.82 can be used to determine the reference values, i_{ds}^*, i_{qs}^*, and ω_2^*, for demanded values of torque, T^*, and rotor magnetizing current, i_{mr}^*. Thus,

$$i_{ds}^* = i_{mr}^* + \tau_r \frac{d i_{mr}^*}{dt} \tag{7.93}$$

$$i_{qs}^* = \frac{T^*}{k i_{mr}^*} \tag{7.94}$$

$$\omega_2^* = \frac{i_{qs}^*}{\tau_r i_{mr}^*} = \frac{T^*}{k \tau_r (i_{mr}^*)^2}. \tag{7.95}$$

These calculations, which are shown in block diagram form in Fig. 7.33, are performed in real time by a microprocessor. The basic implementation of a speed control system for a current-controlled PWM inverter drive is shown in Fig. 7.34. The speed error is fed to

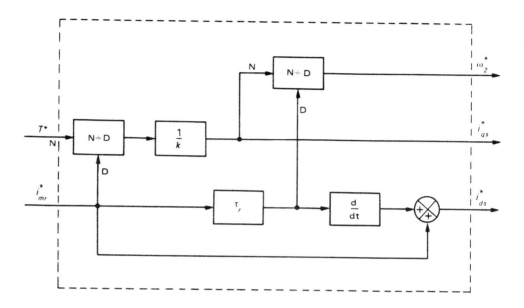

FIG. 7.33. Stator current and slip frequency calculator for indirect field orientation.

the speed controller, which generates the torque command, T^*. As before, the shaft speed is fed to a function generator that demands a constant rotor magnetizing current, i_{mr}, below base speed, and implements field weakening above base speed. The torque and flux commands are used to calculate the reference values, i_{ds}^*, i_{qs}^*, and ω_2^*, as already indicated in Fig. 7.33. The commanded slip frequency, ω_2^*, is integrated to give a slip angular position signal, θ_2^*, which is added to the rotor position signal, ϵ, from the shaft-mounted incremental encoder to determine the rotor flux angle, ρ. These calculations are performed digitally to give the required accuracy and freedom from drift problems. As shown in Fig. 7.34, the angle ρ is then used to implement the vector rotation, $e^{j\rho}$, of i_{ds}^* and i_{qs}^* to stator-based reference currents, just as in direct field orientation. The stator reference currents are reproduced in the motor by the PWM inverter, in the usual manner, and the resulting drive system is capable of full servo performance down to zero speed. As usual, an outer position loop can be implemented, if required.

It is evident from Fig. 7.33 that the stator current and slip frequency reference values are critically dependent on the accuracy of the motor parameter values used in the computation. In particular, if the actual rotor time constant differs from the value used to calculate the reference values, then correct field orientation will not be achieved and the dynamic response of the drive will deteriorate. For correct decoupling of the control, the parameters used in the calculator should track the actual machine parameters, but the temperature-induced variation in rotor resistance is a major obstacle to the achievement of parameter coincidence. De-tuning effects that result from parameter variations have received considerable attention in the literature. The effects of parameter changes on dynamic performance have been studied, and several on-line adaptation schemes have been suggested.[95–105]

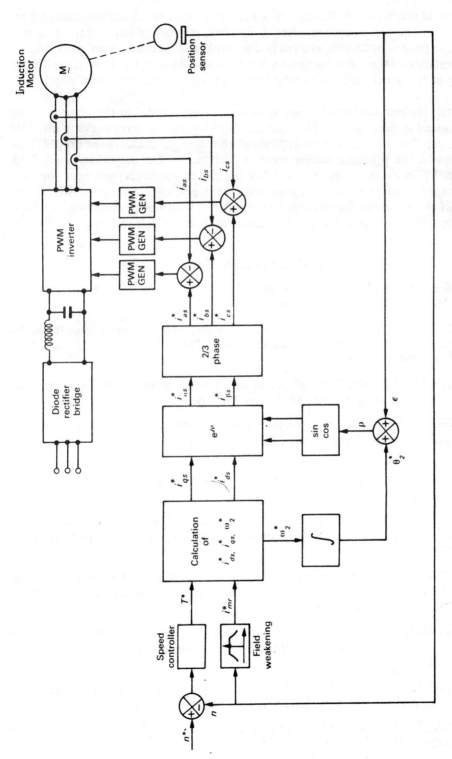

FIG. 7.34. Induction motor speed control by indirect field orientation wih a current-controlled PWM inverter.

Indirect field orientation is readily applied to other types of adjustable-frequency inverter drive. In a current-source inverter (CSI) drive, it is natural to convert the current commands to polar coordinates, as already discussed for direct field orientation. The required orientation of the stator current vector is usually obtained from the integral value of the sum of the demanded slip angular frequency, ω_2^*, and the rotational angular frequency, ω_m.

In general, indirect field orientation systems are very similar to the controlled-slip drives discussed in Section 7.10. The traditional controlled-slip drive seeks to maintain constant airgap flux, but the control implementation does not preserve the proper phase relationships in the machine during transient conditions. On the other hand, field-oriented control maintains constant rotor flux and properly controls the torque angle between the space vectors of rotor flux and stator current, even during transients. The influence of rotor leakage flux cannot be neglected, and it has been shown that field-oriented control on the basis of airgap flux has an inferior dynamic performance.[106]

7.14. REFERENCES

1. MOKRYTZKI, B., The controlled slip static inverter drive, *IEEE Trans. Ind. Gen. Appl.*, **IGA-4**, 3, May/June 1968, pp. 312-317.

2. LANDAU, I.D., Wide-range speed control of three-phase squirrel-cage induction motors using static frequency converters, *IEEE Trans. Ind. Gen. Appl.*, **IGA-5**, 1, Jan./Feb. 1969, pp. 53-60.

3. AGARWAL, P.D., The GM high performance induction motor drive system, *IEEE Trans. Power Appar. Syst.*, **PAS-88**, 2, Feb. 1969, pp. 86-93.

4. RISBERG, R.L., A wide speed range inverter fed induction motor drive, *Conf. Rec. IEEE Ind. Gen. Appl. Group Annual Meeting, 1969*, pp. 629-633.

5. MAAG, R.B., Characteristics and application of current source/slip regulated ac induction motor drives, *Conf. Rec. IEEE Ind. Gen. Appl. Group Annual Meeting, 1971*, pp. 411-416.

6. HASSE, K., Zur Dynamik drehzahlgeregelter Antriebe mit stromrichtergespeisten Asynchron-Kurzschlussläufermaschinen, Techn. Hochsch. Darmstadt, Dissertation, 1969.

7. BLASCHKE, F., The principle of field orientation as applied to the new "transvektor" closed-loop control system for rotating-field machines, *Siemens Rev.*, **39**, 5, May 1972, pp. 217-220.

8. FORSYTHE, J.B., and DEWAN, S.B., Output current regulation with PWM inverter-induction motor drives, *IEEE Trans. Ind. Appl.*, **IA-11**, 5, Sept./Oct. 1975, pp. 517-525.

9. PLUNKETT, A.B., and LIPO, T.A., New methods of induction motor torque regulation, *IEEE Trans. Ind. Appl.*, **IA-12**, 1, Jan./Feb. 1976, pp. 47-55.

10. PLUNKETT, A.B., Direct flux and torque regulation in a PWM inverter-induction motor drive, *IEEE Trans. Ind. Appl.*, **IA-13**, 2, Mar./Apr. 1977, pp. 139-146.

11. PLUNKETT, A.B., D'ATRE, J.D., and LIPO, T.A., Synchronous control of a static ac induction motor drive, *IEEE Trans. Ind. Appl.*, **IA-15**, 4, July/Aug. 1979, pp. 430-437.

12. KAUFMAN, G.A., and PLUNKETT, A.B., A high-performance torque controller using a voltage source inverter and induction machine, *Conf. Rec. IEEE Ind. Appl. Soc. Annual Meeting, 1981*, pp. 863-872.

13. KRISHNAN, R., STEFANOVIC, V.R., and LINDSAY, J.F., Control characteristics of inverter-fed induction motor, *IEEE Trans. Ind. Appl.*, **IA-19**, 1, Jan./Feb. 1983, pp. 94-104.

14. EVANS, R.J., COOK, B.J., and BETZ, R.E., Nonlinear adaptive control of an inverter-fed induction motor linear load case, *IEEE Trans. Ind. Appl.*, **IA-19**, 1, Jan./Feb. 1983, pp. 74-83.

15. SCHAUDER, C.D., CHOO, F.H., and ROBERTS, M.T., High performance torque-controlled induction motor drives, *IEEE Trans. Ind. Appl.*, **IA-19**, 3, May/June 1983, pp. 349-355.

16. JOETTEN, R., and MAEDER, G., Control methods for good dynamic performance induction motor drives based on current and voltage as measured quantities, *IEEE Trans. Ind. Appl.*, **IA-19**, 3, May/June 1983, pp. 356-363.

17. HARASHIMA, F., HASHIMOTO, H., and KONDO, S., MOSFET converter-fed position servo system with sliding mode control, *IEEE Power Electron. Spec. Conf.*, 1983, pp. 73-79.

18. KAZMIERKOWSKI, M.P., and KÖPCKE, H.J., Comparison of dynamic behaviour of frequency converter fed induction machine drives, *Proc. 3rd IFAC Symp. on Control in Power Electronics and Electrical Drives*, 1983, pp. 313-320.

19. SUGI, K., NAITO, Y., KUROSAWA, R., KANO, Y., KATAYAMA, S., and YOSHIDA, T., A microcomputer-based high capacity cycloconverter drive for main rolling mill, *Int. Power Electron. Conf., Tokyo*, 1983, pp. 744-755.

20. BOSE, B.K., Scalar decoupled control of induction motor, *IEEE Trans. Ind. Appl.*, **IA-20**, 1, Jan./Feb. 1984, pp. 216-225.

21. SCHÖNUNG, A., and STEMMLER, H., Static frequency changers with "subharmonic" control in conjunction with reversible variable-speed ac drives, *Brown Boveri Rev.*, **51**, 8/9, Aug./Sept. 1964, pp. 555-577.

22. HEUMANN, K., and JORDAN, K.-G., Das Verhalten des Käfigläufermotors bei veränderlicher Speisefrequenz und Stromregelung, *AEG Mitt.*, **54**, 1/2, 1964, pp. 107-116.

23. STEFANOVIC, V.R., and BARTON, T.H., Static torque characteristics of an induction motor with a variable frequency supply, *IEEE Power Eng. Soc. Winter Meeting*, 1973, Paper C 73 199-7.

24. NÜRNBERG, W., *Die Asynchronmaschine*, Springer-Verlag, Berlin, 1952.

25. RODEWALD, E., Betriebseigenschaften des Asynchronmotors bei Speisung mit niedrigen Frequenzen, *Siemens-Z.*, **40**, Beiheft "Motoren für industrielle Antriebe," Oct. 1966, pp. 131-139.

26. STEFANOVIC, V.R., Static and dynamic characteristics of induction motors operating under constant airgap flux control, *Conf. Rec. IEEE Ind. Appl. Soc. Annual Meeting, 1976*, pp. 436-444.

27. FISCHER, H.-J., Beitrag zum Betriebsverhalten von Asynchronmotoren bei veränderlicher Primärfrequenz, *Elektrie*, **20**, 5, 1966, pp. 205-208.

28. ABBONDANTI, A., Method of flux control in induction motors driven by variable frequency, variable voltage supplies, *IEEE Int. Semicond. Power Converter Conf.*, 1977, pp. 177-184.

29. SLEMON, G.R., and DEWAN, S.B., Induction motor drive with current source inverter, *Conf. Rec. IEEE Ind. Appl. Soc. Annual Meeting, 1974*, pp. 411-417.

30. LIPO, T.A., and CORNELL, E.P., State-variable steady-state analysis of a controlled current induction motor, *IEEE Trans. Ind. Appl.*, **IA-11**, 6, Nov./Dec. 1975, pp. 704-712.

31. CHALMERS, B.J., Torque production in ac motors with constant current supply, *Elect. Mach. and Electromechan.*, **1**, 3, Apr.-June 1977, pp. 237-244.

32. ABRAHAM, L., HEUMANN, K., and KOPPELMANN, F., Wechselrichter zur Drehzahlsteuerung von Käfigläufermotoren, *AEG Mitt.*, **54**, 1/2, 1964, pp. 89-106.

33. ABRAHAM, L., and KOPPELMANN, F., Käfigläufermotoren mit hoher Drehzahldynamik, *AEG Mitt.*, **55**, 2, 1965, pp. 118-123.

34. AMATO, C.J., Variable speed with controlled slip induction motor, *IEEE Ind. Static Power Conv. Conf.*, Nov. 1965, pp. 181-185.

35. SALIHI, J.T., AGARWAL, P.D., and SPIX, G.J., Induction motor control scheme for battery-powered electric car (GM-Electrovair I), *IEEE Trans. Ind. Gen. Appl.*, **IGA-3**, 5, Sept./Oct. 1967, pp. 463-469.

36. SALIHI, J.T., Simulation of controlled slip, variable speed induction motor drive systems, *IEEE Trans. Ind. Gen. Appl.*, **IGA-5**, 2, Mar./Apr. 1969, pp. 149-157.

37. HUMPHREY, A.J., Constant horsepower operation of induction motors, *IEEE Trans. Ind. Gen. Appl.*, **IGA-5**, 5, Sept./Oct. 1969, pp. 552-557.

38. ABBONDANTI, A., and BRENNEN, M.B., Variable speed induction motor drives use electronic slip calculator based on motor voltages and currents, *IEEE Trans. Ind. Appl.*, **IA-11**, 5, Sept./Oct. 1975, pp. 483-488.

39. NABAE, A., Performance of slip-frequency controlled induction machines, *Conf. Rec. IEEE Ind. Appl. Soc. Annual Meeting, 1975*, pp. 852-856.

40. GOSDEN, D., An experimental electric car using an induction motor drive, *Elect. Eng. Trans. (Australia)*, **EE-12**, 1, 1976, pp. 26-34.

41. SEN, P.C., and MacDONALD, M., Slip-frequency controlled induction motor drives using digital phase-locked loop control system, *IEEE Int. Semicond. Power Converter Conf.*, 1977, pp. 413-419.

42. MURPHY, J.M.D., HOFT, R.G., and HOWARD, L.S., Controlled-slip operation of an induction motor with optimum PWM waveforms, IEE Conf. Publ. No. 179, *Electrical Variable-Speed Drives*, 1979, pp. 157-160.

43. STEFANOVIC, V.R., Closed loop performance of induction motors with constant volts/hertz control, *Elect. Mach. and Electromechan.*, **1**, 3, Apr.-June 1977, pp. 255-266.

44. MILES, A.R., NOVOTNY, D.W., and BETRO, J., The effect of volts/hertz control on induction machine dynamic performance, *Conf. Rec. IEEE Ind. Appl. Soc. Annual Meeting, 1979*, pp. 802-809.

45. DIXON, F.F., and TILEY, G.L., Low-frequency performance of a wound-rotor induction motor for mine hoist drives, *Trans. AIEE, Pt. 3, Power Appar. Syst.*, **76**, Dec. 1957, pp. 1140-1145.

46. DE MELLO, F.P., and WALSH, G.W., Reclosing transients in induction motors with terminal capacitors, *Trans. AIEE, Pt. 3, Power Appar. Syst.*, **80**, Feb. 1961, pp. 1206-1213.

47. TSIVITSE, P.J., and KLINGSHIRN, E.A., Optimum voltage and frequency for polyphase induction motors operating with variable frequency power supplies, *IEEE Trans. Ind. Gen. Appl.*, **IGA-7**, 4, July/Aug. 1971, pp. 480-487.

48. JIAN, T.W., NOVOTNY, D.W., and SCHMITZ, N.L., Characteristic induction motor slip values for variable voltage part load performance optimization, *IEEE Trans. Power Appar. Syst.*, **PAS-102**, 1, Jan./Feb. 1983, pp. 38-46.

49. MURPHY, J.M.D., and HONSINGER, V.B., Efficiency optimization of inverter-fed induction motor drives, *Conf. Rec. IEEE Ind. Appl. Soc. Annual Meeting*, 1982, pp. 544-552.

50. KIRSCHEN, D.S., NOVOTNY, D.W., and SUWANWISOOT, W., Minimizing induction motor losses by excitation control in variable frequency drives, *IEEE Trans. Ind. Appl.*, **IA-20**, 5, Sept./Oct. 1984, pp. 1244-1250.

51. PARK, M.H., and SUL, S.K., Microprocessor-based optimal-efficiency drive of an induction motor, *IEEE Trans. Ind. Electron.*, **IE-31**, 1, Feb. 1984, pp. 69-73.

52. KIM, H.G., SUL, S.K., and PARK, M.H., Optimal efficiency drive of a current source inverter fed induction motor by flux control, *IEEE Trans. Ind. Appl.*, **IA-20**, 6, Nov./Dec. 1984, pp. 1453-1459.

53. KUSKO, A., and GALLER, D., Control means for minimization of losses in ac and dc motor drives, *IEEE Trans. Ind. Appl.*, **IA-19**, 4, July/Aug. 1983, pp. 561-570.

54. KIRSCHEN, D.S., NOVOTNY, D.W., and LIPO, T.A., On-line efficiency optimization of a variable frequency induction motor drive, *IEEE Trans. Ind. Appl.*, **IA-21**, 4, May/June 1985, pp. 610-615.

55. LIPO, T.A., Flux sensing and control of static ac drives by the use of flux coils, *IEEE Trans. Magn.*, **MAG-13**, 5, Sept. 1977, pp. 1403-1408.

56. PHILLIPS, K.P., Current-source converter for ac motor drives, *IEEE Trans. Ind. Appl.*, **IA-8**, 6, Nov./Dec. 1972, pp. 679-683.

57. CORNELL, E.P., and LIPO, T.A., Design of controlled current ac drive systems using transfer function techniques, *Proc. 1st IFAC Symp. on Control in Power Electronics and Electrical Drives*, **1**, 1974, pp. 133-147.

58. CORNELL, E.P., and LIPO, T.A., Modeling and design of controlled current induction motor drive systems, *IEEE Trans. Ind. Appl.*, **IA-13**, 4, July/Aug. 1977, pp. 321-330.

59. MacDONALD, M.L., and SEN, P.C., Control loop study of induction motor drives using dq model, *Conf. Rec. IEEE Ind. Appl. Soc. Annual Meeting, 1978*, pp. 897-903.

60. KRISHNAN, R., MASLOWSKI, W.A., and STEFANOVIC, V.R., Control principles in current source induction motor drives, *Conf. Rec. IEEE Ind. Appl. Soc. Annual Meeting, 1980*, pp. 605-617.

61. KRISHNAN, R., LINDSAY, J.F., and STEFANOVIC, V.R., Design of angle-controlled current source inverter-fed induction motor drive, *IEEE Trans. Ind. Appl.*, **IA-19**, 3, May/June 1983, pp. 370-378.

62. WALKER, L.H., and ESPELAGE, P.M., A high-performance controlled-current inverter drive, *IEEE Trans. Ind. Appl.*, **IA-16**, 2, Mar./Apr. 1980, pp. 193-202.

63. KUME, T., and YOSHIDA, Y., Speed transient of induction motor driven by current-source type inverter, *Conf. Rec. IEEE Ind. Appl. Soc. Annual Meeting, 1973*, pp. 865-874.

64. MANN, S., A current source converter for multi-motor applications, *Conf. Rec. IEEE Ind. Appl. Soc. Annual Meeting, 1975*, pp. 980-983.

65. ABBAS, M., and NOVOTNY, D., The stator voltage-controlled current source inverter induction motor drive, *IEEE Trans. Ind. Appl.*, **IA-18**, 3, May/June 1982, pp. 219-229.

66. KOVACS, K.P., and RACZ, I., *Transiente Vorgänge in Wechselstrommaschinen*, Ungarische Akademie der Wissenschaften, Budapest, 1959.

67. NAUNIN, D., The calculation of the dynamic behavior of electric machines by space-phasors, *Elect. Mach. and Electromechan.*, **4**, 1, July/Aug. 1979, pp. 33-45.

68. LEONHARD, W., *Control of Electrical Drives*, Springer-Verlag, Berlin, 1985.

69. STANLEY, H.C., An analysis of the induction machine, *Trans. AIEE*, **57**, 1938, pp. 751-755.

70. WHITE, D.C., and WOODSON, H.H., *Electromechanical Energy Conversion*, J. Wiley, New York, 1959.

71. HUGHES, F.M., and ALDRED, A.S., Transient characteristics and simulation of induction motors, *Proc. IEE*, **111**, 12, Dec. 1964, pp. 2041-2050.

72. KRAUSE, P.C., and THOMAS, C.H., Simulation of symmetrical induction machinery, *IEEE Trans. Power Appar. Syst.*, **PAS-84**, 11, Nov. 1965, pp. 1038-1053.

73. JORDAN, H.E., Analysis of induction machines in dynamic systems, *IEEE Trans. Power Appar. Syst.*, **PAS-84**, 11, Nov. 1965, pp. 1080-1088.

74. JORDAN, H.E., Digital computer analysis of induction machines in dynamic systems, *IEEE Trans. Power Appar. Syst.*, **PAS-86**, 6, June 1967, pp. 722-728.

75. LAWRENSON, P.J., and STEPHENSON, J.M., Note on induction-machine performance with a variable-frequency supply, *Proc. IEE*, **113**, 10, Oct. 1966, pp. 1617-1623.

76. NELSON, R.H., LIPO, T.A., and KRAUSE, P.C., Stability analysis of a symmetrical induction machine, *IEEE Trans. Power Appar. Syst.*, **PAS-88**, 11, Nov. 1969, pp. 1710-1717.

77. ROGERS, G.J., Linearized analysis of induction-motor transients, *Proc. IEE*, **112**, 10, Oct. 1965, pp. 1917-1926.

78. PFAFF, G., Zur Dynamik des Asynchronmotors bei Drehzahlsteuerung mittels veränderlicher Speisefrequenz, *Elektrotech. Z. Ausg. A*, **85**, 22, 1964, pp. 719-724.

79. DE CARLI, A., MURGO, M., and RUBERTI, A., Speed control of induction motors by frequency variation, *IFAC Congress, London, 1966*, pp. 4C1-4C11.

80. WEST, J.C., JAYAWANT, B.V., and WILLIAMS, G., Analysis of dynamic performance of induction motors in control systems, *Proc. IEE*, **111**, 8, Aug. 1964, pp. 1468-1478.

81. JAYAWANT, B.V., and BATESON, K.N., Dynamic performance of induction motors in control systems, *Proc. IEE*, **115**, 12, Dec. 1968, pp. 1865-1870.

82. KASSAKIAN, J.G., Simulating power electronic systems — a new approach, *Proc. IEEE*, **67**, Oct. 1979, pp. 1428-1439.

83. BLASCHKE, F., Das Verfahren der Feldorientierung zur Regelung der Drehfeldmaschine, Techn. Univ. Braunschweig, Dissertation, 1973.

84. NOVOTNY, D.W., and LORENZ, R.D. (Editors), Introduction to field orientation and high performance ac drives, Tutorial course presented at *IEEE Ind. Appl. Soc. Annual Meeting*, 1985.

85. GABRIEL, R., LEONHARD, W., and NORDBY, C., Microprocessor control of induction motors employing field coordinates, IEE Conf. Publ. No. 179, *Electrical Variable-Speed Drives*, 1979, pp. 146-150.

86. GABRIEL, R., LEONHARD, W., and NORDBY, C., Field-oriented control of a standard ac motor using microprocessors, *IEEE Trans. Ind. Appl.*, **IA-16**, 2, Mar./Apr. 1980, pp. 186-192.

87. GABRIEL, R., and LEONHARD, W., Microprocessor control of induction motor, *IEEE Int. Semicond. Power Converter Conf.*, 1982, pp. 385-396.

88. SCHUMACHER, W., Microprocessor controlled ac servo drive, *Conf. on Microelectronics in Power Electronics and Electrical Drives*, Darmstadt, 1982, pp. 311-319.

89. SCHUMACHER, W., LETAS, H.-H., and LEONHARD, W., Microprocessor-controlled ac servo-drives with synchronous and asynchronous motors, IEE Conf. Publ. No. 234, *Power Electronics and Variable-Speed Drives*, 1984, pp. 233-236.

90. SCHUMACHER, W., and LEONHARD, W., Transistor-fed ac-servo drive with microprocessor control, *Int. Power Electron. Conf.*, 2, Tokyo, 1983, pp. 1465-1476.

91. LEONHARD, W., Control of ac machines with the help of microelectronics, *Proc. 3rd IFAC Symp. on Control in Power Electronics and Electrical Drives*, 1983, pp. 769-792.

92. NABAE, A., OTSUKA, K., UCHINO, H., and KUROSAWA, R., An approach to flux control of induction motors operated with variable-frequency power supply, *IEEE Trans. Ind. Appl.*, **IA-16**, 3, May/June 1980, pp. 342-350.

93. SASAKI, Y., and HAYASHI, Y., Application of slip-frequency controlled induction motors for machine tools, *IEEE Power Electron. Spec. Conf.*, 1982, pp. 305-311.

94. BLASCHKE, F., and BÖHM, K., Verfahren der Flusserfassung bei der Regelung stromrichtergespeister Asynchronmaschinen, *Proc. 1st IFAC Symp. on Control in Power Electronics and Electrical Drives*, 1, 1974, pp. 635-649.

95. GARCES, L.J., Parameter adaption for the speed-controlled static ac drive with a squirrel-cage induction motor, *IEEE Trans. Ind. Appl.*, **IA-16**, Mar./Apr. 1980, pp. 173-178.

96. YOSHIDA, Y., UEDA, R., and SONODA, T., A new inverter-fed induction motor drive with a function of correcting rotor circuit time constant, *Int. Power Electron. Conf.*, 1, Tokyo, 1983, pp. 672-683.

97. KOYAMA, M., SUGIMOTO, H., MIMURA, M., and KAWASAKI, K., Effects of parameter change on coordinate control system of induction motor, *Int. Power Electron. Conf.*, 1, Tokyo, 1983, pp. 684-695.

98. OHTANI, T., Torque control using the flux from magnetic energy in induction motors driven by static converter, *Int. Power Electron. Conf.*, 1, Tokyo, 1983, pp. 696-707.

99. OKUYAMA, T., NAGASE, H., KUBOTA, Y., HORIUCHI, H., MIYAZAKI, K., and IBORI, S., High performance ac motor speed control system using GTO converters, *Int. Power Electron. Conf.*, 1, Tokyo, 1983, pp. 720-731.

100. IRISA, T., TAKATA, S., UEDA, R., SONODA, T., and MOCHIZUKI, T., A novel approach on parameter self-tuning method in ac servo system, *Proc. 3rd IFAC Symp. on Control in Power Electronics and Electrical Drives*, 1983, pp. 41-48.

101. KRISHNAN, R., and DORAN, F.C., Study of parameter sensitivity in high-performance inverter-fed induction motor drive systems, *Conf. Rec. IEEE Ind. Appl. Soc. Annual Meeting*, 1984, pp. 510-524.

102. NORDIN, K.B., NOVOTNY, D.W., and ZINGER, D.S., The influence of motor parameter deviations in feedforward field orientation drive systems, *Conf. Rec. IEEE Ind. Appl. Soc. Annual Meeting, 1984*, pp. 525-531.

103. NAGASE, H., MATSUDA, Y., OHNISHI, K., NINOMIYA, H., and KOIKE, T., High-performance induction motor drive system using a PWM inverter, *IEEE Trans. Ind. Appl.*, **IA-20**, 6, Nov./Dec. 1984, pp. 1482-1489.

104. MATSUO, T., and LIPO, T.A., A rotor parameter identification scheme for vector-controlled induction motor drives, *IEEE Trans. Ind. Appl.*, **IA-21**, 3, May/June 1985, pp. 624-632.

105. HARASHIMA, F., KONDO, S., OHNISHI, K., KAJITA, M., and SUSONO, M., Multimicroprocessor-based control system for quick response induction motor drive, *IEEE Trans. Ind. Appl.*, **IA-21**, 3, May/June 1985, pp. 602-609.

106. BAYER, K.H., and BLASCHKE, F., Stability problems with the control of induction machines using the method of field orientation, *Proc. 2nd IFAC Symp. on Control in Power Electronics and Electrical Drives*, 1977, pp. 483-492.

CHAPTER 8

Control Systems for Adjustable-Frequency Synchronous Motor Drives

8.1. INTRODUCTION

This chapter begins with a description of the several types of synchronous machine used with power electronic circuits in adjustable-speed ac motor drives. The operating principles and phasor diagrams for the wound-field machine, the permanent magnet machine, and the synchronous reluctance machine are presented, and the corresponding torque equations are developed. These introductory sections aid in the understanding of the open- and closed-loop control systems described in later sections of the chapter. Some of the earliest applications of adjustable-speed ac drives used synchronous machines and tube-based power electronic converters. Reference 1 describes some of these early applications. References 2 through 4 describe the basic construction and operating characteristics of synchronous machines.

8.2. TYPES OF SYNCHRONOUS MACHINE

Synchronous motors can be of the wound-field, permanent magnet, or reluctance type. The wound-field machine has a distributed polyphase armature winding on the stator and a dc-excited field winding on the rotor. An inverted construction, with a dc-excited stator field winding is possible, but the rotating three-phase armature winding would require four slip-rings (including a neutral connection), and the slip-ring currents and insulation voltage ratings would be quite large. The rotating-field system is preferred because the dc excitation current is relatively small and can be supplied by only two slip-rings. A brushless construction is obtained when the rotor field excitation is provided by permanent magnets. The synchronous reluctance motor is a rugged brushless motor that uses an unexcited rotor having salient or projecting poles.

The wound-field and permanent magnet machines can operate at unity power factor at full load and so minimize the stator current and inverter volt-ampere rating for a given shaft power. These machines can also operate at a leading power factor and thereby provide load commutation for the inverter. The synchronous reluctance motor, on the other hand, always operates at a relatively low lagging power factor. The permanent magnet and reluctance machines are restricted to lower power ratings, but the wound-field motor can be of very high power rating. There are two distinct categories of wound-field machine: (a) those with round or cylindrical rotors and (b) those with salient, or projecting, rotor poles.

339

8.2.1. *Round-Rotor Machine*

The round-rotor, or cylindrical-rotor, synchronous machine has a uniform airgap between a slotted stator and rotor. The stator is composed of iron laminations stacked together. The rotor is a solid forging with rotor slots milled into its surface. A conventional three-phase distributed armature winding is placed in the stator slots, and a single field winding is placed in the rotor slots. For synchronous motor operation, the three-phase armature winding is supplied with balanced three-phase currents, and dc excitation is supplied to the rotor field winding. The balanced three-phase armature currents establish a component flux wave in the airgap, which has an approximately sinusoidal spatial distribution with a constant amplitude and which rotates at synchronous speed. If the rotor also rotates at synchronous speed, the magnetic fields of stator and rotor are stationary relative to one another, and a steady electromagnetic torque is developed because of the tendency of the two magnetic fields to align their axes. The speed n_1 (in revolutions per minute) of the synchronous motor is related to the ac supply frequency, f_1, and the number of pole-pairs, p, by $n_1 = 60 \, f_1/p$. On 50 or 60 Hz ac utility supplies, the synchronous motor has no starting torque; it must be brought up to synchronous speed by induction motor action or by an auxiliary motor. Induction motor torque is developed when the rotor is fitted with a squirrel-cage winding or when an unlaminated steel rotor is used. Figure 8.1 shows a cross-sectional view of an idealized round-rotor machine in which the distributed three-phase stator winding is represented by three concentrated coils.

The rotor field winding can be excited with direct current supplied through slip-rings and brushes from a static phase-controlled rectifier exciter or from a dc generator exciter on the shaft of the main synchronous machine. The field current of the synchronous machine is controlled by varying the field current of the dc generator exciter. A common ap-

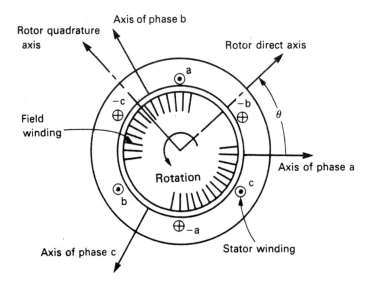

FIG. 8.1. Idealized round-rotor synchronous machine.

proach, up to quite large ratings, is to use a shaft-coupled ac generator exciter with rectification of the ac output. If the exciter armature winding is placed on the rotor and the rectifier diodes are mounted on the rotating shaft, the rectified output can be fed directly to the field winding of the main synchronous machine without any sliding contacts. This is the common brushless excitation system. An alternate approach is to fit a shaft-mounted exciter generator with three-phase windings on both stator and rotor. The three-phase stator winding is energized by the ac power supply system and establishes a rotating airgap field. The rotor is driven against this rotating field so that alternating voltages are generated in the three-phase rotor winding. These ac voltages are rectified by a rotating diode rectifier bridge to provide the dc excitation for the main synchronous machine. The field current of the synchronous machine is varied by controlling the stator voltage of the exciter with a three-phase thyristor voltage controller.

Round-rotor synchronous machines are used for steam and gas turbine-driven generators in utility and industrial generating stations. Ratings can exceed 1500 MW for a single unit.

8.2.2. Salient-Pole Machine

In a salient-pole synchronous machine, the field winding consists of a number of concentrated field coils placed around projecting poles on the rotor. The field winding is excited with direct current to produce alternate north and south rotor poles. The stator has a conventional three-phase distributed armature winding placed in slots, as in the round-rotor machine. As before, synchronous motor torque is developed by the interaction between stator and rotor magnetic fields when the stator winding is energized with balanced three-phase currents.

The salient-pole construction results in a nonuniform airgap, as shown in the idealized two-pole machine of Fig. 8.2. The airgap length is a minimum in the polar, or direct, axis

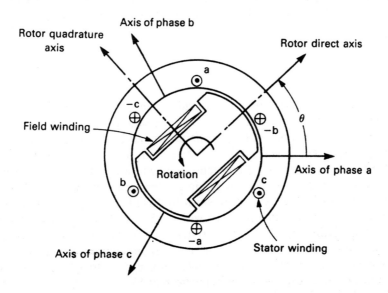

Fig. 8.2. Idealized salient-pole synchronous machine.

and is a maximum in the interpolar, or quadrature, axis. Because of the variation in magnetic circuit reluctance, the stator winding mmf will establish a larger airgap flux when the mmf is centered on the direct axis than when centered on the quadrature axis. This variation in airgap flux implies a corresponding variation in stator phase reactance, leading to a two-axis theory of operation, as discussed later.

The field current is supplied through slip-rings and brushes or by a brushless exciter as previously described. A squirrel-cage damper winding is often embedded in the rotor pole faces to dampen rotor oscillations following sudden load changes and to provide induction motor torque for starting the motor. Salient-pole machines have been built in unit sizes of up to 500 MW for hydrogenerator utility applications.

8.2.3. *Permanent Magnet Machine*

The rotor field flux of a synchronous machine can be produced by permanent magnets rather than by an electromagnet. Various iron-, nickel-, and cobalt-based alloys can be permanently magnetized to provide a source of rotor flux. Ferrite and rare earth (samarium cobalt) materials are also used as permanent magnets. The stator winding is a conventional distributed three-phase winding located in the stator slots and energized by a three-phase ac supply. The physical airgap length is uniform around the circumference of the rotor. However, the magnetic circuit reluctance in the region occupied by the magnet is high because permanent magnet materials have low permeability. If the steel rotor laminations occupy the interpolar space between the magnets, the reluctance in the interpolar region is lower than in the polar region. The permanent magnets can be located either at the airgap (surface permanent magnets) or within the body of the rotor (interior permanent magnets). An elementary surface-mounted permanent magnet machine is illustrated in Fig. 1.8 of Chapter 1. In the interior magnet machine, the flux of adjacent magnets can combine to produce a higher value of airgap flux than can be obtained with a single magnet.

8.2.4. *Synchronous Reluctance Machine*

The synchronous reluctance motor has already been described in Chapter 1, and an elementary two-pole motor is shown in Fig. 1.6. The salient-pole rotor has neither a dc field winding nor permanent magnets, but the unexcited ferromagnetic rotor develops reluctance torque because of the tendency of the salient rotor poles to align themselves with the stator magnetic field axis. If a synchronously rotating stator field is established by means of a conventional polyphase winding excited by a balanced polyphase ac supply, then the rotor runs in exact synchronism with this field as the salient poles seek to maintain the minimum reluctance position with respect to the stator flux. On a fixed-frequency ac supply, the motor is not self-starting unless the rotor is fitted with a squirrel-cage winding to permit starting by induction motor action. When the rotor speed approaches synchronous speed, the salient-pole rotor pulls into synchronism with the stator field, and the rotor cage winding plays no part in the steady-state synchronous operation of the motor. The pull-in torque capability is the maximum load torque that the motor can pull into synchronism with a specified load inertia. Pull-in torque is an important criterion for fixed-frequency operation. The pull-out torque is the load torque required to pull the rotor out of synchronism.

Until the mid-1960s, the reluctance principle had been applied only to fractional horsepower motors, because the advantages of an inexpensive, robust, and reliable construction seemed to be offset by low efficiency, low power factor, poor torque/weight ratio, and high starting current. However, intensive development efforts resulted in better reluctance motors.[5-14] Improved performance characteristics were achieved by the use of a segmented-rotor construction which introduced a flux barrier in each pole,[9,10] or by introducing two flux barriers per pole into a salient-pole rotor.[11,12] In general, the synchronous performance can be improved by increasing the degree of magnetic asymmetry in the rotor. Thus, a reduction in the flux on the interpolar, or quadrature, axis gives a greater pull-out torque, but the pull-in torque of the motor is reduced and the power factor is also affected. For fixed-frequency operation, the final design is a compromise among these factors. Reluctance motors have been constructed that have a power output of 75 to 85 percent of the power of an induction motor of the same size. Power factors of 0.54 to 0.75 have been reported with a full-load efficiency of 85 percent,[10] but better power factors can obviously be achieved with wound-field or permanent magnet machines. The pull-in performance of the synchronous reluctance motor can be improved at the expense of the power factor and efficiency, and segmented-rotor reluctance motors have been developed which can synchronize up to 10 times the rotor inertia against full-load torque, with a starting current of only 4.5 times the full-load current. The pull-out torque is typically about 1.5 times the full-load torque.

The direct-on-line starting performance of the motor is unimportant in adjustable-frequency applications where the motor operates synchronously when accelerating from rest. For such applications, the motor can be designed to give optimum synchronous performance without the trade-offs required to achieve good starting and pull-in characteristics. However, the reluctance motor has a tendency to become unstable at low frequencies and light loads. It has been shown that a stable machine can be obtained by designing the rotor with magnetically saturable bridges between the poles.[13,14] These bridges are saturated at the pull-in torque condition but are unsaturated at light load.

8.2.5. *Inductor Machine*

The inductor machine is a special type of synchronous machine developed for use at high rotational speeds. The armature winding and dc-excited field winding are both placed on the stator, and a homopolar construction is often used. The rotor has many teeth but no winding and can withstand high rotational speeds. When the rotor is rotating, the unidirectional airgap flux at any position pulsates in magnitude as a result of the cyclic variation in magnetic circuit reluctance, and alternating voltages are generated in the stator windings. Inductor machines are used at rotational speeds up to 100 000 rev/min for ac generators on aircraft. A heteropolar construction is also used.

8.3. STEADY-STATE THEORY OF OPERATION

This section reviews the basic theory of operation of the various types of synchronous machine. The relevant phasor diagrams are developed, and torque equations are derived. The analysis assumes sinusoidal ac supply voltages and currents because harmonic effects are usually negligible.

8.3.1. *Round-Rotor Machine*

In a synchronous machine, as explained above, the balanced three-phase currents in the three-phase armature winding establish a synchronously rotating magnetic field in the airgap. The rotor also rotates at synchronous speed, and therefore, the armature mmf wave is stationary with respect to the rotor. If magnetic saturation is neglected, the resultant airgap flux can be regarded as the combination of the flux that results from the dc-excited rotor field winding and that which results from the rotating armature mmf. The influence of the armature mmf on the magnitude and distribution of the resultant airgap flux is termed armature reaction.

It is convenient to regard the two component airgap fluxes as if they existed independently of each other. If the flux waves are assumed to be sinusoidally distributed in space around the airgap, they may be represented by space vectors whose directions show the positions of the maximum values or axes of the flux waves. The airgap flux component, Φ_f, due to the dc-excited field winding generates an excitation emf per phase, E_f. This emf is a maximum in any one phase when the conductors of that phase are opposite the centers of the rotor poles, that is, when the axis of the phase winding is displaced by 90 electrical degrees from the rotor field axis, and the flux linking the phase is zero. This means that the emf phasor, E_f, lags the flux phasor, Φ_f, by 90 degrees, as shown in Fig. 8.3(a).

On the other hand, the armature reaction flux wave, Φ_{ar}, due to the three-phase armature currents is centered on the axis of a particular stator phase when the current in that phase is a maximum. Assume the synchronous machine is operating as a generator and is supplying an armature line current, I_a, which is in phase with the line-to-neutral excitation emf, E_f. Then, at the instant of maximum emf in a particular phase, the current in

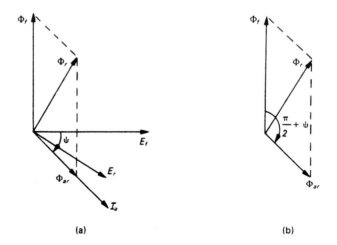

(a)   (b)

FIG. 8.3. Basic flux and voltage relationships in a round-rotor synchronous generator: (a) phasor diagram; (b) space vector diagram.

that phase is also a maximum and the armature reaction flux, Φ_{ar}, is centered on the phase axis. Consequently Φ_{ar} is displaced by 90 electrical degrees from the field flux, Φ_f. In general, if the armature current, I_a, lags the excitation emf, E_f, by an angle ψ, then the armature reaction flux, Φ_{ar}, is displaced in space by an angle $(\pi/2 + \psi)$ from the field flux, Φ_f. The phasor diagram of Fig. 8.3(a) shows the flux and emf phasors, and the space vector diagram for the airgap fluxes is shown in Fig. 8.3(b). The fluxes Φ_f and Φ_{ar} are combined to give the resultant mutual airgap flux, Φ_r, which generates the resultant armature emf, E_r, also known as the airgap voltage. The armature reaction flux phasor, Φ_{ar}, is always parallel to the current phasor, I_a, because the armature reaction flux linking a particular phase is a maximum when the phase current is a maximum. It is clear from Fig. 8.3 that the space vector diagram showing the relative positions of the airgap fluxes, Φ_f, Φ_{ar}, and Φ_r, can be combined with the phasor diagram showing the time variation of the various fluxes linking an armature phase. In future, a single diagram will be used.

When the synchronous machine operates as a motor, the ac supply voltage causes the armature current, I_a, to flow in opposition to the excitation, or back, emf, E_f. Using generator conventions, the phase angle between E_f and I_a exceeds 90 degrees, and the electrical power output is negative, indicating motor operation. However, it is more convenient to adopt motor conventions by reversing the direction of the I_a phasor. The internal power factor angle, ψ, between E_f and I_a is then less than 90 degrees and electrical power input is positive, but the armature reaction flux phasor, Φ_{ar}, must be drawn in antiphase with I_a, as shown in the phasor diagram of Fig. 8.4(a).

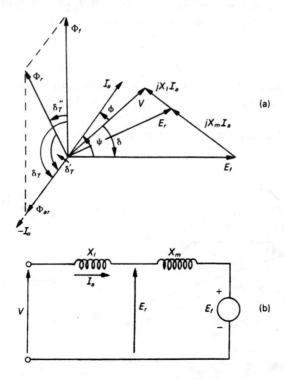

FIG. 8.4. Phasor diagram and equivalent circuit for a round-rotor synchronous motor.

Synchronous Reactance. As stated earlier, the armature reaction flux wave, Φ_{ar}, is aligned with the axis of a particular stator phase when the current in that phase is a maximum. Consequently, Φ_{ar} generates an emf in the phase winding that is proportional to the armature current, I_a, and which lags I_a by 90 degrees. Thus, the effect is the same as that of an inductive reactance, and the armature reaction flux produced by the stator currents may be regarded as being responsible for the per-phase magnetizing inductance, L_m, of the stator winding. The corresponding magnetizing reactance, $X_m = \omega_1 L_m$, where ω_1 is the electrical angular frequency, $2\pi f_1$. The phasor sum of the excitation emf, E_f, and $jX_m I_a$, the voltage drop across the magnetizing reactance, gives the resultant armature emf, E_r, as shown in the phasor diagram of Fig. 8.4(a). Stator leakage flux is also present and its effects are represented in the usual manner by a stator leakage reactance, X_l. The voltage drop, $jX_l I_a$, is added to E_r to give the terminal voltage, V, in Fig. 8.4(a). If the armature resistance voltage drop is significant, an $R_a I_a$ term should also be included. The corresponding equivalent circuit for one phase of the synchronous motor is shown in Fig. 8.4(b).

The sum of the magnetizing and leakage reactances defines X_s, the synchronous reactance per phase of the round-rotor machine. Thus,

$$X_s = X_m + X_l \tag{8.1}$$

and the phasor diagram and equivalent circuit can be simplified as shown in Fig. 8.5. The line-to-neutral voltage, V, is chosen as the reference phasor in Fig. 8.5(a), and the current phasor, I_a, which is perpendicular to the $jX_s I_a$ phasor, leads V by the power factor angle, ϕ. With the circuit conventions indicated in Fig. 8.5(b)

$$V = E_f + jX_s I_a . \tag{8.2}$$

Varying the field excitation of the synchronous motor varies the magnitude of E_f and thereby controls the motor power factor. If E_f is greater than V, the motor is said to be overexcited and operates at a leading power factor (I_a leading V), as shown in Fig. 8.5(a). The underexcited synchronous motor operates at a lagging power factor.

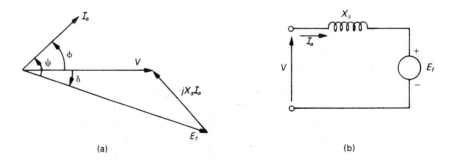

(a)   (b)

FIG. 8.5. Simplified phasor diagram and equivalent circuit for a round-rotor synchronous motor.

On no load, V and E_f are exactly in phase, but when a load torque is applied to the motor shaft, there is a momentary slowing down of the rotor, allowing the phase of the generated emf, E_f, to fall back in relation to the applied voltage, V. Thus, an angle δ appears between the E_f and V phasors, as shown in the phasor diagram. After a brief transient period, steady-state operation at synchronous speed resumes with a value of δ which is appropriate to the value of load power and torque. The angle δ is conventionally termed the load angle. (For generator operation, the rotor is momentarily accelerated by the driving torque of the prime mover, and E_f leads V by the load angle, δ.)

Torque Angle. Torque is developed in a synchronous motor by virtue of the tendency of the stator and rotor airgap fluxes, Φ_{ar} and Φ_f, to align their axes. The space phase angle between these two component airgap fluxes is known as the torque angle, δ_T, and is indicated in Fig. 8.4(a). By inspection

$$\delta_T = \pi/2 + \psi \tag{8.3}$$

where ψ is the internal phase angle by which I_a leads the excitation emf, E_f. (If I_a lags E_f, then ψ must be regarded as negative.)

When the armature current is in phase with E_f, ψ is zero, and δ_T equals 90 degrees. Thus, the axis of Φ_{ar} is spatially displaced by 90 degrees from the rotor field axis. For armature currents which lead or lag E_f by 90 degrees, the axis of Φ_{ar} is coincident with the rotor field axis. In a round-rotor machine, the airgap length is uniform and the armature reaction flux due to a given mmf is independent of the spatial orientation of the armature mmf wave relative to the rotor field poles. Consequently, the motor can be represented by a single value of synchronous reactance, X_s, as in Fig. 8.5.

In general, one can define a torque angle between any two of the three interacting flux waves, Φ_f, Φ_r, and Φ_{ar}. The torque angle, δ_T, between Φ_f and Φ_{ar} has been introduced above, but torque angles δ_T' and δ_T'' can also be defined, as indicated in Fig. 8.4(a). If armature resistance and leakage reactance are both negligible, the torque angle δ_T'' between Φ_f and Φ_r is, in fact, equal to the load angle δ.

Torque/Load-Angle Characteristic. The shaft torque can be expressed in terms of the load angle, δ. Thus, the electrical power input to the synchronous motor for all three phases is given by

$$P = 3\ VI_a \cos\phi. \tag{8.4}$$

An examination of the phasor diagram of Fig. 8.5(a) shows that

$$E_f \sin\delta = X_s I_a \cos\phi. \tag{8.5}$$

Substituting Equation 8.5 in Equation 8.4 gives

$$P = -\frac{3\ VE_f}{X_s} \sin\delta. \tag{8.6}$$

A negative sign is introduced into this equation because δ is negative for motor operation (E_f lagging V), but the electrical power input is positive. For a generator, δ is positive and the input power, P, is negative, indicating an electrical power output.

In a motor, electrical power is converted to mechanical power at the motor shaft. Neglecting stator losses, the airgap torque in newton-meters is obtained by dividing the electrical power input by the synchronous angular velocity, ω_s. Thus

$$T = \frac{P}{\omega_s} = \frac{P}{\omega_1/p} \tag{8.7}$$

where ω_1 is the electrical supply frequency and p is the number of pole-pairs. The useful shaft torque is somewhat less than the airgap torque due to friction and windage torques.

Substituting Equation 8.6 into Equation 8.7 gives

$$T = -\frac{3p}{\omega_1} \cdot \frac{VE_f}{X_s} \sin\delta . \tag{8.8}$$

This equation indicates that, for a fixed field excitation, the airgap torque varies as $\sin\delta$ and is a maximum when δ is 90 degrees. This maximum torque, T_{max}, is termed the pull-out torque because, if the load torque exceeds this value, the motor pulls out of synchronism and stalls. Thus

$$T_{max} = \frac{3p}{\omega_1} \cdot \frac{VE_f}{X_s} . \tag{8.9}$$

Equation 8.8 also defines the counter-torque developed by the synchronous machine when operating as a generator. Figure 8.6 shows the variation in developed torque as a function of δ with E_f/V as parameter. Maximum torque is developed when $\delta = \pm\pi/2$, and stable operation is not possible beyond the steady-state pull-out torque in either the motoring or generating modes. However, when powered by an electronic converter controlled from a shaft position sensor, the synchronous motor can operate in a self-synchronous mode and will not pull out of step. The machine is then approximately equivalent to a dc motor and, on overloads, the synchronous motor remains synchronized with the power converter but slows down until the load torque decreases or the motor stops. This self-synchronous mode of operation will be considered later.

8.3.2. Salient-Pole Machine

As explained above, the spatial alignment of the armature mmf wave with respect to the field poles is a function of the internal power factor angle between phase current, I_a, and generated emf, E_f. In a salient-pole machine, the armature reaction flux due to a given armature mmf is a function of the orientation of the mmf wave with respect to the saliency. Thus, the synchronous reactance of the salient-pole machine is a function of the mmf orientation relative to the rotor poles. This difficulty is overcome by adopting the two-axis theory that separates the effects in the polar or direct axis (or d-axis) and in the interpolar or quadrature axis (or q-axis).

In general, the armature current, I_a, can be resolved into two components, one in phase with, and the other in phase quadrature with, the generated emf, E_f, as shown in

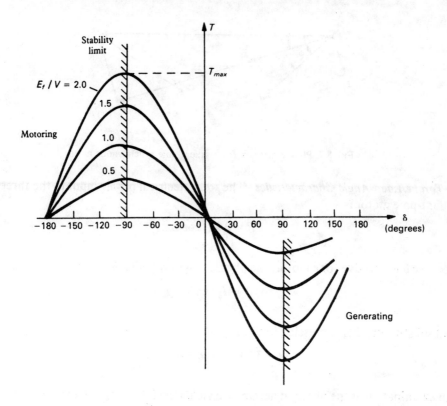

FIG. 8.6. Steady-state torque/load-angle curves for a round-rotor synchronous machine.

the phasor diagram of Fig. 8.7. The q-axis component of armature current, I_q, is in phase with E_f and establishes an armature reaction flux that is centered on the q-axis, or interpolar space between the field poles. The d-axis component of armature current, I_d, is in phase quadrature with the generated emf and establishes an armature reaction flux along the polar axis, or d-axis. Because of the different airgap reluctances in the d and q axes, a different synchronous reactance must be assigned to each axis. Thus, in place of the single synchronous reactance, X_s, of the round-rotor machine, the salient-pole machine has d-axis and q-axis synchronous reactances, X_d and X_q. The fundamental flux due to a given armature mmf is smaller in the q-axis than in the d-axis, and hence X_q is less than X_d. Typically, X_q is about 0.6 X_d. For motor operation, the phasor diagram is completed as shown in Fig. 8.7. As before, armature resistance is assumed negligible. Note that current component I_q establishes a q-axis airgap flux and hence is associated with the q-axis synchronous reactance, X_q, to give the reactance voltage phasor, jX_qI_q. Similarly, current component I_d is associated with X_d, to give the reactance voltage phasor, jX_dI_d. As in the case of the round-rotor machine, the motor power factor, $\cos\phi$, can be controlled by varying the field excitation. In Fig. 8.7, the motor is underexcited and operates with a lagging power factor.

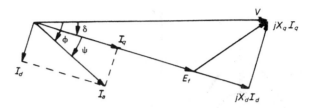

FIG. 8.7. Phasor diagram for a salient-pole synchronous motor.

Torque/Load-Angle Characteristics. The total electrical power input to the three-phase salient-pole motor is

$$P = 3 VI_a \cos \phi . \tag{8.10}$$

Because δ is negative for a motor, the phasor diagram of Fig. 8.7 gives the relationship

$$I_a \cos \phi = I_q \cos \delta + I_d \sin \delta \tag{8.11}$$

and substituting this expression in Equation 8.10 gives

$$P = 3 V(I_q \cos \delta + I_d \sin \delta) . \tag{8.12}$$

An examination of the phasor diagram also yields the following two expressions:

$$V \cos \delta = E_f + X_d I_d \tag{8.13}$$

$$- V \sin \delta = X_q I_q . \tag{8.14}$$

Solving Equations 8.13 and 8.14 for I_d and I_q and substituting in Equation 8.12 results in the following expression for input power:

$$P = -3 \frac{VE_f}{X_d} \sin \delta - \frac{3}{2} V^2 \left(\frac{X_d - X_q}{X_d X_q} \right) \sin 2\delta . \tag{8.15}$$

Neglecting stator losses, the airgap torque in newton-meters is again equal to the input power divided by the mechanical speed in radians/second. Thus

$$T = - \frac{3p}{\omega_1} \left[\frac{VE_f}{X_d} \sin \delta + V^2 \left(\frac{X_d - X_q}{2 X_d X_q} \right) \sin 2\delta \right] . \tag{8.16}$$

The first term in Equation 8.16 is the same as that obtained for the round-rotor machine. The second term is independent of field excitation and represents the reluctance torque, due to the tendency for the salient poles to align themselves with the rotating airgap flux in the position of minimum reluctance. In a round-rotor machine, $X_d = X_q = X_s$ and the reluctance torque term vanishes. The torque/load-angle characteristics for a salient-pole motor are shown in Fig. 8.8.

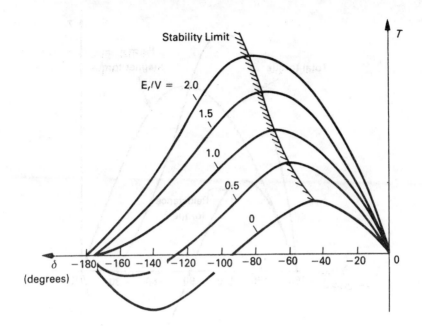

FIG. 8.8. Steady-state torque/load-angle curves for a salient-pole synchronous machine in the motoring region.

8.3.3. *Permanent Magnet Machine*

In a permanent magnet machine operating on a fixed-frequency ac supply, the constant rotor flux produced by the permanent magnets generates a constant value of excitation emf, E_f. The actual value of excitation emf depends on the magnet material, its physical dimensions, the rotor design, and the airgap length. The machine designer must ensure that the magnets are not demagnetized by the application of normal or overload currents. The earlier torque equation for the salient-pole motor, Equation 8.16, can be used for permanent magnet machines if the generated emf, E_f, is set at a constant value. The single value of E_f implies that there is only one torque versus load angle curve rather than a family of curves for different values of E_f/V. Additional torque is produced by the rotor saliency, but in a permanent magnet machine, the direct axis reactance may be less than the quadrature axis reactance. Consequently, the reluctance torque term is opposite in sign to that in the normal salient-pole motor, and it subtracts from the synchronous torque at low values of load angle. The torque versus load angle curve is shown in Fig. 8.9. The stability limit is located in the region from −90 to −135 degrees; the precise value is dependent on the ratio of direct-to-quadrature axis reactance for a particular machine, because this ratio determines the reluctance torque component. If the reluctance torque is comparable to the permanent magnet torque, the composite torque curve of Fig. 8.9 can be negative for small values of load angle.

The phasor diagram of the permanent magnet motor is the same as that for the salient-pole machine (Fig. 8.7). The generated emf, E_f, is constant and has a value such that the motor operates near unity power factor at rated load.

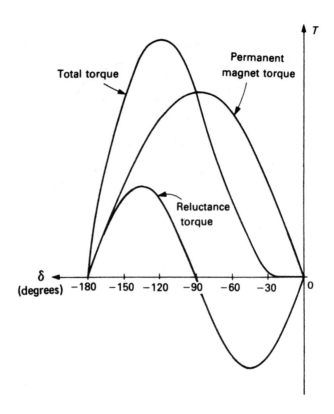

FIG. 8.9. Steady-state torque/load-angle curve for a permanent magnet synchronous machine in the motoring region.

8.3.4. *Synchronous Reluctance Machine*

As described in Section 8.2.4, the synchronous reluctance motor is a salient-pole machine having no source of rotor field excitation. Consequently, there is no generated emf, E_f, and the torque equation is obtained by setting E_f equal to zero in Equation 8.16. Thus,

$$T = - \frac{3p}{2\omega_1} \left(\frac{X_d - X_q}{X_d X_q} \right) V^2 \sin 2\delta . \qquad (8.17)$$

The torque/load-angle curve is shown in Fig. 8.10. The pull-out torque is obtained when $\sin 2\delta$ is unity. Thus, the steady-state stability limit is -45 degrees, and the pull-out torque is

$$T_{max} = \frac{3p}{2\omega_1} \left(\frac{X_d - X_q}{X_d X_q} \right) V^2 . \qquad (8.18)$$

For good torque capability, a high X_d/X_q ratio is required.

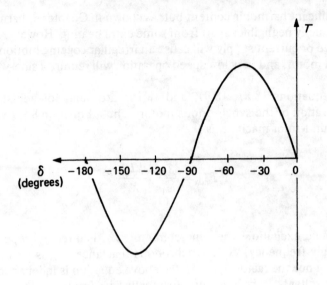

FIG. 8.10. Steady-state torque/load-angle curve for a synchronous reluctance machine in the motoring region.

The phasor diagram is shown in Fig. 8.11 and is obtained by setting E_f equal to zero in Fig. 8.7. Power factor control is now clearly impossible, and the motor operates at a lagging power factor because of the fact that magnetizing current must be drawn from the ac supply.

8.4. ADJUSTABLE-FREQUENCY OPERATION

Motor speed is precisely related to the frequency of the ac supply by the equation $n_1 = 60 f_1/p$, where n_1 is in revolutions per minute. Synchronous motors can therefore provide accurate speed control, particularly when the adjustable-frequency reference is crystal-controlled for stability and freedom from drift. The adjustable-frequency supply

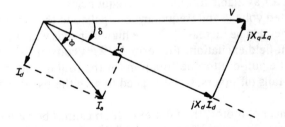

FIG. 8.11. Phasor diagram for a synchronous reluctance motor.

may have a significant harmonic content but, as shown in Chapter 6, harmonic effects in the motor are usually negligible, apart from some extra heating. However, at low speeds, a six-step voltage or current supply will cause an irregular cogging motion, as in the case of the induction motor, and very low speed operation will require a sine-wave PWM supply.

The torque equations (8.8), (8.9), and (8.16) are valid for adjustable-frequency voltage-fed operation of the synchronous motor. Thus, Equation 8.9 gives the pull-out torque of the round-rotor machine as

$$T_{max} = \frac{3p}{\omega_1} \cdot \frac{VE_f}{X_s} .$$

For a fixed dc field excitation, the generated emf, E_f, is directly proportional to shaft speed (and supply frequency). The synchronous reactance, X_s, is also proportional to frequency, and hence the factor E_f/X_s in the above equation is independent of frequency. If the supply voltage, V, is varied proportionally with frequency in the usual constant volts/hertz mode of operation, the term V/ω_1 is also constant. Hence, the pull-out torque of the motor is constant and the torque/load-angle characteristics of Fig. 8.6 are also independent of frequency. This is the usual constant-torque region of operation below base speed. The available shaft power will, of course, increase linearly with speed. These conclusions also apply to the salient-pole wound-field, permanent magnet, and synchronous reluctance motors.

In a synchronous motor, the terminal voltage, V, is approximately equal to the airgap voltage, E_r, assuming the voltage drop across the stator leakage impedance is small. As shown earlier, the emf, E_r, is generated by the resultant mutual airgap flux, Φ_r, and therefore constant applied volts/hertz (V/f_1) implies an approximately constant value of E_r/f_1 and a constant airgap flux, Φ_r. The airgap flux component, Φ_f, due to the rotor field winding is determined by the field excitation and may be significantly greater than Φ_r.

It must be remembered that armature resistance effects have been neglected in the derivation of the various synchronous motor torque equations. With a constant volts/hertz supply, armature resistance plays an important role at low frequencies because the resistive voltage drop is comparable in magnitude to the terminal voltage. As in the case of an induction motor on a constant volts/hertz supply, stator resistance will cause a reduction in the available torque at low frequencies.

At base speed, rated voltage and frequency are applied to the synchronous motor, but the inverter frequency can be increased to give higher speeds with a constant motor voltage, V. For constant field excitation, the term E_f/X_s is constant above base speed, and an examination of the pull-out torque equation for the round-rotor machine shows that the pull-out torque falls off inversely with speed to give the usual constant-horsepower region of operation.

In a permanent magnet motor, the field excitation cannot be varied, but in a wound-field machine, one has the option of varying the field current to give greater flexibility in drive characteristics. Closed-loop control techniques for the wound-field and permanent magnet machines are discussed later in this chapter.

8.4.1. *Controlled-Current Operation*

The discussion to date has assumed that the synchronous motor is supplied from a voltage source, but the power electronic converter may deliver a controlled level of stator current with the supply voltage adjusting itself automatically to suit this condition. As in the case of the induction motor, direct control of motor current gives direct torque control and fast response.

For controlled-current operation, the phasor diagram of the synchronous motor is unchanged, but it is now desirable to express motor torque in terms of stator current rather than stator voltage. For the round-rotor machine, it is clear from Fig. 8.5 that the total input power can be expressed as

$$P = 3 E_f I_a \cos \psi . \tag{8.19}$$

As explained in Section 8.3.1, the rotating armature reaction flux generates an emf in the stator winding of magnitude $X_m I_a$, where X_m is the magnetizing reactance. If the actual dc field current is referred to the armature winding and expressed as an equivalent armature phase current I_f', then

$$E_f = X_m I_f' = \omega_1 L_m I_f' \tag{8.20}$$

and substituting in Equation 8.19 gives

$$P = 3 \omega_1 L_m I_f' I_a \cos \psi . \tag{8.21}$$

As usual, the torque in newton-meters is given by

$$T = \frac{P}{\omega_s} = \frac{P}{\omega_1 / p}$$

and hence.

$$T = 3 p L_m I_f' I_a \cos \psi . \tag{8.22}$$

But the torque angle $\delta_T = \pi/2 + \psi$, and therefore

$$T = 3 p L_m I_f' I_a \sin \delta_T . \tag{8.23}$$

This equation indicates that torque is independent of frequency and speed, and is directly proportional to the product of field current, stator current, and the sine of the torque angle. Thus, if the synchronous motor is controlled so that δ_T is constant and field current is also constant, then a constant-torque load will require the same armature current, I_a, at all speeds.

The main field flux, Φ_f, is proportional to I_f', and consequently, Equation 8.23 can be written as

$$T = K \Phi_f I_a \sin \delta_T \tag{8.24}$$

where K is a constant. This equation shows that torque is proportional to the product of rotor flux, Φ_f, armature current, I_a, and the sine of the torque angle between rotor flux and armature mmf. However, Fig. 8.4(a) shows that

$$\Phi_f \sin \delta_T = \Phi_r \sin \delta_T' \qquad (8.25)$$

and substituting Equation 8.25 in Equation 8.24 gives an alternative form of the torque equation:

$$T = K\Phi_r I_a \sin \delta_T' . \qquad (8.26)$$

Thus, torque is also proportional to the product of the resultant airgap flux, Φ_r, armature current, I_a, and the sine of the torque angle between resultant flux and armature mmf.

Equation 8.24 shows that if Φ_f is constant, as in a permanent magnet machine, the torque per stator ampere is maximized when δ_T is 90 degrees. The developed torque is then proportional to the product of the field flux, Φ_f, and the armature current, I_a, as in a dc machine. Equation 8.26 shows that if Φ_r is maintained constant by field current control of a wound-field machine, the torque angle δ_T' should be 90 degrees to give maximum torque per stator ampere.

Parameter changes due to saturation effects are likely in a current-fed machine because the airgap flux level may vary widely. The resulting variation in L_m must be taken into account when Equation 8.23 is used to evaluate torque. This equation indicates that maximum torque occurs for a torque angle, δ_T, of 90 electrical degrees. In practice, however, due to the influence of saturation on the value of L_m, the torque angle is greater than 90 degrees at the maximum torque condition.[15,16]

8.5. PRINCIPLES OF SYNCHRONOUS MOTOR CONTROL

The converter-fed synchronous motor can be operated under open-loop frequency control or closed-loop self-synchronous control. Voltage-source inverters (or cycloconverters) are used for open-loop frequency control with the inverter frequency controlled by an independent oscillator to give the desired speed. When the inverter firing signals are at a fixed frequency, the synchronous motor operates in the conventional manner: motor speed is constant; load angle, δ, increases with load torque until the pull-out torque is exceeded; and the machine has a tendency to hunt or oscillate for sudden load changes.

In the self-synchronous mode of operation, the inverter frequency is slaved to the motor speed by means of a shaft position sensor from which the inverter firing signals are derived. The synchronous motor and inverter are then equivalent to a dc motor in which the mechanical commutator has been replaced by an electronic commutator, with obvious advantages. The resulting synchronous motor drive is endowed with the flexible adjustable-speed characteristics of the dc brush motor. The torque angle, δ_T, can be set at a constant value that is independent of load torque. Motor speed can then vary with load, as in a conventional dc motor, but the machine can never pull out of step because stator frequency changes automatically with speed. The self-controlled synchronous motor is also known as a dc commutatorless motor (CLM) or electronically commutated motor (ECM).

8.5.1. *Voltage-Source Inverter Drive with Open-Loop Control*

A block diagram of an open-loop synchronous motor drive system with a six-step voltage-source inverter is shown in Fig. 8.12. This system is appropriate for multiple synchronous reluctance or permanent magnet synchronous motors operating from a single voltage-source inverter. These multimotor drives are useful where exact speed synchronism between motors is important as, for example, in textile machinery or in roller drives for a run-out table. If the inverter frequency is determined by a high-stability reference oscillator, precise shaft speeds are obtained which are independent of motor loading. The inverter must also be capable of implementing output voltage control with one of the techniques described in Chapter 4. The voltage control can be open loop as shown in Fig. 8.12, but closed-loop voltage control is also possible with either the inverter output voltage or the dc link voltage as a feedback signal.

For constant-torque applications, the motor voltage is varied proportionally with frequency in the usual constant volts/hertz mode of operation. Under certain conditions of light load and low speed, oscillations in motor speed above and below a mean operating value may occur, particularly in the case of synchronous reluctance motors. If the inverter voltage controller has sufficient bandwidth, the speed oscillations can be compensated by rapid adjustment of the motor voltage. In other cases, the motor voltage may be boosted above the nominal volts/hertz value to reduce the oscillations.

Permanent magnet motors have replaced synchronous reluctance motors in some multiple motor applications to give higher efficiency and improved power factor, but the permanent magnet materials are costly. In general, open-loop frequency control is not suitable for applications with high dynamic performance requirements because the applied frequency must be varied slowly to avoid the risk of pull-out. However, because the self-controlled synchronous motor behaves like a dc motor, a fast dynamic response is feasible with similar closed-loop control techniques. The self-controlled drive system will now be discussed.

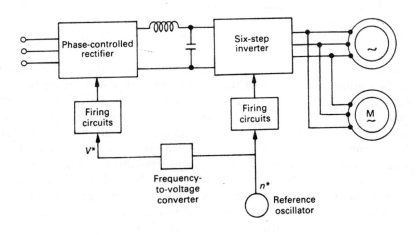

FIG. 8.12. Open-loop drive system for multiple synchronous motors.

8.6. THE SELF-CONTROLLED SYNCHRONOUS MOTOR

In a self-controlled synchronous motor drive, the motor is only a synchronous machine as regards construction, because its mode of operation is like that of a dc motor with an electronic commutator.[17-20]

8.6.1. *Electronic Commutation*

In a conventional dc motor, the armature winding consists of a number of coils distributed around the armature surface and interconnected to form a closed winding. Each interconnection, or tapping, is joined to a segment of the rotating commutator. Current is supplied to the armature winding by stationary carbon brushes which slide on the commutator surface. As a result of this construction, the armature magnetic field is stationary, with its axis in a direction determined by brush position. Stator poles which carry a dc-excited field winding also establish a stationary magnetic field in the airgap. The motor torque is due to the tendency of the armature field to align itself with the main field, and maximum torque is developed when the two fields have a space phase displacement of 90 electrical degrees. The provision of a large number of commutator segments results in a unidirectional armature field that reacts with the main field to produce a steady output torque having negligible ripple. In an electronic commutator, the number of "segments" is reduced, causing the output torque to pulsate above and below its mean value.

When a segment of the mechanical commutator passes under a brush, the corresponding armature tapping is connected to the positive or negative supply terminal. In order to reproduce this behavior in an electronic commutator, each commutator segment is replaced by a pair of power semiconductor devices. One of these devices permits current flow from the positive rail to the armature tapping, and the other device allows current flow from the tapping to the negative rail. The mechanical commutator automatically switches each coil at the correct rotor position, but the switching of the semiconductors must be controlled by a position transducer on the rotor shaft. This shaft-controlled switching ensures that the switching frequency is proportional to the rotor speed and maintains the armature field in space quadrature with the field poles.

When a mechanical commutator is replaced by its electronic counterpart, the number of switching operations is reduced in order to limit the number of semiconductors and improve their utilization. Figure 8.13(a) shows an elementary dc machine with an armature ring winding and a three-segment commutator. Figure 8.14(a) shows the equivalent semiconductor arrangement in which each of the three armature tappings is connected by a pair of thyristors to the positive and negative rails, giving the familiar three-phase bridge converter. This thyristor circuit can utilize natural (or load) commutation because of the generated emfs appearing at the armature slip-rings. If the voltages appearing between armature tappings have a sinusoidal variation with time, the three-phase waveforms of Fig. 8.13(b) and Fig. 8.14(b) are obtained for motor operation. These voltage waveforms show the close correspondence between the two systems. In Fig. 8.13(b) for the mechanical commutator, the brushes are shifted from the neutral axis against the direction of rotation and the brush width, w, is taken into consideration. Figure 8.14(b) is obtained for the electronic commutator with the circuit analysis techniques for the naturally commutated three-phase bridge circuit introduced in Chapter 3.

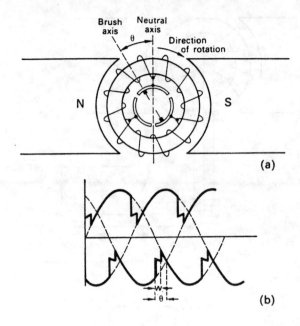

FIG. 8.13. Elementary dc motor with three commutator segments: (a) basic construction; (b) brush voltage waveform.

This diagram shows the voltage waveforms for the three-phase bridge converter operating as a load-commutated inverter with the firing point advanced by an angle, β, to allow sufficient margin for commutation overlap and turn-off.

A large number of commutator segments is required in the conventional dc machine because the voltage between segments is limited by the intersegment insulation. This consideration does not arise in the electronic commutator, and satisfactory operation is obtained with a limited number of commutations per revolution. The commutator in a conventional dc motor is placed on the rotor to avoid the need for rotating brushes, but the construction is inverted when an electronic commutator is employed. The field winding and a pair of slip-rings are then placed on the rotor, and the armature slip-rings of Fig. 8.14(a) are unnecessary. If the armature ring winding is replaced by a phase-wound armature, the machine has the same construction as a three-phase synchronous motor of normal manufacture. Torque pulsations may be troublesome at low speeds, but their effect can sometimes be reduced by fitting a substantial damper winding, which also improves the steady-state and transient characteristics of the motor.[21,22]

As explained above, a three-phase synchronous motor supplied by a bridge inverter is equivalent to a simple dc motor. Thyristors are triggered in the correct sequence by a position transducer on the rotor shaft, and are load commutated by the generated or back emfs of the synchronous motor. As shown in Fig. 8.14(b), the incoming thyristor is fired at an advance angle, β, to ensure that the margin angle, γ, available after overlap is adequate for thyristor turn-off. Increasing the firing advance, β, is equivalent to shifting the brushes against the direction of rotation so that current starts flowing somewhat earlier

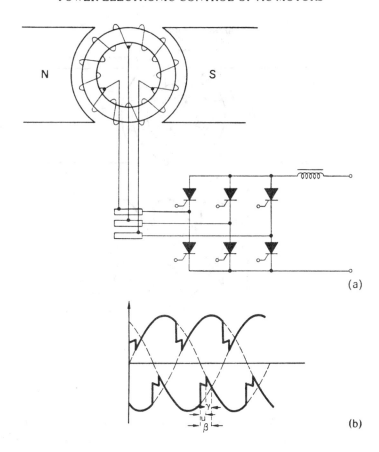

(a)

(b)

FIG. 8.14. Replacement of mechanical commutator of Fig. 8.13 by a thyristor bridge circuit.

than before in an armature phase. This firing advance implies leading power factor operation of the synchronous motor because a phase advance in the thyristor conduction interval results in a corresponding phase advance of the motor line current relative to the motor voltage. Thus, for load commutation of the inverter, the synchronous motor must be overexcited to give the required leading power factor operation for all steady-state conditions (for example, no load, full load, motoring, regenerating) and also during transient changes in torque and speed.

In operation, the self-controlled synchronous motor behaves like a dc motor but has none of the disadvantages of mechanical commutation. The nominal synchronous speed of the motor on a 50 or 60 Hz ac supply is now meaningless, and the upper speed limit is determined by the mechanical limitations of the rotor construction. There is the disadvantage that the rotational emfs required for load commutation of the inverter are not available at standstill and low speeds, but auxiliary forced commutating circuits have been used for these operating conditions.[17,23] Other forms of commutatorless dc motor use forced commutation methods at all speeds; and, in fact, any synchronous motor fed by an adjustable-frequency dc link converter can be operated as a commutatorless dc motor by controlling the inverter frequency from a position transducer on the motor shaft. In Fig. 8.15, the synchronous motor and its inverter can be regarded as a commutatorless dc motor having the characteristics of a dc motor with a mechanical commutator. Thus, the synchronous motor speed is proportional to the direct voltage input to the inverter or electronic commutator just as the speed of the conventional dc motor is proportional to the direct voltage applied to the commutator brushes. Adjustable-speed operation of the commutatorless motor shown in Fig. 8.15 is therefore obtained by variation of the rectifier delay angle, and the no-load speed of the motor adjusts itself to make the armature-generated emf equal to the applied dc link voltage. When load torque is applied to the motor shaft, there is an increase in dc link current; the resulting IR drop in the motor causes a slight drop in shaft speed. Thus, the torque-speed characteristics are like those of a dc shunt motor. Shaft speed also varies approximately as the inverse of the field current.

A cycloconverter, whose output frequency is controlled from the shaft of a synchronous motor, also constitutes a form of commutatorless dc motor in which there is no obvious dc link. Such drives are used for low-speed applications at power ratings up to several megawatts. When powered by a dc link converter, as in Fig. 8.15, or by a cycloconverter, the synchronous motor is equivalent to a dc shunt motor and is termed a shunt commutatorless motor.[24] A series commutatorless motor having the operating characteristics of a dc series motor is obtained by supplying the field winding from a bridge rectifier that is in series with the armature winding; or alternatively, a doubly fed wound-rotor induction motor may be used.[25] A synchronous motor drive of this type, described as a "thyratron motor," was constructed with thyratrons before the development of the thyristor.[26] A 400-horsepower scheme was in operation for many years in the United States as a power station auxiliary drive.[1,27] The increased reliability and efficiency of thyristor circuits have improved the performance of such systems, and they are now practical.[28–32]

For motor ratings below about 25 horsepower, the modern electronically commutated synchronous motor normally utilizes a permanent magnet rotor. In the conventional machine, the magnets are designed to establish an airgap flux wave with a sinusoidal spatial distribution. The permanent magnet motor is used in high-performance servo drives; appropriate control techniques are discussed later in this chapter. The rotor mag-

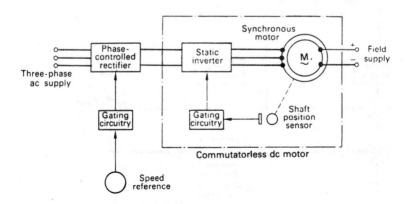

FIG. 8.15. Commutatorless dc motor fed by a phase-controlled rectifier.

nets can also be configured to establish a trapezoidal rather than a sinusoidal flux density wave in the airgap. This trapezoidal flux machine, the so-called "brushless dc motor," is also widely used in servo applications. It is discussed in Chapter 11.

8.7. THE CURRENT-FED, SELF-CONTROLLED SYNCHRONOUS MOTOR

In the earlier discussion of the operating characteristics of the commutatorless dc motor shown in Fig. 8.15, voltage-source operation was assumed. However, it is advantageous to supply the motor with a controlled stator current from a current-fed dc link converter or a current-controlled PWM inverter. This approach gives direct torque control of the commutatorless motor and is analogous to the use of armature current control, rather than armature voltage control, in a converter-fed dc motor drive. The torque equations for controlled-current operation of the synchronous motor have already been derived in Section 8.4.1. The present section considers some practical converter configurations.

8.7.1. *The LCI-Synchronous Motor Drive*

At high power levels, the most common power converter configuration is the current-fed dc link converter of Fig. 8.16. The phase-controlled thyristor rectifier on the supply side of the dc link has a current regulating loop and operates as a controlled-current source. The regulated dc current is delivered through the dc link inductor to the thyristor load-commutated inverter (LCI) which supplies quasi-square-wave line currents to the synchronous motor. As usual, the inverter gating signals are under the control of the shaft position sensor, giving a commutatorless dc motor with armature current control. As already explained, load commutation is ensured by overexcitation of the synchronous motor so that it operates at a leading power factor. The elimination of forced com-

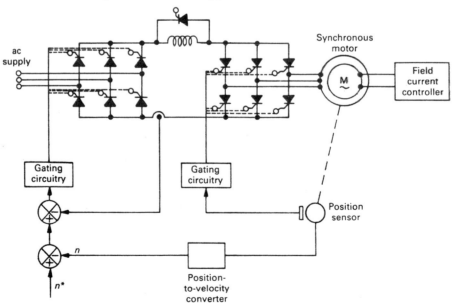

FIG. 8.16. The load-commutated inverter (LCI) synchronous motor drive.

mutating circuitry means fewer components and lower losses and is particularly attractive at high power levels. Consequently, this type of self-controlled LCI-synchronous motor drive is used up to power levels of several megawatts. Various industrial applications have been described in Section 4.9 of Chapter 4.

The drive characteristics are similar to those of a conventional dc motor drive. Motor speed can be increased up to a certain base speed corresponding to the maximum voltage available from the supply-side converter. A further increase in speed is obtained by reducing the field current to give a field weakening region of operation. Regenerative braking is accomplished by shifting the gating signals so that the machine-side converter acts as a rectifier and the supply-side converter functions as an inverter. As explained in Chapter 4, the dc link voltage polarity is reversed, but dc link current continues to flow in the same direction as before, and power is returned to the ac utility network. The direction of motor rotation is also readily reversed by altering the gating sequence of the machine-side converter. Thus, four-quadrant operation is achieved without additional power circuitry.

The generated emfs required for load commutation of the machine-side converter are not available at standstill and low speeds. The inverter can be commutated by pulsing on and off the dc link current. This technique produces large pulsating torques, but many drive applications do not require smooth torque at low speeds. The dc link current is pulsed by phase-shifting the gating signals of the supply-side converter from rectification to inversion and back again. When the current is zero, the machine-side converter is switched to a new conduction pattern and the supply-side converter is then turned on again. The time required for the motor current to fall to zero can be significantly shortened by placing a shunt thyristor in parallel with the dc link inductor, as shown in Fig. 8.16.[33,34] When a current zero is demanded, the line-side converter is phased back to inversion and the auxiliary thyristor is also gated. The dc link inductor is then short-circuited and its current can circulate freely without affecting the motor. When the line-side converter is turned on, the auxiliary thyristor is quickly blocked. This rapid interruption of motor current reduces the effects of the pulsating torques. At about 10 percent of the nominal motor speed, a transition can be made to the normal load-commutated mode of operation. Of course, the starting problem can be avoided by replacing the load-commutated inverter by a forced-commutated current-source inverter. The CSI employs forced commutation at all speeds and will therefore allow synchronous motor operation at unity power factor.

Figure 8.16 also shows a simplified drive control strategy in which the current regulator is the inner loop for an outer speed regulating loop, as in a standard converter-fed dc motor drive. The shaft speed signal is obtained by conversion of the position information. More sophisticated drive control techniques may employ field current control and terminal voltage or airgap flux sensing.[35–47] Examples of these techniques are described in later sections. Approximate transfer function models of the LCI drive have been developed to facilitate the design of the closed-loop control system.[48,49] In modern drives, the control implementation is frequently microprocessor-based.

8.7.2. *The Cycloconverter-Fed Synchronous Motor Drive*

A synchronous motor powered by a cycloconverter can be used as a large low-speed reversing drive with a fast dynamic response.[28,30,31,35] The naturally commutated cycloconverter delivers high-quality sine wave currents at low output frequencies; consequently, the low-speed performance is much superior to that of the LCI-synchronous motor drive with its intermittent dc link current and pulsating torque at standstill and low speeds. The cycloconverter can function as either a voltage- or current-source supply, and it also permits regeneration to the ac supply network, so that four-quadrant operation is available with a smooth transition through zero speed. The speed range is limited by the usual restriction on the maximum usable ratio of cycloconverter output frequency to input frequency. Large cycloconverter drives are used in mine hoists, reversing rolling mills, and low-speed gearless mill drives.

8.7.3. *The PWM Inverter-Fed Synchronous Motor Drive*

If the self-controlled synchronous motor is supplied with balanced sinusoidal currents by a current-controlled PWM transistor inverter, a smooth, nonpulsating torque is developed, and direct torque control is possible down to standstill. When the synchronous motor uses permanent magnet excitation, the resulting brushless drive has all the properties required for servo applications in machine tool feed drives and in industrial robots, as well as in spindle drives. Thus, the drive exhibits a large torque/inertia ratio, a high peak torque capability for fast acceleration and deceleration, insignificant torque ripple at low speeds, and high torsional stiffness at standstill with zero deadband. Most servo applications are in the power range below 10 kW, where power transistors can be used at switching frequencies of 10 kHz, or more, to give accurate current tracking with negligible control delay. Operation outside the audible range of frequencies is also possible, so that high-frequency noise from the motor and inverter can be eliminated.

The PWM inverter is controlled so as to force the phase currents of the machine to track the sinusoidal reference currents generated from the rotor position sensor. As explained in Chapters 4 and 7, the actual current is compared with the reference current and the error controls the inverter switching. Current control can be achieved by means of an independent hysteresis comparator for each phase, or alternatively, a PWM controller can be employed in which the current error for each phase is compared with a common high-frequency triangular wave. The relative merits of the two methods were discussed in Chapter 4, and both of these techniques, and others, have been used in practical synchronous motor drives.[35,50-57]

When the counter emf of the permanent magnet motor approaches the available dc link voltage, the PWM inverter loses its ability to track the reference current, and the inverter operation will then transition to the six-step mode of operation. However, the speed range can be extended by using armature reaction to introduce a field weakening effect, as discussed later.

8.8. TORQUE ANGLE CONTROL OF THE SELF-CONTROLLED SYNCHRONOUS MOTOR

In a conventional dc motor, the torque per ampere is maximized by maintaining a torque angle of 90 electrical degrees between the field, or polar, axis and the armature mmf axis. In a self-controlled synchronous motor, the inverter firing can be controlled so as to emulate the orthogonal spatial orientation of flux and mmf in the dc machine.

The corresponding phasor diagram for a round-rotor machine is shown in Fig. 8.17(a), where, as usual, only fundamental voltage and current components are considered. A torque angle, δ_T, of 90 degrees implies that the internal power factor angle between E_f

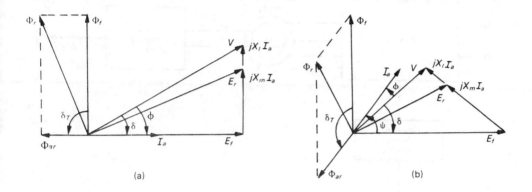

FIG. 8.17. Phasor diagrams for the round-rotor synchronous motor at (a) unity internal power factor ($\delta_T = 90$ degrees) and (b) leading internal power factor ($\delta_T > 90$ degrees).

and I_a is zero, but there is clearly a lagging power factor at the motor terminals which is unacceptable if load commutation of a thyristor inverter is required. In order to create a leading power factor at the terminals, armature current, I_a, must lead voltage, E_f, by a relatively large angle of 40 degrees or more, as shown in Fig. 8.17(b). Consequently, the torque angle, δ_T, must be appreciably greater than 90 degrees. However, a forced-commutated CSI or PWM inverter will allow motor operation with a torque angle of 90 degrees or with a terminal power factor of unity.

The shaft position sensor locates the direct rotor axis and hence locates the Φ_f vector. Using the shaft position information, the inverter firing signals can be adjusted to set the torque angle, δ_T, between Φ_f and Φ_{ar}. As the motor is loaded, the armature current, I_a, increases and establishes an increased armature reaction flux, Φ_{ar}, as shown in Fig. 8.17(b). The resultant airgap flux, Φ_r, is therefore significantly phase advanced relative to Φ_f, and likewise E_r and V are phase advanced with respect to E_f. Thus, although the phase angle, ψ, between I_a and E_f remains unaltered, the phase angle, ϕ, between I_a and V can vary widely as I_a changes in magnitude, and the power factor may even change from leading to lagging. In a load-commutated inverter, with a fixed setting of the shaft position sensor, a commutation failure can occur due to this phase advance of the resultant armature emf, E_r. Commutation overlap also introduces a load-dependent variation in the phase angle, ψ, between the fundamental component of the motor current, I_a, and the generated emf E_f. Consequently, phase shifting of the position sensor signals is needed to maintain a constant torque angle or constant terminal power factor.

In the simplified diagrams of Figs. 8.15 and 8.16, the shaft encoder signals are shown directly controlling the inverter firing, on the assumption that the same firing pulse timing is appropriate for all load torques and speeds. In practice, for the reasons indicated

above, the signals from the position sensor must be phase shifted as a function of armature current, field current, and speed in order to maintain the desired operating conditions. In Fig. 8.18, the position sensor signals are phase shifted by a controlled amount as

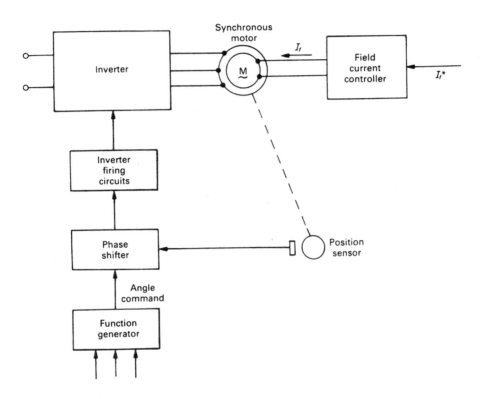

FIG. 8.18. Block diagram for torque angle control by means of a shaft position sensor.

determined by the angle command signal. The output signals from the phase shifter
block are then fed directly to the inverter firing circuits so that the inverter devices are
triggered at selected rotor positions, and the torque angle between the rotor field flux and
armature mmf is appropriately controlled.

Shaft position sensing can be achieved with Hall-effect or optical sensors, but digital
encoders are becoming increasingly popular. The incremental encoder delivers a large
number of pulses, up to several thousand, for a single shaft revolution. Position information is obtained by counting the number of pulses from the initial reference position.
The absolute encoder is more expensive than the incremental encoder, but it gives a
unique digital output signal for each shaft position. Consequently, position information
is available at standstill, whereas the incremental encoder requires an initialization procedure to locate a once-per-revolution index mark. The brushless resolver also gives position information in absolute terms; its operation is based on inductive coupling between stator and rotor windings. High-frequency carrier-type signal techniques are used
to obtain suitable analog signals that are decoded by a resolver-to-digital converter to

give a digital position signal. Electronic circuits can process the position information available from encoders or resolvers to give a velocity signal for closed-loop feedback.

The phasor diagrams of Fig. 8.17 and the accompanying discussion are also valid for self-controlled operation of the permanent magnet synchronous motor, and the control of a permanent magnet machine is similar to the control of a wound-field machine except that the field excitation cannot be varied. Servo control of the permanent magnet motor is discussed in Section 8.10.

8.9. CLOSED-LOOP CONTROL OF THE WOUND-FIELD SYNCHRONOUS MOTOR

In a wound-field machine, it is possible to regulate the field current so that the motor operates with the resultant airgap flux, Φ_r, equal to the maximum permissible value for all loads and speeds. Airgap flux control avoids the extremes of magnetic saturation or underutilization of the iron of the machine. If the armature current, field current, and torque angle are controllable, the power factor at the motor terminals can be controlled. Thus, leading power factor operation can be assured to accomplish reliable load commutation of a thyristor inverter. Alternatively, unity power factor operation can be achieved with a forced-commutated inverter.

In a load-commutated inverter, the thyristors are naturally commutated by the machine voltages and, as discussed in Chapter 3, the margin angle, or extinction angle, defines the interval during which a reverse bias is applied across the outgoing thyristor. A commutation failure will occur if the margin angle becomes so small that the turn-off time provided is less than the thyristor recovery time. As indicated in Fig. 8.14(b), the margin angle $\gamma = \beta - u$, where β is the angle of firing advance and u is the overlap angle. As already explained, when the load on the synchronous motor is increased, armature reaction advances the phase of the machine terminal voltage and reduces its amplitude. Thus, the effective advance angle, β, of the inverter is reduced, and the overlap angle, u, increases. As a result, the margin angle, γ, decreases and a commutation failure may occur.

In simple LCI-synchronous motor drives, the inverter advance angle has a constant value that is large enough to ensure safe commutation for the highest value of motor current. The margin angle is then at the minimum safe value, and the motor is operating at the highest possible power factor, yielding maximum power for a given current. However, on light loads, the armature reaction effect is reduced and the overlap angle decreases. Consequently, the margin angle is unnecessarily large, resulting in a poor power factor, low torque per ampere, and unnecessarily high losses. These disadvantages can be overcome if the angle command signal (Fig. 8.18) implements a load-dependent phase shift by means of a nonlinear function generator that senses armature current, field current, and speed, and then computes the required phase shift. Alternatively, the motor power factor or margin angle can be controlled in a closed-loop system.

8.9.1. *Motor Power Factor Control*

Shaft position sensing can be used in conjunction with voltage and current sensing to allow closed-loop control of motor power factor in forced-commutated or load-commutated inverter drives. Figure 8.19 shows a block diagram of a voltage-fed, self-controlled synchronous motor drive in which the phase relationship between motor voltage and current is automatically varied so that the motor power factor is maintained con-

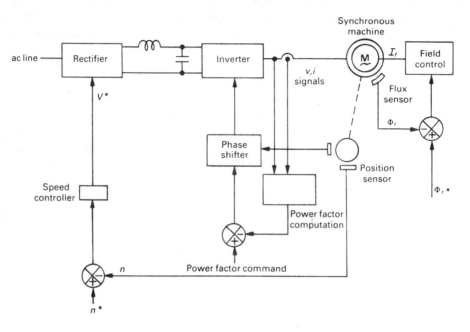

FIG. 8.19. Synchronous motor drive with closed-loop power factor control.

stant at a commanded value. The machine power factor is calculated by using the measured values of motor terminal voltage and current. This computed power factor is compared with the commanded value to give an error voltage that shifts the phase of the inverter firing signals, as shown in Fig. 8.19.

The resultant airgap flux, Φ_r, is independently regulated by closed-loop control. Auxiliary sense windings can be placed in the stator slots close to the airgap to provide a signal that, after integration, is proportional to the airgap flux. Alternatively, the terminal voltage signal can be modified to take account of the stator resistance and leakage reactance voltage drops. The resulting airgap voltage, E_r, is integrated to provide a signal proportional to the airgap flux. The actual value of Φ_r is compared with the desired value, Φ_r^*, and the error signal is used to drive the field current controller. The drive system has an outer speed loop using a shaft speed signal obtained by conversion of the rotor position signal.

With a forced-commutated inverter, the commanded power factor can be unity. A load-commutated thyristor inverter, however, obviously requires a leading power factor command.

8.9.2. *Terminal Voltage Sensing*

In a self-controlled synchronous motor drive, the inverter must receive synchronized firing signals for all steady-state and transient conditions, and, in the discussion to date, it has been assumed that the firing signals are synchronized by means of a position sensor on the shaft. However, because the presence of the shaft sensor may impair the overall ruggedness of the drive, it is attractive to use the machine terminal voltages as synchronizing signals and thereby eliminate shaft position sensing. This approach is analogous to that used in a conventional phase-controlled converter connected to the ac utility network, where the gating signals must be synchronized with the ac line voltages. In the synchronous motor drive, the inverter firing signals can be phase shifted with respect to the terminal voltage, in the usual manner, to control the phase angle between the motor voltage and current, and thereby allow motor operation at leading, lagging, or unity power factor.

Using the machine voltages as synchronizing signals has the advantage over shaft position synchronization that the phase advance of the terminal voltage due to armature reaction is automatically compensated. Thus, the effective angle of inverter advance, β, can be directly controlled. However, the motor voltages are distorted due to the large commutation transients occurring at 60-degree intervals. Consequently, the terminal voltages are usually sensed and clipped to give a square wave signal with the same frequency and phase as the terminal voltage; the voltage-sensing circuit may introduce a frequency-dependent phase shift for which compensation is necessary.

Figure 8.20 is a block diagram illustrating inverter synchronization by means of termi-

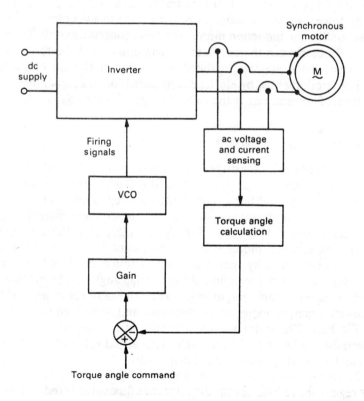

FIG. 8.20. Block diagram for electrical control of torque angle.

nal voltage sensing. Motor voltages and currents are measured with suitable sensors, and this information is processed to determine the torque angle. Thus, as already shown in Equation 8.26, the electromagnetic torque is proportional to the product of the resultant airgap flux, Φ_r, the stator mmf, and the sine of the torque angle, δ_T', between the resultant flux and stator mmf vectors. Hence

$$T = K\,\Phi_r\,I_a \sin\delta_T'.$$

The terminal voltage signal can be corrected for the stator leakage impedance voltage drop and used to determine resultant airgap flux, as already explained. The electromagnetic torque is calculated in terms of the d- and q-axis components of airgap flux and stator current.[40] The resulting torque signal is divided by the product of airgap flux and stator current to produce a signal proportional to $\sin\delta_T'$. The calculated torque angle is compared with an angle command signal, as shown in Fig. 8.20, and the error is amplified and supplied to a voltage-controlled oscillator (VCO) that generates the inverter firing signals. This phase-locked loop maintains the actual torque angle equal to the commanded value.

Motor operation at standstill and very low speeds requires that the voltage and current sensors be able to respond to dc input signals. This requirement can be avoided if open-loop operation from an external oscillator is used to accelerate the motor to a speed at which conventional ac voltage and current transformers can be used as sensors. At no load and unity power factor, the stator current can drop to a very low value, approaching zero, because, unlike the induction motor, the synchronous motor is not magnetized by stator current. Problems can then arise in the computation of the torque angle because the stator current value appears in the denominator of the $\sin\delta_T'$ calculation. This difficulty can be circumvented by electronically switching in a signal that corresponds to a minimum level of current, rather than allowing the signal to go to zero.

8.9.3. *Inverter Margin Angle Control*

In the LCI-synchronous motor drive, the load-commutated inverter should be operated with a minimum margin angle, γ, in order to maximize the motor power factor and efficiency for all loading conditions. This control strategy requires that the inverter advance angle, β, be continuously adjusted so that an adequate margin angle is allowed after commutation overlap (Fig. 8.14). For a given value of β, the overlap angle, u, is a function of dc link current, motor voltage, and motor frequency, but the appropriate value of β can be determined by open-loop computation or by closed-loop control. Thus, an accurate measurement or prediction of the overlap angle can be made, and this value of u is added to the minimum margin angle to determine the advance angle, β. Alternatively, the inverter margin angle can be measured and controlled as shown in the block diagram of Fig. 8.21. The stator voltage and current are sensed, and these signals are used to determine the inverter margin angle. This actual value of γ is maintained equal to the commanded value by phase-locked loop control.

Figure 8.21 also shows an outer speed loop and a flux control loop. In the constant-horsepower region above base speed, the reference flux value is reduced to avoid excessive generated emf at high speeds. The supply-side converter in Fig. 8.21 delivers a regulated dc current to the load-commutated inverter, in the usual manner. If the speed error

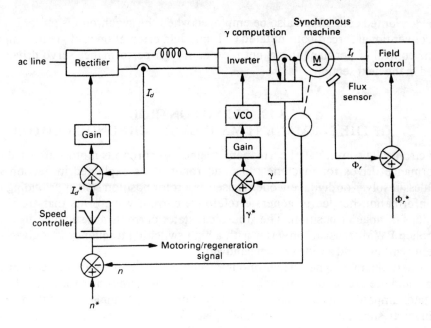

FIG. 8.21. Synchronous motor drive with closed-loop margin angle control of the load-commutated inverter.

is negative, the roles of the two power converters interchange. The drive enters a regenerative mode, and the angle, γ, of the machine-side converter is increased toward 180 degrees — that is, the firing angle is reduced toward zero to give rectifier operation. The supply-side converter now operates in the inverter mode but continues to regulate the dc link current by automatically adjusting its firing angle.

8.9.4. *Field-Oriented Control of the Wound-Field Motor*

Field-oriented control of an ac motor decouples the torque control from the field control, and its implementation is facilitated when the motor currents can be freely controlled, as with a current-controlled PWM inverter. However, in a load-commutated inverter, there are constraints on the inverter firing due to the fact that the safe commutation of the inverter must have top priority, and the minimization of the margin angle is very desirable. Also, the synchronous motor draws current at a leading power factor, and the resulting armature reaction effect has a demagnetizing component that prevents decoupling of the torque and field controls. Consequently, field-oriented control is not appropriate for the LCI-synchronous motor drive.

Wound-field synchronous motors powered by naturally commutated cycloconverters have utilized field-oriented control to give fast response.[35,58–60] The principle of the control is similar to the field-oriented control of an induction motor and, as usual, the practical implementation is facilitated by supplying the motor with impressed sinusoidal stator currents. Consequently, the cycloconverter is operated with fast current control loops that track sinusoidal current references. The resulting four-quadrant drive has a response equivalent to that of a high-performance dc drive.

Field-oriented control can also be employed when the synchronous motor is powered by a conventional forced-commutated CSI, but field orientation is of greater importance for the control of permanent magnet motors in high-performance servo drives, as discussed in the next section.

8.10. SERVO CONTROL
OF THE PERMANENT MAGNET SYNCHRONOUS MOTOR

Electronically commutated permanent magnet synchronous motors are used in high-performance drives for machine tools and robotics, as explained in Section 8.7.3. A brushless resolver or digital encoder is used as a rotor position sensor providing the precise information needed to generate reference current waveforms that are sinusoidal functions of angular position. The sinusoidal stator currents are supplied by a current-controlled PWM transistor inverter with a high switching frequency that gives accurate current synthesis and a wide system bandwidth.

In a permanent magnet synchronous machine, the field excitation cannot be controlled, and the excitation emf, or back emf, E_f, keeps increasing with speed. The torque per stator ampere is maximized by operation with a torque angle, δ_T, of 90 degrees, but the inverter supply voltage will be inadequate for sinusoidal current synthesis above a certain base speed because the motor back emf is excessively large. However, the speed range can be extended above base speed by an increase in the torque angle beyond 90 degrees to give leading power factor operation and a field-weakening effect, as shown in the phasor diagram of Fig. 8.17 (b). The armature reaction flux, Φ_{ar}, now has a demagnetizing effect and the resultant airgap flux, Φ_r, is smaller than the main field flux, Φ_f. The resultant armature emf, E_r, is correspondingly less than the excitation emf, E_f; hence the inverter supply voltage is adequate for proper current synthesis. However, unless the inverter is grossly overrated, saturation of the bus voltage is inevitable at higher speeds. Consequently, the inverter operation transitions from the controlled-current mode to the quasi-square-wave voltage-source mode, and there is a resulting degradation in the quality of the stator current waveform.

Large stator currents are needed to produce the required demagnetizing armature reaction effect, particularly in machines with surface-mounted rare earth magnets, because this magnet material has unity permeability for external fields, and the machine has a wide, effective airgap and low magnetizing reactance. As the speed rises, the torque angle is increased to enhance the field-weakening effect, but the torque per stator ampere is much reduced because of the large torque angle, and the motor copper loss is high due to the large stator currents. Maximum speed is determined by the motor thermal rating or by the inverter current rating or by fears that a failure in the current control could result in destruction of the inverter due to excessive voltage at the motor terminals. High speeds are possible with light loads and large demagnetizing currents, but in practice, the maximum speed of operation does not normally exceed twice base speed.

8.10.1. *Transient Analysis of Permanent Magnet Motor Behavior*[35,51,58]

The modern permanent magnet synchronous motor has samarium cobalt magnet bars attached to the surface of the rotor. As already explained, this rare earth magnet material has unity permeability, and the effective airgap width is constant around the circumference. Consequently, the machine has negligible saliency and can be mathematically modeled for transient conditions by the general differential equations that describe the symmetrical uniform airgap machine of Section 7.11. The latter machine is essentially a three-phase wound-rotor induction motor, but it can be adapted as a model for the permanent magnet synchronous machine. The constant rotor mmf due to the permanent magnet excitation can be modeled by supplying the three-phase rotor winding from two dc current sources, as shown in Fig. 8.22. The constant rotor currents establish a

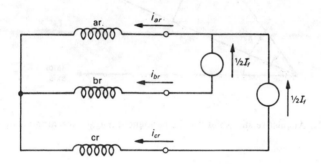

FIG. 8.22. Equivalent wound-rotor circuit for the permanent magnet synchronous motor.

sinusoidal mmf distribution of constant amplitude which is rigidly fixed to the rotor and therefore rotates at the same speed as the rotor.

From Fig. 8.22, the rotor phase currents are

$$i_{ar} = I_f; \ i_{br} = -\frac{I_f}{2}; \ \text{and} \ i_{cr} = -\frac{I_f}{2}. \tag{8.27}$$

Substituting these values in Equation 7.51 gives the fictitious rotor current vector as

$$\bar{i}_r = i_{ar} + a i_{br} + a^2 i_{cr}$$

$$= I_f - \frac{I_f}{2}(a + a^2)$$

$$= \frac{3}{2} I_f \tag{8.28}$$

because $1 + a + a^2 = 0$.

This expression for $\bar{i}_r$ can be substituted in the general torque equation (Equation 7.69) to give

$$T = \frac{2}{3} M \operatorname{Im}\left[\bar{i}_s \left(\bar{i}_r e^{j\epsilon}\right)^*\right]$$

$$= M I_f \operatorname{Im}\left[\bar{i}_s e^{-j\epsilon}\right]$$

$$= \Phi_f \operatorname{Im}\left[\bar{i}_s e^{-j\epsilon}\right] \tag{8.29}$$

where $\Phi_f = MI_f$.

In Equation 8.29, $\bar{i}_s e^{-j\epsilon}$ is the stator current vector in a frame of reference moving with the rotor, as emphasized in Fig. 8.23. The rotor mmf vector, which is now stationary

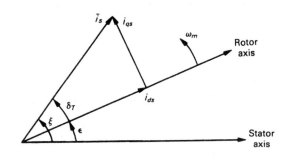

FIG. 8.23. Angular relationships for the permanent magnet synchronous motor.

with respect to the rotor, defines the rotor axis. As usual, the torque angle between the stator and rotor mmfs is denoted by δ_T. From Fig. 8.23

$$\bar{i}_s e^{-j\epsilon} = i_s e^{j(\xi-\epsilon)} = i_s e^{j\delta_T}. \tag{8.30}$$

Again, the direct and quadrature components, i_{ds} and i_{qs}, of the stator current vector can be defined in the rotor-based coordinate system. Thus,

$$\bar{i}_s e^{-j\epsilon} = i_{ds} + ji_{qs}. \tag{8.31}$$

Substituting Equations 8.30 and 8.31 in Equation 8.29 gives

$$T = \Phi_f \, i_s \sin \delta_T$$

$$= \Phi_f \, i_{qs}. \tag{8.32}$$

This equation shows that motor torque is proportional to the quadrature current component, i_{qs}. For a given stator current, i_s, the motor torque is a maximum when the torque angle, δ_T, is 90 degrees, that is, for a purely quadrature stator current. However, at high speeds, the torque angle is increased beyond 90 degrees to give a field-weakening effect, as explained above.

It should be noted that Equation 8.32 has been derived from the general mathematical model which signifies that this torque equation is valid for both transient and steady-state conditions. Because the transfer function between the stator current component, i_{qs}, and the instantaneous torque, T, is a constant, with no inherent delay, it is clear that a fast torque response is obtained if i_{qs} is rapidly varied, as for example, by a current-controlled PWM inverter.

8.10.2. *The Current-Controlled Permanent Magnet Motor Servo Drive*

Figure 8.24 shows a block diagram of a PWM servo controller for a permanent magnet

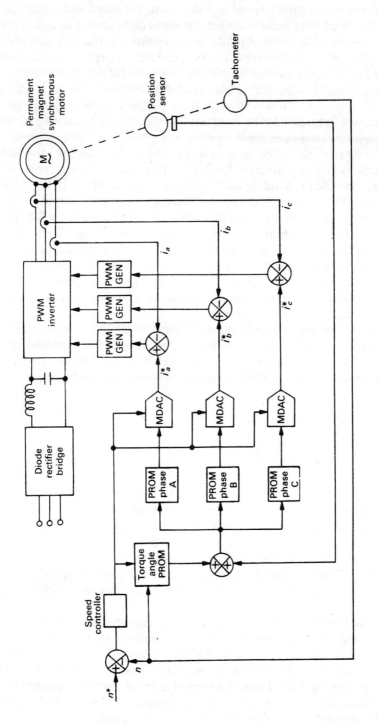

FIG. 8.24. Servo control of a permanent magnet synchronous motor with a current-controlled PWM inverter.

synchronous motor.[54] The input signal is an analog dc voltage whose magnitude and polarity represent the desired motor speed and direction of rotation. A brushless tachometer senses the actual motor speed and direction. Set speed and actual speed are compared, and the speed error defines the set current or commanded torque in the usual manner. A brushless resolver gives absolute rotor position information, and the torque angle is dynamically varied as a function of rotor speed and torque to give optimum drive performance. In Fig. 8.24, the sinusoidal reference currents for the three phases are digitally encoded and stored in a programmable read-only memory (PROM). The instantaneous phase currents are functions of angular position; consequently, the three PROMs are addressed in parallel by the rotor position signal or by a modified version of it that defines the appropriate torque angle.

Each PROM output is converted into an analog demanded current waveform by a multiplying digital-to-analog converter (MDAC). The MDAC output voltage is also proportional in magnitude to an independent reference voltage which, in Fig. 8.24, is the compensated speed error or set current signal. Thus, the sine wave outputs of the three MDACs are the demanded current waveforms for the three phases, and all three currents are controlled in amplitude by the common speed error voltage. These set current values are compared with the actual phase currents in three independent current loops. Each current loop has a conventional PWM control circuit in which a high-frequency triangular wave is mixed with the current error to define a PWM signal that drives one phase leg of the transistor inverter so that the sinusoidal reference current is synthesized in the motor phase winding.

In Fig. 8.24, the spatial orientation of the stator mmf wave relative to the rotor poles is controlled by the torque angle PROM, which is addressed by digitally encoded signals regarding shaft speed and commanded torque. The digitized rotor position signal is added to the output of the torque angle PROM to define the appropriate torque angle.

8.10.3. *Field-Oriented Control of the Permanent Magnet Motor*

Field-oriented control techniques are readily applied to the permanent magnet synchronous motor to realize a high-performance servo drive. As explained above, the permanent magnet rotor is modeled as a three-phase wound rotor with constant current excitation from two dc current sources. Consequently, the rotor voltage equation can be omitted from the machine model. When the stator windings are current fed with impressed sinusoidal currents from a current-controlled PWM transistor inverter, the stator voltage equation is also eliminated from the drive dynamics. The basic field-oriented control scheme is illustrated in the block diagram of Fig. 8.25, which shows the inner torque control loop. This diagram should be compared with the corresponding induction motor diagram of Fig. 7.30, which emphasizes that the basic control principle is the same in both cases.

In general, for field-oriented control, the stator currents are transformed into a frame of reference moving with the rotor flux. In the permanent magnet synchronous machine, the rotor flux is stationary relative to the rotor. The rotor flux position is therefore defined by the mechanical angle of rotation, ϵ, which is obtained from a rotor position sensor, as indicated in Fig. 8.25. Thus, the control is much easier to implement than in the case of the induction motor, which requires a complex parameter-dependent model of the motor to determine rotor flux position down to zero speed.

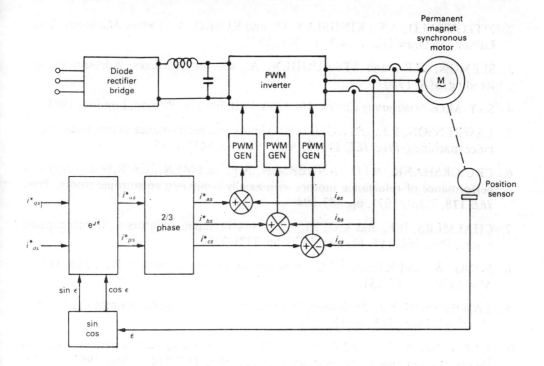

FIG. 8.25. Basic field orientation system for a permanent magnet synchronous motor with a current-controlled PWM inverter.

The reference currents, i_{ds}^* and i_{qs}^*, in Fig. 8.25, represent the commanded flux and commanded torque components of stator current, respectively. These reference currents are in rotor-based coordinates and are transformed to a stator reference frame using the rotor position angle, ϵ. The rotor-to-stator transformation is shown in two stages in Fig. 8.25 with a dq to two-phase vector rotation followed by the usual two-phase to three-phase transformation. The resulting three-phase stator reference currents are at stator frequency and are used as current commands for three current loops, in the usual manner.

An outer speed loop can be superimposed on the inner torque control loop of Fig. 8.25. Below base speed, i_{ds}^* is set at zero, and i_{qs}^* is specified, as usual, by the compensated speed loop error. This approach realizes the optimum torque angle of 90 degrees, because it is clear from Fig. 8.23 that δ_T is 90 degrees when i_{ds} is zero. Above base speed, field weakening is achieved by setting i_{ds}^* at a negative value, which causes the torque angle, δ_T, to exceed 90 degrees.

8.11. REFERENCES

1. OWEN, E.L., MORACK, M.M., HERSKIND, C.C., and GRIMES, A.S., AC adjustable-speed drives with electronic power converters — the early days, *IEEE Trans. Ind. Appl.*, **IA-20**, 2, Mar./Apr. 1984, pp. 298-308.

2. FITZGERALD, A.E., KINGSLEY, C., and KUSKO, A., *Electric Machinery*, Third Edition, McGraw-Hill, New York, NY, 1971.

3. SLEMON, G.R., and STRAUGHEN, A., *Electric Machines*, Addison-Wesley, Reading, MA, 1980.

4. SAY, M.G., *Alternating Current Machines*, Fifth Edition, Pitman, London, 1983.

5. LAWRENSON, P.J., and AGU, L.A., Theory and performance of polyphase reluctance machines, *Proc. IEE*, **111**, 8, Aug. 1964, pp. 1435-1445.

6. CRUICKSHANK, A.J.O., ANDERSON, A.F., and MENZIES, R.W., Theory and performance of reluctance motors with axially laminated anisotropic rotors, *Proc. IEE*, **118**, 7, July 1971, pp. 887-894.

7. CHALMERS, B.J., and MULKI, A.S., New reluctance motors with unlaminated rotors, *Proc. IEE*, **117**, 12, Dec. 1970, pp. 2271-2272.

8. FONG, W., and HTSUI, J.S.C., New type of reluctance motor, *Proc. IEE*, **117**, 3, Mar. 1970, pp. 545-551.

9. LAWRENSON, P.J., Development and application of reluctance motors, *Electron. Power*, **11**, June 1965, pp. 195-198.

10. LAWRENSON, P.J., and GUPTA, S.K., Developments in the performance and theory of segmental-rotor reluctance motors, *Proc. IEE*, **114**, 5, May 1967, pp. 645-653.

11. HONSINGER, V.B., The inductances L_d and L_q of reluctance machines, *IEEE Trans. Power Appar. Syst.*, **PAS-90**, 1, Jan./Feb. 1971, pp. 298-304.

12. HONSINGER, V.B., Steady-state performance of reluctance machines, *IEEE Trans. Power Appar. Syst.*, **PAS-90**, 1, Jan./Feb. 1971, pp. 305-317.

13. HONSINGER, V.B., Stability of reluctance motors, *IEEE Trans. Power Appar. Syst.*, **PAS-91**, 4, July/Aug. 1972, pp. 1536-1543.

14. HONSINGER, V.B., Inherently stable reluctance motors having improved performance, *IEEE Trans. Power Appar. Syst.*, **PAS-91**, 4, July/Aug. 1972, pp. 1544-1554.

15. SLEMON, G.R., DEWAN, S.B., and WILSON, J.W.A., Synchronous motor drive with current-source inverter, *IEEE Trans. Ind. Appl.*, **IA-10**, 3, May/June 1974, pp. 412-416.

16. CHALMERS, B.J., Torque production in ac motors with constant current supply, *Elect. Mach. and Electromechan.*, **1**, 3, Apr./June 1977, pp. 237-244.

17. KÜBLER, E., Der Stromrichtermotor, *Elektrotech. Z. Ausg. A*, **79**, 1958, pp. 15-17.

18. SATO, N., A study of commutatorless motor, *Electr. Eng. in Jpn*, **84**, 8, Aug. 1964, pp. 42-51.

19. KÖLLENSPERGER, D., The synchronous machine as a self-controlled converter-fed motor, *Siemens Rev.*, **35**, 5, May 1968, pp. 195-201.

20. INAGAKI, J., KUNIYOSHI, M., and TADAKUMA, S., Commutators get the brushoff, *IEEE Spectrum*, **10**, 6, June 1973, pp. 52-58.

21. MIYAIRI, S., and TSUNEHIRO, Y., Analysis and the characteristics of a commutatorless motor as a dc motor, *Electr. Eng. in Jpn*, **85**, 9, Sept. 1965, pp. 51-62.

22. MIYAIRI, S., and TSUNEHIRO, Y., Damper winding in a commutatorless dc motor, *Electr. Eng. in Jpn*, **87**, 8, 1967, pp. 67-77.

23. OHNO, E., KISHIMOTO, T., and AKAMATSU, M., The thyristor commutatorless motor, *IEEE Trans. Magn.*, **MAG-3**, 3, Sept. 1967, pp. 236-240.

24. TSUCHIYA, T., Basic characteristics of cycloconverter-type commutatorless motors, *IEEE Trans. Ind. Gen. Appl.*, **IGA-6**, 4, July/Aug. 1970, pp. 349-356.

25. SCHMITZ, N.L., and LONG, W.F., The cycloconverter driven doubly-fed induction motor, *IEEE Trans. Power Appar. Syst.*, **PAS-90**, Mar./Apr. 1971, pp. 526-531.

26. ALEXANDERSON, E.F.W., and MITTAG, A.H., The "thyratron" motor, *Electr. Eng.* (New York), **53**, Nov. 1934, pp. 1517-1523.

27. BEILER, A.H., The thyratron motor at the Logan plant, *Electr. Eng.* (New York), **57**, Jan. 1938, pp. 19-24.

28. STEMMLER, H., Drive systems and electronic control equipment of the gearless tube mill, *Brown Boveri Rev.*, **57**, 3, Mar. 1970, pp. 121-129.

29. HABÖCK, A., and KÖLLENSPERGER, D., Application and further development of converter-fed synchronous motors with self-control, *Siemens Rev.*, **38**, 9, Sept. 1971, pp. 393-395.

30. MAENO, T., and KOBATA, M., AC commutatorless and brushless motor, *IEEE Trans. Power Appar. Syst.*, **PAS-91**, 4, July/Aug. 1972, pp. 1476-1484.

31. TERENS, L., BOMMELI, J., and PETERS, K., The cycloconverter-fed synchronous motor, *Brown Boveri Rev.*, **69**, 4/5, Apr./May 1982, pp. 122-132.

32. MEYER, A., SCHWEICKARDT, H., and STROZZI, P., The converter-fed synchronous motor as a variable-speed drive system, *Brown Boveri Rev.*, **69**, 4/5, Apr./May 1982, pp. 151-156.

33. SATO, N., and SEMENOV, V.V., Adjustable speed drive with a brushless dc motor, *IEEE Trans. Ind. Gen. Appl.*, **IGA-7**, 4, July/Aug. 1971, pp. 539-543.

34. ISSA, N.A.H., and WILLIAMSON, A.C., Control of a naturally commutated inverter-fed variable-speed synchronous motor, *IEE J. Elect. Power Appl.*, **2**, 6, Dec. 1979, pp. 199-204.

35. LEONHARD, W., *Control of Electrical Drives*, Springer-Verlag, Berlin, 1985.

36. DEWAN, S.B., SLEMON, G.R., and STRAUGHEN, A., *Power Semiconductor Drives*, J. Wiley, New York, NY, 1984.

37. SLEMON, G.R., FORSYTHE, J.B., and DEWAN, S.B., Controlled power angle synchronous motor inverter drive system, *IEEE Trans. Ind. Appl.*, **IA-9**, 2, Mar./Apr. 1973, pp. 216-219.

38. LEIMGRUBER, J., Stationary and dynamic behaviour of a speed controlled synchronous motor with cos ϕ or commutation limit line control, *Proc. 2nd IFAC Symp. on Control in Power Electronics and Electrical Drives, 1977*, pp. 463-473.

39. PLUNKETT, A.B., and TURNBULL, F.G., System design method for a load commutated inverter synchronous motor drive, *Conf. Rec. IEEE Ind. Appl. Soc. Annual Meeting, 1978*, pp. 812-819.

40. PLUNKETT, A.B., and TURNBULL, F.G., Load commutated inverter/synchronous motor drive without a shaft position sensor, *IEEE Trans. Ind. Appl.*, **IA-15**, 1, Jan./Feb. 1979, pp. 63-71.

41. HARASHIMA, F., TAOKA, H., and NAITOH, H., A microprocessor-based PLL speed control system of converter-fed synchronous motor, *Proc. IEEE IECI Ind. and Control Appl. Microproc.*, 1979, pp. 272-277.

42. ROSA, J., Utilization and rating of machine commutated inverter-synchronous motor drives, *IEEE Trans. Ind. Appl.*, **IA-15**, 2, Mar./Apr. 1979, pp. 155-164.

43. RICHTER, W., Microprocessor controlled inverter-fed synchronous motor drive, IEE Conf. Publ. No. 179, *Electrical Variable-Speed Drives*, 1979, pp. 161-164.

44. LE-HUY, H., PERRET, R., and ROYE, D., Microprocessor control of a current-fed synchronous motor drive, *Conf. Rec. IEEE Ind. Appl. Soc. Annual Meeting, 1979*, pp. 873-880.

45. LE-HUY, H., JAKUBOWICZ, A., and PERRET, R., A self-controlled synchronous motor drive using terminal voltage sensing, *IEEE Trans. Ind. Appl.*, **IA-18**, 1, Jan./Feb. 1982, pp. 46-53.

46. BOSE, B.K., Adjustable speed ac drives - a technology status review, *Proc. IEEE*, **70**, 2, Feb. 1982, pp. 116-135.

47. NISHIKATA, S., and KATAOKA, T., Dynamic control of a self-controlled synchronous motor drive system, *IEEE Int. Semicond. Power Converter Conf.*, 1982, pp. 357-364.

48. HARASHIMA, F., NAITOH, H., and HANEYOSHI, T., Dynamic performance of self-controlled synchronous motors fed by current-source inverters, *IEEE Trans. Ind. Appl.*, **IA-15**, 1, Jan./Feb. 1979, pp. 36-46.

49. JAKUBOWICZ, A., NOUGARET, M., and PERRET, R., Simplified model and closed-loop control of a commutatorless dc motor, *IEEE Trans. Ind. Appl.*, **IA-16**, 2, Mar./Apr. 1980, pp. 165-172.

50. NASHIKI, M., and DOTE, Y., High performance current-controlled PWM transistor inverter-fed brushless servomotor, *IEEE Int. Semicond. Power Converter Conf.*, 1982, pp. 349-356.

51. LETAS, H. H., and LEONHARD, W., Dual axis servo drive in cylindrical coordinates using permanent magnet synchronous motors with microprocessor control, *Conumel 83*, 1983, pp. II-23 to II-30.

52. SCHWARZ, B., Converter-fed synchronous machine with high performance dynamic behavior for servo-drive application, *Proc. 3rd IFAC Symp. on Control in Power Electronics and Electrical Drives*, 1983, pp. 375-382.

53. PFAFF, G., WESCHTA, A., and WICK, A., Design and experimental results of a brushless ac servo-drive, *IEEE Trans. Ind. Appl.*, **IA-20**, 4, July/Aug. 1984, pp. 814-821.

54. BROSNAN, M.F., and BROWN, B., Closed loop speed control using an ac synchronous motor, IEE Conf. Publ. No. 234, *Power Electronics and Variable-Speed Drives*, 1984, pp. 373-376.

55. SCHUMACHER, W., LETAS, H.H., and LEONHARD, W., Microprocessor-controlled ac servo-drives with synchronous and asynchronous motors, IEE Conf. Publ. No. 234, *Power Electronics and Variable-Speed Drives*, 1984, pp. 233-236.

56. LAJOIE-MAZENC, M., VILLANUEVA, C., and HECTOR, J., Study and implementation of hysteresis controlled inverter on a permanent magnet synchronous machine, *IEEE Trans. Ind. Appl.*, **IA-21**, 2, Mar./Apr. 1985, pp. 408-413.

57. ANDRIEUX, C., and LAJOIE-MAZENC, M., Analysis of different current control systems for inverter-fed synchronous machine, *Proc. European Power Electron. Conf.*, 1985, pp. 2.159-2.165.

58. LEONHARD, W., Control of ac machines with the help of microelectronics, *Proc. 3rd IFAC Symp. on Control in Power Electronics and Electrical Drives, 1983*, pp. 769-792.

59. BAYER, K.H., WALDMANN, H., and WEIBELZAHL, M., Field-oriented closed-loop control of a synchronous machine with the new "transvektor" control system, *Siemens Rev.*, **39**, 5, May 1972, pp. 220-223.

60. NAKANO, T., OHSAWA, H., and ENDOH, K., A high performance cycloconverter fed synchronous machine drive system, *IEEE Int. Semicond. Power Converter Conf.*, 1982, pp. 334-341.

CHAPTER 9

Adjustable-Voltage Induction Motor Drive Systems

9.1. INTRODUCTION

Most ac motors run at a speed that is determined principally by the frequency of the ac supply. The speed of a synchronous motor is completely determined by the stator frequency, but the induction motor runs asynchronously at a speed which is normally within 5 percent of the synchronous speed. On a 50 or 60 Hz fixed-frequency utility supply, the torque developed by the induction motor, at a given slip, varies approximately as the square of the applied voltage, and steady-state operation occurs when the motor torque balances the load torque. Consequently, the rotor slip is determined by the load torque and the applied voltage, and continuous speed control may be obtained by stepless adjustment of the stator voltage without any alteration in the stator frequency.

As the stator voltage is reduced, the rotor speed decreases, but the maximum torque available from the motor also decreases, and the speed range is limited with a constant-torque load. A high-resistance rotor modifies the torque-speed characteristics, and wide-range speed control is possible (Fig 9.1), but the stator and rotor copper losses are excessive and the operating efficiency is low, particularly at reduced speeds. The low-speed performance is poor because the motor current, at a given slip, is proportional to the applied voltage, whereas the electromagnetic torque varies approximately as the square of the voltage. Consequently, the torque per ampere is lower at reduced speeds, and large currents are required to develop appreciable torque. However, in fan and pump drives, the load torque varies approximately as the square of the speed. Hence, the torque required for starting and low-speed running is small, and may often be obtained, without excessive overheating, from a voltage-controlled induction motor with a normal full-load slip of about 10 percent (Fig. 9.2).

The stator voltage can be reduced by connecting variable external impedances between the stator terminals and the ac supply network. Saturable reactors have been used in the past to perform this function, but thyristor circuits are more compact, despite the heat sink requirements, and weigh considerably less. The basic thyristor circuit that is employed is the series ac voltage controller, already described in Section 2.7. This circuit consists of two thyristors connected in inverse parallel, or antiparallel ("back-to-back"), and triggered symmetrically at identical points in their anode-to-cathode voltage cycles. By inserting these circuits into the stator leads and controlling the conduction periods of the thyristors, the effective voltage delivered to the motor can be varied from zero to the full supply voltage.[1-8] The motor is subjected to a chopped sine-wave voltage, and the motor currents also have a high harmonic content, but satisfactory operation has been achieved with small- and medium-sized induction motors up to 100 horsepower, or more.

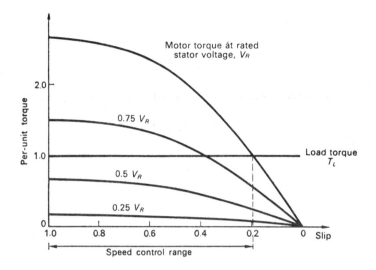

FIG. 9.1. Speed control of a constant-torque load by stator voltage control of an induction motor.

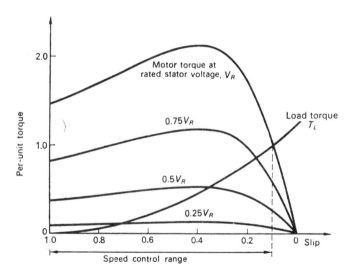

FIG. 9.2. Speed control of a fan load by stator voltage control of an induction motor.

Stator voltage control eliminates the complex circuitry of the adjustable-frequency schemes discussed in earlier chapters and, consequently, is cheaper to install. However, the operating efficiency is poor, and motor derating is necessary at low speeds to avoid overheating due to excessive current and reduced ventilation. Stator voltage control is widely used for fractional horsepower drives, and also for ac-powered cranes and hoists where large torque at high slip is only demanded for intermittent portions of the duty cycle. Adjustable-frequency drives are unnecessarily complex and expensive for such applications.

The chapter concludes with a discussion of the use of stator voltage adjustment, not for induction motor speed control but, rather, for reducing the input power consumption during conditions of light load on the motor. At light loads, the induction motor does not need to be excited with rated voltage corresponding to rated flux. A reduced stator voltage, corresponding to a reduced flux, can drive a lightly loaded motor at low slip. The reduction in stator voltage leads to a reduction in stator core loss and an improvement in the input power factor. If the induction motor operates on a duty cycle that has substantial periods of light or no load operation, an appreciable overall saving in energy can be obtained.

9.2. THEORETICAL PRINCIPLES OF STATOR VOLTAGE CONTROL

Assume that the motor voltages and currents are balanced and sinusoidal, and also regard the rotor resistance as constant and independent of rotor frequency. If ω_s denotes the synchronous angular velocity and ω_m is the rotational angular velocity, the fractional slip is defined, as usual, by $s = (\omega_s - \omega_m)/\omega_s$. The gross mechanical power output, including windage and friction losses, is $P_{mech} = T\omega_m$, where T is the internal motor torque. In accordance with standard theory (see Chapter 6), the total power input to the rotor, across the airgap from the stator, is $P_{ag} = T\omega_s$. The difference between the power input to the rotor and the mechanical power output is dissipated as heat losses in the rotor resistance. Thus, the rotor copper loss is given by

$$P_2 = T(\omega_s - \omega_m) = sT\omega_s = sP_{ag} .\tag{9.1}$$

The basic power division process in the rotor circuit is now evident. A fraction, s, of the rotor input power is dissipated as heat, and the remainder, $(1 - s)$, is the gross mechanical power output, P_{mech}. Hence,

$$P_{mech} = (1 - s) P_{ag} .\tag{9.2}$$

Induction motor speed control by a reduction in the magnitude of the applied stator voltage is therefore inherently inefficient, since the rotor copper loss increases linearly with slip for a constant motor torque. Equation 9.1 also shows that the rotor copper loss is proportional to the product of torque and slip, and the motor heat generated in a given duty cycle may be quickly approximated by evaluating the product of torque, slip, and time, for each portion of the duty cycle.

The stator current, I_1, contains a load component that is proportional to the rotor current, and also a magnetizing component. At reduced voltages, the magnetizing current may be neglected, and hence the stator and rotor currents are proportional. The large rotor currents corresponding to high-torque, high-slip operation will produce correspondingly large stator currents which may cause serious overheating and damage to the heat-sensitive stator winding. Consequently, the motor must be derated for low-speed operation.

Equation 9.1 may be rewritten to give an expression for the motor torque. Thus, as in the earlier Equation 6.6,

$$T = \frac{m_1 I_2^2 R_2}{s\omega_s}$$

where m_1 is the number of stator phases. Therefore,

$$T \propto \frac{I_2^2 R_2}{s} \propto \frac{I_1^2 R_2}{s} \ . \tag{9.3}$$

Stable operation occurs when the motor torque, T, equals the load torque, T_L. For a fan load, $T_L \propto (\omega_m)^2 \propto (1-s)^2$, and equating torques gives the result

$$I_1 \propto I_2 \propto (1-s)\sqrt{(s/R_2)} \ . \tag{9.4}$$

This result indicates that the machine currents are inversely proportional to $\sqrt{R_2}$ and, by differentiation, it is found that the currents have maximum values when $s = 1/3$, that is, when $\omega_m = 2\omega_s/3$. Equation 9.4 can also be used to express the relationship between the maximum stator current, I_{1max}, and the full-load stator current, I_{FL}, at rated slip, s_{FL}. Thus,

$$\frac{I_{1max}}{I_{FL}} = \frac{2}{3\sqrt{3}(1-s_{FL})\sqrt{s_{FL}}} \ . \tag{9.5}$$

If the induction motor has a full-load slip of 15 percent, then Equation 9.5 gives $I_{1max}/I_{FL} = 1.169$, indicating that the maximum stator current is only 16.9 percent greater than the full-load value.

For a constant-torque load, T_L is constant, and equating motor and load torques gives the result

$$I_1 \propto I_2 \propto \sqrt{(s/R_2)} \ . \tag{9.6}$$

The stator and rotor currents are again inversely proportional to $\sqrt{R_2}$, but they now increase steadily as the motor speed is reduced to zero, and are even larger in the plugging region where the slip is greater than unity. In this case, the induction motor with a full-load slip of 15 percent draws a standstill current of 2.6 times the rated current.

These results show that the rotor resistance is an important factor in determining the increase in motor current. If the normal full-load slip of a squirrel-cage motor is small, indicating the presence of a low-resistance rotor, a very large increase in motor current can occur at reduced speeds when stator voltage control is employed. For satisfactory low-speed operation in a fan drive, the rotor should have a high resistance corresponding to a full-load slip of 10 to 15 percent.[5] In a constant-torque drive, the rotor resistance must be even greater. Squirrel-cage rotors with variable-resistance characteristics may also be used. The usual deep-bar or double-cage rotor is not very effective, because the eddy-current loss, on which the operation depends, is proportional to the square of the slip, and hence is not significant unless the slip exceeds about 0.5. An improved performance can be obtained by using a solid-iron rotor,[9] or by placing a bar of permanent magnet material, such as Alnico,* in the top of each rotor slot. The hysteresis loss in the Alnico causes an increase in the effective rotor resistance, thereby increasing the torque per ampere at low speeds and giving the motor a better torque-speed characteristic.[10]

Standard wound-rotor induction motors are commonly used in adjustable-voltage speed-control systems. External rotor resistance is added to modify the torque-speed

---

* Trademark of Magnets and Electronics, Inc.

characteristic so that the torque per ampere at low speeds is increased. Motor heating is reduced, because part of the rotor circuit loss is now dissipated externally.

Speed control may also be obtained by deliberately unbalancing the stator terminal voltages. By the theory of symmetrical components, the unbalanced three-phase voltages can be resolved into positive- and negative-sequence systems. The positive-sequence set establishes balanced currents of the same sequence, and these produce a forward-rotating field which develops forward rotor torque in the usual manner. The negative-sequence voltages produce negative-sequence currents which establish a backward-rotating field and develop a counter-torque opposing motion. The resultant torque is the difference between the positive- and negative-sequence torques, but the positive- and negative-sequence currents both contribute to the motor heating. The low-speed performance is, therefore, even poorer than that obtained with a symmetrical reduction of the stator voltages, and further derating is necessary to limit the temperature rise.[11]

When thyristor switching circuits are used to reduce the stator voltage, either symmetrically or asymmetrically, the resulting voltage and current waveforms are highly distorted, and the additional harmonic losses will further accentuate the overheating problem at low speeds. Because an exact analysis of induction motor behavior under such operating conditions is rather difficult, an analog simulation or a numerical solution on a digital computer is required.[12-18]

9.3. PRACTICAL CIRCUITS [3,5,8,16,19,20]

As already explained in Section 2.7, the effective load voltage in a single-phase ac circuit can be varied by a thyristor controller consisting of a pair of antiparallel thyristors in series with the load. A three-phase version of this circuit is obtained by combining three single-phase circuits in a delta-connection, as in Fig. 9.3. Each pair of antiparallel thyristors controls the voltage delivered to one phase of the stator, and the phase voltage waveform consists of a series of sine wave segments. A phase displacement of 120 degrees is maintained between the sets of gating pulses delivered to each controller, in order to produce a symmetrical reduction of the three-phase voltages. In the full-voltage condition, each thyristor receives a gating signal at the start of the positive half-cycle of its anode-to-cathode voltage. The series thyristor pair is then virtually a short-circuit and the stator phase receives a complete cycle of the ac supply voltage. As the

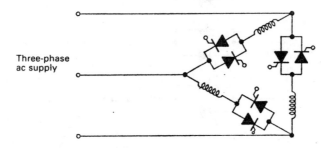

Three-phase
ac supply

FIG. 9.3. Stator voltage control of a delta-connected induction motor by a series antiparallel thyristor pair in each phase.

thyristor firing is delayed, the conduction period of each thyristor is shortened and the effective stator voltage is reduced. When the firing delay is 180 degrees, the thyristor controller is an open circuit and the motor voltage and current are both zero. Using a switching controller in this manner as a nonlinear series impedance has the advantage that the power dissipation in the controller is considerably less than the load power, but the switching action also distorts the stator voltages.

Several alternative power circuits are possible. Thyristor controllers may be inserted in the three ac lines of a wye- or delta-connected load, as in Figs. 9.4(a) and (b). For economy, one thyristor in each antiparallel circuit can be replaced by a rectifier diode. A single triac can replace two antiparallel thyristors.[20] The number of semiconductor devices can be reduced if it is decided to operate with unbalanced stator voltages. Thus, antiparallel thyristors can be inserted into two of the three-phase lines, as in Fig. 9.4(c), with the third line connected directly to the supply. Alternatively, the motor can be operated with a thyristor controller in only one of the three-phase lines. However, these

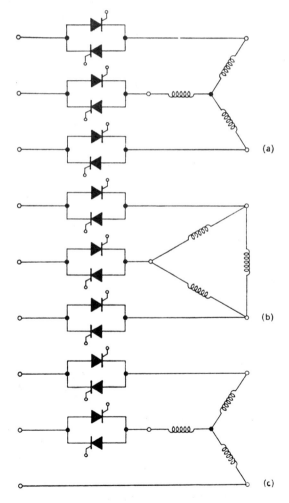

FIG. 9.4. Alternative connections for stator voltage control of a three-phase induction motor.

unbalanced circuits tend to accentuate the severe machine heating problem because some phases are heavily overloaded.

Speed control may be obtained by decreasing or unbalancing the stator voltages of the induction motor, but the direction of rotation can only be reversed by changing the phase sequence of the applied voltages. This sequence change is often achieved by means of a mechanical contactor which interchanges two stator leads. Contactor operation takes place after the stator current has been interrupted by removal of the thyristor gating pulses. Consequently, sparkless operation is obtained and the contactor requires little maintenance. However, the stator phase sequence can also be reversed statically by introducing additional thyristor controllers, as in Fig. 9.5. The extra units, X and Y, are not gated for normal forward rotation of the motor but, rather, only come into operation when A and C are on open-circuit. This arrangement reverses the phase sequence of the applied voltages and thereby changes the direction of rotation. It is, of course, essential that X and Y should not be gated until the gating signals for A and C have been removed; otherwise a short-circuit condition would result. Current sensors in series with A and C inhibit firing of X and Y until zero-current signals are detected. Similarly, the triggering of A and C is inhibited until X and Y have been open-circuited.

Adjustable-speed drives using stator voltage control are normally closed-loop systems.[20,21] In Fig. 9.6, the dc tachometer delivers a voltage proportional to the motor speed, n, and this is compared with the dc reference voltage representing the desired speed, n^*. The difference between the two signals is the error voltage, which is amplified in the speed controller. The resulting signal controls the thyristor firing angles and thereby alters the terminal voltage and motor speed so that the error is reduced. If the reference voltage is greater than the tachometer voltage, the conduction periods of the thyristors are increased. The increased stator voltage allows the development of an

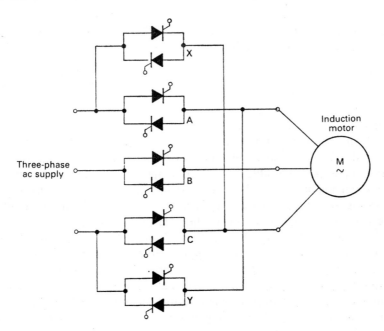

FIG. 9.5. Thyristor configuration for a reversible adjustable-speed induction motor drive.

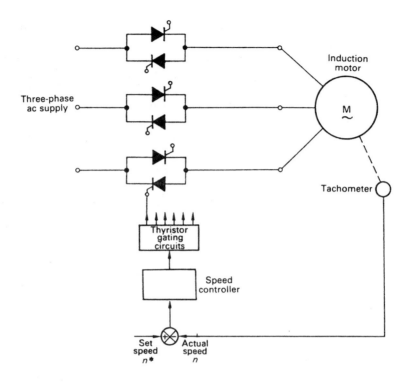

FiG. 9.6. Closed-loop system for induction motor speed control by stator voltage control.

increased motor torque, and hence the speed rises. If the tachometer voltage exceeds the reference voltage, the conduction periods are reduced and the motor torque decreases, causing a reduction in shaft speed. In a high-gain feedback system, the set speed can be accurately maintained and there is no need for the motor to have a flat speed-torque characteristic, because the shaft speed is determined by the reference signal rather than the inherent open-loop characteristic of the motor; stable operation may be obtained at any point on the induction motor characteristic.

In a reversible induction motor drive employing the thyristor configuration of Fig. 9.5, the polarity of the error signal determines whether motoring or braking torque is developed, and operation is possible in the torque-speed quadrants shown in Fig. 9.7. If the reference voltage is positive and exceeds the tachometer voltage, forward torque is produced by gating thyristor controllers A, B, and C, and hence the motor accelerates. If the reference voltage is suddenly reduced, the polarity of the error signal changes, and thyristor units A and C are turned off, and X and Y are gated on. As a result, the phase sequence of the stator voltages is reversed and the motor operates in the plugging region with a rapid reduction in speed. In a crane or hoist application, when the motor is lowering an overhauling load, the actual speed is also greater than the set speed, and a braking counter-torque is developed which permits steady-state lowering of the load. Regenerative braking in the induction generator region is possible for full-speed lowering of an overhauling load.

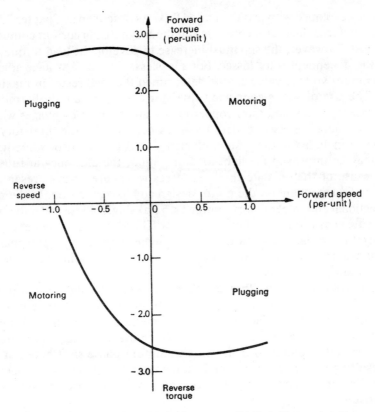

FIG. 9.7. Torque-speed characteristics for a reversible induction motor drive.

9.4. INTEGRAL-CYCLE CONTROL

In the phase-control methods described above, the effective stator voltage is varied by means of antiparallel thyristor circuits which conduct for a portion of each half-cycle. As an alternative, integral-cycle, or time-ratio, control may be employed, using the antiparallel thyristors or triacs as static contactors and alternately switching on and off the supply voltage for a number of cycles at a time.[22] The effective stator voltage is varied by controlling the ratio of on-time to off-time, and speed fluctuations are avoided if the conducting and nonconducting intervals are short compared with the mechanical time constant of the system.[23] Less voltage distortion is produced on the ac supply network, and less radio-frequency interference is propagated, when integral-cycle systems are used rather than the phase-controlled regulators described in this chapter. On the other hand, speed fluctuations may be objectionable, and stepless speed control is not possible when integral-cycle techniques are used.

9.5. PERFORMANCE IMPROVEMENT WITH ADJUSTABLE VOLTAGE

In fixed-speed industrial applications, the circuits shown in Figs. 9.3 and 9.4 can be used to supply a reduced stator voltage to an induction motor to improve its efficiency during operation at light load. The motor voltage is reduced when the required load torque can be developed with less than rated flux. As described in Chapter 7, operation at reduced flux reduces the motor core loss and the magnetizing current component of stator copper loss. However, the slip must increase to develop the load torque, resulting in increased slip-dependent rotor losses. For a lossless, sine wave voltage adjustment system, an optimum voltage value can be determined that will result in maximum motor efficiency. The use of series-connected thyristor circuits results in harmonic currents which increase the stator and rotor copper losses, and harmonic voltages which increase the core loss. There are also on-state losses associated with the thyristors themselves which must be included in the system efficiency calculation. Also, while it is clear that efficiency may be improved for light-load operation, the efficiency at full load will be reduced because of the on-state thyristor losses which are always present. The actual energy saving is dependent on the motor design and its application. In the manufacture of some fractional and integral horsepower (< 5 hp) single-phase and polyphase motors, maximum efficiency is not a primary design goal. These motors can benefit from reduced voltage operation at light load. In the case of higher horsepower, polyphase induction motors that are designed for better efficiency, the energy saving obtained by reduced voltage operation is minimized.

Several types of control algorithm have been developed to find the optimum operating point for a reduced voltage, thyristor phase-controlled system. A popular approach is to maximize the motor power factor.[24–26] This method uses a current transformer to sense the line current and a voltage sensor to determine the voltage zero-crossing points. The control circuit attempts to maintain a minimum phase shift between voltage and current, resulting in maximum power factor. Integrated circuits have been developed that accept the current and voltage inputs, and provide a phase-shifted pulse to gate a triac or thyristor.

Several studies have analyzed the potential for saving energy with the use of the symmetrical thyristor circuits of Figs. 9.3, 9.4(a), and 9.4(b).[27–29] Several theoretical control algorithms have been developed, including minimum input power, maximum power factor, constant power factor, and minimum stator current. The implementation of each algorithm can be related to the firing angle of the thyristors.

This use of stator voltage reduction can best be considered when the motor is operating with a duty cycle that includes a period of light load operation. Graphs can be prepared to show the energy saving as a function of duty cycle for a series of light-load/full-load ratios.[29] The energy saved increases as the percentage of time at light load increases and as the light-load portion of the duty cycle requires lower percentages of rated torque.

Two other main areas of application exist. One is where the motor is rated at higher horsepower than its connected load demands. The oversized motor, driving its maximum connected load, does not develop its rated horsepower. This situation can occur if the maximum load is unknown or is anticipated to increase with time, or if stocking a limited number of motor horsepower ratings is desirable. In these cases, the motor may show reduced energy consumption by the use of an ac voltage controller. The second application is where the motor is connected to a supply voltage that is higher than the rated motor voltage. This situation can occur if the motor is located on a feeder next to power factor correction capacitors, which cause the feeder voltage to rise with load. The ac voltage controller serves as a voltage limiter and prevents motor saturation with its increase in core loss.

Another application for the ac voltage controller system is in the "soft starting" of induction motors.[30] Before the thyristors are gated into conduction, the motor current is zero. Thyristor firing is then slowly advanced, allowing the line current to increase gradually so that the motor accelerates up to its rated speed. Various options can be provided by the starter — for example, a constant-current acceleration or a timed acceleration. In either case, the line current is reduced from the six to eight times full-load current that occurs for a direct-on-line start when full voltage is applied to the induction motor. Some applications maintain the thyristors in the circuit during full voltage operation; others bypass the thyristors with mechanical contacts, thereby eliminating the on-state losses in the starter.

In this chapter, the techniques for reducing induction motor voltage at light load to save energy have been described for an adjustable-voltage, 50 or 60 Hz drive system. The same energy saving principles can be used in adjustable-voltage, adjustable-frequency induction motor drive systems, as previously described in Chapter 7. In this case, the inverter or cycloconverter already has the capability of adjusting the motor voltage, and external voltage controllers are not required.

9.6. REFERENCES

1. ZAGALSKY, N., and SHEPHERD, W., Wide-range reversible voltage controllers for polyphase induction motors, *Trans. AIEE, Pt. 2: Appl. and Ind.*, **81**, Nov. 1962, pp. 272-277.

2. BAILEY, F.M., and KONRAD, C.E., Dynamic characteristics of controlled rectifier alternating current drives, IEE Conf. Publ. No. 11, *Commutation in Rotating Machines*, 1964, pp. 123-126.

3. SHEPHERD, W., and STANWAY, J., The polyphase induction motor controlled by firing angle adjustment of silicon controlled rectifiers, *IEEE Int. Conv. Rec. Pt. 4*, **12**, 1964, pp. 135-154.

4. KORB, F., Einstellung der Drehzahl von Induktionsmotoren durch antiparallele Ventile auf der Netzseite, *Elektrotech. Z. Ausg. A*, **86**, 8, 1965, pp. 275-279.

5. PAICE, D.A., Induction motor speed control by stator voltage control, *IEEE Trans. Power Appar. Syst.*, **PAS-87**, 2, Feb. 1968, pp. 585-590.

6. TAKEUCHI, T.J., *Theory of SCR (Thyristor) Circuit and Application to Motor Control*, Tokyo Electrical Engineering College Press, 1968.

7. GUTT, H.-J., Bemessung stellergespeister Asynchronmaschinen für bürstenlose drehzahlsteuerbare Antriebe, *Siemens-Z.*, **44**, 8, Aug. 1970, pp. 531-534.

8. McMURRAY, W., A comparative study of symmetrical three-phase circuits for phase-controlled ac motor drives, *IEEE Trans. Ind. Appl.*, **IA-10**, 3, May/June 1974, pp. 403-411.

9. FINZI, L.A., and PAICE, D.A., Analysis of the solid-iron rotor induction motor for solid-state speed controls, *IEEE Trans. Power Appar. Syst.*, **PAS-87**, 2, Feb. 1968, pp. 590-596.

10. ALGER, P.L., LUNEBURG, S.E., and MESTER, R.L., Alnico bar capacitor motor with scr speed control, IEE Conf. Publ. No. 11, *Commutation in Rotating Machines*, 1964, p. 128.

11. RAMA RAO, N., and JYOTHI RAO, P.A.D., Rerating factors of polyphase induction motors under unbalanced line voltage conditions, *IEEE Trans. Power Appar. Syst.*, **PAS-87**, 1, Jan. 1968, pp. 240-249.

12. KRAUSE, P.C., A constant frequency induction motor speed control, *Proc. Natl. Electron. Conf.*, **20**, 1964, pp. 361-365.

13. KOVACS, K.P., Über die genaue und vollständige Simulation des am Ständer mit steuerbaren Siliziumtrioden geregelten Drehstrom-Asynchronmotors, *Acta Tech. Acad. Scient. (Hungar)*, **48**, 3/4, 1964, pp. 445-459.

14. NOVOTNY, D.W., and FATH, A.F., The analysis of induction machines controlled by series connected semiconductor switches, *IEEE Trans. Power Appar. Syst.*, **PAS-87**, 2, Feb. 1968, pp. 597-605.

15. SHEPHERD, W., On the analysis of the three-phase induction motor with voltage control by thyristor switching, *IEEE Trans. Ind. Gen. Appl.*, **IGA-4**, 3, May/June 1968, pp. 304-311.

16. LIPO, T.A., The analysis of induction motors with voltage control by symmetrically triggered thyristors, *IEEE Trans. Power Appar. Syst.*, **PAS-90**, 2, Mar./Apr. 1971, pp. 515-525.

17. BEDFORD, R.E., and NENE, V.D., Voltage control of the three-phase induction motor by thyristor switching: A time-domain analysis using the α-β-0 transformation, *IEEE Trans. Ind. Gen. Appl.*, **IGA-6**, 6, Nov./Dec. 1970, pp. 553-562.

18. RAHMAN, S., and SHEPHERD, W., Thyristor and diode controlled variable voltage drives for 3-phase induction motors, *Proc. IEE*, **124**, 9, Sept. 1977, pp. 784-790.

19. PETERSEN, J.G., Thyristor drives for ac cranes and hoists, ASME-IEEE Materials Handling Conf., 1965.

20. KENLY, W.L., and BOSE, B.K., Triac speed control of three-phase induction motor with phase-locked loop regulation, *IEEE Trans. Ind. Appl.*, **IA-12**, 5, Sept./Oct. 1976, pp. 492-498.

21. SHEPHERD, W., and STANWAY, J., An experimental closed-loop variable speed drive incorporating a thyristor driven induction motor, *IEEE Trans. Ind. Gen. Appl.*, **IGA-3**, 6, Nov./Dec. 1967, pp. 559-565.

22. LINGARD, B.W., JOHNSON, R.W., and SHEPHERD, W., Analysis of thyristor-controlled single-phase loads with integral-cycle triggering, *Proc. IEE*, **117**, 3, Mar. 1970, pp. 607-608.

23. ERLICKI, M.S., BEN URI, J., and WALLACH, Y., Switching drive of induction motors, *Proc. IEE*, **110**, 8, Aug. 1963, pp. 1441-1450.

24. NOLA, F.J., Power factor control system for ac induction motor, U.S. Patent 4,052,648, Oct. 4, 1978.

25. Save power in induction motors, *NASA Tech. Brief,* MFS-2328, Summer 1977.

26. NOLA, F.J., Power factor controller — an energy saver, *Conf. Rec. IEEE Ind. Appl. Soc. Annual Meeting, 1980,* pp. 194-198.

27. MOHAN, N., *Evaluation and Comparison of State-of-the-Art Techniques for Energy Conservation by Reduced Losses in AC Motors,* Electric Power Research Institute Report EM-2037, Sept. 1981.

28. JIAN, T.W., NOVOTNY, D.W., and SCHMITZ, N.L., Characteristic induction motor slip values for variable-voltage part load performance optimization, *IEEE Trans. Power Appar. Syst.,* **PAS-102,** 1, Jan./Feb. 1983, pp. 38-46.

29. ROWAN, T.M., and LIPO, T.A., A quantitative analysis of induction motor performance improvement by SCR voltage control, *IEEE Trans. Ind. Appl.,* **IA-19,** 4, July/Aug. 1983, pp. 545-553.

30. MUNGENAST, J., Design and application of a solid state ac motor starter, *Conf. Rec. IEEE Ind. Appl. Soc. Annual Meeting, 1974,* pp. 861-866.

CHAPTER 10

Speed Control of Wound-Rotor Induction Motors

10.1. INTRODUCTION

The wound-rotor induction motor is not as rugged as the squirrel-cage motor and cannot operate at speeds as high. However, in a slip-ring machine there is little wear or sparking at the brushes, and hence the maintenance requirements are considerably less than in a dc motor. Consequently, the slip-ring induction motor is an acceptable alternative to the squirrel-cage machine in many industrial applications, and the availability of the rotor terminals introduces new possibilities for wide-range speed control. The most important method of speed control involves the removal of rotor slip power by a converter cascade circuit.[1-18] The principle of slip-energy recovery offers high efficiency and high reliability, and the drive has good control characteristics. Static converter cascades have found wide acceptance in recent years, particularly for a limited subsynchronous range of speed variation, when the converter rating can be significantly less than the motor rating. Subsynchronous static converter cascades have been used in pump drives with motor output ratings of up to 20 MW. Drive efficiencies of over 95 percent are obtained in the megawatt range.

10.2. THEORETICAL PRINCIPLES OF SLIP-ENERGY RECOVERY

The division of power in the rotor circuit has already been discussed in Sections 6.2.2 and 9.2. The fundamental power delivered to the rotor across the airgap, P_{ag}, is divided between the mechanical power output, P_{mech}, and the rotor copper loss, P_2. The following equations have been previously derived:

$$P_2 = sP_{ag}$$

and

$$P_{mech} = (1 - s) P_{ag} .$$

Also

$$P_{ag} = T\omega_s$$

where T is the electromagnetic torque developed by the motor, and ω_s is the synchronous angular velocity.

From these equations, it is obvious that the airgap power is constant when the induction motor drives a constant-torque load, and hence the rotor copper loss is proportional to the slip. Speed control of a wound-rotor motor connected to the ac utility network by the introduction of external rotor resistance is, therefore, inherently inefficient. At half synchronous speed, the airgap power is divided equally between mechanical power output and rotor I^2R loss, giving an overall efficiency, when stator losses are taken into consideration, of less than 50 percent. In general, at slip s, mechanical power is obtained from the airgap power with a per-unit conversion efficiency of $(1 - s)$, and the overall motor efficiency is even lower than this. At low speeds, the airgap power is almost completely dissipated in the rotor circuit, and the efficiency is very poor. The rotor resistance method of speed control is, therefore, uneconomical except for a very small subsynchronous speed range. However, it is not essential that the slip power, sP_{ag}, be dissipated in resistance losses, because it can be removed from the rotor circuit and utilized externally, thereby improving the overall efficiency of the drive system. In these cascade connections, the slip power is either returned to the ac supply network or is used to drive an auxiliary motor which is mechanically coupled to the induction motor shaft.

The operation of the cascade connection may also be regarded as speed control by emf injection in the rotor circuit. Assume the motor is operating normally at slip s, and an external voltage is applied at the slip-rings in phase opposition to the rotor emf. The resultant decrease in rotor current causes a reduction in motor torque, because torque is proportional to in-phase rotor current, assuming constant airgap flux. The motor speed therefore decreases due to the braking action of the load. However, as the slip increases, the rotor emf and current also increase, and stable operation is obtained at some reduced speed when the motor torque again equals the load torque.

When rotor resistance control is adopted, the slip-frequency voltage drop in the external resistor constitutes the injected voltage, but an external emf source of low impedance is equally effective and does not introduce excessive heat loss. The main problem in providing a suitable emf source is that the frequency of the injected emf must match the rotor slip frequency at all speeds. Historically, mechanical frequency-conversion methods have been employed. In the ac commutator motor, the frequency-changing action of the commutator was utilized to match frequencies and give a self-contained adjustable-speed machine. At higher power levels, a conventional slip-ring induction motor was employed, and auxiliary rotating machines in the cascade circuit performed the frequency-changing function. In the traditional Scherbius system shown in Fig. 10.1, a rotary converter rectifies the slip power, and the rectified output drives a dc motor which is mechanically coupled to a squirrel-cage induction generator. The induction generator is driven at supersynchronous speeds and returns the slip power to the ac supply. In the traditional Kramer system, the slip power is also rectified by a rotary converter, and the output supplies a dc motor which is mechanically coupled to the main induction motor. The slip power is thus converted to mechanical power at the induction motor shaft.

Static frequency converters have now replaced the auxiliary machines in the Scherbius system, and this change has resulted in a more compact adjustable-speed drive with an improved operating efficiency and a better dynamic response. The Kramer drive can also be modified with the use of a diode rectifier bridge in place of the rotary converter, but a dc motor is still required to convert the rectified slip power to mechanical power. Consequently, the Kramer drive with a conventional dc brush motor is now obsolete. How-

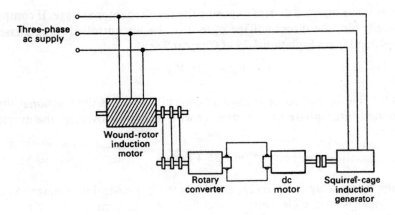

FIG. 10.1. The traditional Scherbius system for speed control of a wound-rotor induction motor.

ever, the reader should note that, in the literature, the modern Scherbius drive is occasionally termed a static Kramer drive.

10.3. SUBSYNCHRONOUS STATIC CONVERTER CASCADE

For subsynchronous speed control, the auxiliary machines of the Scherbius drive are replaced by a static converter, as shown in Fig. 10.2. This dc link converter consists of a three-phase diode rectifier bridge which operates at slip frequency and feeds rectified slip power through the smoothing inductor to the phase-controlled thyristor inverter. The inverter returns the rectified slip power to the ac supply network. The rectifier and inverter are both naturally commutated by the alternating emfs appearing at the slip-rings and supply busbars, respectively.

The rectification of the slip-ring voltages eliminates the problem of matching the frequencies of the injected emf and the rotor emf, because an adjustable dc back emf can now be used as the injected voltage for speed control. In Fig. 10.2, the average back emf

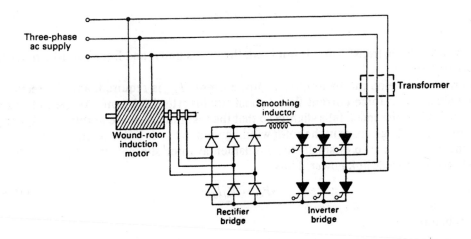

FIG. 10.2. Induction motor and subsynchronous static converter cascade.

of the inverter is the injected dc emf opposing the rectified rotor voltage. If commutation overlap is negligible, the direct voltage output of the uncontrolled three-phase bridge rectifier is obtained from Equation 3.25 of Chapter 3 as

$$V_{do} = 1.35 \ V_r s \tag{10.1}$$

where V_r is the line-to-line rotor voltage at standstill, and s is the fractional slip. For a line-commutated three-phase bridge inverter with negligible overlap, the average back emf is given by

$$V_d = 1.35 \ V_L \cos \alpha \tag{10.2}$$

where α is the firing delay ($\alpha \geq \pi/2$), and V_L is the ac line-to-line voltage. At no load, the motor torque is negligible and the rectified rotor current is almost zero. Consequently, the two direct voltages of Equations 10.1 and 10.2 must balance. Thus,

$$1.35 \ V_r s + 1.35 \ V_L \cos \alpha = 0 \tag{10.3}$$

and hence

$$s = - (V_L/V_r) \cos \alpha = -a \cos \alpha = a|\cos \alpha| \tag{10.4}$$

where a is the effective stator-to-rotor turns ratio of the motor.

Speed control is, therefore, obtained by a simple variation of the inverter firing angle. If a is unity, the no-load speed of the motor can be controlled from near standstill to full speed, as $|\cos \alpha|$ is varied from almost unity to zero.

In practice, the motor turns ratio, a, is usually greater than unity, resulting in a low rotor voltage. Consequently, a transformer is often required between the ac supply network and the inverter in order to step down the utility voltage to a level that is appropriate for the slip-ring circuit. This transformer is indicated in outline in Fig. 10.2. If the transformer turns ratio for the utility side relative to the inverter side is denoted by a_T, then the ac line-to-line voltage applied to the inverter terminals is V_L/a_T, and Equation 10.4 has the modified form

$$s = \frac{a}{a_T}|\cos \alpha| . \tag{10.5}$$

Clearly, if a cascade transformer is not used, then a_T is unity, and Equation 10.5 reduces to Equation 10.4.

In order to develop motor torque, a rotor current, I_{2r}, is required, and the rectified rotor voltage must force current flow against the inverter back emf. As the induction motor is loaded, the speed falls slightly so that the resulting increase in rotor voltage can overcome the voltage drops in the rotor windings and in the dc link circuit.

If the rotor resistance is small, the fundamental rotor slip power, sP_{ag}, is approximately equal to the dc link power. Thus,

$$sP_{ag} = V_d I_d \tag{10.6}$$

but, as before,

$$P_{ag} = T\omega_s$$

and hence

$$T = \frac{V_d I_d}{s \omega_s}. \tag{10.7}$$

If the speed droop on load is neglected, Equation 10.4 for the no-load slip can be substituted in Equation 10.7. Substituting also for V_d from Equation 10.2 gives the torque expression

$$T = \frac{1.35 \, V_L I_d}{a \omega_s}. \tag{10.8}$$

This equation is also valid when a transformer is present in the cascade circuit. Thus, the steady-state torque is proportional to the rectified rotor current, I_d, which, in turn, is equal to the difference between the rectified rotor voltage and the average back emf of the inverter divided by the resistance of the dc link inductor. For a fixed firing angle, the inverter emf is constant, and hence the rotor slip increases linearly with load torque, giving a torque-speed characteristic similar to that of a separately excited dc motor with armature-voltage control. In practice, the complete open-loop torque-speed characteristics have the form shown in Fig. 10.3.

10.3.1. *Power Factor*

The principal disadvantage of the subsynchronous cascade drive is its low fundamental power factor, or displacement factor, particularly at reduced speeds. If the system is designed for wide-range speed control, the full-load power factor at maximum speed may be as low as 0.5, decreasing to 0.3 or less as speed is reduced. This low power factor is partly due to the commutating reactive power that is drawn through the induction motor from stator to rotor by the three-phase bridge rectifier.[6–8] However, the reactive

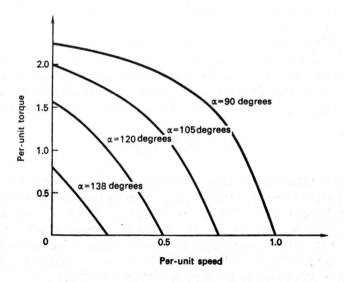

FIG. 10.3. Open-loop torque-speed characteristics for the induction motor and static converter cascade.

power consumption of the line-commutated inverter is largely responsible for the low power factor of the cascade drive. The average back emf of the inverter is a maximum at the lowest controllable speed. Ideally, under these conditions, the inverter firing angle is 180 degrees, but, as usual, commutation overlap and thyristor recovery time are significant, and hence the inverter firing must be advanced to prevent a commutation failure. This firing advance causes the inverter output current to lead the corresponding phase voltage, and the inverter acts as a generator of leading reactive power — that is, a consumer of lagging reactive power (Section 3.3.4). The reactive power consumption increases when the inverter firing point is further advanced to reduce the back emf and motor slip. At full speed, the inverter firing angle is 90 degrees, and the inverter kVA is almost completely reactive. At a given firing angle, the net power drawn from the ac supply is the difference between the power delivered to the stator and the power returned by the inverter. However, the total reactive kVA is the sum of the reactive powers absorbed by the motor and inverter, and consequently, the system power factor is poor at low speeds when the active power consumption is small.

As explained in the next section, the cascade drive is frequently operated with a reduced subsynchronous range of speed control. In order to maintain the drive power factor at the highest possible value throughout the speed range, the inverter delay angle should be a maximum at the lowest controllable speed. Thus, at the maximum controllable slip, s_{max}, the rectifier bridge has zero delay angle, and the inverter bridge has a delay angle of almost 180 degrees. Consequently, when the inverter is connected directly to the ac supply network without a voltage-matching transformer, the slip-ring voltage at the greatest controllable slip, $s_{max} V_r$, is approximately equal to the ac network voltage, V_L. Hence

$$s_{max} = \frac{V_L}{V_r} = a \ . \tag{10.9}$$

Thus, for optimum operation, the stator-to-rotor transformation ratio, a, should be equal to s_{max}; for example, if subsynchronous speed control is required down to a slip of 0.6, the effective stator-to-rotor turns ratio should also be 0.6.

As already explained, the alternating voltage applied to the inverter terminals is often reduced by means of a step-down transformer because the slip-ring voltage is limited. The transformer ratio, a_T, is now chosen in conjunction with the motor turns ratio, a, so that the maximum controllable slip again requires full inversion. Thus, the maximum slip-ring voltage, $s_{max} V_r$, is equal to the inverter supply voltage, V_L/a_T, so that in general

$$s_{max} = \frac{V_L}{V_r a_T} = \frac{a}{a_T} \ . \tag{10.10}$$

Numerous circuit modifications have been devised to improve the power factor of the cascade drive. A special type of "through-pass" inverter has been developed to replace the phase-controlled inverter.[19] This inverter circuit can be line commutated or forced commutated; it returns slip power to the ac supply network without drawing large amounts of reactive power. Modified gating of conventional line-commutated bridge inverters can be adopted to give half-controlled operation or "controlled flywheeling" with a resulting improvement in drive power factor and efficiency.[20-24] A modified line-commutated inverter bridge configuration has also been devised in which two auxiliary series-connected thyristors are placed across the dc link with the midpoint joined to the ac supply neutral.[25] This eight-thyristor inverter circuit also gives improved drive performance.[26] Another circuit variation involves the introduction of a chopper thyristor across the dc link of a conventional six-thyristor bridge inverter.[27] By suitably gating the chopper thyristor, the ac line commutation of the inverter bridge is assisted, and a significant improvement in drive power factor is achieved. Motor-winding tap-changing has also been suggested as a means of improving the power factor and reducing the converter rating in cascade drives with fan-type loads,[28] but if a voltage-matching transformer is present in the cascade circuit, its transformation ratio can be altered in several steps over the speed range to maintain optimum conditions. Capacitive compensation has also been suggested as a method of improving the power factor and giving increased torque at low speeds.[29]

10.3.2. *Operation with a Limited Subsynchronous Speed Range*

As shown above, the slip power, $P_2 = sP_{ag}$, is carried by the cascade converter circuit, and, for a restricted speed range below synchronous speed, the slip power can be substantially less than the full-load power rating of the motor. Thus, if the lowest motor speed is 80 percent of synchronous speed, the slip power for a constant-torque load is limited to 20 percent of the stator input power.

The kVA ratings of the cascade rectifier and inverter are a function of the speed range. In a subsynchronous cascade drive that is controlled from standstill to full speed, the static converter must have the same kVA rating as the induction motor. When the subsynchronous speed range is limited, however, the converter rating can be reduced proportionately and the system power factor is improved. The induction motor must be run up from standstill in the usual manner with a starting resistor in the rotor circuit. At the lowest speed in the control range, the cascade converter is automatically switched in by a speed detector, and the starting resistor is then disconnected.

The three-phase bridge rectifier has a voltage rating that is determined by the maximum working slip of the motor, because rotor emf and slip are proportional. The current rating of the rectifier bridge is determined by the maximum rotor current corresponding to the maximum output torque. With a constant-torque load, the rotor current, I_{2r}, and the airgap power, P_{ag}, are approximately constant. Maximum rotor voltage occurs at the greatest controllable slip, s_{max}, when, neglecting rotor losses, the cascade converter transfers the slip power, $s_{max}P_{ag}$, from the rotor to the ac supply network. With a fan load, the torque varies as the square of the speed, and hence the rotor current is a maximum at full speed. The slip-ring voltage, however, is greatest at the lowest controllable speed, and the rectifier bridge must be rated for the maximum alternating voltage and current, even though they do not occur simultaneously.

The inverter bridge is rated for the applied ac voltage and the alternating current corresponding to the maximum output torque of the motor. In order to minimize the ratings of rectifier and inverter, the transformation ratios of the motor and cascade transformer must be chosen such that the lowest controllable speed requires full inversion. This design procedure also optimizes the power factor of the system, as already explained. The inverter transformer is, of course, only rated for the inverter kVA.

When the subsynchronous speed range is limited, the power factor of the cascade drive is improved.[20] This is illustrated in Fig. 10.4, which is an approximate phasor dia-

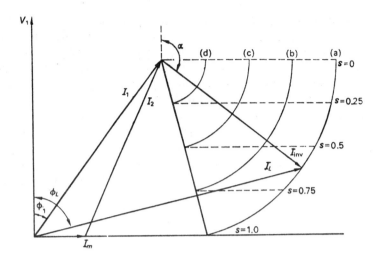

FIG. 10.4. Locus diagram of supply current for constant-torque operation and a speed control range: (a) from 0 to 100 percent of rated speed; (b) from 25 to 100 percent of rated speed; (c) from 50 to 100 percent of rated speed; (d) from 75 to 100 percent of rated speed.

gram for a constant-torque loading of the motor. All harmonic currents are neglected, and the current phasors in the diagram represent the fundamental components of the nonsinusoidal waveforms, referred to the stator. The fundamental stator current, I_1, is obtained by combining I_m, the magnetizing component, and $I_2 (= I_{2r}/a)$, the rotor current referred to the stator, as in Fig. 10.4. The stator current lags the phase voltage, V_1, by the phase angle, ϕ_1, and the tip of the I_1 phasor lies inside the normal circle diagram of the motor with a short-circuited rotor. This is due to the commutating reactive power which is drawn by the diode rectifier bridge in the rotor circuit.[6-8] The fundamental component of inverter current, I_{inv}, lags the supply voltage, V_1, by the firing angle, α, if commutation overlap is neglected. The phasor sum of I_1 and I_{inv} gives the total line current, I_L, with the phase angle, ϕ_L.

In Fig. 10.4, a constant load torque is assumed and, consequently, the referred rotor current, I_2, and the stator current, I_1, do not vary appreciably. The inverter firing angle is assumed to be 165 degrees at the lowest controllable speed, thus allowing 15 degrees for reliable thyristor commutation. For speed control down to standstill, the current locus is given by curve (a). At standstill, the slip power returned to the supply network is equal to the stator input power, if all losses are neglected. The total line current, I_L, then lags the supply voltage by 90 degrees. At a slip of 0.5, the slip power is half the stator power, and the active component of I_{inv} is half the active component of I_1. This simplified analysis yields the loci (a), (b), (c), and (d) of Fig. 10.4, showing the reduction in reactive current and reactive power consumption as the speed range is limited.

A significant improvement in the system power factor and a reduction in the rating and cost of the cascade converter are achieved, therefore, when the subsynchronous speed range is limited. The converter rating is then appreciably less than that of the static converter in an armature-controlled dc motor drive for the same horsepower. In the dc drive, the converter is rated for the maximum armature voltage and current, corresponding to full-load operation at maximum speed. In a static cascade drive with a speed control range from 70 to 100 percent of full speed, the converter kVA rating is only 30 percent of the motor rating. The induction motor and static converter cascade system is, therefore, particularly suitable for drives requiring a limited speed range below synchronism. Typical applications are fan and pump drives, for which the power output varies as the cube of the speed; hence, a limited subsynchronous speed range is adequate.

10.3.3. *The Influence of Harmonics*

In the cascade converter, the slip power is rectified to dc, filtered, and inverted to ac. The rectification process produces nonsinusoidal rotor currents which induce associated harmonic currents in the stator winding. These current harmonics cause additional harmonic torques and losses which necessitate some derating of the induction motor. An asynchronous harmonic torque is produced by the interaction between a time harmonic rotor flux and the corresponding induced stator current. The asynchronous torques are usually small in magnitude.[9] Space harmonic fluxes also produce synchronous locking torques at certain speeds, but their effect is negligible in a good machine design.

The smoothing reactor in the dc link is necessary for satisfactory operation of the inverter, and a large inductance is required to give continuous conduction over most of the load range. The harmonic currents in the stator and rotor cause harmonic copper losses which slightly reduce the motor efficiency. In order to limit these harmonic losses to a given percentage of the machine losses, the value of the filter inductance should be increased as the rated dc link current of the system decreases.[9] Converter losses must also be taken into consideration for calculating the overall system efficiency. In a static converter cascade system, the converter losses are considerably lower than in a traditional Scherbius drive, which has three rotating machines in the cascade circuit. The full-load efficiency near maximum speed in a static converter cascade drive system is only a few percentage points less than the normal motor efficiency on full load with a short-circuited rotor.

The harmonic losses cause an increased temperature rise in the induction motor and reduce its rating. The commutating reactive power, which is drawn through the motor by the bridge rectifier, also reduces the useful power rating of the motor. As a result, the induction motor must be derated by 10 to 15 percent when it is employed in a static cascade connection.

At operating speeds close to synchronous speed, the rotor frequency and generated emf are both low. This condition imposes difficulties in the commutation of the diode rectifier bridge because the overlap period may be appreciably prolonged due to the small rotor emfs and the relatively large rotor leakage reactance. If the overlap period exceeds 60 degrees, then four diodes conduct simultaneously, and the dc link is short-circuited by two conducting diodes in the same rectifier leg. Under these conditions, the dc link voltage cannot be established and slip power cannot be transferred. Consequently, the maximum speed is limited to 93 to 95 percent of synchronous speed. Contactors can short out the slip-rings for operation at the lowest possible level of slip.

10.3.4. *Closed-Loop Control*

The cascade drive, like the separately excited dc motor, has a linear torque-current relationship, and speed control is normally obtained with a double-loop control system

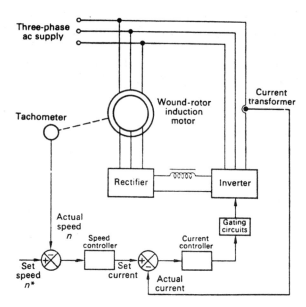

FIG. 10.5. Block diagram of control system for closed-loop operation.

which is similar to that used in single-converter dc motor drives. Figure 10.5 shows a block diagram of the control scheme. The inner control loop adjusts the current in the cascade circuit by variation of the inverter firing angle, and this current determines the motor torque. The actual cascade current is measured by current transformers on the ac side of the inverter, and the desired current value is set by the outer speed loop, which measures the difference between the desired speed and the actual tachometer value. Thus, as usual, a speed error produces a motor torque that reduces the error. Current limitation is readily incorporated by limiting the reference current value delivered by the speed controller. In this manner, the cascade current may be limited to 120 percent, for example, of rated current. If the desired speed greatly exceeds the measured value, the motor accelerates with maximum current and torque until the set speed is approached. The current and torque are then reduced automatically to the values required by the load. For a sudden reduction in demanded speed, the current is reduced to a low level and the braking action of the load reduces the motor speed.

In the cascade drive, the control system is obviously much simpler than in other adjustable-speed induction motor drives. The double-loop control system is readily stabilized, and the speed regulation on load may be reduced by means of a high-gain speed controller. In this manner, a speed droop from no load to full load of 0.1 percent or less may be achieved.

10.4. SUPERSYNCHRONOUS STATIC CONVERTER CASCADE

10.4.1. *DC Link Converter Cascade* [30-34]

The static converter circuit of Fig. 10.2 achieves subsynchronous speed control by the removal of slip power from the rotor circuit. In fact, supersynchronous speeds may also be obtained by reversing the direction of power flow in the cascade circuit, thereby feeding additional power into the rotor. This power reversal is readily achieved with rotating machinery because all electrical machines are inherently capable of motor or generator operation. In order to reverse the power flow in the converter circuit of Fig. 10.2, the rectifier must operate as an inverter and vice versa. For speed control on either side of synchronism, the converter cascade therefore needs two phase-controlled thyristor bridges, one operating at slip frequency as a rectifier or inverter, while the other operates at network frequency as an inverter or rectifier. The circuit diagram is shown in Fig. 10.6. The replacement of six diodes by six thyristors increases the converter cost and also necessitates the introduction of slip-frequency gating circuits. As already mentioned, difficulty is experienced near synchronism when the slip-frequency emfs are insufficient for natural commutation, and special circuit configurations employing forced commutation or devices with a self-turn-off capability are necessary for the passage through synchronism. Thus, the provision of supersynchronous speed control complicates the static converter cascade system and nullifies the advantages of simplicity and economy which are inherent in a purely subsynchronous drive.

10.4.2. *Cycloconverter Cascade* [35–40]

A second approach for achieving bidirectional power flow in the cascade circuit is through use of a line-commutated cycloconverter (Chapter 4). An 18-thyristor circuit is shown in Fig. 10.7, but other circuit configurations can be employed, and a voltage-matching transformer may be introduced between the cycloconverter and the ac supply network. The cycloconverter must be controlled so that the frequency of the injected slip-ring voltages tracks the rotor slip frequency. As before, for subsynchronous motoring, the slip-frequency rotor emfs deliver slip power to the cycloconverter, and this power is returned to the ac utility supply. At supersynchronous speeds, the slip-ring voltages have opposite phase sequence, and the cycloconverter delivers slip power to the rotor. Thus, for supersynchronous motoring, the ac utility network must supply both the stator input power and the slip power.

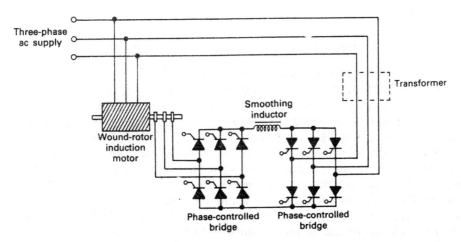

FIG. 10.6. Induction motor and supersynchronous static converter cascade using a naturally commutated dc link converter.

Three-phase ac supply

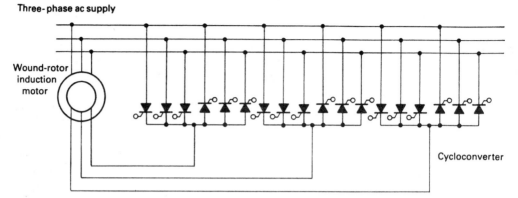

FIG. 10.7. Induction motor and supersynchronous static converter cascade using a line-commutated cycloconverter.

The line-commutated cycloconverter can achieve a minimum step-down frequency ratio of approximately three to one with near-sinusoidal low-frequency voltage and current. For a 60 Hz utility supply to the stator, the maximum rotor frequency is approximately 20 Hz, corresponding to operation at 67 percent of synchronous speed. With rotor voltages of 20 Hz and opposite phase sequence, the motor speed is 133 percent of synchronous speed. For synchronous operation, the rotor current is at zero frequency, and the cycloconverter supplies dc excitation to the rotor, thereby causing the induction motor to function as a synchronous motor. The overall power factor of the drive is maximized if the transformation ratios of motor and transformer are such that the largest injected voltage required at the slip-rings corresponds to the maximum output voltage available from the cycloconverter.

The cycloconverter cascade is expensive, and it introduces additional control complexity, but the near-sinusoidal rotor currents minimize harmonic heating effects and low-frequency torque pulsations. Commutation problems near synchronous speed are also eliminated.

10.5. STATIC CONTROL OF ROTOR RESISTANCE [41–45]

Intermittent low-speed operation of a wound-rotor induction motor can be simply obtained by the introduction of external rotor resistance to dissipate the slip power. Conventionally, speed control is obtained by mechanical variation of the external resistance in discrete steps, and the resulting drive is suitable for crane and hoist applications where sustained low-speed running is not required. Nowadays, a high-frequency chopper circuit allows the external rotor resistance to be varied statically and steplessly, and provides a low-cost adjustable-speed drive with a good dynamic response.

Figure 10.8 shows a practical circuit in which the rotor slip power is rectified in a diode rectifier bridge and is then fed through a smoothing reactor to a resistor, R. A single thyristor, or GTO, in parallel with the resistor is switched on and off, at a frequency of about 1 kHz. The duty cycle, or ratio of on-time to cycle-time, determines the effective value of rotor resistance and thus controls the motor speed by altering its torque-speed characteristic. By introducing a capacitor in series with the external resistor, it is possible

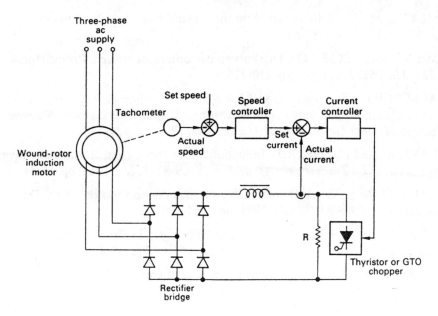

FIG. 10.8. Induction motor speed control by static variation of external rotor resistance.

to obtain a variation in the effective resistance from zero to infinity, thus permitting a wider range of speed control.[46,47] Static control of external rotor resistance may also be used in conjunction with the stator-voltage-control circuits of Chapter 9.[48,49] When the load torque is small, speed control is obtained by variation of the stator voltage, and rotor resistance control is used in the high torque range. Alternatively, the duty cycle of the chopper can be controlled as a predetermined function of rotor speed so that a high value of torque is available at all speeds without excessive stator current.

10.6. REFERENCES

1. MEYER, M., The single-range induction motor and static converter cascade system — an excellent drive for small speed control ranges, *Siemens Rev.*, **29**, 5, May 1962, pp. 175-178.

2. ANSCHÜTZ, H., Der Einsatz von Stromrichterkaskaden für industrielle Antriebe, *BBC Nachr.*, **47**, 1, Jan. 1965, pp. 13-20.

3. ELGER, H., and WEISS, M., A subsynchronous static converter cascade for variable-speed boiler feed pump drives, *Siemens Rev.*, **35**, Oct. 1968, pp. 405-407.

4. SHEPHERD, W., and STANWAY, J., Slip power recovery in an induction motor by the use of a thyristor inverter, *IEEE Trans. Ind. Gen. Appl.*, **IGA-5**, 1, Jan./Feb. 1969, pp. 74-82.

5. BLAND, R.J., HANCOCK, N.N., and WHITEHEAD, R.W., Considerations concerning a modified Kramer system, *Proc. IEE*, **110**, 12, Dec. 1963, pp. 2228-2232.

6. MEYER, M., Über die untersynchrone Stromrichterkaskade, *Elektrotech. Z. Ausg. A*, **82**, 19, 1961, pp. 589-596.

7. MIKULASCHEK, F., Die Ortskurven der untersynchronen Stromrichterkaskade, *AEG Mitt.*, **52**, 5/6, 1962, pp. 210-219.

8. ALBRECHT, S., and GAHLEITNER, A., Bemessung des Drehstrom-Asynchronmotors in einer untersynchronen Stromrichterkaskade, *Siemens-Z.*, **40**, *Beiheft Motoren für industrielle Antriebe*, Oct. 1966, pp. 139-146.

9. LAVI, A., and POLGE, R.J., Induction motor speed control with static inverter in the rotor, *IEEE Trans. Power Appar. Syst.*, **PAS-85**, 1, Jan. 1966, pp. 76-84.

10. ERLICKI, M.S., Inverter rotor drive of an induction motor, *IEEE Trans. Power Appar. Syst.*, **PAS-84**, 9, Nov. 1965, pp. 1011-1016.

11. WALLACH, Y., ERLICKI, M.S., and BEN-URI, J., Transients in a Kramer cascade, *IEEE Trans. Ind. Gen. Appl.*, **IGA-2**, 2, Mar./Apr. 1966, pp. 158-162.

12. HORI, T., and HIRO, Y., The characteristics of an induction motor controlled by a Scherbius system, *Conf. Rec. IEEE Ind. Appl. Soc. Annual Meeting, 1972*, pp. 775-782.

13. NODA, J., HIRO, Y., and HORI, T., Brushless Scherbius control of induction motors, *Conf. Rec. IEEE Ind. Appl. Soc. Annual Meeting, 1974*, pp. 111-118.

14. KUSKO, A., and SOMUAH, C.B., Speed control of a single-frame cascade induction motor with slip-power pump-back, *IEEE Trans. Ind. Appl.*, **IA-14**, 2, Mar./Apr. 1978, pp. 97-105.

15. WEISS, H.W., Adjustable speed ac drive systems for pump and compressor applications, *IEEE Trans. Ind. Appl.*, **IA-10**, 1, Jan./Feb. 1974, pp. 162-167.

16. WAKABAYASHI, T., HORI, T., SHIMIZU, K., and YOSHIOKA, T., Commutatorless Kraemer control system for large-capacity induction motors for driving water service pumps, *Conf. Rec. IEEE Ind. Appl. Soc. Annual Meeting*, 1976, pp. 822-828.

17. TSUCHIYA, T., Suboptimal control of a static Scherbius induction motor system using a microprocessor, *IEEE Trans. Ind. Appl.*, **IA-16**, 5, Sept./Oct. 1980, pp. 686-699.

18. BROWN, J.E., DRURY, W., JONES, B.L., and VAS, P., Analysis of the periodic transient state of a static Kramer drive, *IEE Proc.*, **133, Pt. B**, 1, Jan. 1986, pp. 21-30.

19. MILJANIC, P.N., The through-pass inverter and its application to the speed control of wound rotor induction machines, *IEEE Trans. Power Appar. Syst.*, **PAS-87**, 1, Jan. 1968, pp. 234-239.

20. STÖHR, M., Vergleich zwischen Stromrichtermotor und untersynchroner Stromrichterkaskade, *Elektrotech. Maschinenbau*, **57**, 49/50, 1939, pp. 581-591.

21. DRURY, W., FARRER, W., and JONES, B.L., Performance of thyristor bridge converters employing flywheeling, *IEE Proc.*, **127, Pt. B**, 4, Apr. 1980, pp. 268-276.

22. DRURY, W., JONES, B.L., and BROWN, J.E., Application of controlled flywheeling to the recovery bridge of a static Kramer drive, *IEE Proc.*, **130, Pt. B**, 2, Mar. 1983, pp. 73-85.

23. DEWAN, S.B., and DUNFORD, W.G., Improved power factor operation of a three phase rectifier bridge through modified gating, *Conf. Rec. Ind. Appl. Soc. Annual Meeting*, 1980, pp. 830-837.

24. RAO, N.N., DUBEY, G.K., and PRABHU, S.S., Slip-power recovery scheme employing a fully controlled convertor with half-controlled characteristics, *IEE Proc.*, **130, Pt. B**, 1, Jan. 1983, pp. 33-38.

25. STEFANOVIC, V.R., Power factor improvement with a modified phase-controlled converter, *IEEE Trans. Ind. Appl.*, **IA-15**, 2, Mar./Apr. 1979, pp. 193-201.

26. OLIVIER, G., STEFANOVIC, V.R., and APRIL, G.E., Evaluation of phase-commutated converters for slip-power control in induction drives, *IEEE Trans. Ind. Appl.*, **IA-19**, 1, Jan./Feb. 1983, pp. 105-112.

27. TANIGUCHI, K., and MORI, H., Applications of a power chopper to the thyristor Scherbius, *IEE Proc.*, **133, Pt. B**, 4, July 1986, pp. 225-229.

28. PAICE, D.A., Speed control of large induction motors by thyristor converters, *IEEE Trans. Ind. Gen. Appl.*, **IGA-5**, 5, Sept./Oct. 1969, pp. 545-551.

29. SHEPHERD, W., and KHALIL, A.Q., Capacitive compensation of thyristor-controlled slip-energy-recovery system, *Proc. IEE*, **117**, 5, May 1970, pp. 948-956.

30. STÖHR, M., Stromrichterkaskade für Doppelzonenregelung, *Elektrotech. Maschinenbau*, **58**, 17/18, 1940, pp. 177-186.

31. OHNO, E., and AKAMATSU, M., Secondary excitation of an induction motor using a self-controlled inverter (super-synchronous thyristor Scherbius system), *Electr. Eng. in Jpn*, **88**, 10, 1968, pp. 76-86.

32. KAZUNO, H., A wide-range speed control of an induction motor with static Scherbius and Krämer systems, *Electr. Eng. in Jpn*, **89**, 2, 1969, pp. 10-19.

33. ZIMMERMANN, P., Super-synchronous static converter cascade, *Proc. 2nd IFAC Symp. on Control in Power Electronics and Electrical Drives*, 1977, pp. 559-566.

34. JOETTEN, R., and ZIMMERMANN, P., Sub- and super-synchronous cascade with current-source inverter, *Proc. 3rd IFAC Symp. on Control in Power Electronics and Electrical Drives*, 1983, pp. 337-343.

35. MAYER, C.B., High response control of stator watts and vars for large wound rotor induction motor adjustable speed drives, *IEEE Trans. Ind. Appl.*, **IA-19**, 5, Sept./Oct. 1983, pp. 736-743.

36. LONG, W.F., and SCHMITZ, N.L., Cycloconverter control of the doubly fed induction motor, *IEEE Trans. Ind. Gen. Appl.*, **IGA-7**, 1, Jan./Feb. 1971, pp. 95-100.

37. CHATTOPADHYAY, A.K., An adjustable speed induction motor drive with a cycloconverter type thyristor-commutator in the rotor, *IEEE Trans. Ind. Appl.*, **IA-14**, 2, Mar./Apr. 1978, pp. 116-122.

38. CHATTOPADHYAY, A.K., Digital computer simulation of an adjustable speed induction motor drive with a cycloconverter type thyristor commutator in the rotor, *IEEE Trans. Ind. Electron. and Control Instrum.*, **IECI-23**, 1, Feb. 1976, pp. 86-92.

39. SMITH, G.A., Static Scherbius system of induction-motor speed control, *Proc. IEE*, **124**, 6, June 1977, pp. 557-560.

40. LEONHARD, W., Field oriented control of a variable speed alternator connected to the constant frequency line, *IEEE Conf. on Control in Power Systems*, 1979, pp. 149-153.

41. SEN, P.C., and MA, K.H.J., Rotor chopper control for induction motor drive: TRC strategy, *IEEE Trans. Ind. Appl.*, **IA-11**, 1, Jan./Feb. 1975, pp. 43-49.

42. SEN, P.C., and MA, K.H.J., Constant torque operation of induction motors using chopper in rotor circuit, *IEEE Trans. Ind. Appl.*, **IA-14**, 5, Sept./Oct. 1978, pp. 408-414.

43. KELKAR, S.S., and PILLAI, S.K., A modified rotor chopper for speed control of slip ring induction motors, *Proc. 2nd IFAC Symp. on Control in Power Electronics and Electrical Drives*, 1977, pp. 567-574.

44. RAMAMOORTY, M., and WANI, N.S., Dynamic model for a chopper-controlled slip-ring induction motor, *IEEE Trans. Ind. Electron. and Control Instrum.*, **IECI-25**, 3, Aug. 1978, pp. 260-266.

45. VAN WYK, J.D., Variable-speed ac drives with slip-ring induction machines and a resistively loaded force commutated rotor chopper, *IEE Proc., Electric Power Appl.*, 2, 6, Dec. 1979, pp. 149-160.

46. ABRAHAM, L., and PATZSCHKE, U., Pulstechnik für die Drehzahlsteuerung von Asynchronmotoren, *AEG Mitt.*, **54**, 1/2, 1964, pp. 133-140.

47. GOLDE, E., Asynchronmotor mit elektronischer Schlupfregelung, *AEG Mitt.*, **54**, 11/12, 1964, pp. 666-671.

48. KOPPELMANN, F., and MICHEL, M., Kontaktlose Steuerung der Drehzahl von Asynchronmotoren mit Hilfe antiparalleler Thyristoren, *AEG Mitt.*, **54**, 1/2, 1964, pp. 126-132.

49. CROWDER, R.M., and SMITH, G.A., Induction motors for crane applications, *IEE Proc., Electric Power Appl.*, 2, 5, Oct. 1979, pp. 149-160.

CHAPTER 11

Brushless DC Motor, Stepping Motor, and Variable-Reluctance Motor Drives

11.1. INTRODUCTION

The power electronic control of the conventional induction motor and synchronous motor has been discussed at length in earlier chapters. Now, increasing emphasis is being placed on the integrated design of the ac motor and its electronic controller. This system approach to adjustable-speed drives is aimed at the development of enhanced drive systems with reduced cost and improved performance. Notable examples of this approach are the modern brushless dc motor drive and the variable-reluctance, or switched-reluctance, motor drive, as described in this chapter. In these drive systems, the stator winding currents are switched synchronously with rotor position by means of a shaft position sensor. This self-controlled operation is equivalent to the replacement of the mechanical commutator and brushes of a conventional dc machine by an electronic commutator, and these synchronously commutated machines have the versatility and controllability of a dc machine, without the limitations imposed by mechanical commutation. As a result, brushless dc motor drive systems can be used in general-purpose applications or in more demanding high-performance servo applications.

The conventional stepping motor is also brushless and can be used at low power levels in open-loop positioning systems. In recent years, there has been a very rapid growth in the use of stepping motors and small brushless dc motors, due primarily to the fact that they have found widespread application in the rapidly growing computer peripheral market. The treatment of the brushless dc motor and stepping motor in the early sections of this chapter serves as an ideal introduction to the final section on the variable-reluctance motor drive. The industrial variable-reluctance, or switched-reluctance, motor has the basic construction and torque-producing mechanism of the variable-reluctance stepping motor, but it can be operated as a brushless dc motor to give a flexible adjustable-speed drive, and it is also possible to realize servo performance.

In the past, conventional dc armature windings have been used with electronic commutators that simulated the progressive switching action of the multisegment mechanical commutator.[1,2] However, as shown in Chapter 8, each commutator segment is equivalent to two power semiconductor switches, and if a conventional dc motor with a multisegment commutator is to be converted to an equivalent brushless dc motor, the cost of the semiconductor devices is far too high. For economic reasons, it is essential to minimize the number of semiconductors; the drive systems described in this chapter clearly demonstrate that it is possible to realize a high-performance brushless dc motor drive with a limited number of semiconductor devices.

413

11.2. BRUSHLESS DC MOTORS

In a conventional permanent magnet dc motor, the torque-speed characteristic is linear, except at high torque levels, where armature reaction effects may become significant. The term "brushless dc motor" is used to identify the combination of ac machine, solid-state inverter, and rotor position sensor that results in a drive system having a linear torque-speed characteristic, as in a conventional dc machine.[3-7] The ac motor has a polyphase winding on the stator and permanent magnets on the rotor. Motor operation is made self-synchronous by the addition of a rotor position sensor that controls the firing signals for the solid-state inverter. In response to these firing signals, the inverter directs current through the stator phase windings in a controlled sequence.

If the stator is fitted with a conventional three-phase winding, the motor has the construction of a standard permanent magnet synchronous machine and operates as a self-controlled synchronous motor, or inverted dc motor with an electronic commutator, as explained in Chapter 8. The terminology used in this area of drive technology is far from consistent, but it is now becoming accepted practice to reserve the term "brushless dc motor" for a self-synchronous machine in which the airgap flux distribution and counter emf, or back emf, waveform are approximately trapezoidal, as in a conventional dc machine. However, the standard permanent magnet synchronous motor, in which the airgap flux distribution and back emf waveform are both sinusoidal, is sometimes classified as a sinusoidal brushless dc motor when operating in a self-controlled mode. Indeed, the drive characteristics and control methods are very similar for the trapezoidal and sinusoidal machines, and in both cases, the motor must be energized with controlled currents that are synchronized with rotor position. However, a distinction is now being drawn between the two drives because of the differences in machine construction and because the standard synchronous motor requires sinusoidal current excitation, whereas the trapezoidal machine is energized with square-wave or quasi-square-wave currents. The rotor position sensor for the trapezoidal machine usually consists of a number of simple position detectors such as Hall-effect devices that can sense the rotor magnetic field and so determine the phase switching points. The sinusoidal machine requires more precise position information to allow accurate synthesis of the sinusoidal current waveforms.

Self-controlled operation of the permanent magnet synchronous machine has been studied in detail in Chapter 8, but it is instructive to consider the operation of the trapezoidal and sinusoidal brushless motors from the same viewpoint. Consequently, in this chapter the self-controlled permanent magnet synchronous motor drive will be briefly reexamined.

In general, the torque contribution of a particular stator phase is a function of phase current and rotor position. If a constant direct current is supplied to one stator phase and the rotor is rotated by an external force, the developed torque due to the interaction between the winding current and the magnet flux will vary periodically with shaft position. This characteristic is known as the torque function or static torque/angle characteristic of the motor. In the brushless dc motor, the torque function is trapezoidal, whereas in the permanent magnet synchronous motor, the torque function is sinusoidal.

11.2.1. *The Three-Phase Half-Wave Brushless DC Motor*

Figure 11.1 shows a basic three-phase half-wave brushless dc motor with its electronic

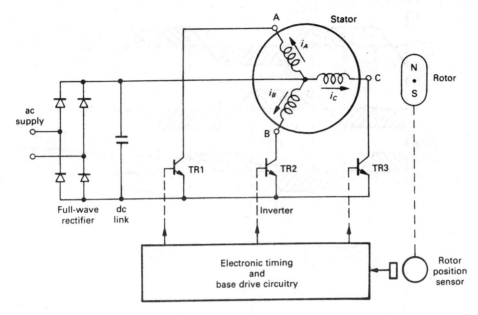

FIG. 11.1. A three-phase, half-wave brushless dc motor drive.

controller. The three stator phases are wye-connected with the neutral point joined to the positive terminal of the dc supply. Transistors TR1, TR2, and TR3 deliver unidirectional phase currents in response to base drive signals that are under the control of the rotor position sensor. This simple half-wave circuit is unusual in that there are no free-wheeling or feedback diodes to provide an alternative path for the inductive winding current when a transistor is turned off. Nevertheless, the circuit is satisfactory for small, low-cost systems in which the inductive winding energy is so small that it cannot cause destructive breakdown of the transistor at turn-off.

The idealized static torque/angle characteristics for the individual phases are shown in Fig. 11.2(a), (b), and (c). This trapezoidal torque function has a 120-degree flat-topped region that can be utilized to produce useful shaft torque. Thus, if the motor phases are energized sequentially with a constant current during each 120-degree interval of constant positive torque, the motor develops a steady positive torque that is independent of shaft position. The corresponding transistor, or phase, currents are shown in Fig. 11.2(d), (e), and (f). Each motor phase has a conduction angle of 120 electrical degrees, and the current amplitude is I, giving a resultant motor torque of

$$T = K_T I \qquad (11.1)$$

where the torque constant, K_T, is the torque per ampere over the flat portion of the trapezoid.

An examination of the static torque/angle characteristic shows that a torque reversal is achieved when the transistor conduction interval is delayed by 180 electrical degrees. The current direction is unchanged, but each phase now conducts during the 120-degree interval of negative torque, as shown in Fig. 11.2(g), (h), and (i). Thus, a torque rever-

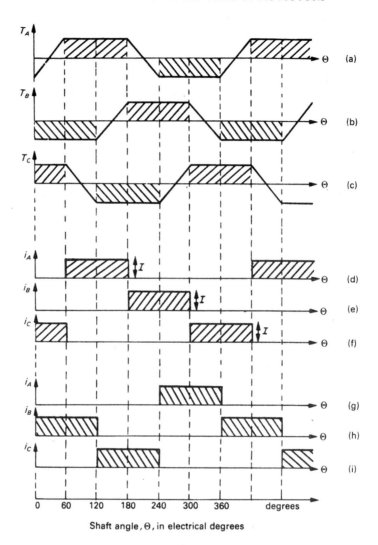

FIG. 11.2. Idealized waveforms for the three-phase, half-wave brushless dc motor: (a), (b), (c) static torque/angle characteristics; (d), (e), (f) phase currents for positive torque; (g), (h), (i) phase currents for negative torque.

sal is obtained in a brushless dc motor by shifting the firing signals by 180 electrical degrees, whereas in a conventional brush-type dc motor, a reversal of the supply voltage polarity and armature current direction is necessary.

The three-phase half-wave configuration of Fig. 11.1 is suitable for small brushless dc motors in a wide variety of applications at power levels from a few watts to 100 W. Practical imperfections in the idealized torque functions and current waveforms will result in some fluctuation in shaft torque during a revolution. A four-phase half-wave system having a 90-degree conduction angle is used when the torque ripple content must be reduced. Typical applications for these small brushless dc motors include turntable drives in record players, spindle drives in hard disk drives for computers, and various applications in low-cost instruments and computer peripheral equipment.

When the drive power level is increased, the stored energy in the motor winding inductance is more significant, and it must be returned to the dc supply to prevent destructive breakdown of the transistor at turn-off. This energy feedback can be accomplished in a half-wave system by using a center-tapped dc supply and feedback diodes, as shown later in Section 11.5.1 for the variable-reluctance motor drive. However, a full-wave system permits the use of a single dc supply and gives improved winding utilization.

11.2.2. The Three-Phase Full-Wave Brushless DC Motor

Figure 11.3 shows a three-phase full-wave brushless motor system with the familiar feedback diodes for inductive energy recovery. The wye-connected stator winding has no neutral connection, and two of the three phases are active at all times. For line-to-line dc currents from A to B, B to C, and C to A, the motor has the idealized trapezoidal torque functions of Fig. 11.4(a), (b), and (c), showing a 60-degree flat-topped region. Motor operation is always in this constant-torque region if the phases are supplied with quasi-square-wave currents of amplitude I, as shown in Fig. 11.4(d), (e), and (f). These currents are obtained by firing the transistors at 60-degree intervals in the sequence in which they are numbered in Fig. 11.3 and allowing a 120-degree conduction angle. Each transistor switching occurs in response to the rotor position sensor. Figure 11.4 also shows that each motor phase conducts for two 120-degree periods in each cycle, giving twice the winding utilization of the previous three-phase half-wave system. A steady nonpulsating torque is developed of magnitude $T = K_T I$, and as before, a torque reversal is achieved by phase-shifting the transistor base drive signals by 180 degrees. The idealized quasi-square-wave currents of Fig. 11.4 imply instantaneous switching of current from one phase combination to the next. In a practical voltage-fed system, the inductive load will delay the build-up of current and will also prolong conduction after the theoretical turn-off instant. These aspects of inverter operation have already been studied in Chapter 4. In an actual motor, the torque function will also depart somewhat from the ideal trapezoidal waveshape. Despite these practical imperfections, the commercial brushless dc motor can achieve a very low torque ripple and is eminently suitable for use in a high-performance servo drive, as discussed later.

11.2.3. The Sinusoidal Type of Brushless DC Motor

As already explained, the sinusoidal brushless dc motor is a self-controlled permanent magnet synchronous motor of normal manufacture. The usual design procedures result in a sinusoidal torque function. For convenience, consider a two-phase motor in which the torque functions for the two phases have a sine-cosine relationship, as shown in Fig. 11.5. The instantaneous torques developed by phases A and B are, respectively

$$T_A = i_A K_T \sin \Theta$$
$$T_B = -i_B K_T \cos \Theta \tag{11.2}$$

where i_A and i_B are instantaneous phase currents, and K_T is the torque constant of each winding.

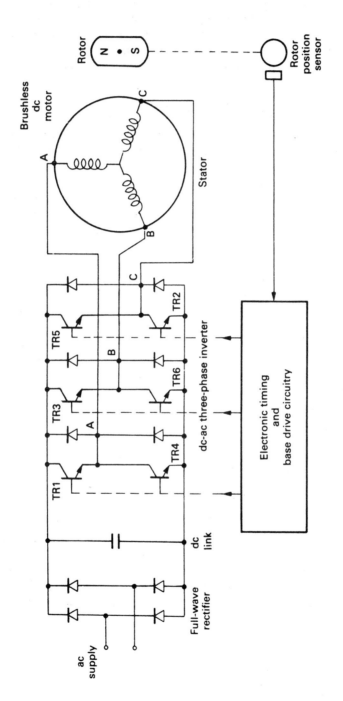

FIG. 11.3. A three-phase, full-wave brushless dc motor drive.

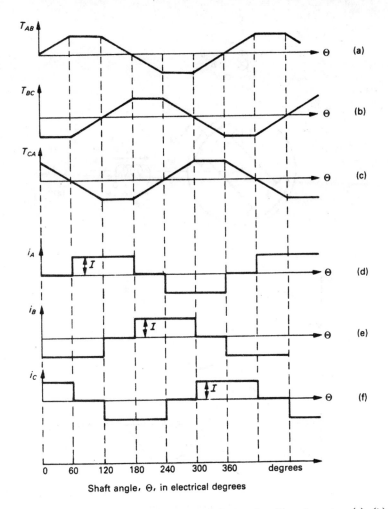

FIG. 11.4. Idealized waveforms for the three-phase, full-wave brushless dc motor: (a), (b), (c) static torque/angle characteristics; (d), (e), (f) phase currents for positive torque.

Now assume that the phase currents i_A and i_B are varied sinusoidally as a function of shaft position, Θ, so that

$$i_A = I_m \sin \Theta$$

$$i_B = -I_m \cos \Theta \qquad\qquad (11.3)$$

where I_m is the amplitude of the phase current. The torque contributions of the two phases are then

$$T_A = K_T I_m \sin^2 \Theta$$

$$T_B = K_T I_m \cos^2 \Theta$$

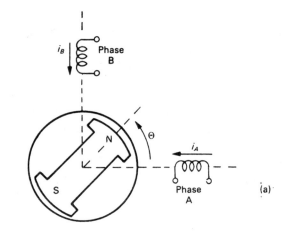

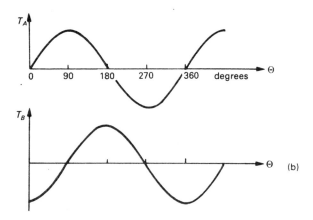

FIG. 11.5. A two-phase sinusoidal type of brushless dc motor: (a) basic construction; (b) static torque/angle characteristics.

and the resultant torque at any angle Θ is given by

$$T = T_A + T_B$$

$$= K_T I_m (\sin^2 \Theta + \cos^2 \Theta)$$

$$= K_T I_m . \tag{11.4}$$

Thus, the shaft torque is proportional to the current amplitude, I_m, and is independent of rotor position.

. The same principle is readily extended to the more common three-phase synchronous motor. In this case, the sinusoidal torque functions for each of the three phases are mutually displaced by 120 electrical degrees, as shown in Fig. 11.6. The instantaneous torques developed by the instantaneous phase currents i_A, i_B, and i_C are now

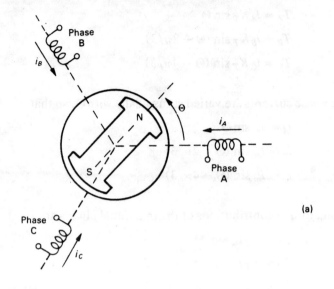

(a)

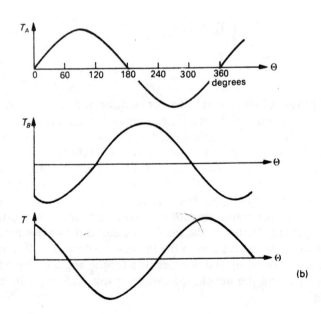

(b)

FIG. 11.6. A three-phase sinusoidal type of brushless dc motor: (a) basic construction; (b) static torque/angle characteristics.

$$T_A = i_A K_T \sin \Theta$$
$$T_B = i_B K_T \sin (\Theta - 2\pi/3)$$
$$T_C = i_C K_T \sin (\Theta - 4\pi/3) \, . \tag{11.5}$$

Assume that the phase currents are varied sinusoidally with Θ, so that

$$i_A = I_m \sin \Theta$$
$$i_B = I_m \sin (\Theta - 2\pi/3)$$
$$i_C = I_m \sin (\Theta - 4\pi/3) \, . \tag{11.6}$$

The corresponding torque contributions of the individual phases are

$$T_A = K_T I_m \sin^2 \Theta$$
$$T_B = K_T I_m \sin^2 (\Theta - 2\pi/3)$$
$$T_C = K_T I_m \sin^2 (\Theta - 4\pi/3) \tag{11.7}$$

and the resultant torque is

$$T = T_A + T_B + T_C$$
$$= \frac{3}{2} K_T I_m \, . \tag{11.8}$$

Again, the shaft torque is independent of rotor position and is linearly related to current amplitude, as in a conventional dc machine. This result is hardly surprising because it simply confirms the fact, already established in Chapter 8, that a conventional three-phase permanent magnet synchronous motor can be operated as a brushless dc motor by supplying the stator winding with balanced three-phase currents that are varied sinusoidally as a function of shaft position, Θ. If the sinusoidal variation in phase current is synchronized with the corresponding torque function for that phase, as assumed above, then the torque per stator ampere is maximized, and examination of Fig. 11.6 shows that the torque angle between the resultant stator mmf wave and the rotor magnet axis is fixed at 90 electrical degrees, as in the brush-type dc motor. When the phase conduction intervals are delayed by 180 degrees, the torque angle is altered from +90 degrees to −90 degrees, and the developed torque changes from positive (motoring) to negative (braking).

11.2.4. *Position Sensors*

The rotor position sensor is an integral part of the brushless dc motor system. Hall-effect sensors, electro-optical sensors, resolvers, and digital encoders are the most commonly used devices. As explained above, the Hall-effect sensor can detect the magnitude and direction of a magnetic field. The three-phase half-wave motor of Fig. 11.1 requires three of these sensors symmetrically mounted on the stator to detect the magnetic field due to the main rotor magnets or due to separate shaft-mounted magnets. The output signals from the sensors are processed to provide the logic signals required for the base drive circuits.

The electro-optical sensor consists of a light-emitting diode (LED) and phototransistor, which act as a light transmitter and detector, respectively. A slotted wheel is mounted on the motor shaft with a number of stator-mounted sensors around its periphery. Shaft rotation produces a shutter action so that the sensor outputs are switched high and low. Again, these output signals are processed to provide the logic signals required for electronic commutation.

Simple Hall-effect or electro-optical sensors are appropriate for the trapezoidal type of brushless motor with its abrupt phase-to-phase current switchings. In a sinusoidal type of motor, however, the phase current is a sinusoidal function of rotor position, and an absolute encoder or resolver or other high-resolution sensor is necessary to obtain position information with the required resolution. In the absolute optical encoder, an accurately patterned disk rotates between a light source and a detector, giving a unique digital output signal for every shaft position. Standard encoders are available with up to 16-bit resolution and with natural binary, gray code, or binary coded decimal output formats. However, each bit in the digital word represents an independent track on the encoder disk, resulting in a complex and costly sensor.

Brushless resolver operation is based on inductive coupling between stator and rotor windings. The resolver with its resolver-to-digital (R/D) converter also gives precise absolute digital position information, but again, the cost is often prohibitive.

In a sinusoidal type of brushless motor, absolute rotor position information is required to at least a 9- or 10-bit resolution.[8] Sinusoidal reference current waveforms are generated with this precise position information, and the actual phase currents track the reference currents in a current-controlled PWM inverter, as described in Chapter 8.

The provision of a high-resolution sensor adds significantly to the cost of the sinusoidal system. On the other hand, the trapezoidal motor has a rugged and inexpensive sensor and simple control logic. It also has excellent performance characteristics and is the preferred brushless motor in a wide variety of applications. Consequently, the remaining discussion on the brushless dc motor will focus on the trapezoidal machine.

11.2.5. Drive Characteristics and Control Principles

Since the brushless dc motor is an inverted permanent magnet dc motor with an electronic commutator, it is not surprising that the basic equations and terminal characteristics of the two machines are very similar. Ignoring second-order effects, the trapezoidal brushless dc motors of Fig. 11.1 and Fig. 11.3 have the following general voltage equation:

$$V = RI + L \, dI/dt + K_E \omega \tag{11.9}$$

where V is the motor terminal voltage, I is the motor current, R and L are the resistance and inductance of a phase winding, ω is the angular velocity of the rotor, and K_E is the voltage constant, or back emf constant, of the motor over the conduction interval. In Equation 11.9, $K_E \omega$ is the usual generated emf or back emf term.

As shown earlier, the trapezoidal machine develops a torque of $K_T I$, where K_T is the torque constant. The dynamic equation is

$$T = K_T I = J \, d\omega/dt + D\omega + T_F + T_L \tag{11.10}$$

where J is the total system inertia, D is the viscous damping coefficient, T_F is the frictional torque, and T_L is the load torque. Equations 11.9 and 11.10 are the usual dc machine equations, and within the international system of units (SI), the values of K_E and K_T are numerically equal.

The inverter and rotor position sensor of the brushless dc motor constitute an electronic commutator in which the inverter dc link voltage and current correspond to the armature voltage and current of the brush motor. Consequently, the brushless dc motor can employ standard dc drive techniques for speed and torque control. Thus, the average dc link voltage for the inverter can be controlled by a series transistor acting as a dc chopper, or PWM regulator. In this manner, the voltage supplied to the electronic commutator is varied and motor speed is controlled. This approach is clearly analogous to speed control by armature voltage regulation of a dc brush motor.

It is more usual to operate a dc motor with an inner current loop that gives direct torque control, and in the brushless dc system, a series transistor regulator in the dc link can operate in a current-controlled PWM mode. However, this external transistor is not really necessary because the main inverter transistors can regulate the amplitude of the motor current by PWM control as well as commutating the current from phase to phase at the appropriate shaft positions. Current sensing is required in the motor leads or in the dc link, and the current feedback signal is used in a conventional PWM current loop. These controllers are now available in integrated circuit form for small brushless dc motors in low-cost high-volume applications. For continuous current ratings of 2 A, the controller chip can incorporate six output power transistors, a PWM current control circuit, and a Hall logic decoder, together with thermal and undervoltage protection circuitry.[9]

11.3. SERVO CONTROL OF THE BRUSHLESS DC MOTOR

The trapezoidal type of brushless dc motor is widely used in high-performance servo systems because of its inherent advantages over conventional dc brush servo drives and other brushless ac servo drives employing the cage-rotor induction motor or permanent magnet synchronous motor. In particular, the brushless dc servo has the following main attributes:

- Exceptional torque/inertia ratio and high peak torques, permitting rapid acceleration and deceleration

- Excellent high-speed capability

- Improved heat dissipation

- Excellent controllability

- Low torque ripple, giving smooth rotation at all speeds

The rapid advances in magnet technology have had a significant impact on brushless dc motor design. Modern high-performance brushless dc motors use samarium cobalt rare earth material, which has a very high energy product and high coercivity. Consequently, relatively thin sections of material can be positioned on the rotor surface to form the magnetic poles. The small size of the magnets results in a compact low-inertia rotor with a torque/inertia ratio which is several times greater than that of the dc brush motor.[10] In a brushless machine, the speed and power limitations imposed by the mechanical commutator, and its maintenance requirements, are eliminated. As a result, very high speed operation is possible, and the absence of a commutation power limit allows higher peak torques and faster acceleration times.

The permanent magnet brushless motor has no heat-producing components on the rotor, unlike the conventional dc motor and the induction motor, in which significant rotor heating occurs. This rotor heat must flow along the shaft or across the airgap to the stator and thence to the external surface of the motor. Because of this poor thermal path, rotor currents must be limited to prevent excessive rotor winding temperatures. The brushless dc motor uses an inverted, or "inside-out," construction as compared with the dc brush motor. This construction permits increased heat dissipation in the motor because the heat-generating windings are close to the stator surface, and external cooling is facilitated. Consequently, the brushless dc motor has a higher current rating and a greater continuous torque output than a dc brush motor of the same frame size. It has been demonstrated that the brushless dc motor can produce over 30 percent more torque per unit volume than a high-quality rare earth magnet dc brush motor.[11]

Because of the close analogy between brush and brushless dc motors which has been repeatedly emphasized in this chapter, it must be clear that the familiar control methods used in conventional dc motor servo drives can be used in brushless dc systems. Thus, the motor torque is determined by the impressed stator currents, and the resulting stator mmf is decoupled from the rotor magnet flux. Unlike the induction motor servo, the control variables are readily available, and the control is unaffected by variations in motor resistance and inductance parameters.

Many of the advantages listed above are common to the trapezoidal and sinusoidal brushless dc servo drives, but the cost advantage of the trapezoidal type of motor has already been emphasized because a low-resolution sensor will suffice to detect the phase-switching points. In addition, for a given shaft torque, the peak current demand is less in a trapezoidal system than in a sinusoidal system, and therefore less current capacity is required in the transistor inverter.[12]

The trapezoidal brushless dc motor might be expected to have a much greater torque ripple than the sinusoidally based system or the induction motor. However, the careful magnetic circuit design of the trapezoidal machine and the use of rare earth materials result in satisfactory torque smoothness over a revolution.[10] The effects of any residual torque ripple can be suppressed by the closed-loop control action of the velocity feedback loop, to give excellent low-speed performance. Torque smoothness and static stiffness are perfectly satisfactory for industrial servo applications in machine tools and robotics.

In summary, because of its simplicity, low cost, and good performance characteristics, the trapezoidal brushless dc motor is a major contender in the field of high-performance servo drives.

11.3.1. *The Current-Controlled Brushless DC Motor Servo Drive*

The techniques used in a servo controller for a brushless dc motor are very similar to those used in a conventional PWM transistor servo controller, or servo amplifier, for a dc brush motor. Figure 11.7 shows a block diagram of a servo controller for the three-phase full-wave trapezoidal motor of Fig. 11.3. Servo control of the sinusoidal brushless, or permanent magnet synchronous, motor has already been discussed in Chapter 8, but the control electronics in Fig. 11.7 are noticeably simpler than in the earlier Fig. 8.24.

In Fig. 11.7, precise control of motor speed is achieved by a classical double-loop control scheme with an outer high-gain speed loop and an inner current loop. As usual, the magnitude and polarity of the set speed command represent the desired motor speed and direction of rotation, respectively. This signal is compared with the tachometer signal to produce a speed error that is fed to the current loop. Naturally, the tachometer used in a brushless drive is itself brushless.[8,13] The speed loop is compensated in the normal manner, with lead and lag networks, to ensure stable operation and to allow dynamic matching of the drive system to loads with varying friction and inertia. As in the classical dc servo drive, the compensated speed error is the demanded current and hence the demanded torque in the motor. A negative value of speed error signifies a negative torque demand and, in the brushless dc motor, this torque reversal is achieved by energizing each motor phase in the negative-torque region of the static torque/angle characteristic. In Fig. 11.7, set current and actual current are compared in a single current loop by multiplexing the current feedback signals from the three phases into one loop. Selection of the correct phase current is controlled by the position information derived from the shaft-mounted sensor.

The current error signal is used to generate base drive signals for the inverter transistors so that the current error is reduced. This feedback loop is implemented by the usual current-controlled PWM technique, in which the amplified current error is fed to a comparator circuit with a fixed-frequency triangular wave of several kilohertz. The comparator functions as a PWM generator, as described in Chapter 4, and delivers PWM waveforms whose duty cycle varies with the current error. The PWM signals are fed to a programmable read-only memory (PROM), as shown in Fig. 11.7. This phase-control PROM stores a table containing the correct state (on or off) of each transistor for each position of the shaft and for positive and negative torques. The PROM is also supplied with the rotor position information and delivers six output signals to apply the appropriate base drive currents for a particular position. Thus, positive or negative torque can be developed with a magnitude determined by the demanded current.

When an incoming motor phase with zero current is multiplexed into the current loop, the large current error results in a large duty cycle and the sudden application of maximum voltage across the motor winding. As a result, there is a rapid increase in current which minimizes torque ripple at low speeds. It has also been found advantageous to supply the phase-control PROM with a polarity signal from the tachometer to indicate the direction of motor rotation, as shown in Fig. 11.7. This signal allows the introduction

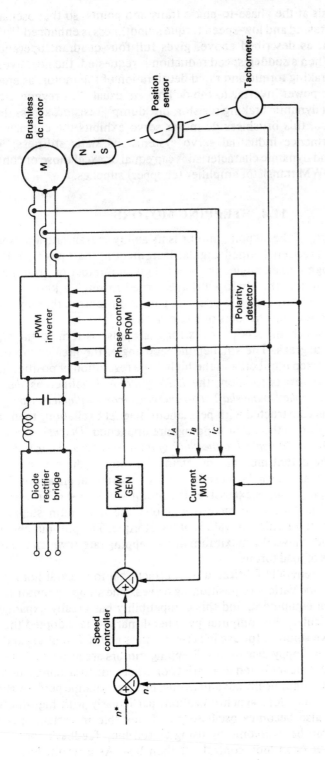

FIG. 11.7. Block diagram of a high-performance brushless dc motor servo drive.

of directional hysteresis at the phase-to-phase transition points, so that oscillations between phases are suppressed and low-speed torque smoothness is enhanced.[10,11]

The control system, as described above, gives full four-quadrant operation of the brushless dc motor. When a sudden speed reduction is requested, the negative speed error results in a large braking torque and rapid deceleration of the motor, as energy is regenerated through the power inverter to the dc link. As usual, this regenerated energy must be dissipated in a dynamic braking resistor, or "dump" resistor, across the dc link. In overall performance, this brushless dc motor servo exhibits the characteristics required in high-performance industrial servo systems: its static stiffness, low-speed torque smoothness, and dynamic characteristics can equal or excel those of conventional dc brush servos with PWM transistor amplifier (chopper) supplies.

11.4. STEPPING MOTORS

The essential property of the stepping motor is its ability to translate a pulsed square-wave excitation into a precisely defined angular increment in shaft position. Each pulse moves the rotor through a fixed angle, and when a given number of drive pulses have been supplied to the motor, the shaft will have turned through a known angle. This means that the motor is ideally suited to open-loop position control, thereby eliminating expensive feedback loops using encoders or tachometers.

The angle through which the shaft moves for each pulse is termed the step angle, which is expressed in degrees. The smaller the step angle, the greater the number of steps per revolution, or step number, and the higher the resolution of positioning can be. At standstill, the motor resists rotation; the *holding torque* is defined as *the maximum steady torque that can be applied to the shaft of an energized motor without causing continuous rotation.* If the motor is constructed with permanent magnet excitation, then the motor shaft can resist rotation when the stator windings are unexcited. *Detent torque* is *the maximum steady torque that can be applied to the shaft of an unenergized motor without causing continuous rotation.* The detent mechanism retains the rotor at the step position in the event of a power failure, and it can be a useful feature in certain applications. If the pulsing frequency, or stepping rate, is excessive, the motor will lose synchronism with the drive pulses. The pull-in torque characteristic shows the maximum stepping rate at which the motor will start for different values of load torque. The pull-out characteristic, or slewing characteristic, shows the maximum fixed stepping rate at which the motor will run for different values of load torque.

Stepping motors are generally fabricated in subfractional to integral horsepower ratings and are used in a wide variety of positioning drives. The stepping motor is compatible with modern digital equipment, and this compatibility has greatly expanded its potential market. In particular, the computer peripheral market has adopted the stepping motor as a cost-effective solution for use in serial printers and X-Y plotters and for magnetic head positioning in floppy disk drives. Stepping motors are also widely used in numerical controls for machine tools and in a variety of other industrial applications.

Open-loop control is economically advantageous, but the dynamic performance is limited because the motor may lose synchronization, particularly with high-inertia loads. The rotor movement also becomes oscillatory and unstable in certain speed ranges. These disadvantages can be overcome by using closed-loop feedback control, but the economic advantages of open-loop control are then lost. As a result, both brush and brushless dc servo drives are performance- and cost-competitive with the closed-loop stepping motor drive.

The three major types of stepping motor are described in subsequent sections. References 14 to 19 present further information on stepping motor construction, characteristics, and controls.

11.4.1. *The Variable-Reluctance Stepping Motor*

The variable-reluctance (VR) motor is the most basic type of stepping motor. It typically consists of salient-pole stator and rotor laminations assembled in a single stack. The stator and rotor pole numbers are different so as to ensure self-starting capability and bidirectional rotation. The stator winding consists of a number of concentrated coils placed over the stator poles. There are no windings on the rotor. Figure 11.8 shows a variable-reluctance motor with 8 stator poles and 6 rotor poles. Windings on diametrically opposite stator poles are connected in series (or parallel) so that one stator pole acts as a north pole and the other acts as a south pole. Thus, there are four phases, A, B, C, and D, giving a four-phase 8/6-pole motor.

When phase A of the stator is energized with direct current, a singly excited magnetic system is formed. In such a system, torque is developed by the tendency for the magnetic circuit to adopt the configuration of minimum reluctance. This reluctance torque,

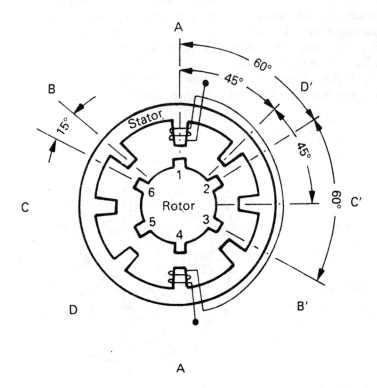

FIG. 11.8. A single-stack, four-phase, 8/6-pole, variable-reluctance stepping motor. Only the phase A winding is shown, for clarity.

which is independent of current direction, causes the rotor to turn until a pair of its poles (1 and 4) are exactly aligned with the stator poles of phase A, as shown in Fig. 11.8. The rotor is stable in this position and cannot move until phase A is de-energized. If phase B is then energized, the rotor turns clockwise through the step angle of 15 degrees until rotor poles 3 and 6 are aligned with phase B. Energization of phases C and D in sequence causes further clockwise rotation through two more 15-degree steps. After the sequence ABCDA, the rotor has turned through four steps, or 60 degrees, and hence six cycles of this sequence are required for one revolution of the rotor. The phase sequence for clockwise rotation is ABCDA and for anticlockwise rotation is ADCBA. Thus, the rotor steps around in the opposite direction to the stepped rotation of the stator field, and one revolution is completed in 24 steps.

A three-phase reluctance motor can have 6 stator poles and 4 rotor poles, giving a step angle of 30 degrees. In Fig. 11.9, the number of poles is doubled, giving a three-phase 12/8-pole machine and halving the step angle to 15 degrees.

Smaller step angles can be achieved if the current energization is gradually shifted from one phase to the next in a discrete number of stages, rather than abruptly, as assumed above. Thus, if phases A and B in Fig. 11.8 are energized simultaneously, the rotor will move to a mid-position resulting in a half-step. A logical extension of this technique is to control the currents in the individual phase windings so that several stable equilibrium positions are created. Normally, the step angle is reduced by factors of 1/2, 1/5, 1/10, or 1/16. This technique is known as microstepping.[19]

Another technique for providing smaller step angles is to use a multistack variable-reluctance motor. In this configuration, the motor is divided into a number of magnetically isolated sections or stacks along its axial length. Each stack corresponds to a phase, and both stator and rotor have the same number of poles; hence they have the same pole pitch. Figure 11.10 shows a cut-away view of a three-stack (three-phase) variable-reluctance machine. In each of the three stacks, the stator and rotor laminations have 12

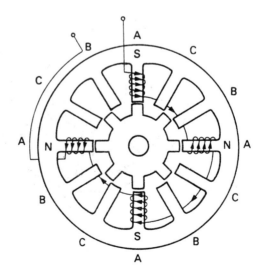

FIG. 11.9. A single-stack, three-phase, 12/8-pole, variable-reluctance stepping motor.

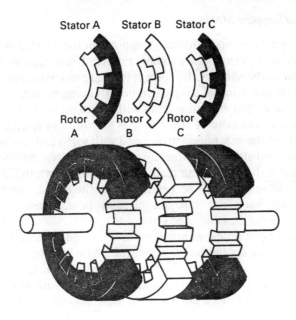

Fig. 11.10. A three-stack, three-phase, variable-reluctance stepping motor.

poles. The poles of the stator laminations are in line, but those of the rotor laminations are offset from each other by one-third of a pole pitch. In the multistack machine, all the stator pole windings in a given stack or phase are energized simultaneously, unlike the single-stack motor, where only the windings on one pair of poles are energized. The three stator stacks in Fig. 11.10 are energized sequentially to produce a step angle of one-third of a pole pitch, or 10 degrees. Step angles are typically in the range of 2 to 15 degrees.

The variable-reluctance stepping motor has a high torque/inertia ratio, giving high rates of acceleration and fast response. A possible disadvantage is the absence of detent torque.

11.4.2. *The Permanent Magnet Stepping Motor*

The permanent magnet (PM) stepping motor has a salient-pole stator construction similar to the single-stack variable-reluctance machine, but the rotor is cylindrical and is composed of radially magnetized permanent magnets. When a particular stator phase is energized, the rotor magnetic poles move into alignment with the excited stator poles. The polarities of the winding currents determine the direction of rotation. For bidirectional operation, positive and negative currents must be provided.

Due to its permanent magnet excitation, the PM stepping motor has a low power requirement and a high detent torque. However, the torque per unit volume is usually poor and, due to the difficulty of manufacturing a small permanent magnet rotor with a large number of poles, the step size is relatively large, being in the range of 30 to 90 degrees. A recent variant of the permanent magnet motor uses a disk rotor which is magnetized axially to give a small step size and low inertia.[18]

11.4.3. *The Hybrid Stepping Motor*

The hybrid (or synchronous inductor) stepping motor combines the features of the variable-reluctance and permanent magnet motors. Rotor flux is produced by a permanent magnet and is directed by the rotor teeth to appropriate parts of the airgap. Figure 11.11 shows a cross section of a typical hybrid stepping motor. The stator structure is similar to that of the earlier machine types with a coil on each pole. A concentric permanent magnet lies in the core of the rotor, as shown, and is magnetized in the axial direction. Each pole of the magnet is surrounded with soft-steel toothed laminations, but the teeth on the two end stacks are offset by a half tooth-pitch, so that a tooth at one end coincides with a slot at the other. The main flux path, as shown in Fig. 11.11, is from the north pole of the magnet, into the end stack, across the airgap, through the stator pole, axially along the stator back iron, through the stator pole, across the airgap, and back to the magnet south pole via the other end stack.

There are typically eight poles, numbered 1 to 8 around the stator periphery, and each pole has between two and six teeth. There are two phase windings: winding A consists of the coils placed on stator poles 1, 3, 5, and 7, and winding B is formed by the coils on poles 2, 4, 6, and 8. Energizing phases A and B alternately with positive or negative currents will cause the motor to step clockwise or anticlockwise, depending on the current polarities.

Hybrid motors have a small step length (typically 1.8 degrees), which is advantageous for high-resolution angular positioning. The torque per unit volume is greater than in the variable-reluctance motor, and the permanent magnet flux provides some detent torque when the stator windings are unexcited.

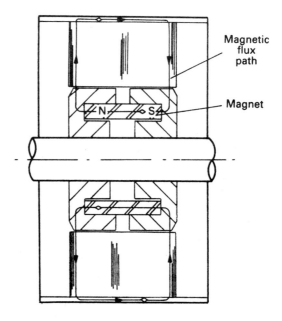

Magnetic
flux
path

Magnet

FIG. 11.11. Cross section of a hybrid stepping motor showing the axial permanent magnet and flux path.

11.4.4. *Drive Circuits*

The stepping motor is normally operated from a dc power supply obtained by ac rectification. The phase windings are energized with square-wave currents by drive circuits in which power transistors operate as solid-state switches. The base drive signals for the transistors can be generated by digital integrated circuits or by a microprocessor, followed by amplification to the required level.

In the variable-reluctance motor, the current polarity is irrelevant to torque production, and a simple unipolar drive circuit — so called because it produces unidirectional currents — is sufficient. Typically, a power transistor is placed in series with each phase winding to switch current on and off. As explained in Chapter 2, a free-wheeling diode must be connected across the inductive winding. This diode provides an alternative path for the winding current when the transistor is turned off and so prevents transistor destruction due to inductive overvoltages.

In general, the natural time constant of the phase winding is long, and the build-up and decay of current is too slow for satisfactory operation at high speeds. Consequently, a "forcing" resistor is placed in series with the motor winding to reduce its time constant. This is a cheap and simple solution, but the losses in the resistor are high and circuit efficiency is poor. A resistor or Zener diode can be placed in series with the free-wheeling diode to quicken the decay of current. The motional, or back, emf in the phase winding also distorts the current waveform at high speeds, but increasing the dc supply voltage (and forcing resistance) will overcome this problem.

A more sophisticated drive circuit requiring two series transistors and two feedback diodes per phase can be used in high-performance applications. This circuit, which will be discussed in the next section in the context of a power converter circuit for the adjustable-speed variable-reluctance motor drive, is shown in Fig. 11.12(b). At switch-off, the inductive energy stored in the winding is returned to the dc supply, giving a rapid decay of current. In this circuit, current-controlled chopping or PWM operation of the transistors can be used to generate the required square-wave currents with high efficiency.

In the permanent magnet and hybrid stepping motors, the direction of rotation is determined by the polarities of the phase currents. Consequently, bidirectional (or bipolar) current excitation is necessary and a bipolar drive circuit is required for each phase winding. The single-phase full-bridge dc-ac inverter circuit of Fig. 4.1(b) is suitable. A number of manufacturers now market plug-in driver cards rated up to 10 A, and beyond. Single-chip integrated circuit drive modules can be obtained for small motors.

11.5. VARIABLE-RELUCTANCE MOTOR DRIVES

The variable- or switched-reluctance motor has a doubly salient construction with projecting poles on both stator and rotor. It is, therefore, a member of the family of single-stack variable-reluctance stepping motors, as described in Section 11.4.1, but motor design procedures and control strategies are modified to give an efficient adjustable-speed drive system. The variable-reluctance motor should not be confused with the synchronous reluctance motor. The latter machine, as described in Chapters 1 and 8, has a cylindrical, or nonsalient, stator and is essentially a salient-pole synchronous motor without a rotor field winding.

In the doubly salient variable-reluctance motor, various combinations of stator and rotor pole numbers are possible. A four-phase machine with eight stator poles and six rotor poles is most common, but a three-phase 6/4-pole machine is also used. Simple, concentrated windings are placed over the stator poles, and diametrically opposite windings are connected to form a stator phase, as in the 8/6-pole stepping motor of Fig. 11.8. As explained in Section 11.4.1, excitation of a stator phase causes the most adjacent pair of rotor poles to be attracted into alignment with the magnetized stator poles. However, in the adjustable-speed variable-reluctance drive, the stator phase currents are switched on and off in synchronism with rotor position to ensure continuous rotation, as in a brushless dc motor drive. Shaft-mounted position sensors are usually of the Hall-effect or optical type. Machine torque is produced solely by reluctance variation and is therefore independent of current direction. Consequently, unidirectional (unipolar) phase currents will suffice.

The variable-reluctance motor has a number of inherent advantages that have sparked interest in its use as an adjustable-speed drive. These features are —

- Simple, low-cost machine construction due to the absence of rotor windings and magnets and the use of a small number of concentrated stator coils similar to the field coils of a dc machine

- Efficient motor cooling because all windings are on the stator

- Suitability for sustained high-speed operation because of the robust rotor construction and absence of brushes

- Low rotor inertia and high torque/inertia ratio

- Simplified power electronic converter giving greater economy and reliability because bipolar (bidirectional) currents are not required.

- Motor phases operate almost independently of each other, so that the loss of one phase of the motor or converter does not prevent drive operation at reduced power.

These major advantages have stimulated research activity aimed at utilizing the variable-reluctance principle in high-power industrial drives. This activity has been facilitated by the advent of powerful computer-aided machine design facilities, fast and efficient power semiconductor switches, and sophisticated analog and digital integrated circuits. As a result, the variable-reluctance motor drive has now emerged as a candidate for general-purpose adjustable-speed drive applications.[20-28] The variable-reluctance drive also has excellent controllability, and recent development effort has resulted in a high-performance servo drive.[29-31]

11.5.1. Power Converter Circuits

As explained above, each stator phase of the variable-reluctance motor must be energized with a unidirectional current pulse while the rotor is appropriately positioned relative to the stator. A rapid response is necessary for positive and negative changes in the demanded phase current level. This function requires a two-quadrant power converter that is capable of applying equal positive and negative phase voltages to produce approximately equal rates of current increase and decrease. The required converter characteristic is provided by each of the circuit configurations of Fig. 11.12. Because the inverter phases are completely independent of each other, Fig. 11.12 shows the circuitry required for just one phase of the reluctance motor. Thus, a four-phase machine will require four of these circuits. The dc supply is usually obtained from a diode rectifier bridge with an LC filter. Power transistors are shown in Fig. 11.12, but gate turn-off thyristors and MOSFETs can also be used. Conventional thyristors will require auxiliary forced commutating circuitry, but economies can be achieved with the use of a single commutating circuit for all the main power devices.

Each circuit topology shown in Fig. 11.12 has the advantage that there is always a motor winding in series with a main power semiconductor device. Consequently, there is no shoot-through fault current path across the dc bus, giving added ruggedness and reliability to the power electronic converter. In Fig. 11.12(a), the phase current is controlled by a single power transistor in conjunction with a closely coupled bifilar motor winding and feedback diode. When the transistor is switched off, free-wheeling current continues to flow through the secondary winding and the feedback diode, and inductive energy is returned to the dc supply as the current collapses. At the instant of turn-off, the transistor collector-to-emitter voltage rises to twice the dc link voltage, assuming a turns ratio of unity and perfect coupling between the bifilar windings. In practice, there is always some uncoupled leakage inductance giving rise to higher voltage transients and requiring large snubber circuits to protect the semiconductors. The bifilar winding also results in a poor utilization of copper in the motor and doubles the number of motor connections as compared with alternative circuits.

The converter circuit of Fig. 11.12(b) dispenses with the bifilar winding but requires two power transistors and two feedback diodes in each phase. The upper and lower power transistors are turned on and off simultaneously. At turn-off, the two diodes conduct, thereby reversing the voltage across the motor phase and allowing the return of inductive energy to the dc supply. In Fig. 11.12(c), only one power transistor and one diode are used per phase, but a center-tapped dc supply is required. Note that the dc link voltage must be doubled in this connection, as compared with the other circuits, in order to apply the same voltage across the motor phase winding. The dc center-tap can be created by a split capacitor circuit, as shown in Fig. 11.12(c). It is important to maintain a bal-

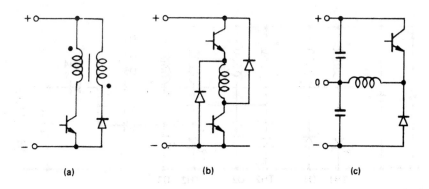

(a) (b) (c)

FIG. 11.12. Power converter configurations for one phase of a variable-reluctance motor drive.

anced loading on the two halves of the dc supply to preserve the correct voltage levels, and therefore, the motor must have an even number of phases.

In conventional ac motor drives, two switching devices per phase are necessary, but the circuits of Fig. 11.12(a) and Fig. 11.12(c) require only one device per phase, giving a more compact and economic converter with lower forward conduction losses. For a given winding voltage rating, the power transistor current rating is the same for each of the three configurations in Fig. 11.12, but the voltage rating of the devices in Figs. 11.12(a) and (c) is double that in Fig. 11.12(b). Since the latter circuit requires twice as many devices, the total device volt-ampere rating is the same for each configuration.

Practical power converters can be based on any of the three configurations shown in Fig. 11.12, but the circuits in (a) and (c) are usually preferred because of the reduced number of devices. Figure 11.13 shows a thyristor converter for a three-phase bifilar-wound motor with a common commutating circuit for all three phases.[20] However, the bifilar winding is normally undesirable, except at low supply voltages, because of the large voltage spikes at turn-off. In the center-tapped dc supply circuit of Fig. 11.12(c), there may be continuous regeneration or "pump-up" of energy to one side of the dc supply for certain operating conditions, but a novel circuit that overcomes this problem has been described in Reference 29. An improved power converter has also been developed with a unipolar dc supply and a capacitor which stores the trapped inductive energy at turn-off, before returning it to the dc source.[32]

11.5.2. Torque Production in the Variable-Reluctance Motor

In the simplified doubly salient reluctance motor of Fig. 11.14(a), only one pair of a larger number of stator poles is shown. As usual, the stator coils on the diametrically opposite poles are connected in series or parallel to form a stator phase, and when current flows in the phase winding, torque is exerted on the rotor, tending to align the stator and rotor poles. In the fully aligned position, the stator winding inductance is a maximum and the reluctance of the magnetic circuit is a minimum. The direction of the winding current is irrelevant because machine torque always acts to increase the inductance and reduce the reluctance. Consequently, as already explained, there is no need for bipolar current excitation.

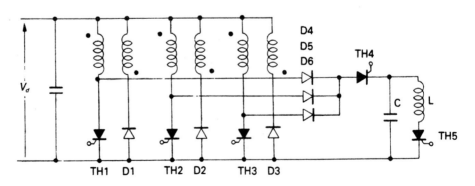

FIG. 11.13. Thyristor power converter for a three-phase, bifilar-wound variable-reluctance motor.

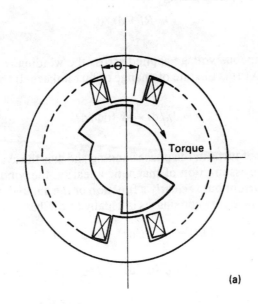

(a)

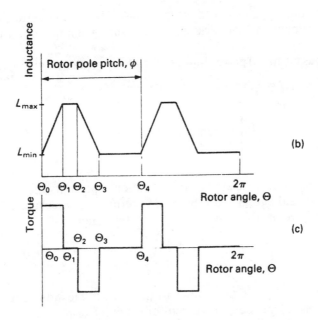

FIG. 11.14. Torque production in a variable-reluctance motor: (a) elementary doubly salient motor showing only one stator and rotor pole pair; (b) idealized phase inductance variation with rotor angle, Θ; (c) static torque/angle characteristic.

In general, for a stator phase winding

$$v = Ri + d\lambda/dt \tag{11.11}$$

where v is the instantaneous voltage applied across the winding, i is the instantaneous winding current, and $d\lambda/dt$ is the rate of change of flux linkage. If the winding resistance is negligible

$$v = d\lambda/dt = N\,(d\Phi/dt) \tag{11.12}$$

where N is the number of turns in the phase winding and Φ is the average flux per turn.

Making the simplistic assumption of magnetic linearity, the winding self-inductance, L, is independent of current but is clearly a function of the angular position of the rotor, Θ. Also, mutual inductance between phases is usually negligible, and therefore

$$v = \frac{d}{dt}\,(Li) = L\,(di/dt) + i\,(dL/dt)$$

$$= L\,(di/dt) + i\,(dL/d\Theta) \cdot d\Theta/dt$$

$$= L\,(di/dt) + i\,\omega\,(dL/d\Theta) \tag{11.13}$$

where $\omega = d\Theta/dt$, the angular speed of rotation. The term $i\omega\,(dL/d\Theta)$ in Equation 11.13 is the motional emf, or back emf, and is zero when the motor is stationary or $dL/d\Theta$ is zero.

The instantaneous electrical power input to the phase winding is given by

$$vi = Li\,(di/dt) + i^2\omega\,(dL/d\Theta)$$

$$= \frac{d}{dt}\left[\frac{1}{2}\,Li^2\right] + \frac{1}{2}\,i^2\omega\,\frac{dL}{d\Theta}\;. \tag{11.14}$$

The first term on the right-hand side of this equation represents the rate of increase in the stored magnetic field energy, and the second term is the mechanical power output, $T\omega$, where T is the developed torque in newton-meters. Hence,

$$T = \frac{1}{2}\,i^2\,\frac{dL}{d\Theta}\;. \tag{11.15}$$

Clearly, the sign of the torque is determined by the slope of the inductance variation with rotor position and is independent of current direction. The precise nature of the L/Θ variation is determined by machine parameters such as stator and rotor pole numbers and pole arc to pole pitch ratios. An idealized phase inductance profile for the stator poles of Fig. 11.14(a) is shown in Fig. 11.14(b), where inductance is assumed to be independent of current magnitude. In an actual motor, the number of cycles of inductance variation per revolution is determined by the number of rotor pole-pairs.

The corresponding static torque/angle characteristic, or torque function, for a constant phase current is shown in Fig. 11.14(c). As predicted by Equation 11.15, positive torque is produced when $dL/d\Theta$ is positive, and negative torque is produced when $dL/d\Theta$ is negative. Torque cannot be developed when $dL/d\Theta$ is zero. In order to produce a unidirectional forward motoring torque, the stator phases must be energized sequentially with single current pulses that coincide with the angular periods for which $dL/d\Theta$ is positive. Similarly, a backward torque for braking or reverse rotation is developed when the current pulses coincide with the periods of negative $dL/d\Theta$. A rotor position sensor is obviously required to initiate and terminate conduction in each phase at times that are appropriate for maximum torque production.

The ideal current waveform for positive torque production is a constant-height pulse occurring for the duration of the increasing inductance interval, as shown in Fig. 11.15(a). However, in practice, the current pulses are switched from a dc voltage source and the winding inductance delays the build-up of current. Typical current, voltage, and flux waveforms are shown in Fig. 11.15(b), where a dc voltage, V_d, is applied to

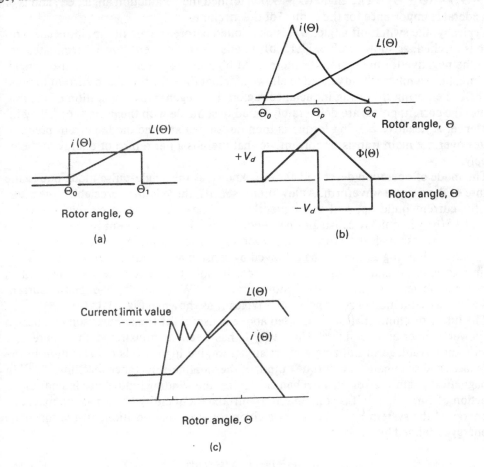

FIG. 11.15. Operation of a magnetically linear variable-reluctance motor: (a) ideal rectangular current pulse; (b) typical phase current, phase voltage, and flux waveforms in the single-pulse mode; (c) typical phase current waveform in the low-speed chopping mode.

the winding at the switch-on angle, Θ_o, and an equal but opposite voltage is applied at the switch-off angle, Θ_p. Equation 11.12 shows that the flux, Φ, increases at a uniform rate when a constant positive voltage, V_d, is applied to the stator phase winding; and conversely, the flux decreases linearly when $-V_d$ is applied, as shown in Fig. 11.15(b). Typically, the power semiconductor is turned on at an angle, Θ_o, in advance of the rising inductance region. The winding inductance is then small, allowing the current to build up rapidly and ensuring that a substantial phase current is established before the rotor enters the torque-producing zone. Subsequently, the rising inductance and the motional emf cause the current to fall, until the switch is opened at angle Θ_p. Thereafter, the negative voltage applied to the winding causes the current to fall more rapidly until it collapses to zero at the extinction angle, Θ_q. Energy flows from the source into the motor during the transistor conduction angle $(\Theta_p - \Theta_o)$, and energy is returned to the source via the diode, during the diode conduction angle $(\Theta_q - \Theta_p)$. The maximum flux always occurs at the instant of transistor switch-off, as defined by Θ_p, and in general $(\Theta_p - \Theta_o) = (\Theta_q - \Theta_p)$. The angle $(\Theta_q - \Theta_o)$ is termed the conduction angle, Θ_c, and is of considerable importance for the control of the machine.

Typically, the switch-off angle, Θ_p, is positioned before the maximum inductance region is reached, as shown in Fig. 11.15(b), to ensure that current flow is terminated before the negative torque region is entered. At high motor speeds, the switch-on angle, Θ_o, must be significantly advanced to allow sufficient time for the phase current to build up before entering the rising inductance region. For regenerative operation of the machine, the current pulses are deliberately timed to coincide with the periods of falling inductance. A negative braking torque is then developed and the diode current predominates over the main transistor current, so that there is a net return of energy to the dc supply.

The mode of operation described above is known as the single-pulse mode due to the nature of the current waveform. At low motor speeds, there is an appreciable time interval for current build-up, and consequently, the linear increase in current shown in Fig. 11.15(b) will quickly cause an unacceptably high level of current in the motor and inverter. It is essential, therefore, to incorporate a current limit in the power converter. This current-limiting action can be achieved by sensing phase current and switching the main transistor on and off repeatedly in order to hold the current between prescribed upper and lower levels. In this "chopping," or PWM, mode, the phase current waveform as a function of rotor position is typically as shown in Fig. 11.15(c).

The linear machine analysis presented above is useful for the comparison and design of power converter circuits.[22] However, it has been demonstrated that there is a significant advantage in utilizing a short airgap to give high levels of saturation in the pole faces and so enhance the torque output of the variable-reluctance machine.[33–35] In a magnetically saturated electromechanical device, the winding inductance is a nonlinear function of current, and it is convenient to introduce a quantity known as the magnetic coenergy of the system.[36,37] Thus, for a given rotor angular position, Θ, the magnetic coenergy is defined by

$$W_f'(\Theta, i) = \int_0^i \lambda(\Theta, i) \, di \qquad (11.16)$$

where the flux linkage, λ, is now a nonlinear function of current.

It can be shown that the machine torque, T, is given by the angular rate of change of coenergy.[36,37] Thus

$$T(\Theta, i) = \frac{\delta W'_f(\Theta, i)}{\delta \Theta}.$$ (11.17)

Magnetic saturation also has the beneficial effect of linearizing the torque/current characteristic which, in the absence of magnetic saturation, obeys a square law relationship, as shown by Equation 11.15. However, it is impossible to maintain magnetic saturation at low current levels, and consequently, the torque per ampere is always reduced at light loads.

11.5.3. *Drive Characteristics and Control Principles*

The fundamental theory underlying the variable- (or switched-) reluctance motor drive has now been introduced. These basic concepts have been applied in a range of general-purpose industrial drives that are cost- and performance-competitive with other ac and dc motor drives.[25] In particular, the variable-reluctance motor has a high specific power output and a torque per unit volume that is comparable with that of the induction motor, while its torque/inertia ratio is appreciably greater than that of the induction motor.[28] The adjustable-speed reluctance motor drive has been shown to offer a high system efficiency over a wide operating range of torque and speed,[21,25,27] and the converter kVA rating for a given shaft power is comparable with that required in a PWM inverter-fed induction motor drive.[27,28,38]

On the other hand, the variable-reluctance motor, unlike the induction motor, cannot operate without its power electronic converter, and there are a larger number of interconnections between motor and converter, particularly with bifilar windings. The pulsating nature of the electromagnetic forces in the variable-reluctance machine are likely to cause problems of acoustic noise and vibration due to the excitation of structural resonances. In general, in one revolution of the motor, each stator phase conducts as many pulses of current as there are rotor poles. Thus, the basic four-phase 8/6-pole machine has 6 current pulses per phase in each revolution and there are 24 torque pulses per shaft revolution. The three-phase 6/4-pole machine has 12 torque pulses per revolution. These torque pulsations are significant at low speeds, unless the drive system is specifically designed to suppress them. At high speeds, the switching frequency of the phase currents influences the core losses in the motor and the switching losses in the converter. For a given shaft speed, the switching frequencies are higher in a reluctance motor drive than in a six-step inverter-fed induction motor drive. At a speed of 3000 rev/min, the switching frequency for a four-phase 8/6-pole machine is 300 Hz as compared with 50 Hz for a two-pole induction motor drive. At 3000 rev/min, the three-phase 6/4-pole machine has a switching frequency of 200 Hz, and clearly, if efficiency is important, the number of phases and the number of rotor poles should be kept as low as possible.

The switch-on and switch-off angles of the reluctance motor must be accurately synchronized with rotor position. The shaft position sensor typically consists of a rotating slotted disk and stationary optical sensors that generate a three-phase or four-phase position reference signal for the three-phase or four-phase machine. The finer resolution of position required for switching angle control can be obtained with a high-frequency clock signal derived from the position sensor signals with a phase-locked loop.

The natural characteristic of the reluctance motor is that which occurs when the dc supply voltage and switching angles are fixed and independent of speed. Under these conditions, the torque-speed characteristic is the same as that of a dc series motor because, as motor speed, ω, is reduced, the length of time during which a phase winding is energized increases inversely with the reduction in speed, and so also does the peak flux (Fig. 11.15b). In a linear singly excited magnetic device, torque is proportional to the square of the flux, and hence motor torque is inversely proportional to ω^2, or

$$T = \frac{K}{\omega^2} \qquad\qquad (11.18)$$

as in a dc series motor. Also, the mechanical power output is given by

$$P = T\omega = \frac{K}{\omega} . \qquad\qquad (11.19)$$

At a given speed, the peak motor flux is also proportional to the dc supply voltage, V_d, and so motor torque varies as V_d^2, giving a family of series torque-speed characteristics for different supply voltages. The upper limiting characteristic is set by the maximum rated voltage. As the speed falls, the flux and current rise, and base speed, ω_b, is the speed at which maximum flux and current occur at maximum voltage. As usual, this base speed is the lowest speed at which maximum power can be obtained and is the highest speed for maximum torque.

Control of speed below ω_b can be achieved by variation of the applied voltage, as in a dc series motor drive. Adjustment of the dc link voltage may be employed, but it is more usual to operate with a fixed dc link voltage and to employ pulse-width modulation or chopping of the main converter devices. If the PWM switching is under current control, as already discussed, a controlled current is delivered to the motor and a constant-torque characteristic is implemented below base speed, as in a dc motor drive. In this chopping mode, the motor current and torque can obviously be varied by controlling the PWM modulation index or duty cycle. (Single-pulse operation at constant voltage is possible below base speed if the conduction angle, Θ_c, is shortened, but this entails high peak fluxes and currents for constant average torque.)

Above base speed, the winding back emf is sufficient to limit the current, and the motor operates in the single-pulse mode. The supply voltage cannot be increased and Equation 11.18 shows that torque decreases as $1/\omega^2$. However, this equation assumes natural operation with fixed switching angles. By increasing the conduction angle as a function of speed, the natural fall in torque can be offset, and a variety of characteristics can be obtained. A constant-horsepower characteristic is frequently required, particularly in traction applications and in machine-tool spindle drives. To obtain this characteristic, torque must fall as $1/\omega$, and hence flux must fall as $1/\sqrt{\omega}$. By increasing Θ_c appropriately, this condition can be satisfied to give a wide speed range of operation at constant horsepower.[23]

In general, in the single-pulse mode of operation the switch-on and switch-off angles, Θ_o and Θ_p, are available as control inputs to determine the duration and location of the conduction angle, Θ_c. By controlling the size of the current pulse, the torque and power produced at a given speed can be controlled. A wide variety of current waveforms can produce the same torque, but careful testing will determine the optimum pair of switching angles for each torque-speed condition to give the most torque-effective current pulse that maximizes inverter efficiency with acceptable current ratings for the inverter components.

In the adjustable-speed industrial drive, set speed and actual speed are compared in an outer speed loop. The shaft speed signal can be obtained from the rotor position sensor by using a frequency-to-voltage converter. The speed error, which represents the torque demand, is used to control the current level in the chopping mode, and controls the switching angles in the single-pulse mode. As already explained, regeneration to the dc link occurs naturally when the current pulses are positioned where the inductance slope is negative. As usual, this energy must be dissipated in a dynamic braking resistor in the dc link.

The flexibility of the variable-reluctance motor drive is noteworthy, and also its high efficiency at partial loads. It is both easy and inexpensive to modify the switching angles to give a variety of different torque-speed characteristics. Thus, constant-torque, constant-horsepower, or series characteristics can be provided, or a combination of these. The electronic control system can employ standard digital electronic circuits, or it can be microprocessor based. [22,39,40]

11.6. SERVO CONTROL OF THE VARIABLE-RELUCTANCE MOTOR

The variable-reluctance motor has the controllability of a dc machine and can be regarded as a brushless dc motor. However, the sequential energization of the stator phases results in a stepping, or cogging, motion at low speeds, making the variable-reluctance machine an unlikely candidate for a high-performance wide-range adjustable-speed drive. Nevertheless, as will be explained, very smooth low-speed operation *can* be achieved, together with the other demanding characteristics required in a servo drive, such as high peak torque for acceleration and good static stiffness.

The torque developed by a stator phase is determined by the stator and rotor geometries, as explained in Section 11.5.2. In Fig. 11.14(c), the static torque/angle characteristic showing the developed torque per phase for a constant dc current, has an idealized quasi-square waveform. In practical machines, magnetic saturation and fringing will have a significant influence on this static characteristic. Magnetic saturation also results in an approximately linear relationship between torque magnitude and current, except at low current levels.

In general, the resultant output torque of the variable-reluctance motor is obtained as the sum of the torque contributions of the energized phases. By appropriate design of the stator and rotor laminations, the torque-producing region of the outgoing phase overlaps with that of the incoming phase. In the torque overlap region, the two phases can contribute torque simultaneously, and if the phase currents are correctly programmed as a function of rotor position, then the two torque contributions sum to a constant. Thus, it is possible to obtain a smooth torque transition from one phase to the next, and the resultant motor torque can be practically independent of shaft position, giving smooth rotation, even at very low speeds.

An extension of the concept of torque overlap in the four-phase 8/6-pole motor, results in a machine design with sinusoidal torque/angle characteristics for the four stator phases, as shown in Fig. 11.16(a).[29,30] If the phases are supplied with half-sinusoids of current that are synchronized with rotor position, a constant torque is developed in all rotor positions. Figure 11.16(b) shows the phase current variation required to develop a smooth positive torque. Maximum current in a phase coincides with the peak of the torque/angle characteristic for that phase. In any rotor position, two phases are energized to make torque contributions proportional to $\sin^2 \Theta$ and $\cos^2 \Theta$, respectively, as in the two-phase sinusoidal type of brushless dc motor of Section 11.2.3. Because $\sin^2 \Theta + \cos^2 \Theta = 1$, the resultant torque output is constant and independent of shaft position. In order to reverse the torque direction, each motor phase must be energized in the negative-torque half-cycle of its static torque/angle curve. This condition is realized if the phase currents are as shown in Fig. 11.16(c). Clearly, a torque reversal is achieved by shifting phase conduction by 180 electrical degrees, which is exactly the same technique as used in the brushless dc motors of Section 11.2. In general, the magnitude of the shaft torque is controlled by the amplitudes of the half-sinusoids of phase current.

The phase current waveforms shown in Fig. 11.16 are ideal in that no allowance is made for the inherent torque/current nonlinearity at low current levels when the machine comes out of saturation. However, this difficulty may be overcome by a compensating nonlinear amplifier in the circuit of the reference current waveform generator. The outer velocity loop can also be expected to suppress the effects of small irregularities in motor torque.

11.6.1. *The Current-Controlled Variable-Reluctance Motor Servo Drive*

A block diagram of the variable-reluctance servo drive is shown in Fig. 11.17. The reader should compare this diagram with Fig. 8.24 for the permanent magnet synchronous motor servo drive and Fig. 11.7 for the brushless dc motor servo. As in Fig. 8.24, the digitally encoded reference current waveforms are stored in programmable read-only memory (PROM) and are accessed under the control of the rotor position sensor. Absolute position information must be available with a sufficiently fine resolution to allow accurate synthesis of the programmed current waveforms. A resolver or optical encoder or other high-resolution sensor can be used.

The digital output of the PROM is converted to an analog reference current waveform by the Multiplying Digital-to-Analog Converter (MDAC). An examination of Fig. 11.16 shows that, in a four-phase machine, phases 1 and 3 do not conduct simultaneously, and likewise, phases 2 and 4 are mutually exclusive. The control circuit of Fig. 11.17 takes advantage of these factors to minimize memory and MDAC requirements by storing the reference currents for a pair of phases in one PROM and using a demultiplexer to switch the reference current waveform to the correct phase output.

The outer control loop in Fig. 11.17 is a conventional speed loop in which the set speed signal is compared with the tachometer signal. The speed-error voltage is fed via the speed controller to a precision rectifier to obtain its absolute value. This voltage represents the desired torque magnitude and is used as a reference voltage for the MDACs to control the amplitudes of the reference current waveforms, which are the set current signals for four conventional PWM current loops. These feedback loops cause the actual phase currents to track the reference current waveshapes using power transistors switching at several kilohertz for accurate current synthesis.

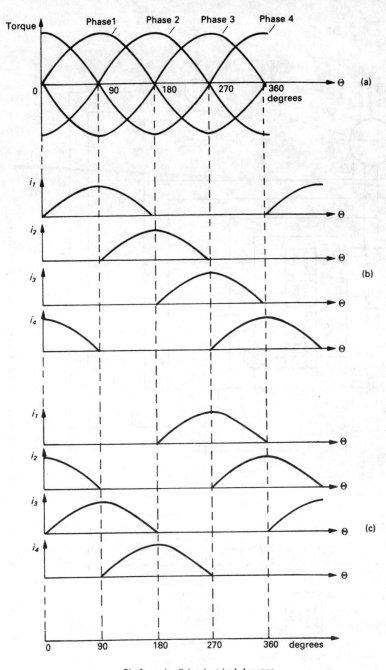

FIG. 11.16. Idealized waveforms for the four-phase variable-reluctance servo motor: (a) static torque/angle characteristics; (b) phase currents for positive torque; (c) phase currents for negative torque.

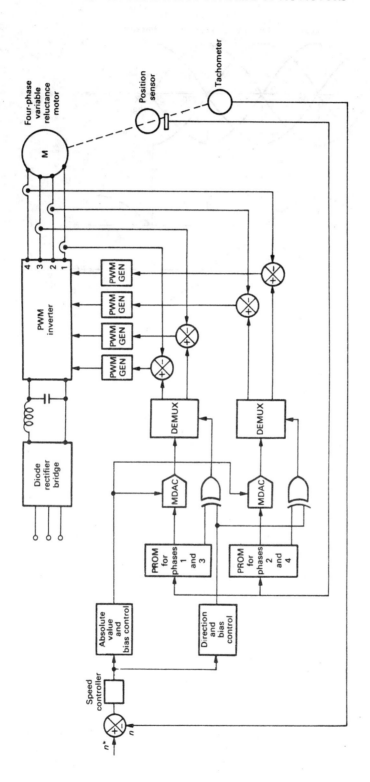

Fig. 11.17. Block diagram of a high-performance variable-reluctance motor servo drive.

The compensated speed-error voltage is also fed to a polarity detector which determines the sign of the error signal and uses it to control the direction of the motor torque by altering the phase excitation pattern. The direction of the motor torque can be controlled by simply inverting the demultiplexer control bits with the use of two EX-OR gates, under the control of the torque direction signal, as shown in Fig. 11.17.

As already noted, modern variable-reluctance motors employ high levels of magnetic saturation in the overlapping pole faces to enhance the specific power output, but the motor comes out of saturation on light load and the torque per ampere is reduced. This nonlinear torque/current characteristic gives rise to an inherent deadband effect, which is unacceptable in a high-performance servo system. Bias control circuitry overcomes this effect by offsetting the torque magnitude signal and processing the torque direction signal so that vestigial bias currents are present in all the machine windings for low values of speed error. Thereby, small positive and negative torques are developed. This technique gives excellent static torsional stiffness allowing a smooth transition through zero speed with zero deadband.

The variable-reluctance servo drive displays extremely smooth low-speed rotation and has a wide bidirectional speed range with rapid reversal of rotation. Overall drive performance equals or excels that of a conventional permanent magnet dc servo drive powered by a four-quadrant transistor PWM converter. Performance is somewhat inferior to that of the best brushless dc motor servos, but the elimination of rotor magnets gives reduced cost and increased ruggedness.

11.7. REFERENCES

1. LAMB, C. St. J., Commutatorless alternating-voltage-fed variable-speed motor, *Proc. IEE*, **110**, 12, Dec. 1963, pp. 2221-2227.

2. TUSTIN, A., Electric motors in which commutation is by switching devices such as controlled silicon rectifiers, *Proc. IEE*, **111**, 1, Jan. 1964, pp. 151-152.

3. *Engineering Handbook: DC Motors, Speed Controls, Servo Systems*, Electro-Craft Corp., Fifth Edition, 1980.

4. KENJO, T., and NAGAMORI, S., *Permanent-Magnet and Brushless DC Motors*, Oxford University Press, 1985.

5. HOWLETT, J.F., Brushless dc motors, *Proc. Motor-Con Conf., San Francisco, 1982*, pp. 594-604.

6. PERSSON, E.K., and MESHKAT, S., Brushless motors and controls, *Proc. Motor-Con Conf., Geneva, 1983*.

7. BROSNAN, M.F., and BARBER, N.T., Rare-earth high performance brushless drives and their application, *Proc. Motor-Con Conf., Hannover, 1985*, pp. 75-85.

8. TOMASEK, J., Velocity and position feedback in brushless dc servo systems, *Proc. Motor-Con Conf., Hannover, 1985*, pp. 61-74.

9. de SAE SILVA, C., Brushless dc motors get a controller IC that replaces complex circuits, *Electron. Des.*, Sept. 19, 1985, pp. 149-156.

10. STONE, A.C., and BUCKLEY, M.G., Novel design and control of a trapezoidal back emf motor: The smooth transition from brush to brushless dc, *Proc. Motor-Con Conf., Hannover, 1985*, pp. 86-95.

11. STONE, A.C., and BUCKLEY, M.G., Ultra high performance brushless dc drive, *Proc. Conf. Drives/Motors/Controls, 1984*, pp. 86-91.

12. PERSSON, E.K., and MESHKAT, S., Brushless servo system with expanded torque-speed operating range, *Proc. Motor-Con Conf., Hannover, 1985*, pp. 96-106.

13. LANGLEY, L.W., and FISHER, R.L., Precision brushless tachometry, *Proc. Conf. Drives/Motors/Controls, 1983*, pp. 26-30.

14. KUO, B.C., *Theory and Application of Step Motors*, West Publishing, St. Paul, MN, 1974.

15. ACARNLEY, P.P., *Stepping Motors: A Guide to Modern Theory and Practice*, Peter Peregrinus Ltd., Second Edition, 1984.

16. KENJO, T., *Stepping Motors and Their Microprocessor Controls*, Oxford University Press, Oxford, 1984.

17. KORDIK, K., The step motor — what it is and what it does, *Proc. Third Annual Symp. Incremental Motion Control Systems and Devices*, 1974, pp. A-1 — A-49.

18. DETTMER, R., Stepping motors, variations on a theme, *Electron. Power*, **31**, 6, June 1985, pp. 467-471.

19. CAREY, T., The technology of microstepping, *OEM Des.*, Oct. 1983, pp. 69-71.

20. RAY, W.F., and DAVIS, R.M., Inverter drive for doubly salient reluctance motor: its fundamental behaviour, linear analysis and cost implications, *IEE J. Electr. Power Appl.*, **2**, 6, Dec. 1979, pp. 185-193.

21. LAWRENSON, P.J., STEPHENSON, J.M., BLENKINSOP, P.T., CORDA, J., and FULTON, N.N., Variable-speed switched reluctance motors, *IEE Proc.*, **127, Pt. B**, 4, July 1980, pp. 253-265.

22. DAVIS, R.M., RAY, W.F., and BLAKE, R.J., Inverter drive for switched reluctance motor: Circuits and component ratings, *IEE Proc.*, **128, Pt. B**, 2, Mar. 1981, pp. 126-136.

23. BYRNE, J.V., and McMULLIN, M.F., Design of a reluctance motor as a 10 kW spindle drive, *Proc. Motor-Con Conf., Geneva, 1982*, pp. 10-24.

24. LAWRENSON, P., Switched-reluctance motor drives, *Electron. Power*, **29**, 2, Feb. 1983, pp. 144-147.

25. REED, J., Switched-reluctance drive systems, *Proc. Conf. Drives/Motors/Controls, 1983*, pp. 118-121.

26. POWELL, B., A low cost, efficient 1 kW motor driver, *Proc. Motor-Con Conf., Atlantic City, 1984*, pp. 229-237.

27. RAY, W.F., LAWRENSON, P.J., DAVIS, R.M., STEPHENSON, J.M., FULTON, N.N., and BLAKE, R.J., High performance switched reluctance brushless drives, *Conf. Rec. IEEE Ind. Appl. Soc. Annual Meeting, 1985*, pp. 1769-1776.

28. HARRIS, M.R., FINCH, J.W., MALLICK, J.A., and MILLER, T.J.E., A review of the integral horsepower switched reluctance drive, *Conf. Rec. IEEE Ind. Appl. Soc. Annual Meeting, 1985*, pp. 783-789.

29. EGAN, M.G., MURPHY, J.M.D., KENNEALLY, P.F., LAWTON, J.V., and McMULLIN, M.F., A high performance variable reluctance drive: Achieving servomotor control, *Proc. Motor-Con Conf., Chicago, 1985*, pp. 161-168.

30. BYRNE, J.V., McMULLIN, M.F., and O'DWYER, J.B., A high performance variable reluctance drive: A new brushless servo, *Proc. Motor-Con Conf., Chicago, 1985*, pp. 147-160.

31. BYRNE, J.V., and DEVITT, F., Design and performance of a saturable variable reluctance servo motor, *Proc. Motor-Con Conf., Chicago, 1985*, pp. 139-146.

32. BASS, J.T., EHSANI, M., MILLER, T.J.E., and STEIGERWALD, R.L., Development of a unipolar converter for a variable reluctance motor drive, *Conf. Rec. IEEE Ind. Appl. Soc. Annual Meeting, 1985*, pp. 1062-1068.

33. BYRNE, J.V., Tangential forces in overlapped pole geometries incorporating ideally saturable materials, *IEEE Trans. Mag.*, **8**, 1, Mar. 1972, pp. 2-9.

34. BYRNE, J.V., and LACY, J.G., Characteristics of saturable stepper and reluctance motors, IEE Conf. Publ. No. 136, *Small Electrical Machines*, 1976, pp. 93-96.

35. LAWRENSON, P.J., HODSON, D.P., and HARRIS, M.R., Electromagnetic forces in saturated magnetic circuits, IEE Conf. Publ. No. 136, *Small Electrical Machines*, 1976, pp. 89-92.

36. FITZGERALD, A.E., KINGSLEY, C., and KUSKO, A., *Electric Machinery*, McGraw-Hill, New York, NY, 1971.

37. SLEMON, G.R., and STRAUGHEN, A., *Electric Machines*, Addison-Wesley, Reading, MA, 1980.

38. MILLER, T.J.E., Converter volt-ampere requirements of the switched reluctance motor drive, *Conf. Rec. IEEE Ind. Appl. Soc. Annual Meeting, 1985*, pp. 813-819.

39. CHAPPELL, P.H., RAY, W.F., and BLAKE, R.J., Microprocessor control of a variable reluctance motor, *IEE Proc.*, **131, Pt. B**, 2, Mar. 1984, pp. 51-60.

40. BOSE, B.K., MILLER, T.J.E., SZCZESNY, P.M., and BICKNELL, W.H., Microcomputer control of switched reluctance motor, *Conf. Rec. IEEE Ind. Appl. Soc. Annual Meeting, 1985*, pp. 542-547.

CHAPTER 12

Power Semiconductor Application Techniques

12.1. INTRODUCTION

In power electronics, as distinct from signal electronics, ultraconservative design is seldom feasible, and power semiconductor devices may be required to operate close to their absolute maximum ratings. The important parameters of power transistors and thyristors are reviewed in this chapter and rating procedures are discussed.

Power semiconductor devices are also inherently susceptible to damage or destruction from transient overcurrents and overvoltages. These transients may cause immediate breakdown or may progressively degrade the device so as to shorten its useful life. For reliable operation of power electronic equipment, it is essential to employ effective and secure techniques for protection against overcurrent and overvoltage. These aspects are also treated in this chapter.

In general, correct application of a power semiconductor requires careful observation of its voltage and current ratings so that device damage due to excessive voltage and junction overheating due to excessive current are avoided. The latter aspect requires an appreciation of thermal design considerations, and so the chapter concludes with a general treatment of power semiconductor cooling.

12.2. APPLICATION OF POWER TRANSISTORS[1–5]

The power transistor and power Darlington are now serious competitors to the thyristor for inverter applications up to 200 kVA or more. This section reviews the basic rating techniques for the transistor and discusses the use of snubber networks for transistor stress relief. The important concept of transistor safe operating area is also explained.

12.2.1. *Transistor Current Ratings*

Transistor junction temperature must be limited to avoid destructive overheating of the semiconductor. For a silicon transistor, the rated junction temperature is typically 150 °C. Cooling methods are discussed later in this chapter, but it is obvious that in the on-state, the continuous, or dc, current ratings of the device must be specified so that the rate of energy dissipation at the junction does not exceed the rate at which heat can be removed. This constraint defines thermal limits for the continuous collector, emitter, and base currents. Peak or pulsed current ratings are also specified, based on the fusing current of bonding wires within the device package. The maximum power that the transistor can dissipate and the continuous-current ratings are often specified for a transistor case temperature of 25 °C. This temperature is unrealistic for most practical applications, and the transistor must be derated for operation at higher case temperatures.

The maximum usable value of collector current may be limited by factors other than junction temperature. The dc current gain, h_{FE}, falls off at high collector currents and the on-state saturation voltage, $V_{CE(sat)}$, increases rapidly, as shown in Fig. 12.1. To reduce $V_{CE(sat)}$ and the associated on-state losses by overdriving the base requires a large base current because of the low h_{FE}; hence, the maximum allowable base current is quickly reached. High values of h_{FE} should be available at rated collector current to provide low values of $V_{CE(sat)}$ with reasonable values of forced gain $\beta_F (= I_C/I_B)$, and to permit optimization of turn-on time by overdriving the base.

On the basis of a gain limit, the continuous collector current rating is that current which results in a minimum acceptable value of h_{FE}. The circuit designer must be careful when comparing devices because some manufacturers specify a collector current rating based on a dc current gain of 10, while other manufacturers use higher values. Indeed, "specmanship" is widely practiced in semiconductor data sheets, with the result that the maximum usable continuous collector current at normal operating junction temperatures may be only 60 or 70 percent of the headlined continuous rating.

Junction temperature affects most transistor parameters. At high collector currents, an increase in junction temperature raises the saturation voltage and increases the switching times. Data sheet values for a junction temperature of 25 °C are unduly optimistic for normal operating junction temperatures of 150 °C to 200 °C. Thus, the saturation voltage specification is meaningless without the corresponding values of collector current, base current, and junction temperature.

12.2.2. Transistor Voltage Ratings

Like all semiconductor devices, the power transistor cannot be subjected to excessive voltage without suffering breakdown. In general, a breakdown of either transistor junction occurs when the applied reverse voltage is increased to a value that causes the leakage current to dislodge additional carriers, thereby initiating avalanche breakdown. For the emitter-base junction, the breakdown voltage, designated by $V_{(BR)EBO}$, is typically about 6 V. When the collector-base junction is reverse biased with the emitter open-circuited, there is a small collector leakage current, I_{CBO}, until the breakdown voltage, $V_{(BR)CBO}$, is exceeded. This is the highest voltage rating for the transistor.

In the usual common emitter configuration, there are a number of collector-to-emitter voltage ratings, each one being specified for particular base-emitter conditions. Most of the applied collector-to-emitter voltage appears as a reverse bias across the collector-base junction, producing the leakage current, I_{CBO}. If the base terminal is open-circuited, this current flows across the internal base-emitter junction, producing, by transistor action, a collector current of $h_{FE} I_{CBO}$. The total collector leakage current is therefore

$$I_{CEO} = I_{CBO} (1 + h_{FE}) \qquad (12.1)$$

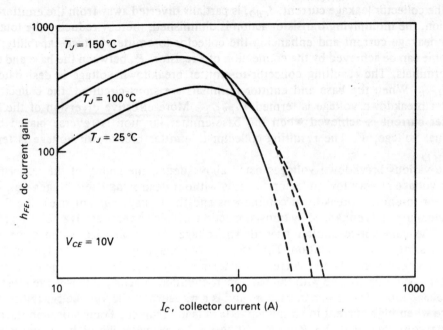

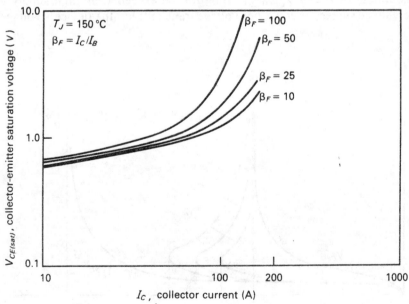

FIG. 12.1. Variation in dc current gain, h_{FE}, and on-state saturation voltage, $V_{CE(sat)}$, as a function of collector current for a General Electric D67DE power Darlington.

where $(1 + h_{FE})$ is much greater than unity. Consequently, the junction power limit is reached at a lower voltage than before, and the collector-to-emitter breakdown voltage for this condition is $V_{(BR)CEO}$.

If the collector leakage current, I_{CBO}, is partially diverted away from the emitter-base junction, the multiplying transistor action is diminished, thereby reducing the total collector leakage current and enhancing the collector-to-emitter voltage capability. This outcome can be achieved by the connection of a resistor, R, between the base and emitter terminals. The resulting collector-to-emitter breakdown voltage is designated by $V_{(BR)CER}$. When the base and emitter terminals are short-circuited, the collector-to-emitter breakdown voltage is termed $V_{(BR)CES}$. More effective diversion of the base-emitter current is achieved when the base-emitter junction is reverse biased by an external voltage, V. The resulting collector-to-emitter breakdown voltage is termed $V_{(BR)CEV}$.

The various breakdown voltages listed above define the ability of the transistor to block voltage at very low collector currents without destroying itself. In general, when collector-to-emitter breakdown commences and the leakage current rises, the flow of collector current is enhanced by transistor action, causing a regenerative build-up of current. The collector-to-emitter breakdown voltage then collapses to the same value regardless of the base condition. This breakdown voltage, which is also relatively insensitive to changes in collector current, as shown in Fig. 12.2, is termed the collector-to-emitter sustaining voltage with the base open-circuited, $V_{CEO(sus)}$. This is the most critical voltage rating for power transistors and is a measure of the voltage capability of the device when high current in an inductive load is interrupted. Transistor manufacturers often recommend that the $V_{CEO(sus)}$ rating not be exceeded by high supply voltages, inductive voltage spikes, or any other voltage transient; but the device's voltage switch-

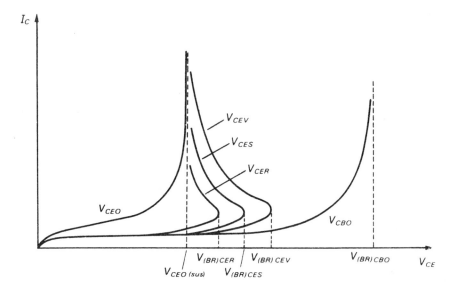

FIG. 12.2. Collector-emitter avalanche characteristics of a transistor.

ing capability can be extended beyond $V_{CEO(sus)}$ with proper control of the operating locus and appropriate base drive conditions.

In addition to the junction thermal limits and the operating limits on voltage and current, there is a further restriction on the operating region of the power transistor due to the onset of secondary breakdown. This is caused by localized junction heating, and in conjunction with the voltage and current constraints, leads to the definition of a safe operating area, as discussed later.

12.2.3. *Inductive Load Switching*

Transistor switching of a resistive load has been studied in Chapter 2. In practice, however, the collector load is usually inductive, at least to the extent that load current remains constant for the duration of the switching interval. Such behavior is characteristic of the simplified circuit of Fig. 12.3, in which a dc supply feeds an inductive load through an n-p-n transistor switch. A continuous load current, I_L, is assumed, and when the transistor is turned off, I_L is diverted into the free-wheeling, or clamping, diode, D1.

In the switching waveforms of Fig. 12.4, it is assumed that the time constant of the load circuit is large compared with the switching period, T. Consequently, the load current is practically constant and is circulating in the free-wheeling diode at time zero when the transistor is turned on. As the current I_L transfers from diode to transistor, the transistor is subjected to the full collector supply voltage, V_{CC}, as shown in Fig. 12.4, because the conducting diode has a small on-state voltage drop. Thus, the transistor experiences maximum voltage and maximum current simultaneously, and the power dissipation at turn-on is greater than that with a resistive load. When the diode current has totally transferred to the transistor, the diode turns off, and the transistor voltage, V_{CE}, decreases rapidly to the saturation voltage, $V_{CE(sat)}$. In practice, the transient characteristics of the free-wheeling diode play an important part in circuit behavior at transistor turn-on because the diode reverse recovery current is carried by the transistor and a fast-recovery diode should be used. Transistor turn-on dissipation can be quite

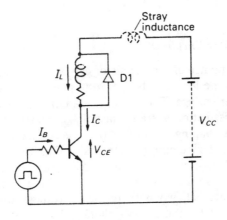

Fig. 12.3. Basic transistor switching circuit for an inductive load.

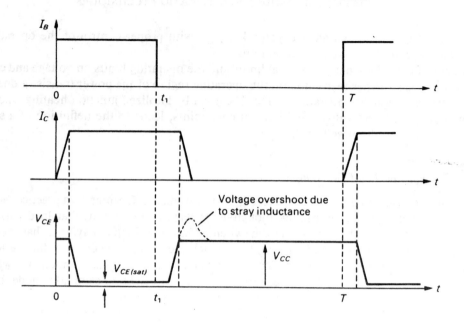

FIG. 12.4. Transistor switching waveforms for the circuit of Fig. 12.3.

high with a slow-recovery diode, but stray circuit inductance, or added series inductance, will ease the problem by delaying the build-up of transistor current.

At turn-off, as shown in Fig. 12.4, the base drive is removed at time t_1, but conduction continues due to free carriers within the base region. After this storage time, the collector-emitter voltage increases rapidly to the dc supply voltage, V_{CC}, before the transistor current starts to transfer to the free-wheeling diode. Again, this means that turn-off dissipation in the transistor is higher than that with a resistive load. Stray circuit inductance will now cause the collector-emitter voltage to overshoot the supply voltage, V_{CC}, as shown in Fig. 12.4, thereby increasing the instantaneous power dissipation and accentuating the risk of transistor breakdown. In general, the transistor losses at turn-off are much more significant than at turn-on.

12.2.4. Transistor Snubber Circuits

The switching performance of the transistor can be displayed in the form of a dynamic load line that indicates the locus of its operating point in the plane of collector current and collector-to-emitter voltage. The dynamic load line highlights the existence of high instantaneous power dissipation or transient voltage overshoot. Stress relief is achieved by appropriate load-line-shaping techniques that use auxiliary snubber circuitry. [1,6,7]

In the earlier, idealized switching waveforms of Fig. 2.13 for a purely resistive load, the variations of I_C and V_{CE} are linear at turn-on and turn-off. The instantaneous power dissipation is zero at the start and finish of the switching interval, and is at maximum midway through the interval, when current and voltage are at half their maximum values. The switching locus in the I_C-V_{CE} plane is the normal linear load line (shown in Fig. 12.5), which is rapidly traversed in both directions.

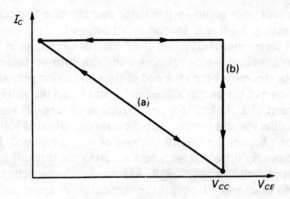

FIG. 12.5. Ideal transistor load lines: (a) purely resistive load; (b) clamped inductive load.

Switching conditions are different with an inductive load, however, and the transistor operating point traverses a dynamic load line that may entail a transient excursion into a region of excessively high voltage or excessive power dissipation. Switching into a clamped inductive load, as discussed above, results in the current I_C increasing to its maximum value, while the transistor sees the dc supply voltage, V_{CC}. Then the free-wheeling diode recovers and the transistor voltage falls to $V_{CE(sat)}$. These conditions imply a dynamic load line that is approximately rectangular, as shown in Fig. 12.5. Peak current and voltage now occur simultaneously in the transistor, and instantaneous power dissipation is high.

Again, at turn-off, the load line is rectangular, because collector current stays practically constant until V_{CE} increases to V_{CC}; I_C then decreases rapidly to the cut-off value. The presence of supply inductance or added series inductance creates different load lines for turn-on and turn-off. Figure 12.6 shows the beneficial effect at turn-on: the load line is now in a region of low power dissipation, indicating a significant reduction in turn-on losses. However, with series inductance, the turn-off locus has a voltage overshoot that causes the collector-emitter voltage to exceed the supply voltage, V_{CC}, as shown in

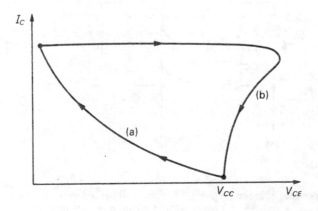

FIG. 12.6. Typical transistor load lines for the circuit of Fig. 12.3: (a) turn-on locus; (b) turn-off locus.

Fig. 12.6. This overshoot may result in transistor damage due to excessive voltage or excessive localized heating, leading to secondary breakdown.

In general, a small series inductor, L_s, in the collector circuit, as in Fig. 12.7(a), will function as a turn-on snubber, which reduces transistor turn-on losses by limiting the rate of increase of collector current. At the end of the conduction period, the shunt resistor, R_s, permits discharge of the energy stored in L_s, and limits the voltage overshoot.

The main component of a snubber for the reduction of turn-off losses is a capacitor connected between collector and emitter. Load current is diverted into the capacitor as the transistor turns off; hence, the rate of increase of V_{CE} is reduced. If the capacitor is large, the transistor has very little voltage across it during the turn-off interval and so the power dissipation in the device is very small. At turn-on, the capacitor is charged to the dc supply voltage and discharges through the transistor. A series resistor is normally introduced to limit this discharge current, giving the conventional nonpolarized turn-off snubber of Fig. 12.7(b). However, the presence of the series resistor is detrimental at turn-off, but it may be shunted by a diode, as in Fig. 12.7(c), to give a polarized snubber. At turn-off, the capacitor, C, is now effectively across the transistor, because the resistor, R, is shunted by the forward-biased diode. At turn-on, the diode is reverse biased, and R limits the discharge current of the capacitor, as required.

A diode may also be placed in series with R_s of Fig. 12.7(a) to give a polarized version of the turn-on snubber. Figure 12.7(d) shows the transistor complete with polarized turn-on and turn-off snubber protection. Figure 12.8 gives typical turn-off load lines with, and without, snubber protection, indicating that the overvoltage surge has been suppressed by the snubber and the transistor switching loss has been substantially reduced.

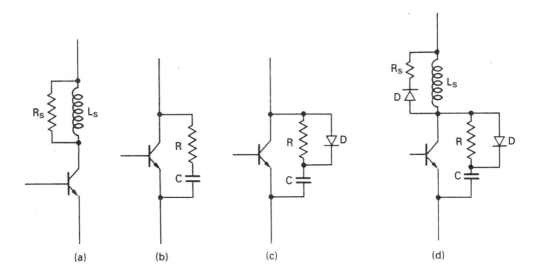

(a)   (b)   (c)   (d)

FIG. 12.7. Transistor snubber circuits: (a) turn-on snubber inductor with discharge resistor; (b) nonpolarized turn-off snubber; (c) polarized turn-off snubber; (d) transistor with polarized turn-on and turn-off snubbers.

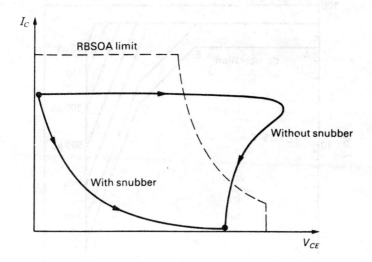

FIG. 12.8. Transistor turn-off load lines with, and without, snubber protection.

During each transistor switching cycle, energy is stored in the snubber components L_s and C_s, and is then discharged. The resulting power loss is proportional to the switching frequency. However, it should be noted that the device power loss is not merely transferred to the snubber resistors, because the total power loss in a snubber-protected system can be lower than that in the unprotected circuit, giving a higher overall efficiency.[6,8]

12.2.5. *Safe Operating Area*

Apart from switching loss reduction, a prime object in the application of snubbers in transistor converters is the avoidance of secondary breakdown. This is a failure mode that occurs when the transistor is in the active region between on- and off-states with appreciable collector current and collector-emitter voltage. Because of device imperfections, current is unevenly distributed over the cross-section of the junction, and localized hot spots may be created, resulting in destruction or irreversible deterioration of the semiconductor.

In general, the capability of the transistor is represented in the form of a safe operating area (SOA or SOAR) in the plane of collector current and collector-emitter voltage. The operating locus of collector current and voltage must always lie within this area, even under fault conditions. When the base-emitter junction is forward biased to turn on the transistor, the relevant area is known as the forward-bias safe operating area (FBSOA). Figure 12.9 shows typical limits for dc and single-pulse operation with the usual logarithmic scales for the I_C and V_{CE} axes. Boundary AB is determined by the maximum current capability of the device, with the continuous-current rating as the limit for dc operation and the peak-current rating setting the limit for pulsed operation. Boundary BC represents the maximum permissible power dissipation at the stated case temperature. This constraint ensures that rated junction temperature is not exceeded. Boundary CD

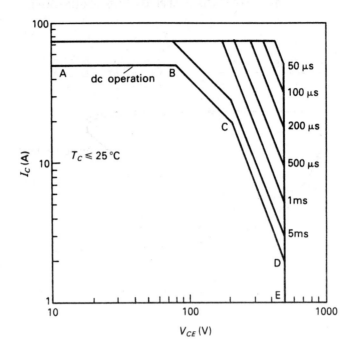

FIG. 12.9. Typical forward-bias safe operating area (FBSOA) for a power transistor.

defines the secondary breakdown limit, which is related to the formation of localized hot spots. Boundary DE is the maximum voltage capability as defined by the $V_{CEO(sus)}$ rating.

Reliable operation of the forward-biased transistor requires that its operating point always lie within the FBSOA. This requirement is readily satisfied when the transistor is in the saturated state and operates close to the I_C axis. However, the dynamic load line at turn-on must also lie within the FBSOA. For pulsed operation, the FBSOA is expanded, as shown in Fig. 12.9, because of the small but finite thermal capacity of the transistor. Modern transistors have turn-on times of a few microseconds, and for these short switching times, the FBSOA is essentially the rectangle bounded by the maximum voltage and current limits.

It should be noted that FBSOA curves are usually presented by the manufacturer for a case temperature of 25 °C and for dc and single-pulse operation. For normal operation at more realistic case temperatures, and with repetitive pulses, the circuit designer must modify the curves with the help of the transient thermal impedance characteristic of the device.[9]

Transistors are usually more severely stressed at turn-off, when they are particularly susceptible to secondary breakdown. It is now common practice to specify a reverse-bias safe operating area (RBSOA) which is applicable at turn-off, when the base current is removed or the base-emitter junction is reverse biased (Fig. 12.10). The RBSOA is smaller than the FBSOA and is usually the limiting factor when a transistor converter is designed. For safe operation under normal or fault conditions, the converter must be designed so that the dynamic load line at turn-off lies within the RBSOA. If this condition is not satisfied, a more rugged transistor is required, or the load line must be modified by means of a snubber network. As explained in Chapter 2, the application of a

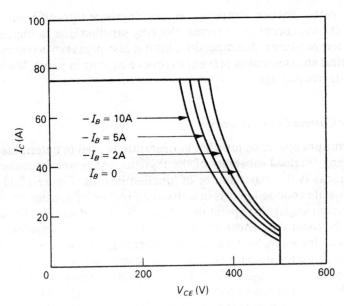

FIG. 12.10. Typical reverse-bias safe operating area (RBSOA) for a power transistor.

reverse base-emitter bias minimizes transistor storage time and enhances the off-condition by reducing the collector leakage current. However, if this reverse bias is applied during the turn-off interval, there is a reduction in size of the RBSOA, as shown in Fig. 12.10, and a greater possibility of secondary breakdown.

In summary, the circuit designer must ensure that turn-on and turn-off load lines always lie within the relevant safe operating areas. Figure 12.8 shows turn-off load lines superimposed on the RBSOA. Without snubber protection, stray inductance causes voltage overshoot and an excursion outside the RBSOA, but a turn-off snubber modifies the load line appropriately. Ideally, the load should appear capacitive when the transistor is switched off (so that current is quenched before the device voltage rises significantly); the load should appear inductive when the transistor is switched on (so that the device is fully on before appreciable current flows). These objectives can be achieved with suitable snubber circuits, as already discussed; but if the transistor has a very wide RBSOA, the snubber circuitry can be minimized or eliminated altogether.

12.3. APPLICATION OF THYRISTORS [10–12]

This section considers the conventional thyristor, or silicon controlled rectifier (SCR), but much of the material is applicable to other thyristor devices such as the gate turn-off thyristor (GTO) and the triac.

12.3.1. *Junction Temperature Limitations*

The maximum operating junction temperature of a thyristor is normally in the range of 100 °C to 150 °C. An upper temperature limit must be imposed because of the temperature dependence of certain critical parameters. Thus, the forward breakover voltage is reduced at elevated junction temperatures, and thyristor turn-off time is increased. Thermal instability may occur at a reverse-blocking junction due to increased leakage currents at high temperatures. A temperature limit is also necessary to restrict thermally induced mechanical stresses and to prevent excessive heating in joints, leads, and interfaces within the device package.

12.3.2. On-State Current Characteristics

The rated maximum operating junction temperature is used to determine the steady-state and recurrent overload capability of the thyristor. At power frequencies, the on-state conduction loss is the major source of junction heating. Figure 12.11(a) shows a typical plot of on-state conduction loss in watts as a function of average on-state current at various conduction angles, for operation between 50 Hz and 400 Hz. These curves are included in the thyristor data sheet and are based on a current waveform that is the remainder of a half sine wave. Such current waveforms are obtained when delayed firing is used in a single-phase resistive load circuit. The power dissipation curves for a specific device are derived from the on-state voltage-current characteristic by the integration of the product of the instantaneous anode current and the corresponding on-state voltage drop. Similar power dissipation curves are presented in the data sheet for rectangular current waveforms with different duty cycles. Such waveforms occur in polyphase rectifier circuits with highly inductive loads.

The heat developed at the junction flows through the silicon chip to the thyristor case and then to the heat sink and on to the surrounding ambient fluid. The junction temperature rises above the stud, or case, temperature in direct proportion to the amount of heat flowing from the junction and the thermal resistance presented to the flow of heat. Power semiconductor cooling is considered in more detail in Section 12.6 of this chapter, but for the present, this simple picture will suffice.

Average current rating. Because the maximum operating junction temperature is fixed, the power loss curves can be used to relate on-state current to maximum allowable case temperature for the specified junction-to-case thermal resistance. The resulting curves in Fig. 12.11(b) plot the maximum allowable case temperature as a function of the average on-state current for sine wave segments of current with various conduction angles. This presentation helps the user by specifying the current rating in terms of the easily measured case temperature rather than the difficult-to-measure operating junction temperature. It can be seen in Fig. 12.11(b) that the maximum average current for a given case temperature is significantly reduced at small conduction angles, and the allowable current approaches zero as the case temperature approaches the maximum permissible junction temperature of 125 °C. Similar characteristics are published for rectangular current waveforms with duty cycle as a parameter.

In practice, the semiconductor has a small but finite thermal capacity — that is, it has the ability to store heat and requires time to heat up and cool down. At power frequencies, the periodic variation in instantaneous power dissipation in the thyristor causes an appreciable fluctuation in junction temperature during a cycle. The simplest method of establishing current ratings is to neglect these temperature variations and use average junction temperature as the maximum operating temperature rating. However, this

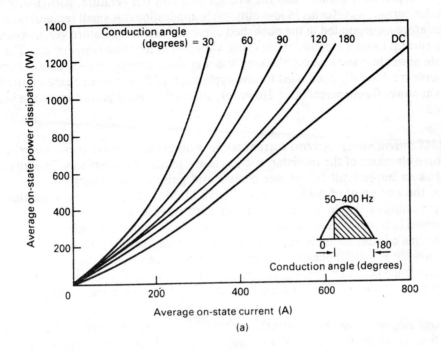

(a)

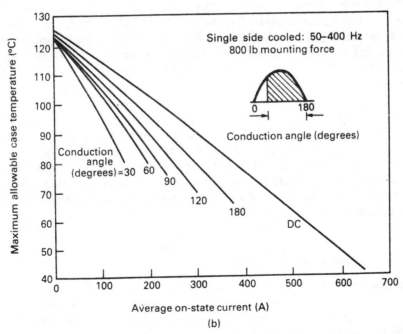

(b)

FIG. 12.11. On-state characteristics of a General Electric C430 phase control thyristor: (a) on-state power dissipation, (b) maximum allowable case temperature.

approach is unsatisfactory because the maximum instantaneous junction temperature may be appreciably greater than the average junction temperature, particularly for rectangular current waveforms in low-duty-cycle applications. A small temperature margin is therefore incorporated in the published curves of case temperature versus average on-state current to take account of the cyclic variation in junction temperature. The effects of gate power loss and reverse blocking loss may also be incorporated. These characteristics have traditionally been used for the application of thyristors in phase-controlled circuits at power frequencies.[10-13] However, alternative rating procedures have been suggested.[14]

RMS current rating. An rms current rating is necessary to prevent excessive heating in resistive elements of the thyristor, such as leads, joints, and interfaces. The rms rating is used as an upper limit for all steady-state and pulse current ratings for the thyristor. Thus, the end point of each curve in Fig. 12.11(b) corresponds to the maximum rms current rating of the device, and each end point represents the same rms current value expressed in terms of average current for the particular current waveform.

The rms current rating is important for the application of thyristors in high peak current, low-duty-cycle applications. Although the average value of the current is quite low, the user must take care that the rms rating is not exceeded. It should also be noted that the headlined current rating of the thyristor is usually its rms current rating.

Surge current rating. For unusual, nonrepetitive overloads or short circuits, the rated junction temperature of the thyristor can be exceeded for a brief instant, thereby allowing additional overcurrent rating. Nonrepetitive surge current values are specified for half-wave sinusoidal current waveforms in a 50 or 60 Hz ac circuit. A curve is usually presented by plotting the peak allowable sine wave current for a specified number of half-cycles. For very short duration (subcycle) overloads, an I^2t rating may also be specified, where I is the rms value of the fault current and t is the fault duration in seconds. A single I^2t value may be quoted, or the I^2t capability may be plotted as a function of overload fault duration in milliseconds. These ratings allow the semiconductor to be coordinated with circuit protective devices such as circuit breakers and fuses, as discussed in Section 12.4.2. The number of current surges permitted over the life of the thyristor must be limited to prevent device degradation.

As discussed in Chapter 2, the thyristor may be damaged by the rapid rate of rise of anode current, and consequently, a maximum di/dt rating is specified. The di/dt rating may also be incorporated as part of a more comprehensive high-frequency current rating procedure, as discussed in the next section.

12.3.3. *High-Frequency Current Ratings*

When the thyristor is turned on, the off-state blocking voltage falls to the on-state value, while the anode current increases rapidly. As a result, the instantaneous power dissipation during the turn-on interval is very high, but the average dissipation is usually negligible at power frequencies. However, as the switching frequency is increased, these turn-on losses form an increasingly significant part of the total average power dissipation of the thyristor and are concentrated in a small area near the gate electrode. The instantaneous power dissipation also peaks during the turn-off interval, and reverse recovery loss may be high if turn-off dv/dt is large. However, turn-on losses are more critical, particularly if the rate of rise of on-state current (di/dt) is high. Proper application of the thyristor in short-pulse and high-switching-frequency inverter applications requires that these switching losses be considered.

The high-frequency current capability of the thyristor is specified in the form of plots of maximum allowable peak current as a function of pulse width for various pulse repetition rates.[15,16] Characteristics are usually provided for sinusoidal and trapezoidal current waveforms and for several stated values of case temperature. A typical set of curves for sinusoidal current waveforms is shown in Fig. 12.12(a). Unlike the earlier power-frequency current rating curves, these high-frequency characteristics are *not* determined by average power dissipation and maximum junction temperature limitations. This difference is explained by the fact that at low frequencies, temperature is uniform across the silicon wafer, whereas, in high-frequency and pulse applications, the entire junction area is not conducting. Consequently, device losses are concentrated in the turn-on region, resulting in high local power densities, even at relatively low average dissipation levels.

The high-frequency current ratings are based on sinusoidal and trapezoidal current waveforms because these are typically present in thyristor inverter circuits. Forced commutating circuits usually employ auxiliary commutating thyristors and LC components which force a near-sinusoidal current through the auxiliary thyristor. In medium- to high-frequency inverters, the main thyristor current is usually sinusoidal. In dc chopper circuits and in low- to medium-frequency PWM and current-source inverters, the main thyristor has a current waveform that is basically rectangular, but finite di/dt values give it steeply sloping sides and a trapezoidal waveshape.

Average device power dissipation must be known if the heat sink is to be sized properly to maintain the case temperature at the value specified in Fig. 12.12(a). This average dissipation is determined from energy-per-pulse data for sinusoidal and trapezoidal current pulses. Figure 12.12(b) plots constant-energy-per-pulse contours for sinusoidal current pulses on a grid of peak on-state current versus pulse base-width. Switching losses due to the RC snubber discharge have been included. In general, the joules-per-pulse value for the particular current waveshape is multiplied by the pulse repetition rate to give average power dissipation.

12.3.4. *Thyristor Voltage Ratings*

Thyristor voltage ratings were discussed briefly in Chapter 2. In the reverse direction, the conventional thyristor and some GTOs behave like rectifier diodes and have similar reverse voltage ratings. These are the repetitive peak reverse voltage rating, V_{RRM}, and the transient or nonrepetitive rating, V_{RSM}. These limits are specified to avoid operation in the avalanche region, where there is a sharp increase in leakage current that may destroy the device.

In the forward direction, the thyristor has repetitive and nonrepetitive off-state voltage ratings, V_{DRM} and V_{DSM}, respectively. The repetitive rating, V_{DRM}, must be observed to ensure that forward breakover does not occur in normal operation, and hence the thyristor remains in the off-state until triggered at the gate. The forward break-

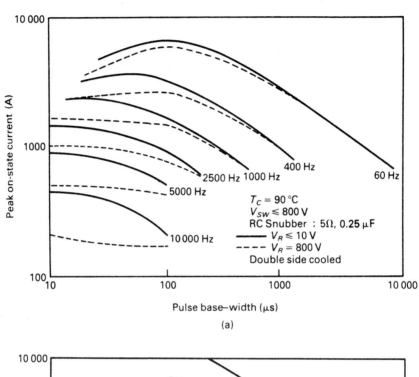

(a)

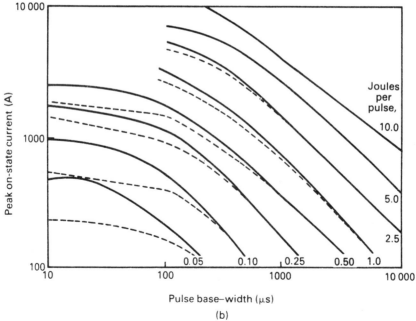

(b)

FIG. 12.12. High-frequency ratings for the General Electric C376 inverter-grade thyristor: (a) peak on-state current; (b) constant energy-per-pulse contours.

over voltage decreases with increasing junction temperature, and consequently, V_{DRM} must be specified at the maximum allowable junction temperature.

Thyristor triggering due to excessive rate of rise of off-state (anode-to-cathode) voltage has also been discussed in Chapter 2, and the static dv/dt rating is listed in the thyristor data sheet. Again, it should be noted that the dv/dt withstand capability of the thyristor decreases with increasing junction temperature. The circuit designer can limit the maximum dv/dt applied to the thyristor by connecting a snubber network between the anode and cathode terminals, as discussed in the next section.

12.3.5. *Thyristor Snubber Circuits*

Snubber networks have already been discussed in connection with transistors, but the snubber is also an essential element in thyristor circuits. However, the turn-off snubber for the thyristor is primarily intended to limit the rate of rise of voltage, whereas the transistor snubber is designed to reduce switching losses and to provide turn-on and turn-off stress relief by appropriate load-line shaping.

The simplest method of dv/dt suppression is a series RC snubber circuit connected across the thyristor, as shown in Fig. 12.13(a). The inductance, L, is the effective circuit inductance, and its presence limits the initial di/dt of the thyristor at turn-on. A series inductor, either linear or saturable, may be introduced to augment the supply circuit inductance. Normally, the LCR circuit is slightly underdamped, and when a forward voltage step is applied to it, the peak voltage appearing across the thyristor and its rate of change are both limited to acceptable values. Typical R and C values are 50 Ω and 0.1 μF, respectively, and snubber losses can be appreciable, particularly at high switching frequencies.

Another important function of the snubber circuit is to limit the amplitude and dv/dt of the reverse voltage generated at turn-off of a power diode or thyristor due to the hole storage effect. As shown in Chapter 2, when the semiconductor turns off, there is a brief pulse of reverse recovery current that rises to a peak value, at which time the device blocks. In the absence of an RC snubber, the abrupt interruption of the reverse recovery current in the series inductance, L, will cause transient $L\,di/dt$ overvoltages that may destroy the device itself or other semiconductors in the converter circuit.

If an RC snubber is connected across the thyristor, the reverse recovery current can transfer to the RC path when the device blocks, as shown in Fig. 12.13(b). The voltage across the RC path appears as an oscillatory reverse voltage across the semiconductor. A correctly designed snubber will limit the amplitude of this reverse recovery voltage and will also limit its rate of rise. By delaying the build-up of reverse voltage across the semiconductor, the recovery losses in the device are also reduced. If the snubber resistance is too small, excessive ringing will occur in the LCR circuit, and the reverse blocking voltage capability of the semiconductor may be exceeded. Appropriate snubber design procedures have been described in the literature.[17,18]

As already explained, thyristor forward voltage must be reapplied at a controlled rate following conduction because the thyristor will fail to commutate if the reapplied dv/dt is unduly high. The presence of a snubber is vital to suppress external forward voltage transients during this critical recombination period. Reverse recovery dv/dt in a commutating diode or thyristor has been discussed in the preceeding paragraphs. In many practical circuits, the diode or thyristor which is turning off has another thyristor con-

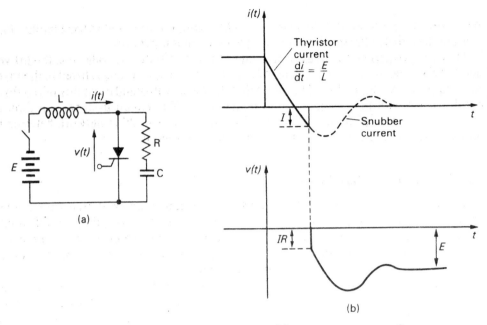

FIG. 12.13. (a) Thyristor with snubber circuit; (b) reverse recovery waveforms.

nected inversely across it. Consequently, the reverse recovery dv/dt of the commutating device appears in the forward direction across the inverse parallel thyristor and may cause spurious triggering or a commutation failure. In the McMurray inverter circuit, for example, the reverse recovery of a feedback diode occurs just after commutation of the inverse parallel thyristor. Thus, the reverse dv/dt of the recovering diode, D1 in Fig. 12.14(a), constitutes forward reapplied dv/dt on thyristor TH1. A thyristor commutation failure will occur if this reapplied dv/dt is excessive.

When thyristor TH1 in Fig. 12.14(a) turns on, the snubber capacitor, C, discharges into the thyristor, but resistor R limits the discharge current and prevents excessive di/dt at turn-on. However, at turn-off, a forward IR voltage drop suddenly appears across the thyristor due to reverse recovery of the diode, as indicated in Fig. 12.13(b). Thus, the presence of the resistor impairs the forward dv/dt-limiting performance of the snubber. If the primary function of the snubber is to limit forward dv/dt on the thyristor, the polarized snubber of Fig. 12.14(b) is often used. Resistor R1 is a small damping resistor giving enhanced forward dv/dt protection, while R2 is a much larger resistor for limiting the snubber discharge current when the thyristor is gated.

12.4. OVERCURRENT PROTECTION OF POWER SEMICONDUCTORS

Fault currents can produce catastrophic overheating of semiconductors in a fraction of a second because semiconductor devices have a very small mass, and junction temperature changes rapidly with changes in current. These abnormal currents may arise from excessive loading of the power converter or an accidental short-circuit at the load terminals. A particularly severe fault condition can arise in the common voltage-source

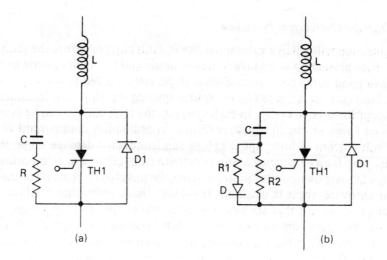

FIG. 12.14. Snubber circuits for a McMurray inverter: (a) conventional nonpolarized snubber;
(b) polarized snubber.

inverter circuit, where each inverter half-bridge has two devices in series across the dc
supply voltage. Normally, the devices conduct alternately, but incorrect triggering due to
electrical noise or interference can result in a "shoot-through" fault, with both devices
conducting simultaneously. This action short-circuits the dc supply and destroys both
power semiconductors unless appropriate overcurrent protection is incorporated.

The current overload capability of the transistor is markedly inferior to that of the conventional thyristor and the GTO. Because thyristor devices are self-latching, the on-state
voltage drop remains relatively low for currents far in excess of the normal operating
current. But, if the base current of a transistor is inadequate for a particular collector current, the transistor comes out of saturation, resulting in high power dissipation and rapid
overheating of the device. This desaturation effect occurs at collector current levels that
are not appreciably greater than the normal operating current.

12.4.1. Transistor Overcurrent Protection

Fuse protection of the delicate power transistor is virtually impossible, but effective
overcurrent protection can be achieved by utilizing its fast turn-off capability. This active
protection technique can be based on overcurrent sensing or $V_{CE(sat)}$ sensing. In the first
category, the collector current is continuously monitored; when it exceeds a preset limit,
the base drive current is rapidly removed to switch off the transistor. The second
category involves monitoring of the collector-emitter voltage of the transistor to detect a
rise in $V_{CE(sat)}$ when the transistor starts to come out of saturation following an abnormal
rise in collector current. Again, the protective circuit initiates rapid cut-off of the transistor by removing its base drive. Because the storage time of the transistor causes a delay in
the turn-off process, it may be necessary to introduce series inductance to limit the rate
of rise of fault current and so allow the protection circuit time to function.[19]

12.4.2. *Thyristor Overcurrent Protection*

Fuse protection will entail a substantial rise in fault current before the fault is cleared. However, fuse protection is feasible in power diode and thyristor circuits because these devices have good surge current capability. If the circuit is fed by a weak ac supply network, the fault current is limited by the source impedance; thus, semiconductor damage may not occur for several cycles. In such systems, the fault current can be interrupted by conventional fuses or circuit breakers. Proper coordination is important to guarantee that the fault current is interrupted before semiconductor damage takes place and to ensure that only faulty branches of the circuit are isolated. In a line-commutated thyristor circuit, adequate overcurrent protection may be possible by removal of the gating signals when an overcurrent is detected. However, these protection methods are inadequate in ac or dc circuits that are powered by an electrically stiff supply system. In such applications, the fault current and junction temperature rise excessively within a few milliseconds, destroying the semiconductor. Special high-speed current-limiting fuses are required for fuse protection of the solid-state devices in these circuits.

Current-limiting fuses. Because these fuses have thermal properties similar to those of the power diode or thyristor, the coordination of fuse and semiconductor is simplified. The fusible link consists of one or more fine silver ribbons or wires of small thermal capacity mounted between heavy metal end-pieces. This arrangement is thermally analogous to a solid-state device mounted on a cooling metal heat sink. The light silver elements have a very short fusing time, giving the fuse a current-limiting action by forcing the available fault current to zero before it rises to its maximum value. [20,21]

In a sine wave ac circuit without fuse protection, the available fault current quickly rises to a high peak value. This behavior is indicated in Fig. 12.15, where a fault occurs at time zero. In a fused circuit, the current increases as before, until a time, t_m, when the fuse melts. An arc is then struck, but the current may continue to rise for a brief interval depending on the circuit parameters and fuse design. The maximum current, I_p, called the peak let-through current, is considerably less than the peak available current. The lower current results from the current-limiting action of the fuse. The fault current decreases from its peak value as the arc resistance increases. The energy stored in the circuit inductance during the build-up of fault current is dissipated in the arc loss until, eventually, at time t_c, all the stored energy has been dissipated and the arc is extinguished. The total clearing time, t_c, is the sum of the melting time, t_m, and the arcing time, $(t_c - t_m)$, as in Fig. 12.15.

Fuse melting occurs when the energy dissipated in the fusible element raises its temperature to the melting value. At any instant, let i denote the fault current and R the resistance of the fusible link. The instantaneous power input to the fuse is $i^2 R$, and the total energy input during the melting time is $\int_0^{t_m} i^2 R \, dt$. The value of $\int_0^{t_m} i^2 \, dt$ is an important parameter in fuse behavior and is usually termed the melting $I^2 t$ of the fuse. The total clearing $I^2 t$ is given by $\int_0^{t_c} i^2 \, dt$. This value is also described as the let-through $I^2 t$ of the fuse, and is a measure of the total thermal energy which the semiconductor device must withstand before the fault current is interrupted. The fuse manufacturer normally provides data in the form of separate graphs of peak let-through current and let-through $I^2 t$ plotted as a function of prospective fault current.

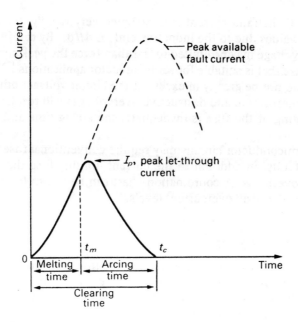

FIG. 12.15. The limitation of fault current in an ac circuit by means of a current-limiting fuse.

Fuse selection. As stated above, the current-limiting fuse and semiconductor device have similar thermal properties. Consequently, for times in the millisecond range, the semiconductor is also described in terms of its peak current capability and fusing I^2t value. A fuse is usually placed in series with each semiconductor. Obviously, the fuse must be rated to carry the full-load current indefinitely. Overcurrent protection is provided by the selection of a fuse whose let-through I^2t rating is less than the I^2t rating of the semiconductor which it protects. However, these I^2t values for fuse and semiconductor vary considerably with pulse width and current waveshape. The thyristor data sheet specifies surge current capability in the form of a graph showing nonrepetitive I^2t rating and peak current rating for pulse durations from about 1 ms to 10 ms. For testing convenience, these curves are based on half sine wave current pulses, whereas the fuse let-through current, as shown in Fig. 12.15, is approximately triangular for short clearing times. Using published I^2t data for fuse and semiconductor coordination is difficult, because the thyristor I^2t rating for a triangular current wave is substantially lower than the published I^2t value for a half-sinusoid current pulse. However, test results and analytical studies have shown that matching the published thyristor peak current capability with the fuse let-through current capability provides a conservative basis for thyristor protection by means of current-limiting fuses.[22,23] From these two sets of data, peak current can be plotted against surge duration on a single graph for a range of fuses and for the thyristor itself. It is then a simple matter to select a fuse whose let-through current curve lies below the curve showing the thyristor surge current rating up to the maximum prospective fault current.[22,24] For fuse-thyristor coordination in dc circuits, additional fuse information is required on dc peak let-through current values.[10]

The voltage across the fuse during the arcing period is known as the arcing, or recovery, voltage; it equals the sum of the source voltage and the induced emf in the circuit inductance. If the fault current is interrupted very rapidly, the arcing voltage may attain excessive values due to the inductive emf, $L \, di/dt$. By careful fuse design, however, the arcing voltage may be limited to less than twice the peak supply voltage, to produce a type of fuse that is suitable for semiconductor applications.[25] The voltage rating of the fuse should not be greatly in excess of the circuit voltage; otherwise, excessively abrupt current interruption and destructive overvoltages will result. On the other hand, if the voltage rating of the fuse is inadequate, the arcing time and let-through I^2t are both increased.

In general, semiconductor circuits may require conventional fuses or circuit breakers for less severe faults, in addition to the current-limiting fuse that operates positively only on severe overloads. A coordination chart can be drawn to ensure that adequate protection is provided at all overcurrent levels.

12.4.3. *Electronic Crowbar Protection*

Because of its high surge current capability, the thyristor can be used in an electronic crowbar circuit for overcurrent protection of power electronic converters using delicate semiconductor devices. The crowbar circuit is used to achieve a rapid shutdown of the power converter before device damage ensues.

Figure 12.16 shows a basic dc-powered converter circuit in which a crowbar thyristor is shunt-connected across the input dc terminals. A detection circuit senses the converter current; if it exceeds a preset limit, the crowbar thyristor is gated and switches fully on within a few microseconds. This action short-circuits the input terminals of the converter and shunts away the converter fault current. The crowbar thyristor continues to receive a fault current from the dc supply, which is limited only by the source impedance, until the fuse (shown in Fig. 12.16) interrupts the fault current. The fuse may be replaced by a circuit breaker if the crowbar thyristor has an adequate surge current rating. Proper operation of this circuit requires a careful wiring layout to minimize stray circuit inductance. A power diode may be required in antiparallel with the crowbar thyristor to prevent a negative voltage transient being applied across the power converter.[26]

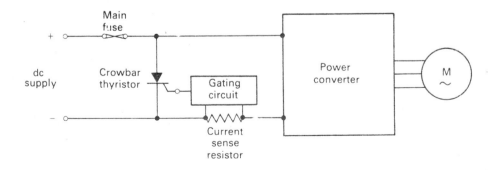

FIG 12.16. Basic electronic crowbar circuit.

12.5. OVERVOLTAGE PROTECTION OF POWER SEMICONDUCTORS

Most semiconductor devices are very intolerant of excessive voltage: destructive breakdown or damage can occur with transient overvoltages that persist for only a few microseconds. These overvoltage transients are probably the main cause of unreliability in power electronic equipment, and reliable circuit operation requires the use of effective overvoltage suppression techniques.

12.5.1. *Origin of Voltage Transients*

Voltage transients may originate within the power electronic circuit, or they may be superimposed on the normal ac network voltage due to system switching operations and occasional lightning surges. [27,28] If a diode or thyristor converter obtains its supply from a transformer, a voltage transient is generated whenever the primary is suddenly energized or de-energized. When the primary is energized, a ringing oscillation occurs on the secondary side due to the sudden application of voltage to the oscillatory circuit formed by the secondary leakage reactance and the winding capacitance. If the primary is energized at a peak of the supply voltage, an overvoltage of twice the normal peak secondary voltage can be generated. The voltage transient following sudden interruption of the primary magnetizing current is usually the most severe transient occurring in practice, because the main transformer flux is forced to decay rapidly to zero, producing a transient overvoltage that may rise to more than ten times the normal working voltage.

In general, the interruption of current flow in an inductive circuit produces dangerous overvoltages unless alternative low-resistance discharge paths are provided. If unsuppressed, these voltage transients will appear within the circuit itself and may also affect neighboring circuits. As already explained, the commutation of a diode or thyristor generates overvoltages which are capable of damaging the device itself or other semiconductor devices in the circuit. The reverse recovery current rapidly decays at rates of up to 10^6 or 10^7 A/s and significant overvoltages are generated if this current is flowing in a series inductance such as a stray circuit inductance or transformer leakage inductance. The rapid current change also generates considerable radio-frequency interference (RFI).

12.5.2. *Suppression of Voltage Transients*

The use of semiconductor devices which are capable of withstanding all transient overvoltages would be uneconomic and impractical; consequently, voltage suppression techniques are essential. A simple RC snubber circuit has traditionally been used to provide overvoltage protection. The snubber is connected across the device to be protected, as shown earlier in Fig. 12.13(a) and again in Fig. 12.17(a). The slow charging rate of the capacitor limits the magnitudes of fast voltage transients and reduces the rate of voltage rise (dv/dt). A small resistor is required in series with the capacitor to damp ringing oscillations between the capacitance and stray circuit inductance. As explained earlier, this series resistor also limits the initial discharge current of the capacitor when the semiconductor device is triggered. In ac-utility-powered circuits, snubber networks are often connected between the ac supply lines to limit incoming utility-borne transients. [29] RC snubbers are also connected across transformer secondary windings and other potential

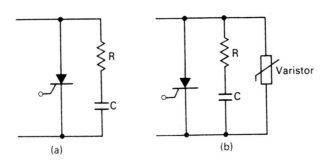

FIG. 12.17. Transient voltage suppression: (a) with RC snubber; (b) with RC snubber and varistor.

sources of inductive voltage transients. However, snubber networks are only partially effective in suppressing transient overvoltages, and hence large safety margins must be applied to the semiconductor blocking voltages. Voltage-clamping devices are more effective in limiting transient overvoltages than RC snubbers and have lower losses for normal operating conditions.

A voltage-clamping device is a nonlinear voltage-dependent resistive element with electrical behavior similar to that of a pair of series-connected, back-to-back Zener diodes. At normal voltages, below the clamping level, the device has a high resistance and draws a small leakage current. Large transient voltages are limited when the device operates in the low-resistance region above its clamping voltage. The increased current produces an increased voltage drop in the source impedance, which effectively clamps the transient voltage at a safe level. The potentially destructive surge energy is dissipated in the voltage-clamping device, but its operating current returns to the normal value after the surge.

Voltage-clamping devices such as the avalanche diode suppressor, the metal oxide varistor (MOV), and the selenium surge suppressor are now available.[27,30] The avalanche diode suppressor consists of two p-n junction diodes in one silicon wafer. The diodes are series-connected back to back and have an avalanche reverse breakdown characteristic which is nondestructive. Consequently, the suppressor draws a large current in either direction when the avalanche voltage is exceeded.

The metal oxide varistor (MOV) consists of a pair of metal contact plates separated by a thin layer of small zinc oxide granules bonded in an amorphous mixture of other oxides. The boundaries of the granules have p-n junction characteristics, so that the device has a symmetrical I-V characteristic similar to that of the avalanche diode suppressor. Varistors are now available with ac operating voltages from 6 V to 2800 V and with a capability extending up to 50 000 A and 6500 J for the larger units.[27] The selenium surge suppressor does not have the clamping ability of the modern avalanche diode suppressor or varistor, and consequently, its field of application is diminishing.

As stated above, the RC snubber is only partially effective, and high-voltage, high-energy transients are not adequately suppressed. Conversely, the voltage-clamping device has no effect on voltages that do not exceed the clamping level, and hence cannot attenuate dv/dt at the operating voltage. Combined RC and voltage-clamping protection is therefore recommended for ac-line-operated equipment, as shown in Fig. 12.17(b). A smaller RC snubber can be used when a voltage-clamping device is present, and hence reduced snubber losses are also achieved.[30]

12.6. COOLING OF POWER SEMICONDUCTORS

In the analysis of power semiconductor switching circuits, device losses are often neglected because they are usually a small fraction of the load power. In practical circuit design, however, the heat dissipated within the semiconductor device cannot be ignored, and good thermal design is vital to ensure that the specified maximum operating junction temperature is not exceeded. As already explained, junction temperature is a critical parameter because of thermal stability considerations and the temperature dependence of device characteristics. Power semiconductor current ratings are determined by the need to operate within specified thermal limits, thereby avoiding permanent device damage and ensuring long-term reliability. The maximum junction temperature is typically 125 °C for conventional thyristors and GTOs and between 150 °C and 200 °C for most silicon power diodes, transistors, and MOSFETs.

The power dissipated in the junction region of a semiconductor device has the following components:

- On-state conduction loss

- Off-state loss due to leakage current

- Switching losses at turn-on and turn-off

- Triggering losses.

The on-state conduction loss is usually the major component, but switching losses will dominate at high operating frequencies. Triggering losses are usually negligible in the case of a power MOSFET or pulse-gated thyristor, but base drive power is often significant in a power transistor.

Electrical losses in the semiconductor device produce thermal energy or heat that must be conducted away from the junction region. Heat dissipation is achieved by mounting the semiconductor on a heat sink. The heat generated within the silicon wafer flows to the case or stud of the device and then through the heat sink and into the ambient fluid, which is usually air or water. The junction temperature rises until the rate of heat dissipation from the heat sink equals the electrical losses in the device; consequently, the current ratings of a power semiconductor are often based on thermal considerations.

The heat sink material is usually aluminum or copper, and the cooling fluid is air in most industrial applications. Heat is removed from the heat sink predominantly by convection. Natural convection cooling is common, and its effectiveness is increased by fitting the heat sink with protruding peripheral fins to enlarge the surface area. Heat loss by radiation is also present and is enhanced if the metal has a black anodized finish.

The heat dissipation of a natural convection finned heat sink is primarily a function of the volume contained within the envelope of the heat sink. If the naturally cooled system is excessively large, a more compact heat sink can be achieved by employing forced air cooling, with a fan blowing cooling air over the fins. Forced convection air cooling increases the heat-removing capability of a naturally cooled heat sink by a factor of two to three and is widely used. In high-power installations, where device losses are large, water cooling gives a very compact assembly.

12.6.1. Thermal Resistance

In general, a flow of thermal energy takes place from a region of high temperature to a region of lower temperature. Similarly, in an electric circuit, current flows from a point of high potential to a point of lower potential. Heat transfer is therefore analogous to current flow, as shown in Fig. 12.18. Thermal power in watts corresponds to current in amperes, temperature difference corresponds to voltage difference, and thermal resistance is analogous to electrical resistance. The thermal resistance is denoted by R_θ and is a measure of the temperature difference per watt of heat flow (°C/W), just as electrical resistance, R, is a measure of the voltage difference per ampere of current flow in volts/ampere or ohms.

As indicated in Fig. 12.18, by analogy with Ohm's law, the steady-state temperature difference when a constant thermal power of P watts flows through a thermal resistance of R_θ °C/W is given by

$$T_1 - T_2 = PR_\theta. \tag{12.2}$$

Using the electrical analogy, heat flow from semiconductor junction to ambient fluid can be represented by the series dc circuit of Fig. 12.19 in which the thermal power of P watts at the junction is analogous to a constant-current source. The steady-state temperature difference between junction and ambient is given by

$$T_J - T_A = P(R_{\theta JC} + R_{\theta CS} + R_{\theta SA}) \tag{12.3}$$

where $R_{\theta JC}$, $R_{\theta CS}$, and $R_{\theta SA}$ are the thermal resistances from junction to case, case to sink, and sink to ambient, respectively. Also,

$$P = \frac{T_J - T_C}{R_{\theta JC}} = \frac{T_C - T_S}{R_{\theta CS}} = \frac{T_S - T_A}{R_{\theta SA}} = \frac{T_J - T_A}{R_{\theta JA}} \tag{12.4}$$

where $R_{\theta JA} = R_{\theta JC} + R_{\theta CS} + R_{\theta SA}$, the total thermal resistance between junction and ambient.

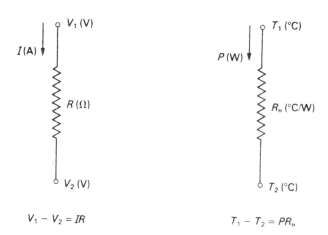

$$V_1 - V_2 = IR \qquad\qquad T_1 - T_2 = PR_H$$

Fig. 12.18. Electric circuit analogy of the heat transfer problem.

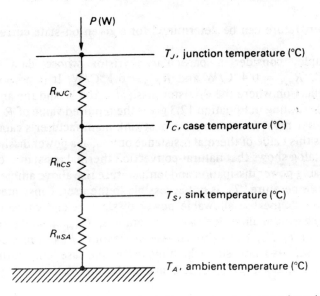

P (W)

T_J, junction temperature (°C)

$R_{\theta JC}$

T_C, case temperature (°C)

$R_{\theta CS}$

T_S, sink temperature (°C)

$R_{\theta SA}$

T_A, ambient temperature (°C)

FIG. 12.19. Thermal equivalent circuit for a semiconductor on a heat sink.

The junction-to-case thermal resistance, $R_{\theta JC}$, is specified in the semiconductor data sheet. The thermal resistance across the case-to-sink interface, $R_{\theta CS}$, depends on the size of the semiconductor case because size affects the contact area between the device and its heat sink. The quality of the contact, also important, is dependent on the flatness of the surfaces, the clamping pressure, and whether a thermally conducting grease is used between the surfaces. Practical values of $R_{\theta CS}$ vary between 0.05 °C/W and 0.5 °C/W, and the semiconductor data sheet may specify a typical value, assuming correct installation procedures and the use of interface thermal grease. If the case is electrically isolated from the heat sink with a thin insulating wafer, $R_{\theta CS}$ may be substantially increased.

The sink-to-ambient thermal resistance, $R_{\theta SA}$, is measured from the device mounting area, where the heat sink is hottest, to the cooling fluid. Its value depends on parameters such as heat sink material, surface area and finish, volume occupied, and air flow. For a natural-convection heat sink, the thermal resistance can be as low as 0.5 °C/W, but for lower values, heat sink size is excessive and forced convection cooling or liquid cooling is necessary.

12.6.2. Heat Sink Specification

The semiconductor data sheet specifies the maximum allowable junction temperature, $T_{J(max)}$, and the junction-to-case thermal resistance, $R_{\theta JC}$. When the case-to-sink thermal resistance, $R_{\theta CS}$, is specified or estimated, Equation 12.3 can be used to determine the largest permissible heat sink thermal resistance, $R_{\theta SA}$, and hence the size of heat sink required for a given steady-state power dissipation, P, and ambient temperature T_A. The continuous dc current rating of the device can then be evaluated with the use of the on-state voltage-current characteristic. Conversely, if the heat sink has been specified, the junction temperature can be determined for a given on-state current and ambient temperature.

As an example, consider a particular thyristor whose data sheet specifies $T_{J(max)} = 125\,°C$, $R_{\theta JC} = 0.4\,°C/W$ and $R_{\theta CS} = 0.1\,°C/W$. It is proposed to use this device in an application where the thyristor dissipates 100 W and the ambient temperature is 35 °C. Substituting in Equation 12.3 gives the required value of $R_{\theta SA}$ as 0.4 °C/W. The circuit designer must now examine a heat sink manufacturer's catalog and select a heat sink that has this value of thermal resistance or less, at a power dissipation of 100 W. The heat sink catalog shows that natural-convection thermal resistance decreases somewhat with increasing power dissipation and temperature rise above ambient.

The junction temperature, T_J, is not accessible to the user; consequently, the device manufacturer may define the allowable power dissipation and current capability as a function of the maximum allowable case temperature, $T_{C(max)}$, which results in a junction temperature of $T_{J(max)}$. For a given power dissipation, P, and ambient temperature, T_A, the user now selects a heat sink which maintains the case temperature at, or below, the specified value. Neglecting $R_{\theta CS}$, the heat sink thermal resistance is given by

$$R_{\theta SA} = \frac{T_C - T_A}{P}. \tag{12.5}$$

As an example, assume a power semiconductor dissipates 100 W for operating conditions for which the data sheet specifies a maximum case temperature of 95 °C. If the ambient temperature is 35 °C, the maximum permissible heat sink thermal resistance is given by Equation 12.5 as 0.6 °C/W.

The above equations and calculations have assumed dc current flow in the semiconductor device and steady-state temperature conditions. If the semiconductor is subjected to cyclic operation, the average power dissipation can be used in these equations to determine the average junction temperature. As already stated, for phase control applications, a thyristor data sheet specifies maximum average power dissipation for square wave and rectified sine wave currents for operation up to 400 Hz. The average power dissipation permitted is less than the steady-state dc power dissipation. This derating is necessary because the instantaneous junction temperature varies cyclically above and below the average value due to the pulsating nature of the power loss. Consequently, the average junction temperature must be reduced if the peak allowable junction temperature of 125 °C is not to be exceeded. At normal power frequencies of 50 or 60 Hz, the peak junction temperature may exceed the average temperature by 5 °C, or more. Precise calculation of peak junction temperature for cyclic or transient currents uses the concept of transient thermal impedance.

12.6.3. Transient Thermal Impedance

For transient conditions, the thermal storage capacity of the semiconductor must be taken into consideration for calculating junction temperature. Thermal capacity is the heat energy stored per degree of temperature rise, and its value is much smaller for a semiconductor than for an electric motor or transformer. This means that the semiconductor device heats up very rapidly on overcurrents, but the small thermal capacity can be utilized to give a significant short term overcurrent capability and thereby avoid unduly conservative circuit design.

The effect of thermal capacity in a power semiconductor is that heat generated by electrical losses is partly stored in the device as its temperature increases, while the remainder of the heat loss at the junction flows to the heat sink. Thermal capacity is analogous to capacitance in the electrical circuit, and the transient thermal behavior of the semiconductor can be modeled approximately by a distributed RC network, as shown in Fig. 12.20. In this diagram, the case of the device is taken as reference.

Assume a constant heating power, P, is suddenly applied to the junction at time zero. For this step input in heating power, there is a corresponding step input of current in the electrical analog of Fig. 12.20. This current must charge up the network capacitance before a voltage can appear at input terminal J. Consequently, there is a delay in the build-up of voltage at the input terminals which is analogous to the delay in the build-up of junction temperature. Similarly, there is a time lag in the decay of junction temperature when the input P is removed.

Figure 12.21 plots the growth of junction temperature as a function of time for a step function power loss, P. The junction temperature increases at a rate which is dependent on the response of the thermal network. If ΔT_J is the rise in junction temperature at some time, t, after the application of P, then $\Delta T_J / P$ is the rise in temperature per unit power. This quantity is termed the transient thermal impedance, $Z_{\theta JC}$, which has the same units as thermal resistance (°C/W). Thus, for a step input of power, P, at time zero, the junction temperature rise at some subsequent time, t, is

$$\Delta T_J = P Z_{\theta JC} \qquad (12.6)$$

where $Z_{\theta JC}$ is the transient thermal impedance at time t.

When steady-state conditions are achieved, the capacitors in Fig. 12.20 are fully charged. Their presence can be ignored when calculating the steady-state temperature rise of the junction above case temperature. Thus,

$$T_J - T_C = P \left[R_{\theta 1} + R_{\theta 2} + \cdots R_{\theta n} \right]$$

$$= P R_{\theta JC} \qquad (12.7)$$

where $R_{\theta JC}$ is the series thermal resistance between junction and case, as defined earlier for dc conditions. The transient thermal impedance, therefore, has an initial value of

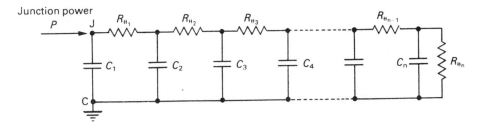

FIG. 12.20. Simplified equivalent network for transient temperature response of the semiconductor junction.

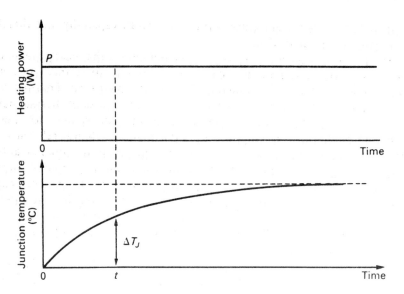

FIG. 12.21. Thermal response of a semiconductor junction to a step input of heating power.

zero, and for long time intervals, it approaches the steady-state thermal resistance, $R_{\theta JC}$.

The semiconductor data sheet includes a plot of $Z_{\theta JC}$, the transient thermal impedance from junction to case as a function of pulse duration. Figure 12.22 shows a typical plot which must represent the highest value that can be expected from the manufacturing distribution of the product. This characteristic can be used to determine instantaneous peak junction temperature for single or repetitive pulse operation and for

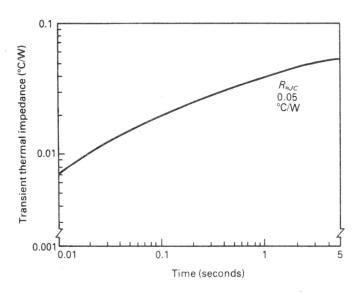

FIG. 12.22. Transient thermal impedance as a function of pulse duration.

transient overcurrents and fault conditions. Conversely, the surge current capability can be evaluated for the specified maximum allowable junction temperature. Thus, if a thyristor has $T_{J(max)} = 125\,°C$ and $Z_{\theta JC} = 0.02\,°C/W$ at $t = 0.01$ s, and if the initial case and junction temperatures are 25 °C, it means that a single step-function power pulse of $P = \Delta T_J/Z_{\theta JC} = 100/0.02 = 5000$ W can be tolerated for 0.01 s without exceeding the maximum allowable junction temperature.

For more complex waveforms, the circuit designer can calculate intermittent and pulse current ratings by applying the principle of superposition. This calculation involves replacing the actual junction power waveform by an equivalent series of superimposed positive and negative step functions. Thus, a typical train of heating pulses, as in Fig. 12.23(a), can be replaced by the positive and negative step functions of

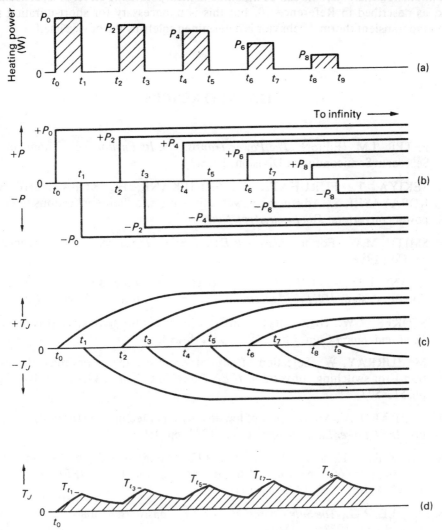

FIG. 12.23. Use of superposition to determine peak junction temperature: (a) junction heating power; (b) equivalent heat input by superposition; (c) junction temperature response to step power inputs; (d) actual junction temperature.

Fig. 12.23(b), and the peak junction temperature can be calculated in terms of the transient thermal impedance. For a pulse train of uniform, equally spaced rectangular pulses, the earlier pulses may be represented to a good approximation by the average value, and the last one or two pulses can be used to determine the peak instantaneous junction temperature. Irregularly shaped power pulses are replaced by equivalent rectangular pulses.[11,31]

When the semiconductor is mounted on a heat sink, the transient thermal impedance characteristic is modified for long-duration pulses. However, the heat sink arrangement is not under the control of the device manufacturer, and hence the data sheet characteristic for transient thermal impedance does not include the influence of the heat sink. The published characteristic can be modified to incorporate the effect of a particular heat sink, as described in Reference 10, but this is unnecessary for short-duration pulses when the transient thermal behavior is determined solely by the device itself.

12.7. REFERENCES

1. PETER, J.M. (Editor), *The Power Transistor in Its Environment*, Thomson — CSF Sescosem Semiconductor Division, 1978.

2. SEKIYA, T., FURUHATA, S., SHIGEKANE, H., KOBAYASHI, S., and KOBAYASHI, S., Advancing power transistors and their applications to electronic power converters, *Fuji Electr. Appl. Note*, 1981.

3. SMITH, M.W. (Editor), *Electronic Data Library, Transistors - Diodes*, General Electric Co., 1982.

4. EVANS, P.D., and HILL-COTTINGHAM, R.J., Some aspects of power transistor inverter design, *IEE Proc., Electr. Power Appl.*, **2**, 3, June 1979, pp. 73-80.

5. STOKES, R.W., High voltage transistor inverters for ac traction drives, *IEEE Int. Semicond. Power Conv. Conf.*, 1977, pp. 270-294.

6. McMURRAY, W., Selection of snubbers and clamps to optimize the design of transistor switching converters, *IEEE Trans. Ind. Appl.*, **IA-16**, 4, July/Aug. 1980, pp. 513-523.

7. FERRARO, A., An overview of low-loss snubber technology for transistor converters, *IEEE Power Electron. Spec. Conf.*, 1982, pp. 466-477.

8. CALKIN, E.T., and HAMILTON, B.H., Circuit techniques for improving the switching loci of transistor switches in switching regulators, *IEEE Trans. Ind. Appl.*, **IA-12**, 4, July/Aug. 1976, pp. 364-369.

9. NOBLE, P.G., *The Safe Operation of Power Transistors*, Mullard Technical Information, No. 68, 1978.

10. GRAFHAM, D.R., and GOLDEN, F.B. (Editors), *SCR Manual*, Sixth Edition, General Electric Co., 1979.

11. LOCHER, R.E., and SMITH, M.W. (Editors), *Electronic Data Library, Thyristors — Rectifiers*, General Electric Co., 1982.

12. HOFT, R.G. (Editor), *SCR Applications Handbook*, International Rectifier Corp., 1974.

13. READ, J.S., and DYER, R.F., Power thyristor rating practices, *Proc. IEEE*, **55**, 8, Aug. 1967, pp. 1288-1301.

14. NEWELL, W.E., Dissipation in solid-state devices — the magic of I^{1+N}, *IEEE Trans. Ind. Appl.*, **IA-12**, 4, July/Aug. 1976, pp. 386-396.

15. DYER, R.F., The rating and application of SCRs designed for power switching at high frequencies, *IEEE Trans. Ind. Gen. Appl.*, **IGA-2**, 1, Jan./Feb. 1966, pp. 5-15.

16. LOCHER, R.E., Characterization of high-frequency high-current reverse blocking triode thyristors for trapezoidal current waveforms, *IEEE Trans. Ind. Gen. Appl.*, **IGA-4**, 2, Mar./Apr. 1968, pp. 228-233.

17. RICE, J.B., Design of snubber circuits for thyristor converters, *Conf. Rec. IEEE Ind. Gen. Appl. Group Annual Meeting, 1969*, pp. 485-489.

18. McMURRAY, W., Optimum snubbers for power semiconductors, *IEEE Trans. Ind. Appl.*, **IA-8**, 5, Sept./Oct. 1972, pp. 593-600.

19. EVANS, P.D., and SAIED, B.M., Fault-current control in power-conditioning units using power transistors, *IEE Proc.*, **128, Pt. B**, 6, Nov. 1981, pp. 335-337.

20. GUTZWILLER, F.W., The current-limiting fuse as fault protection for semiconductor rectifiers, *Trans. AIEE, Pt. 1, Commun. and Electron.*, **77**, Nov. 1958, pp. 751-755.

21. HOWE, A.F., and NEWBERY, P.G., Semiconductor fuses and their applications, *IEE Proc.*, **127, Pt. B**, 3, May 1980, pp. 155-168.

22. GOLDEN, F.B., Take the guesswork out of fuse selection, *The Electronic Engineer* (USA), July 1969.

23. SCHONHOLZER, E.T., Fuse protection for power thyristors, *IEEE Trans. Ind. Appl.*, **IA-8**, 3, May/June 1972, pp. 301-309.

24. KING, K.G., Which fuse will protect your thyristor?, *Electron. Eng.* (GB), May 1973, pp. 59-61.

25. JACOBS, P.C., Fuse application to solid-state motor and power control, *IEEE Trans. Ind. Gen. Appl.*, **IGA-2**, 4, July/Aug. 1966, pp. 281-285.

26. EVANS, P.D., and SAIED, B.M., Protection methods for power-transistor circuits, *IEE Proc.*, **129, Pt. B**, 6, Nov. 1982, pp. 359-363.

27. SMITH, M.W., and McCORMICK, M.D. (Editors), *Electronic Data Library, Transient Voltage Suppression*, General Electric Co., 1982.

28. MARTZLOFF, F.D., The coordination of transient protection for solid-state power conversion equipment, *IEEE Int. Semicond. Power Conv. Conf.*, 1982, pp. 97-105.

29. MERRETT, J., Practical transient suppression circuits for thyristor power-control systems, *Mullard Technical Communications*, No. 104, Mar. 1970, pp. 82-88.

30. DE BRUYNE, P., and WETZEL, P., Improved overvoltage protection in power electronics using active protection devices, *IEE Proc., Electr. Power Appl.*, 2, 1, Feb. 1979, pp. 29-36.

31. GUTZWILLER, F.W., and SYLVAN, T.P., Power semiconductor ratings under transient and intermittent loads, *Trans. AIEE, Pt. 1, Commun. and Electron.*, 79, Jan. 1961, pp. 699-706.

CHAPTER 13

The Adjustable-Speed AC Motor Drive

13.1. INTRODUCTION

The basic circuitry and control techniques for the solid-state ac motor drive have been described in preceding chapters. Adjustable-frequency control offers greatest flexibility and the line-commutated cycloconverter is attractive for large, low-speed drives, but the dc link converter is more versatile and economical and has been more widely used in industry. In this chapter, the selection or specification of a voltage-source inverter is studied, and the selection and design procedures for an inverter-fed ac motor are considered. In the low-to-medium power range, the dc link converter usually feeds a cage-rotor induction motor; the most common forms of inverter-fed induction motor drive are compared in this chapter. The importance of computer modeling and simulation techniques in the design of adjustable-frequency ac motor drives is also emphasized. Finally, the advantages of adjustable-speed ac motor drives are listed, and major application areas are identified.

13.2. STATIC INVERTER RATING

As explained in Chapter 12, the semiconductor has a maximum allowable junction temperature which cannot be exceeded without damaging the device. Consequently, care must be exercized in specifying or selecting a converter for an ac motor drive.

The static inverter has several current ratings, each of which must be strictly observed. The continuous-current rating is determined by the ambient temperature, the thermal resistance of the heat sink, and the maximum allowable rise in junction temperature. The small thermal capacity of a semiconductor device causes it to heat up rapidly on overcurrents, and the overload capability of a static inverter is much less than that of an electric motor, which has a thermal time constant measured in minutes. Consequently, in a motor, large overcurrents may be carried for short intervals of time without causing overheating, but in a semiconductor, destructive overheating occurs in a fraction of a second, and the overcurrent rating based on the transient thermal characteristics must be carefully respected. A third current rating is also specified for the forced-commutated thyristor inverter. This is the instantaneous peak current delivered by the inverter, and it is limited by the commutating capability of the system. If the current at the instant of commutation exceeds this limit, a commutation failure occurs, which may blow fuses or trip out the inverter. The maximum current to be commutated is decided at the design stage, and the commutating circuit is constructed on this basis. In subsequent service, the built-in commutating capability of the inverter cannot be exceeded without causing a shutdown. Similarly, the GTO inverter has a maximum controllable current capability which must not be exceeded. It is determined primarily by device characteristics and snubber circuit design.

In general, therefore, the static inverter has three kVA ratings—the continuous, momentary, and commutating ratings. The *continuous-kVA rating* is the maximum output which the inverter can deliver indefinitely at maximum ambient temperature, while remaining within the frequency and voltage specifications. The *momentary-kVA rating* is the maximum inverter output for a specified time, following operation at rated continuous current and maximum ambient temperature. The *commutating-kVA rating* is the peak instantaneous output which the inverter can deliver without exceeding the commutating or turn-off capability of the system. A typical momentary rating for a thyristor inverter is 150 percent of the continuous rating for 1 min and 200 percent of the continuous rating for 0.5 s. Unless the commutating rating is specified independently, it is assumed to be the same as the momentary rating.

The fact that the momentary rating of a static inverter is considerably less than that normally associated with rotating power supplies is a serious disadvantage in ac motor applications. When started at rated voltage and frequency, a typical cage-rotor induction motor draws six times full-load current, while a synchronous reluctance motor may draw up to eight times rated current. If the static inverter is to supply these large transient currents for the duration of the run-up period, it must have a continuous-current rating of this order, because of the limited thermal capacity of the power semiconductors. Oversizing the inverter to this extent would render the solid-state ac drive highly uneconomical. In a single-motor drive, it is essential to adopt an appropriate control strategy which ensures that motor and inverter ratings are closely matched. A current limit is usually incorporated into the system to keep the motor current within the inverter rating, but good starting performance is achieved if the voltage and frequency are reduced at starting and then increased gradually to speed up the motor. However, in voltage-fed multiple-motor drives it is not always feasible to start the motors simultaneously from standstill. If the motors are run up individually, the inverter must have a momentary current rating equal to the standstill current of one motor, plus the full-load current of the remainder.

It must also be borne in mind that the continuous-current rating of the inverter varies with frequency. At very low frequencies, each semiconductor device carries the full input current for an appreciable interval, and at zero frequency the direct-current rating of the inverter equals the continuous-current rating of one device. As the operating frequency increases, the conduction periods become shorter and the current rating increases. Even at frequencies of 1 or 2 Hz, the current rating of a thyristor inverter is about 40 or 50 percent greater than the dc rating. However, circulating currents are also present in some inverter circuits due to the forced commutation method employed. The mean circulating current increases with output frequency, and consequently the continuous-current rating is reduced at high frequencies.

Most inverter circuits generate a nonsinusoidal output voltage, and the harmonic distortion of the inverter current must be taken into consideration. The presence of harmonics increases both the rms and peak values of the inverter current. The increase in rms value causes additional heating and may necessitate somewhat larger cables and heat sinks. The increase in peak inverter current imposes a more severe commutating duty on the inverter. As already explained in Chapter 6, the harmonic currents in a voltage-fed machine are largely determined by the leakage reactance of the motor and are practically independent of the loading conditions and speed over the constant volts/hertz range when the static inverter is designed to generate a six-step voltage waveform. The total rms inverter current at full load as a function of the per-unit reactance of the motor is as shown in Fig. 6.5 of Chapter 6. The ratio of the peak inverter current to the peak fundamental current at full load is shown in Fig. 6.8. The inverter rating is usually expressed in rms units based on a sine wave current, as described in the next section.

13.2.1. *Voltage-Source Inverter Specification*[1-2]

The induction motor and synchronous reluctance motor operate at lagging power factor, and the reactive power requirements must be taken into consideration in specifying the inverter rating. The rectifier supplies dc power to the inverter via the dc link; clearly, the rectifier must be rated to deliver the maximum continuous dc power required by the inverter. The reactive power and current requirements of the ac motor are supplied by the inverter, which must therefore be rated for the apparent power output. The inverter frequency range is determined by the desired speed range and the number of poles in the ac motor. In a general-purpose industrial drive, the minimum frequency is often about 10 Hz, to prevent overheating due to reduced ventilation and also to avoid the speed pulsations and motor instability which may occur at lower frequencies. The maximum inverter frequency with standard ac motors is often limited to 150 or 200 Hz, because higher frequencies introduce excessive core loss in the motor and cause the rotor speed to rise excessively.

The ac motor usually delivers a constant-torque output, and the stator voltage increases linearly with frequency, while the stator current is roughly constant over the frequency range. Thus, the motor kVA is greatest at the highest speed of operation, and the continuous-kVA rating of the inverter is calculated at the maximum output frequency. In a multiple-motor drive application, a voltage-source inverter is usually employed, and in order to determine the inverter rating, the number of motors must be specified, and their full-load and locked-rotor characteristics on a sine wave supply must be known. The continuous-kVA rating of the inverter is determined by multiplying the kVA per motor at maximum frequency by the number of motors employed in the drive, and also by the appropriate harmonic factor, obtained from Fig. 6.5. This factor takes account of the increase in rms inverter current due to the presence of harmonics. The continuous-kVA rating should also be checked at the minimum frequency, particularly if the inverter has a voltage boost for low-frequency operation.

The following example illustrates the specification procedure. Specify the kVA ratings of a six-step voltage-source inverter to supply 24 three-phase motors, each having a full-load rating at maximum frequency of 2 A at 200 V and 0.6 power factor. The corresponding locked-rotor current of the motor at rated voltage is 16 A at 0.5 power factor. Four motors are required to start simultaneously.

The per-unit reactance of the motor is calculated by substitution in Equation 6.45. Thus,

$$X_{pu} = \frac{I_{FL}}{I_s} \sin \phi_s = \frac{2}{16} \sin(\text{arc} \cos 0.5) = 0.108 .$$

Figure 6.5 shows that the total rms current at full load is 1.09 times the fundamental rms value. The continuous-kVA rating of the inverter is, therefore

$$kVA_{cont} = 24(\sqrt{3} \times 200 \times 2 \times 10^{-3}) \times 1.09 = 18.1 \text{ kVA}.$$

The momentary rating of the inverter must be sufficient to supply the standstill current of four motors together with the full-load current of the remainder. Hence,

$$kVA_{mom} = 4(\sqrt{3} \times 200 \times 16 \times 10^{-3}) + 20(\sqrt{3} \times 200 \times 2 \times 10^{-3}) \times 1.09 = 37.3 \text{ kVA}.$$

The time for which the momentary rating must be maintained depends on the starting duty. In this calculation, the locked-rotor current is not multiplied by the harmonic factor, because the fundamental motor current at standstill is eight times rated current; but the harmonic currents are practically the same at full load and at standstill. Consequently, the locked-rotor current is mainly a fundamental current.

The peak or commutating rating must be sufficient to deliver the maximum instantaneous motor current. This condition also occurs when four motors are started with all other motors already running, but the commutating rating is greater than the momentary rating because the commutating rating takes account of the peak inverter current resulting from harmonic distortion. From Fig. 6.8, the peak inverter current is 1.78 times the peak fundamental current. An equivalent sine wave inverter current with the same peak value has an rms value of 1.78 times the fundamental rms current. Thus, the commutating-kVA rating is expressed in rms units based on a sine wave current as

$$kVA_{comm} = 4(\sqrt{3} \times 200 \times 16 \times 10^{-3}) + 20(\sqrt{3} \times 200 \times 2 \times 10^{-3}) \times 1.78 = 46.8 \text{ kVA}.$$

Again, in this calculation, the locked-rotor current is not multiplied by the harmonic factor, because the current waveform is nearly a true sine wave. Each of the above ratings is normally multiplied by a factor of 1.1 or 1.2 to allow a 10 or 20 percent safety margin for manufacturing tolerances and unbalance between phases.

The continuous-kilowatt rating of the rectifier is determined from a knowledge of the continuous-kVA rating of the inverter and its output power factor and efficiency. The product of inverter kVA and power factor gives the kilowatt output, which is divided by the efficiency in order to determine the input power to the inverter. As a conservative estimate, a full-load efficiency of 85 percent can be assumed for a thyristor inverter and 90 percent for a transistor or GTO circuit. When several inverters are operating from a common rectified supply, the rectifier rating is less than the sum of the peak loads of all inverters, if each inverter does not operate at maximum frequency and full load at the same time.

A more detailed description of specification procedures can be found in the literature. Reference 3 describes a general method for sizing the PWM inverter and induction motor in a traction application. Reference 4 discusses the converter rating for an induction motor slip-energy recovery scheme, and Reference 5 gives a comprehensive treatment of the converter design for a self-controlled, load-commutated synchronous motor drive.

13.3. MOTOR SPECIFICATION AND DESIGN

Solid-state ac motor drives usually employ cage-rotor induction motors or synchronous motors of the wound-rotor, permanent magnet, or reluctance type. A standard induction motor is frequently used in an open-loop system without tachometer feedback, but the proper coordination of motor and converter provides the required operating characteristics and ensures system stability for all operating conditions. Parasitic torque and speed pulsations can also be minimized by an appropriate choice of system parameters. When a good dynamic performance is important, closed-loop control techniques are employed.

Synchronous motors are used mainly in large, single-motor drives, in small servo drives, and in low-to-medium-power, multiple-motor drives. DC-field-excited synchronous machines span the power range from a few kilowatts to tens of megawatts and permit power factor improvement by the control of field excitation. A completely brushless synchronous motor is obtained when permanent magnet excitation is employed. Such motors are now available in the power range from 0.5 to 30 kW, or more. Although they are more expensive than synchronous reluctance motors, they are very compact and have a better power factor. They also eliminate the excitation loss of the conventional synchronous motor and give higher efficiency. The permanent magnet synchronous motor is tending to replace the synchronous reluctance motor in multiple-motor drives.

13.3.1. Motor Characteristics[6-10]

The torque-speed characteristic of the voltage-fed ac motor is determined by the output voltage-frequency characteristic of the static converter, as explained in Chapter 7. A constant volts/hertz ratio is applied to give an approximately constant airgap flux density close to the saturation level of the iron. The induction motor is then capable of developing rated torque at rated current over the entire speed range with a constant slip speed in revolutions/minute. At a particular frequency, the synchronous motor has a constant-speed characteristic up to the pull-out torque of the motor. Constant volts/hertz operation gives a constant-torque capability, as in the case of the induction motor.

The terminal voltage of an ac motor is the phasor sum of motor emf and resistive and reactive voltage drops in the stator winding. The IR drop at rated current is a constant, and at rated frequency is a small percentage of the applied voltage. This percentage value increases progressively as voltage and frequency are reduced. To compensate for this effect, and sustain the airgap flux, a volts/hertz boost is usually applied at low frequencies to give increased starting torque. In commercial units, the voltage boost may be adjustable, but over-excitation should be avoided because the magnetizing current is large, even on light loads, and motor ventilation is poor at low speeds. Consequently, excessive heating may occur during low-speed running of the motor.

As explained above, the stator voltage of a synchronous or induction motor is increased linearly with frequency to develop rated torque at rated current. However, the applied voltage is limited by the motor insulation and by the peak voltage rating of semiconductor devices in the converter. The maximum voltage available from the converter at rated volts/hertz usually defines the base speed of the drive. If the motor is operated above base speed by a further increase in frequency, both maximum voltage and current are constrained to remain constant. This maximum allowable power input implies that output torque capability is inversely proportional to speed, giving a constant-horsepower region of operation.

In a current-source converter drive, the induction or synchronous motor is also operated at rated flux density close to the saturation level, to give a constant-torque range of operation. This may be achieved in an induction motor drive by a closed-loop control system, which regulates stator current and rotor slip frequency so that the rated airgap flux level is maintained. Above base speed, a constant-horsepower region of operation at reduced flux can also be implemented. The characteristics of a current-fed motor are therefore closely analogous to those of a voltage-fed machine.

With a constant-torque output, the power rating of the machine is proportional to speed, and large power is available at high speed in a compact high-frequency motor. Thus, a motor which is rated for 10 kW at 1000 rev/min will develop 40 kW at 4000 rev/min without any increase in frame size. The current and torque ratings at a particular speed depend on the cooling method and type of enclosure used, and improved ventilation is obtained at high speed because of the increased effectiveness of the cooling fan. However, an increase in current rating may not be feasible because of the higher core loss. In a standard machine, the core loss is usually excessive at frequencies in excess of 150 or 200 Hz, and a torque derating is required unless forced cooling is used. This derating sets a limit to the maximum usable frequency of operation. In addition, standard bearings are unsuitable for high-speed operation, and centrifugal force increases as the square of the speed. These mechanical considerations also restrict the high-speed capability of a standard motor construction. However, with special motor designs using high-speed bearings and thin, low-loss laminations, high power is available at high speed from an inverter-fed machine.

13.3.2. *Motor Design for Inverter Supplies*

For optimum performance with an inverter supply, special motor designs are necessary, and these machines may differ considerably in their design parameters from standard motors intended for fixed-voltage, fixed-frequency operation on a sine wave ac supply.[11-14] In an inverter drive, the induction motor or synchronous motor is accelerated from rest by a gradual increase in frequency, and motor design can ignore the direct-on-line starting conditions or pull-in performance, which greatly influence the design of general-purpose motors. In the case of the induction motor, the leakage reactance, rotor-cage construction, and skewing practice are released from the restrictions imposed by the demand for good starting performance. The leakage inductance of a standard induction motor is a compromise value which limits the starting current but permits the development of the required starting and breakdown torques. Double-cage or deep-bar rotors are used to enhance the starting torque without increasing the rated slip. Slot skewing is adopted to eliminate parasitic torques during motor acceleration.

Starting considerations are unimportant in an inverter-supplied machine, and motor leakage reactance can be chosen to suit the inverter supply. When the motor is fed by a voltage-source inverter, a high leakage reactance is beneficial, because it limits harmonic current flow and therefore minimizes harmonic losses, pulsating torques, and peak currents. When a current-source inverter is used, a low motor leakage reactance is desirable to limit the commutating voltage spikes in the output voltage. In general, deep-bar or double-cage rotors should not be used: a simpler rotor construction minimizes rotor skin effect and the associated increase in rotor resistance at harmonic frequencies. Harmonic I^2R loss in the rotor is often the dominant harmonic loss in the motor, and minimization of skin effect in the cage rotor is important, particularly in the case of PWM systems with large-amplitude, high-frequency harmonics. A rotor resistance that is smaller than usual also helps reduce harmonic I^2R loss and gives a reduced speed drop from no load to full load in open-loop, voltage-fed induction motor drives.

Slot skewing has been traditionally used in standard induction motors to avoid parasitic torques at low speeds. However, because skew-leakage flux can contribute significantly to time-harmonic core loss, as discussed in Chapter 6, a nonskewed motor design should be used in both voltage-source and current-source inverter drives. The replacement of a skewed rotor by a nonskewed design also reduces the leakage reactance of the motor. This reduction in leakage is particularly significant in smaller motors and is clearly advantageous if the motor is powered by a current-source inverter.

A special-purpose induction motor for inverter drives should also have high-grade winding insulation and enhanced cooling. However, if one proposes excessive diversification and variety in motor construction to suit different inverters, the low-cost advantage of the mass-produced induction motor may be impaired. Consequently, the introduction of external reactance between motor and inverter has been suggested as a means of modifying motor leakage reactance to suit different inverter types and waveforms.[13,15]

Similar design considerations apply to the synchronous motor. Thus, in a synchronous reluctance or synchronous induction motor for fixed-frequency operation, a cage-type rotor winding is fitted to allow the motor to run up by induction motor action to a speed that is very close to synchronous speed. The load inertia is then accelerated or pulled into synchronism by the motor for subsequent operation at synchronous speed. Pull-in performance is enhanced by fitting a substantial low-impedance cage winding, but its introduction reduces the power factor and efficiency of the motor, and impairs the synchronous performance. For fixed-frequency supplies, motor design is a compromise between these factors. If, however, the motor is supplied with an adjustable frequency which is reduced for starting and then increased gradually, the motor can be designed for optimum synchronous performance without regard to the pull-in requirements.

13.3.3. Harmonic Effects and Motor Derating

For adjustable-frequency operation, it is not feasible to filter the output of the static converter; consequently, harmonic effects may be significant. As shown in Chapter 6, harmonic currents have a negligible effect on the average torque developed by the motor, but pulsating torques are also developed, which produce a nonuniform stepping, or cogging motion, of the rotor at low speeds. Harmonics in the motor current may also cause increased acoustic noise, particularly with PWM inverter supplies, because the switching frequency is normally in the range from 500 Hz to 10 kHz and may coincide with a mechanical resonance of the structural parts of the motor.

The additional harmonic losses in the motor must be taken into consideration for operation on a nonsinusoidal supply. As shown in Section 6.5, the reduction in motor efficiency is usually not excessive: in the case of a six-step inverter supply, the increase in full-load loss is about 20 percent. When the motor is powered by a PWM inverter, motor loss may be greater or less than this, depending on the modulation strategy used and the number of switchings per cycle. High-order harmonics can contribute significantly to rotor I^2R loss because of the pronounced skin effect in the cage rotor. In the stator winding, skin effect has much less influence, but overheating of the stator winding insulation is seriously detrimental to motor life. Stator heating is primarily due to I^2R loss and consequently is determined by the total rms stator current, including harmonics.[16] However, in totally enclosed machines, rotor losses also have an appreciable effect on the temperature rise of the stator winding.

At rated speed, with a six-step voltage or current supply, the stator temperature rise at full load is typically 20 percent greater than that with a sine wave supply. For waveforms with a higher degree of harmonic distortion, the temperature rise is correspondingly greater; to avoid excessive overheating, it is necessary to reduce the continuous power and torque ratings of the motor. This derating compensates for the heating effect of the harmonic loss by a reduction in fundamental loss. If limiting the total motor loss to the normal full-load value is required, a reduction in power output of approximately 10 times the harmonic loss is necessary for typical motors. Thus, a motor which develops 100 kW on a sine wave supply, with a fundamental loss of 10 kW, can deliver only 90 kW when fed from an inverter supply, which causes a harmonic motor loss of 1 kW. Derating of this magnitude may be unnecessary because some motors can tolerate the increased heating of a six-step supply, particularly energy-efficient motors, which run cooler than standard designs. However, the precise derating required should be checked with the motor manufacturer.

As motor speed is reduced, the effectiveness of the internal cooling fan decreases rapidly. If the load has a constant-torque characteristic, motor current and I^2R losses remain relatively constant over the entire speed range, and, even with a sine wave supply, low-speed derating is mandatory to avoid serious overheating. The harmonic losses associated with an inverter supply obviously accentuate this problem. Derating may be minimized or eliminated by the adoption of forced cooling. If the motor drives a fan-type load, torque output varies as the square of the speed. Consequently, motor loading is light at reduced speeds, and low-speed derating may be unnecessary.

In addition to motor insulation damage from stator overheating, it has been suggested that the large, steep-fronted voltage pulses which are characteristic of PWM inverter supplies can cause voltage stresses in the stator winding, resulting in insulation damage and reduced motor life.[17] However, ac motors using standard grades of insulation have been operating satisfactorily for many years without any evidence of this failure mechanism.

13.4. SIMULATION OF ADJUSTABLE-FREQUENCY AC DRIVES

For the complex dynamic behavior of the adjustable-frequency ac drive, computer simulation is a valuable design tool, because the effects of parameter variation are readily examined, and different open-loop and closed-loop designs may be tested without costly and time-consuming experiments with hardware. The dynamic performance of a drive can also be studied without the sophisticated instrumentation required in a real-world system, while the steady-state characteristics are obtained as a by-product of the general simulation. System response to various fault conditions can also be investigated.

Analog computer modeling has been used extensively because of its advantages with respect to speed of computation,[18-24] but the greater availability and versatility of the digital computer, and continuing advances in the state of the art of digital simulation, ensure the digital computer's wide use.[25-31] Analog and digital computers may be combined in a hybrid simulation, with a digital approach being adopted for the control, particularly when it is microcomputer-based. A more recent modeling technique uses the parity simulator.[32] This simulator is a synthetic breadboard which electronically models the physical terminal characteristics of each network element.

With the help of modeling or simulation techniques, it is possible at the design stage to determine the extent to which low-frequency instability and harmonic torque pulsations are a problem, to evaluate various closed-loop control strategies, and to predict system performance. Simulation also facilitates the selection of dc link filter components and a study of their influence on system behavior,[30] the investigation of different inverter PWM strategies,[31] and the selection and dimensioning of the inverter circuit. When properly used, a simulation study saves time and money.

A digital simulation of the six-step, voltage-source, inverter-fed induction motor drive is described in Reference 28. The basic drive system is shown in Fig. 13.1 with current and voltage variables labeled. The four components of the drive—the rectifier, filter, inverter, and ac motor—are simulated in separate Fortran subroutines. This modular arrangement gives flexibility and facilitates the checking or substitution of individual elements of the drive. The main program contains the numerical integration procedures and output routines.

13.4.1. *Rectifier Simulation*

To avoid excessive computation time, particularly when studying slow electrome-chanical transients, it is necessary to make certain simplifying approximations. The thyristors in the rectifier circuit are treated as ideal static switches having zero forward voltage drop during conduction, zero leakage current in the off-state, and negligible switching times. If the ac supply is assumed to have zero impedance, then the transfer of current from one thyristor to the next will take place instantaneously without commutation overlap. As usual, in the phase-controlled rectifier circuit of Fig. 13.1, commutation can be delayed by retarding the gating pulse of the incoming thyristor. During transient operating conditions, the delay angle changes rapidly, but the corresponding variation in

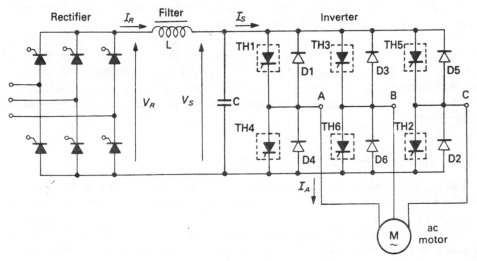

FIG. 13.1. Basic circuit of a dc link converter feeding an ac motor (auxiliary forced commutating circuits not shown).

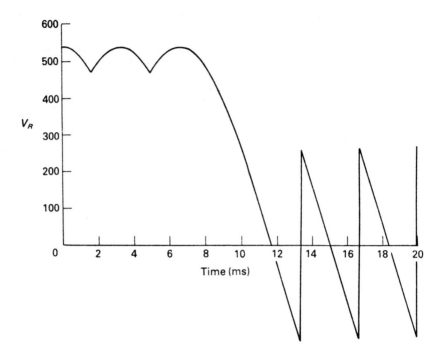

FIG. 13.2. Transient variation of rectified voltage for a sudden increase in rectifier delay angle.

the rectified output voltage does not occur instantaneously, because the new value of delay angle does not become evident until the next commutation occurs. In some drive studies, the rectifier simulation may need to take account of this phenomenon, and Fig. 13.2 is a plotter output showing the variation of the instantaneous rectified voltage, as computed by the subroutine for a highly inductive load, when the delay angle is suddenly increased from zero (maximum mean rectified voltage) to 90 degrees (zero mean rectified voltage) after 7 ms. A time lag in the response of the rectifier is evident.

The rectifier subroutine must also take account of the nature of the loading imposed by the LC filter. Because the thyristor carries current in one direction only, the rectifier current, I_R, in Fig. 13.1, cannot reverse. When the capacitor voltage, V_S, exceeds the rectified voltage, V_R, the current I_R is forced to zero, and the terminal voltage of the rectifier is equal to the capacitor voltage, V_S, and is no longer a function of the delay angle alone.

13.4.2. *Filter and Inverter Simulation*

The LC filter is simulated by writing equations for the rate of change of inductor current and capacitor voltage. These values are then returned to the numerical integration procedure in the main program.

The simulation of the three-phase bridge inverter also uses an idealized model of the thyristor in which switching effects are neglected and the thyristors are simply regarded as ideal switches. Obviously, a more detailed inverter model is required for dimensioning the inverter components. The inverter subroutine must also correctly model the transient operating conditions when the demanded output frequency is varying.

13.4.3. AC Motor Simulation

For a complete transient study of the drive, the simulation of the ac motor must represent both electrical and mechanical transients. A general set of equations which is valid for both transient and steady-state conditions can be derived from a simplified linear model of the motor, as described in Chapter 7. The resulting nonlinear differential equations can be somewhat simplified by a mathematical transformation of phase voltage and current variables to give the dq, or two-axis, model of the ac machine. However, with certain types of inverter drive, the motor may have one or two phases on open circuit for a portion of each cycle. In these circumstances, it is preferable to dispense with the transformation of variables and to model the machine directly in terms of actual phase quantities. The resulting motor equations are programmed in the motor subroutine, which returns the various current derivatives and the speed derivative to the main program for numerical integration.

13.4.4. Simulation Runs

Figure 13.3 shows the plotter output from a simulation run of the inverter-fed induction motor drive for steady-state operating conditions. The variables plotted are

(a) Rectifier output voltage V_R
(b) Rectifier output current I_R
(c) Inverter input voltage V_S
(d) Inverter input current I_S
(e) Motor phase voltage V_A
(f) Motor line current I_A
(g) Electromagnetic torque TQ
(h) Motor speed N

The dc link converter is operated from a three-phase, 380 V, 50 Hz ac utility network, and the induction motor is a typical 7.5 kW, three-phase, 380 V, 50 Hz, four-pole, wye-connected machine. In Fig. 13.3, the motor is running unloaded at a speed of 1000 rev/min with an inverter frequency of 33-1/3 Hz. Time is in seconds and all other plotted variables are expressed in normalized or per-unit form using as base quantities the peak values of the rated phase voltage and phase current for the motor and also its rated torque and speed.

In Fig. 13.3, the rectifier output voltage, V_R, exhibits the usual ripple at six times the supply frequency. The LC filter reduces the harmonic content, and V_S is the filtered dc supply voltage for the inverter. The rectifier current, I_R, is discontinuous, and the voltage at the rectifier terminals is equal to the capacitor voltage during intervals of zero current. The motor phase voltage waveform, V_A, shown in Fig. 13.3, is the basic six-step waveform, but more complex PWM waveshapes can also be generated by the inverter subroutine.[31] Because of the harmonic content of the six-step voltage waveform, a torque pulsation at six times the fundamental stator frequency is developed, as is evident in the electromagnetic torque waveform, TQ, of Fig. 13.3.

Figure 13.4 shows the corresponding drive waveforms when the motor is required to develop rated torque at a speed of 1200 rev/min, corresponding to an inverter frequency of about 40 Hz. The application of a shaft load increases the rectifier current, which is

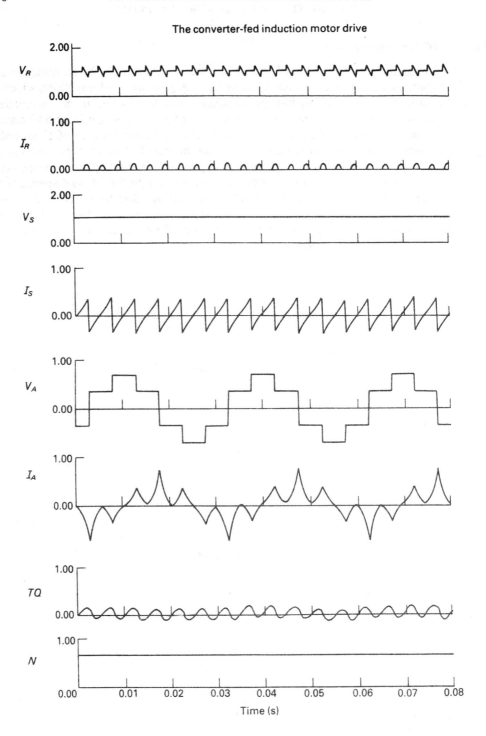

The converter-fed induction motor drive

FIG. 13.3. Simulation results for steady-state operation at 33 1/3 Hz with zero load torque.

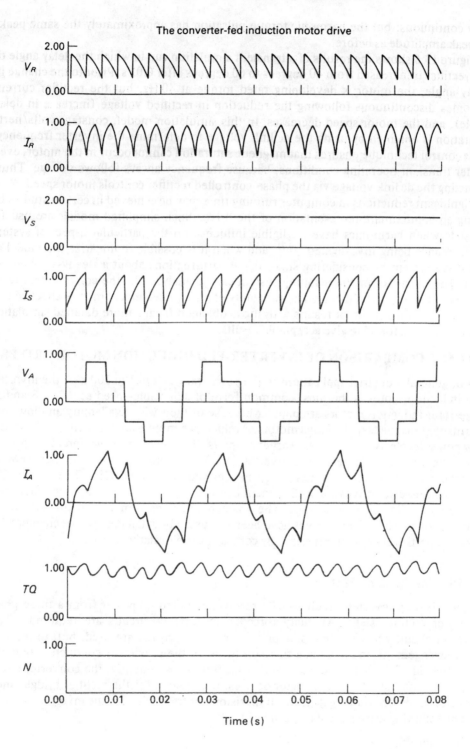

The converter-fed induction motor drive

FIG. 13.4. Simulation results for steady-state operation at 40 Hz with rated load torque.

now continuous; but the harmonic torque pulsation has approximately the same peak-to-peak amplitude as before.

Figure 13.5 shows the results of a transient simulation run in which the delay angle of the rectifier is increased from 40 degrees to 60 degrees after 0.03 s. Prior to the change in delay angle, the motor is developing rated torque at 40 Hz, but the rectifier current becomes discontinuous following the reduction in rectified voltage (increase in delay angle), and the motor speed decreases. In this simulation model, constant volts/hertz operation is obtained by using the dc link voltage to determine the inverter frequency. This control technique ensures that magnetic saturation cannot occur in the motor, even under transient operating conditions, because frequency always follows voltage. Thus, adjusting the dc link voltage via the phase-controlled rectifier controls motor speed.

Significant reductions in computer running time may be achieved in certain studies by using an approximate representation of the drive. Such simplified models are usually feasible when harmonics have negligible influence on the particular aspect of system performance being investigated,[20,21] and when it is possible to linearize the model of the drive system by considering small signal perturbations about a steady-state condition. The dynamics of the resulting system can be analyzed by the conventional techniques of linear control theory. The suitability of a simplified model may be checked by a preliminary comparison of results with those obtained from a more detailed simulation which has been found to give acceptable results.

13.5. A COMPARISON OF INVERTER-FED INDUCTION MOTOR DRIVES

For general industrial applications in the power range from 1 to 500 kW, the inverter-fed induction motor is the most common form of adjustable-speed ac drive. Standard cage-rotor induction motors are popular because of their wide availability and low cost. Permanent magnet synchronous motors or reluctance motors are less commonly used at low power levels. As explained in earlier chapters, three inverter power circuits are available, the six-step voltage-source inverter (VSI), the pulse width-modulated (PWM) voltage-source inverter, and the six-step current-source inverter (CSI). The six-step VSI and PWM inverter drives are predominant below 100 kW with transistorized units being cost effective in this power range. The six-step CSI is used mainly in thyristor-based drives above 20 kW. These drives have been discussed in detail in previous chapters, but their properties are now summarized by comparing their relative merits.

13.5.1. The Six-Step VSI Drive

The six-step inverter circuit usually receives rectified dc power from a three-phase thyristor rectifier stage. Auxiliary forced commutating circuits are required in the inverter circuit when conventional or asymmetric thyristors are used, but, in the simplest arrangement, the incoming thyristor turns off the previously conducting thyristor with the aid of auxiliary inductors and capacitors. Consequently, the converter circuit requires a total of 12 semiconductor devices, 6 thyristors for the rectifier bridge, and 6 inverter-grade thyristors, gate turn-off thyristors, or transistors for the inverter circuit.

The following drive advantages can be listed:

- High efficiency
- Suitability for standard induction motors

The converter-fed induction motor drive

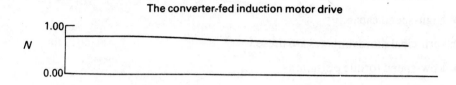

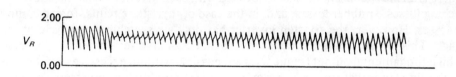

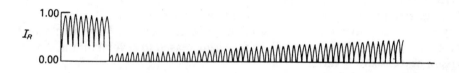

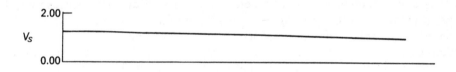

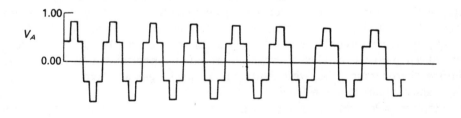

Fig. 13.5. Transient behavior of the drive for a sudden increase in rectifier delay angle.

- Good reliability

- High-speed capability

The principal disadvantages are these:

- Low-speed torque pulsations

- Possible low-speed instability

There is an inevitable energy loss associated with each inverter switching due to device switching losses, snubber losses, and, in the case of thyristor circuits, forced commutation losses. However, the six-step inverter has only six switchings per cycle of output voltage. This low switching frequency means that high inverter efficiencies are achieved and high fundamental output frequencies are possible without excessive inverter losses.

In the six-step waveform, each harmonic voltage amplitude is inversely proportional to the order of the harmonic. Because there are no pronounced high-order harmonics, motor losses are small. Induction motors with normal values of leakage inductance can be used on a six-step voltage supply.

The six-step inverter has a high reliability because of its unsophisticated control circuitry and the low dynamic stressing of the inverter devices. Very high speed motor operation is possible if the output frequency is increased to several hundred hertz with a thyristor or transistor inverter.

On the other hand, the low-order harmonic content of the six-step waveform results in a pronounced torque pulsation at six times the fundamental output frequency. The open-loop drive may also become unstable at frequencies in the 10 to 20 Hz range. However, the lower frequency limit is generally quoted as 5 or 6 Hz, because instability effects, if present, can be overcome by appropriate feedback control techniques, and low-speed torque pulsations can be reduced by flux weakening.

Other disadvantages are inherent in the six-step inverter drive. The phase-controlled rectifier presents a low power factor to the ac supply network. The dc link filter capacitor is large and reduces the speed of response of the system. Because of the variable dc link voltage, back-up protection cannot be provided by a battery for continuity of operation in the event of a failure in the ac utility supply. If the adjustable-speed motor is overhauled by the load, regeneration occurs and power is returned to the dc link through the feedback diodes of the static inverter. This regenerated energy cannot be returned to the ac supply by the phase-controlled bridge circuit unless a second thyristor bridge rectifier is introduced in inverse parallel. However, this double-converter connection requires six additional thyristors together with appropriate control logic, at a significant increase in the cost and complexity of the drive.

The input power factor and response time of the drive can be improved by replacing the phase-controlled rectifier with a diode rectifier circuit feeding a dc chopper that regulates the input voltage to the inverter. The chopper thyristor or transistor is required to handle the full dc current, but the fixed dc voltage input to the chopper can be buffered by a battery for back-up protection. Regeneration to this fixed-voltage link is possible if a regenerative chopper is used. This configuration is advantageous in a multiple-motor drive in which a single rectifier feeds several choppers simultaneously, because regenerated power that is returned to the dc link by one motor will help reduce the net load on the rectifier.

13.5.2. *The PWM Inverter Drive*

The PWM voltage-source inverter circuit has the same configuration as the six-step inverter circuit but employs a more complex switching sequence. With conventional or asymmetric thyristors, there is usually an auxiliary commutating thyristor associated with each of the six main thyristors. The PWM inverter therefore requires 12 inverter-grade thyristors or 6 devices with a turn-off capability. A diode rectifier bridge is used to provide a constant dc link voltage.

The drive has the following advantages:

- Low-speed capability

- Good transient performance

But the main disadvantage is this:

- High motor losses with simple PWM strategies

If the PWM inverter develops a near-sinusoidal waveform with negligible low-order harmonics, then low-speed cogging torques are eliminated. Because open-loop instability is rarely a problem at reduced speeds, the PWM inverter is the ideal solution in low-speed drive applications. And because voltage and frequency control are both performed within the PWM inverter, a quick response is obtained for changes in demanded voltage and frequency. The dc link filter can be appreciably smaller than that in a six-step inverter drive, giving an attractive reduction in size, weight, and cost. The diode rectifier circuit presents a high power factor to the ac supply network, and a battery can be floated across the dc link to give a ride-through capability during power system outages. The PWM inverter also regenerates naturally to the dc link, and if a single rectifier bridge feeds several inverters, it is possible for one inverter to regenerate into another via the dc link, but regeneration to the ac line is not possible.

On the negative side, the high switching frequency in a PWM inverter means that inverter devices and commutating components are subjected to high dynamic stressing, particularly in higher voltage systems. Consequently, reliability may not be as good as with a six-step inverter. Control circuit complexity can be minimized and reliability improved by using large-scale integration (LSI) or microprocessor techniques, which have now been introduced in many commercial inverter drives.

As the fundamental output frequency is increased, inverter losses become excessive because of the high switching frequency. The upper frequency limit may be raised by making a transition to a less sophisticated PWM waveform with a lower switching frequency and ultimately to a six-step waveform if the application demands it. However, difficulties can arise at the transitions between different modulation strategies due to discontinuities in the fundamental output voltage and a significant increase in the harmonic content, with associated motor losses, but the use of digital control and preprogrammed look-up tables can minimize these problems. Elementary PWM strategies such as multiple-pulse modulation within a six-step envelope also produce excessive harmonic losses in the motor, but more sophisticated modulation strategies can now be implemented with custom-made LSI and preprogrammed ROM or by a microprocessor-based approach.

In a PWM waveform, pronounced harmonics are present at the switching frequency and multiples thereof, and the harmonic content is high during the transition to six-step operation. In general, a motor with high leakage reactance and low skin effect is required to ensure that harmonic I^2R losses are not excessive.

The power transistor and GTO have significant advantages in PWM inverter applications because of their fast switching capability and ease of turn-off. The elimination of forced commutating circuits gives an inverter configuration which is less complex than that obtained with conventional thyristors. Sophisticated PWM techniques can also be implemented with a higher switching frequency and better efficiency than are possible with a conventional thyristor inverter. Further growth in transistor applications is being fueled by the availability of higher voltage devices which can provide a cost-competitive system for operation on the 380 to 460 V three-phase ac utility networks that are prevalent world-wide. The GTO also has this voltage capability, of course, and is being used increasingly in commercial PWM inverters.

13.5.3. *The Six-Step CSI Drive*

The dc link converter is given the characteristics of a current source by removing the shunt capacitor of the dc link filter and employing a current regulation loop to control the output dc current from the phase-controlled rectifier. The rectifier and filter inductor feed a regulated dc current to the machine windings via the particular inverter thyristors which are gated into conduction at any instant. In the six-step CSI, each thyristor conducts for 120 degrees, resulting in an ac line current with a quasi-square waveshape similar to the ac line voltage waveform of a six-step VSI.

The following drive advantages can be listed:

- Simple circuit
- Inherent four-quadrant operation
- Short-circuit proof

The main disadvantages are these:

- High voltage stresses
- Low-speed torque pulsations

The simplicity of the converter circuit is attractive. Six thyristors are needed for the rectifier stage, and six more thyristors are required for the usual autosequentially commutated inverter (ASCI). However, a fast turn-off capability is not necessary in the inverter devices and relatively inexpensive slow-turn-off thyristors are satisfactory. Despite its simplicity, the converter is inherently capable of regeneration to the ac utility network. Because of the large dc link inductor and the controlled-current mode of operation, there is a slow rise of fault current following an inverter fault. Consequently, the converter can recover from an inverter commutation failure or an accidental short-circuit at the output terminals by suppression of the rectifier gating signals. This feature gives the converter an inherent ruggedness and an ability to recover from maloperation.

Assuming comparable full-load losses in a six-step VSI drive and a CSI drive, the efficiency of the CSI drive will be higher under partial load conditions because losses decrease with load current, whereas in the VSI drive the losses are essentially independent of motor loading. On the other hand, the large dc link inductor gives the CSI drive a somewhat sluggish response, and the inductor and commutating capacitors make the converter bulky and expensive. In addition, the sudden changes in current that are inherent in the output current waveform cause large transient overvoltage spikes at the machine terminals. Consequently, high-voltage semiconductor devices are required in the inverter circuit. Although auxiliary voltage-clamping circuits are often used, they also introduce extra losses.

The frequency range of the six-step CSI is lower than that of the six-step VSI. The upper frequency limit is determined by the onset of the current bypass effect when inverter commutation is occurring simultaneously in the upper and lower thyristor groups. This occurrence causes a loss of power in the motor and affects the efficiency and stability of the drive. For a typical 50 or 60 Hz induction motor, the upper frequency limit might be about 150 Hz, but higher frequencies can be achieved if the motor has an unusually low leakage reactance. Forced commutation techniques with auxiliary thyristors may be employed to speed up commutation and raise the upper frequency limit, but the inverter circuit then loses some of its previous simplicity. In general, CSI and motor behavior are very interdependent, and the CSI design is, in fact, linked to the motor parameters, especially the leakage or subtransient reactance. A particular inverter design is valid for a certain range of motor parameters. Low-speed torque pulsations are also present in a CSI drive because of the six-step current waveform. Elimination of troublesome lower order harmonics by PWM techniques is possible but cannot be achieved as readily as in the case of a VSI.

The CSI drive is very attractive for single-motor applications where regeneration to the ac line is important. For stable operation in the magnetically unsaturated region of the induction motor torque-speed characteristics, closed-loop operation is essential. This requires the introduction of a shaft-coupled tachometer. For a multiple-motor drive it is necessary to regulate the output voltage by controlling the current delivered to the motors. This type of voltage-regulating system can be difficult to stabilize.

The CSI drive also has a phase-controlled rectifier input with the attendant disadvantages of low power factor and susceptibility to maloperation because of line notching and disturbance. In some drives, the phase-controlled rectifier has been replaced by a diode rectifier bridge and a dc chopper that regulates the current delivered to the inverter. However, this modification eliminates the ability of the drive to regenerate naturally into the ac utility network.

13.5.4. Summary

In the field of inverter drives it is obvious that there is no single "best buy" for all applications. The six-step VSI drive has been widely used because of its cost and reliability advantages, its high-frequency capability, and the flexibility which provides for single-motor operation in open- and closed-loop configurations and also permits open-loop, multiple-motor operation. High-performance requirements cannot be satisfied with a six-step VSI drive, but it has been popular as a general-purpose drive in low-to-medium-power industrial applications when the speed range is limited to 10:1.

For a single-motor reversing drive with controlled operation through zero speed, the CSI drive is ideally suited, provided low-speed torque smoothness is achieved by pulse-width modulation of motor current or programmed variation of dc link current. It is also attractive for less demanding applications because of its ruggedness, but it is not as tolerant of machine parameter variations as the six-step VSI.

A high-performance low-speed drive requires a PWM inverter. The operation of a number of PWM inverters on a common dc bus provides an economic multiple-motor system with adequate regenerative capability and with continuity of operation assured by a single back-up battery across the dc link. However, a careful choice of PWM strategy is vital, and careful motor selection or design is necessary to ensure high efficiency. PWM inverter control is now possible with a custom-made LSI chip that can implement complex modulation strategies.

The continuing advances in power semiconductor technology are boosting the status of the PWM inverter drive in both general-purpose and high-performance drive applications. This trend commenced in small general-purpose drives when high-voltage power transistors and GTOs became available, resulting in PWM inverters of small size, light weight, and low cost. LSI technology has now been successfully applied to the development of new power semiconductors with improved characteristics, but the six-step VSI and CSI are nonevolutionary in that they cannot take advantage of these new devices. By contrast, the PWM inverter can capitalize on the improved switching speeds to give better waveform quality and faster inverter response. Power transistors are now being used in PWM inverter drives rated up to 200 kW or more. The GTO is used at all power levels, but its main impact is in drive applications with power ratings in excess of 100 kW.

13.6. INDUSTRIAL APPLICATIONS OF ADJUSTABLE-SPEED AC DRIVES

Adjustable-speed electrical drives are now used at all power levels in a variety of industrial applications. An adjustable-speed drive is mandatory for many conveyor lines, roller table drives, and traction applications, and in industrial processes where adjustment of running speed, controlled acceleration, or multiple-motor synchronization are essential features. However, adjustable-speed drives are also being introduced in new application areas where speed adjustment is not essential but offers advantages in energy conservation or productivity. The potential energy savings are particularly significant in the area of centrifugal pump and fan drives in which fixed-speed ac motors have traditionally been used with flow controlled by a throttling valve or damper, which introduces appreciable energy losses. When an adjustable-speed drive is introduced, throttling is avoided and the extra expenditure on the drive is quickly repaid in reduced running costs.[33-35] In existing fixed-speed process lines, the introduction of adjustable speed can result in increased productivity and improved product quality by allowing compensation for variations in the prevailing conditions. Speed can also be accurately and rapidly controlled in response to a command signal, so that precise automatic regulation is possible with minimum supervision, and automation of the process is facilitated.

Dramatic innovation has occurred in ac drive technology during the past 20 years. Fast semiconductor switching devices are now available, improved converter configurations have been devised, and sophisticated drive control strategies have been developed. As a result of these developments, ac motor drives are now prominent in traditional application areas and in the new, expanding sectors of the adjustable-speed drive market. The inherent advantages of the ac drive include

- The use of highly reliable ac motors
- Improved speed accuracy
- Easy synchronization of multiple-motor loads
- High-speed capability
- High-power capability
- Continuity of operation
- Favorable supply loading

13.6.1. *Motor Reliability*

The advantages of brushless ac motors are well known. The absence of commutators, slip rings, and brushes eliminates the maintenance requirements associated with dc motors and ac commutator motors. The periodic maintenance required by these commutator machines is often inconvenient, causing undesirable service interruptions. Regular maintenance is impossible if the application demands prolonged motor operation in unattended or inaccessible locations. When driving a submersible pump, for example, an adjustable-speed induction motor runs completely submerged. It also operates reliably in the high-temperature environment of a hot rolling mill with its severe impact loading. And it can be used in chemical or petrochemical plants in the presence of explosive gas mixtures. In short, the rugged ac motor can be placed in abrasive, corrosive, or explosive atmospheres and in wet or dust-laden environments without detriment to its performance. The inherent advantages of a brushless construction also provide the incentive for the development of cost-competitive ac motor drives for general industrial applications in nonhostile environments.

In addition to its superior reliability, the cage-rotor induction motor also has a higher torque/inertia ratio than the dc machine and, for a given rating, has about 60 percent of the weight and volume of the dc motor. The smaller size can significantly ease problems of siting the motor on or in the driven machine. Thus, the ac motor has advantages in traction applications that use axle-mounted traction motors where space is limited, unsprung weight must be reduced, and maintenance requirements minimized.[36]

13.6.2. *Speed Accuracy*

Open-loop adjustable-frequency control of the cage-rotor induction motor gives adequate speed accuracy for many applications, as motor speed normally lies within a few percentage points of synchronous speed. If the induction motor has a rotor resistance that is smaller than usual, the speed regulation from no load to full load can be reduced to 2 percent. The speed-holding accuracy can be further improved to 1 percent, or less, by employing an open-loop slip compensation technique, in which the stator frequency is increased as load is applied to the motor. For higher accuracy in an induction motor drive, closed-loop methods of speed control may be used. In multiple-motor conveyor drives, high-slip induction motors with 5 to 8 percent slip at full load are sometimes employed to improve load-sharing between individual motors.

As explained in Chapter 4, the inverter frequency can be determined solely by the reference oscillator and is then unaffected by changes in inverter loading and by normal variations in ac supply voltage and frequency. An inverter-powered permanent magnet synchronous motor or synchronous reluctance motor runs at the precise synchronous speed that corresponds to the applied frequency, and so provides very precise speed control in a simple open-loop system which is ideal for applications with low dynamic performance requirements. Analog oscillators can provide a long-term speed accuracy of ± 0.05 percent of set speed. Digital control methods using a crystal oscillator and divider can provide speed-holding accuracies of ± 0.001 percent or better.

13.6.3. *Multiple-Motor Drives*

The ability of the static frequency converter to deliver a precise output frequency independent of load is highly advantageous for multiple-motor applications. A static inverter feeding a number of cage-rotor induction motors in an open-loop, multiple-motor drive, a system now widely used in the steel industry for roller table drives, has excellent load-sharing characteristics and high reliability. Alternatively, each motor may be driven by its own converter with all inverter frequencies locked to a common reference oscillator.

In many continuous production processes, the quality of the product is dependent on the long-term speed stability of a multisection process line. For such applications, the adjustable-frequency ac drive has outstanding advantages. A multiple reluctance motor or permanent magnet synchronous motor drive can provide precise speed synchronization between motors without feedback tachometers. Several motors can be operated with precise preset speed ratios, or small frequency differences can be readily introduced between a number of digitally controlled inverters in order to introduce tension control in a process line. Because the inverter has zero load regulation, a motor can be started or stopped without other machines in the group being affected.

Such multiple-motor drives are ideally suited to conveyors, packaging equipment, plate glass manufacture, and other continuous production processes where reliability and exact speed coordination are essential. In the man-made fiber industry, in particular, such drives have proved highly successful and are widely used. In this application, synthetic fiber spinning requires large numbers of small motors for pumps and godets, for example, to be precisely controlled for uniform fiber thickness. Inverter-powered synchronous reluctance and permanent magnet motors have provided the reliability and precision essential to quality control.

These multiple-motor drives are also readily extended by installing an additional converter which is synchronized with the existing equipment. This method provides exact synchronization or speed ratio control between items of equipment which are physically remote, without the use of complicated mechanical links.

13.6.4. *High-Speed Capability*

Because the speed of an ac motor increases linearly with supply frequency, a high-frequency inverter and brushless ac motor can be used to give a reliable high-speed drive. Thus, a two-pole machine will run at 6000 rev/min on a 100 Hz supply. Heavier bearings and thinner low-loss laminations permit higher speeds, and special machine designs may be used to increase the speed capability to 50 000 rev/min or more.

The solid-rotor construction of the ac motor provides a well-balanced machine, capable of prolonged operation without excessive noise or vibration. In contrast, dynamic balance is more difficult to achieve and maintain in the dc motor, and commutation difficulties restrict its high-speed operation. Consequently, the ac motor has significant advantages for high-speed applications in the machine tool industry. In high-speed grinders, a consistent quality in the machined product is obtained as a result of the excellent speed-holding accuracy and freedom from vibration of the adjustable-frequency ac drive. Stepless adjustment of speed is available to compensate for grinding wheel wear by maintaining a constant linear surface speed.

13.6.5. *High-Power Capability*

The power output and speed capability of the dc motor are restricted by commutation difficulties and centrifugal forces on the commutator. Thus, the actual power rating is limited to about 500 kW at a speed of 5000 rev/min and 2.5 MW at 1000 rev/min. The ac machine can supply a significantly greater power output, and very large drives must use ac motors.

When the drive rating increases above 1 MW, forced-commutated inverters are unacceptable because of the inordinately large commutating capacitors required. Forced commutation losses also pose cooling problems at high power levels. Thus, a 5 MW inverter with an efficiency of 90 percent has losses of 0.5 MW and an obvious cooling problem. High converter efficiency can be achieved in a naturally commutated circuit such as a cycloconverter or load-commutated inverter.

The self-controlled synchronous motor with its load-commutated inverter has the high efficiency required at high power levels; it has been used up to several tens of megawatts for large fans, compressors, extruders, and test stands. It is also used for start-up applications in gas turbine and pumped storage hydrogenerating stations, where large synchronous generators must be run up to speed prior to synchronization with the ac utility grid.[37,38] The generator is run up as a synchronous motor by a load-commutated inverter, which is then disconnected.

Cycloconverter-fed induction or synchronous motors are also used in very high power drives. Typical applications include gearless cement or ball mill drives, which use a directly coupled, slow-running induction or synchronous motor.[39,40] The cycloconverter has also been used in large synchronous motor drives for rolling mills, and a 2.5 MW drive system has been described.[41]

The wound-rotor induction motor with a subsynchronous converter cascade has an efficient naturally commutated dc link converter for slip energy recovery. It is used in fan and pump drives in the power range from 100 kW to 15 or 20 MW, particularly when a limited speed range below base speed is adequate. The converter rating is then less than the machine rating, and the total drive cost is reduced. Speed control above and below synchronism is possible when a cycloconverter is used to feed slip power to or from the rotor.[4,42] The cycloconverter is naturally commutated by the ac network voltages and also has a reduced rating when the drive has a limited speed range above and below synchronism.

13.6.6. *Continuity of Operation*

Continuity of operation is essential in many critical process industries to avoid production losses and plant damage. In plate glass manufacture or in synthetic fiber spinning, for example, even brief interruptions in the ac utility supply could completely disrupt production if continuity of operation was not assured. In such applications, it is vital that the drive should ride through ac supply disturbances and interruptions; in this respect, the adjustable-frequency ac drive has a major advantage over other drive systems.

The voltage-source inverter has a large capacitor across the dc terminals to filter the input voltage. When the dc link supply is interrupted as a result of an ac power failure, the smoothing capacitor rapidly discharges, and the inverter output is maintained for only a few cycles of the supply frequency. Auxiliary capacitor banks across the dc link will extend the inverter shutdown time and so provide continuity of supply for a short-duration ac power loss. For long-term operation, however, it is necessary to use an auxiliary standby battery connected across the dc link, as in Fig. 13.6. This battery provides continuity of supply until the ac power is restored or until a standby generator is run up and can take over the load.

13.6.7. *Favorable Supply Loading*

For a normal direct-on-line start, at rated voltage and frequency, a typical cage-rotor induction motor draws a starting current of five or six times rated current. This large inrush current can be eliminated in an adjustable-frequency system by reducing the stator voltage and frequency for starting. This reduction enhances the torque per ampere of the motor and limits the standstill current.

In the dc link converter, an uncontrolled diode rectifier bridge can be used to supply a PWM inverter or a chopper that regulates the dc link voltage. The absence of firing delay in the rectifier circuit means that a high input power factor, of almost unity, is presented to the ac supply network over the full speed range. By contrast, in the thyristor-controlled dc motor drive, the power factor decreases as the armature voltage and motor speed are reduced by delayed firing. The harmonic current loading of the ac supply network is also somewhat lower in the case of a diode rectifier circuit.

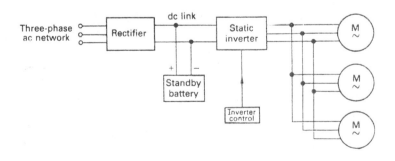

FIG. 13.6. DC link converter with standby battery for continuity of operation.

13.6.8. *Further Applications*

Because of the advantages detailed above, the ac motor drive is opening up new vistas in the adjustable-speed drive market and is also a serious contender for industrial applications where the dc motor drive previously dominated. For robotic applications and feed drives for machine tools, the exacting performance requirements have, in the past, been met by dc servo drives using permanent magnet motors. In these applications, four-quadrant operation is required with very fast response and exceptionally smooth low-speed rotation at speeds of less than 1 rev/min. Torque is also required at zero speed to give torsional stiffness to the stationary shaft. The dc servo drive has satisfied these requirements, but the inherent advantages of a brushless ac system have led to a major development effort to meet these demanding specifications in an ac servo drive. For drive ratings up to 20 kW, one can now choose between the permanent magnet synchronous motor, the permanent magnet brushless dc motor, and the cage-rotor induction motor. Each drive system can provide a wide constant-torque region of operation with rapid transient response and excellent torque smoothness. The induction motor drive requires the highest degree of sophistication in its control, and the permanent magnet motor drives are usually the preferred choice. If the application demands a wide constant-power range of operation at high speed, as in a spindle drive for a machine tool, then the induction motor is the obvious choice because of its inherent field-weakening capability when the motor frequency is increased with the applied voltage held constant. For high-performance servo drives above 20 kW, the permanent magnet motors are expensive, and the induction motor drive is normally used for economic reasons. Nowadays, the complex signal processing required for field-oriented control can be readily handled by the microcomputer.

Open-loop constant volts/hertz control is used in low-performance drives such as energy-saving induction motor drives and in multiple-motor drives with several ac motors operating on the same inverter supply. In an age of soaring energy costs, it is essential to maximize process efficiency and, as already stated, adjustable-speed ac drives are currently being introduced into many industrial installations employing centrifugal pumps, blowers, and compressors. In the past, these units were driven by fixed-speed ac motors with flow reduction obtained by a throttling valve or adjustable damper. The pump or fan unit is rated for the peak delivery, which may rarely be required, and throttling losses are high during most of the operating time of the unit. Pump or fan efficiency is also poor for the high-pressure, low-delivery operating condition. The energy losses due to throttling are eliminated when an adjustable-speed drive is used to reduce motor speed at times of reduced demand. Pump efficiency also remains high at partial loads. Overall energy savings as high as 40 or 50 percent have been quoted, which quickly repay the extra capital cost involved. [33-35]

These economic considerations and the long-term reliability of the ac motor have led to the use of adjustable-frequency control of ac motors for pump drives in municipal water and sewage treatment plants and in general industrial pumping applications. The adjustable-frequency drive has good efficiency, even under partial load conditions, and at low speed, and is readily adapted to provide automatic control of pressure, flow, or level. In an installation containing several motors, it may be possible to run the fully loaded machines direct-on-line and utilize a single static frequency converter for efficient adjustable-speed operation of the partially loaded unit.

13.6.9. *Summary*

Adjustable-speed ac drive systems have now proven technically and economically attractive for many industrial applications. For operation under very arduous conditions or at very high power levels, the ac drive is clearly the only viable solution. In areas where fixed-speed ac machines have traditionally been used, the adjustable-speed ac drive will clearly be favored, particularly for retrofitting adjustable-speed control to upgrade performance or conserve energy. For a wide spectrum of more general industrial applications, both adjustable-speed dc and ac drive solutions are feasible. In multiple-motor systems, the ac drive solution has been more economical for many years because of the significant saving in machine costs achieved by the use of ac rather than dc motors.[43] In the past, however, a single-motor ac drive was often more expensive than a dc drive, because the lower cost of the ac motor was offset by the higher cost of the power and control electronics. But motor technology is now at a mature stage, and motor costs are increasing, whereas the cost of drive electronics is falling. Thus, the ac drive with its higher content of power and control electronics is being affected much more favorably by the rapid evolution in electronics technology. Advances in integrated electronics, microprocessors, and power semiconductors are dramatically reducing the size, complexity, and cost of the static frequency converter so that the single-motor ac drive is often less expensive than its dc counterpart.[44] As a result, there is now a spectacular growth in industry application of ac drives.

The enhanced economic position of the ac drive is being complemented by a growing recognition of its performance capabilities. The microcomputer permits a universal hardware design with flexible software control; the resulting ac drive is readily integrated into a computer-based system for overall process automation. In the future, the computational capability of the microcomputer will be used for the implementation of sophisticated modern control theory in ac motor drives, together with the usual diagnostic and self-check features. The integration of the power semiconductor and its signal electronics is already possible at low power levels. Further advances in this area will give size and cost reductions with improved reliability. It appears inevitable that adjustable-speed ac drives will find even more widespread application in the future.

13.7. REFERENCES

1. HELMICK, C.G., How to specify adjustable-frequency drives, (ii) selecting the inverter, *Control Eng.*, **14**, May 1967, pp. 83-84.

2. BECK, C.D., and CHANDLER, E.F., Motor drive inverter ratings, how to specify, how to analyze, how to compare, *IEEE Trans. Ind. Gen. Appl.*, **IGA-4**, 6, Nov./Dec. 1968, pp. 589-595.

3. PLUNKETT, A.B., and PLETTE, D.L., Inverter-induction motor drive for transit cars, *IEEE Trans. Ind. Appl.*, **IA-13**, 1, Jan./Feb. 1977, pp. 26-37.

4. WEISS, H.W., Adjustable speed ac drive systems for pump and compressor applications, *IEEE Trans. Ind. Appl.*, **IA-10**, 1, Jan./Feb. 1974, pp. 162-167.

5. ROSA, J., Utilization and rating of machine-commutated inverter-synchronous motor drives, *IEEE Trans. Ind. Appl.*, **IA-15**, 2, Mar./Apr. 1979, pp. 155-164.

6. HELMICK, C.G., How to specify adjustable-frequency drives, (i) selecting the motor, *Control Eng.*, **14**, Apr. 1967, pp. 69-71.

7. BARNES, E.C., Performance and characteristics of induction motors for solid state variable frequency drives, *Conf. Rec. IEEE Ind. Gen. Appl. Group Annual Meeting, 1969*, pp. 653-659.

8. CHALMERS, B.J., Design of induction motors supplied by thyristors, *Electr. Rev.*, **184**, Apr. 11, 1969, pp. 525-528.

9. AUINGER, H., Einflüsse der Umrichterspeisung auf elektrische Drehfeldmaschinen, insbesondere auf Käfigläufer-Induktionsmotoren, *Siemens—Energietechnik*, **2**, 1981, pp. 46-49.

10. CONNORS, D.P., JARC, D.A., and DAUGHERTY, R.H., Considerations in applying induction motors with solid-state adjustable frequency controllers, *IEEE Trans. Ind. Appl.*, **IA-20**, 1, Jan./Feb. 1984, pp. 113-121.

11. DE JONG, H.C.J., *AC Motor Design with Conventional and Converter Supplies*, Oxford University Press, Oxford, England, 1976.

12. SATTLER, P.K., and BALSLIEMKE, G., Experimental and theoretical investigations concerning the design of induction machines, especially machines with high power density with respect to the inverter supply and speed regulation, *Proc. 2nd IFAC Symp. on Control in Power Electronics and Electrical Drives, 1977*, pp. 387-398.

13. DE BUCK, F.G.G., Leakage reactance and design considerations for variable-frequency inverter-fed induction motors, *Conf. Rec. IEEE Ind. Appl. Soc. Annual Meeting, 1979*, pp. 757-769.

14. LEVI, E., Design considerations for motors used in adjustable-speed drives, *IEEE Trans. Ind. Appl.*, **IA-20**, 4, July/Aug. 1984, pp. 822-826.

15. DE BUCK, F.G.G., Design adaptation of inverter-supplied induction motors, *IEE Proc. Electr. Power Appl.*, **1**, 2, May 1978, pp. 54-60.

16. DE BUCK, F.G.G., Losses and parasitic torques in electric motors subjected to PWM waveforms, *IEEE Trans. Ind. Appl.*, **IA-15**, 1, Jan./Feb. 1979, pp. 47-53.

17. PHILLIPS, G.R., Induction motor behavior on switched ac drives, *Proc. Motorcon, 1981*, Paper 3B.2.

18. STIEBER, M., Nachbildung des Verhaltens umrichtergespeister Asynchronmotoren auf dem Analogrechner, *Fifth International Congr. Assoc. for Analog Computation (AICA), 1967*, pp. 719-728.

19. KRAUSE, P.C., and WOLOSZYK, L.T., Comparison of computer and test results of a static ac drive system, *IEEE Trans. Ind. Gen. Appl.*, **IGA-4**, 6, Nov./Dec. 1968, pp. 583-588.

20. KRAUSE, P.C., and LIPO, T.A., Analysis and simplified representations of a rectifier-inverter induction motor drive, *IEEE Trans. Power Appar. Syst.*, **PAS-88**, 5, May 1969, pp. 588-596.

21. KRAUSE, P.C., and LIPO, T.A., Analysis and simplified representations of rectifier-inverter reluctance-synchronous motor drives, *IEEE Trans. Power Appar. Syst.*, **PAS-88**, 6, June 1969, pp. 962-970.

22. MAYER, C.B., and LIPO, T.A., The use of simulation in the design of an inverter drive, *Conf. Rec. IEEE Ind. Appl. Soc. Annual Meeting, 1972*, pp. 745-752.

23. LOCKWOOD, M., Simulation of inverter/induction machine systems including discontinuous phase currents, *IEE Proc. Electr. Power Appl.*, **1**, 4, Nov. 1978, pp. 105-114.

24. LIPO, T.A., Simulation of a current source inverter drive, *IEEE Trans. Ind. Electron. and Control Instrum.*, **IECI-26**, May 1979, pp. 98-103.

25. BELLINI, A., and DE CARLI, A., Digital simulation of inverter-fed ac drives, *AICA Symp. on Simulation of Complex Systems, 1971*, pp. F13/1-F13/8.

26. RAJAN, S.D., JACOVIDES, L.J., and LEWIS, W.A., Digital simulation of a high-performance ac drive system, *IEEE Trans. Ind. Appl.*, **IA-10**, 3, May/June 1974, Part I, pp. 391-396, Part II, pp. 397-402.

27. CHATTOPADHYAY, A.K., Digital computer simulation of an adjustable-speed induction motor drive with a cycloconverter-type thyristor-commutator in the rotor, *IEEE Trans. Ind. Electron. and Control Instrum.*, **IECI-23**, 1, Feb. 1976, pp. 86-92.

28. MURPHY, J.M.D., Digital simulation of adjustable-frequency ac motor drives, *Simulation*, **27**, July 1976, pp. 39-44.

29. AL-NIMMA, D.A.B., and WILLIAMS, S., Modelling a variable-frequency induction motor drive, *IEE Proc. Electr. Power Appl.*, **2**, 4, Aug. 1979, pp. 132-134.

30. KAMPSCHULTE, B., Comparison of an asynchronous machine drive with current-source inverter and voltage-source inverter, *Conf. Rec. IEEE Ind. Appl. Soc. Annual Meeting, 1979*, pp. 945-951.

31. BOWES, S.R., and CLEMENTS, R.R., Digital computer simulation of variable speed PWM inverter-machine drives, *IEE Proc.*, **130, Pt. B**, 3, May 1983, pp. 149-160.

32. KASSAKIAN, J.G., Simulating power electronic systems—a new approach, *Proc. IEEE*, **67**, Oct. 1979, pp. 1428-1439.

33. MORTON, W.R., Economics of ac adjustable speed drives on pumps, *IEEE Trans. Ind. Appl.*, **IA-11**, 3, May/June 1975, pp. 282-286.

34. ALLEN, J.P.C., and BENTZEN-BILQVIST, I., Two views of adjustable frequency drives in the cement industry, *Conf. Rec. IEEE Ind. Appl. Soc. Annual Meeting, 1976*, pp. 575-579.

35. CHAUPRADE, R., and ABBONDANTI, A., Variable speed drives: Modern concepts and approaches, *IEEE Int. Semicond. Power Converter Conf., 1982*, pp. 20-37.

36. VAN WYK, J.D., SKUDELNY, H.-CH., and MÜLLER-HELLMAN, A., Power electronics, control of the electromechanical energy conversion process and some applications, *IEE Proc.*, **133, Pt. B**, 6, Nov. 1986, pp. 369-399.

37. PETERSON, T., and FRANK, K., Starting of large synchronous motor using static frequency converter, *IEEE Trans. Power Appar. Syst.*, **PAS-91**, 1, Jan./Feb. 1972, pp. 172-179.

38. MUELLER, B., SPINAGER, T., and WALLSTEIN, D., Static variable frequency starting and drive system for large synchronous motors, *Conf. Rec. IEEE Ind. Appl. Soc. Annual Meeting, 1979*, pp. 429-438.

39. STEMMLER, H., Drive system and electronic control equipment of the gearless tube mill, *Brown Boveri Rev.*, **57**, 3, Mar. 1970, pp. 120-128.

40. ALLAN, J.A., WYETH, W.A., HERZOG, G.W., and YOUNG, J.A.I., Electrical aspects of the 8750 hp gearless ball-mill drive at St. Lawrence Cement Company, *IEEE Trans. Ind. Appl.*, **IA-11**, 6, Nov./Dec. 1975, pp. 681-687.

41. NAKANO, T., OHSAWA, H., and ENDOH, K., A high performance cycloconverter fed synchronous machine drive system, *IEEE Int. Semicond. Power Converter Conf., 1982*, pp. 334-341.

42. HORI, T., and HIRO, Y., The characteristics of an induction motor controlled by a Scherbius system (application to pump drive), *Conf. Rec. IEEE Ind. Appl. Soc. Annual Meeting, 1972*, pp. 775-782.

43. POLLACK, J.J., Some guidelines for the application of adjustable-speed ac drives, *IEEE Trans. Ind. Appl.*, **IA-9**, 6, Nov./Dec. 1973, pp. 704-710.

44. STEFANOVIC, V.R., Present trends in variable speed ac drives, *Int. Power Electron. Conf., Tokyo, 1983*, pp. 438-449.

INDEX

AC
  motor
    adjustable speed, 1, 485
      selection, 2
    harmonic behavior, 225
    operation
      adjustable frequency, 9
      nonsinusoidal supply waveforms, 217
    simulation, 495
  power
    control, 51
    regulator, 51
  DC converter circuits, 69
Acoustic noise, 136, 139, 441
Active region, 32
Adjustable-frequency
  drive, 3
  operation, 9
    ac motors, 9
Adjustable-speed drives, 16, 263, 299, 302
  ac motors, 485
  voltage-controlled system, 383
Advance angle, 82, 359, 367, 370
Airgap
  emf, 217
  flux, 4
    constant operation, 272
    control, 297
    sensing, 297, 319
  mmf harmonics, 220
Analog simulation, 242, 313, 492
Angle
  advance, 82, 359, 367, 370
  characteristic, 414, 415, 439
  conduction, 440
  control schemes, 301
  delay, 51, 73, 74, 82, 90, 150, 200, 402
  displacement, 82
  extinction, 440
  firing see Delay angle
  load, 347
  margin, 359, 367, 370, 82
    control, 370
  overlap, 78
  step, 430, 431, 438
Antisaturation circuit, 34
Application of
  power transistors, 451
  thyristors, 461

Approximate equivalent circuit, 154
Arcing
  time, 470
  voltage, 471, 472
Armature reaction, 344, 348, 365, 367, 369, 372
Asymmetric thyristor (ASCR), 55
Autosequentially commutated inverter (ASCI), 148, 150
Avalanche
  breakdown, 27, 44, 452, 474
  diode, 27
  voltage, 474
Average current rating, 462

Backward rotating waves, 223
Base speed, 18, 275, 285, 286, 442, 489
Bifilar winding, 435, 436, 441
Bipolar drive circuit, 433
Blocking diodes, 151
Breakdown torque, 10, 268, 269, 270, 273, 277, 279, 285, 286
Brushless
  dc servo, 424, 426
  excitation, 341
  motor, 414
    dc, 413, 425, 426, 509
      full wave, 417
      half wave, 414

CSI see Current-source inverter
Cage-rotor, 2
  induction motor, 1, 9, 217
Carrier
  ratio, 125, 128, 131, 133, 137
  wave, 123, 124
Chopper circuit, 21, 95, 117, 408
Circuit
  ac/dc converter, 69
  antisaturation, 34
  bipolar drive, 433
  chopper, 21, 22, 95, 117, 408
  commutation
    forced, 23, 47, 48, 148, 165, 167
    impulse, 167
    input, 175
  dc chopper, 22, 95, 117, 408
  drive, 433

515